AVENUE21. Planning and Policy Considerations
for an Age of Automated Mobility

Mathias Mitteregger · Emilia M. Bruck ·
Aggelos Soteropoulos · Andrea Stickler ·
Martin Berger · Jens S. Dangschat ·
Rudolf Scheuvens · Ian Banerjee

Editors

AVENUE21. Planning and Policy Considerations for an Age of Automated Mobility

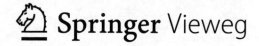

Editors

Mathias Mitteregger
TU Wien
future.lab Research Center
Vienna, Austria

Aggelos Soteropoulos
future.lab Research Center and Research Unit
Transport System Planning (MOVE)
TU Wien
Vienna, Austria

Martin Berger
Research Unit Transport System Planning
(MOVE) TU Wien
Vienna, Austria

Rudolf Scheuvens
future.lab Research Center and Research Unit of
Local Planning (IFOER)
TU Wien
Vienna, Austria

Emilia M. Bruck
future.lab Research Center and Research Unit of
Local Planning (IFOER)
TU Wien
Vienna, Austria

Andrea Stickler
future.lab Research Center and Research Unit
Sociology (ISRA)
TU Wien
Vienna, Austria

Jens S. Dangschat
Research Unit Sociology (ISRA)
TU Wien
Vienna, Austria

Ian Banerjee
Research Unit Sociology (ISRA)
TU Wien
Vienna, Austria

ISBN 978-3-662-67006-4   ISBN 978-3-662-67004-0   (eBook)
https://doi.org/10.1007/978-3-662-67004-0

Translation from the German language edition: AVENUE21. Politische und planerische Aspekte der automatisierten Mobilität by Mathias Mitteregger, et al., © Der/die Herausgeber bzw. der/die Autor(en) 2021. Published by Springer Viewer. All Rights Reserved.

INTO THE FUTURE WITH KNOWLEDGE

FOREWORD BY THE DAIMLER AND BENZ FOUNDATION

The digital transformation affects practically all areas of life. Whether in medicine and care, teaching and learning, cultural activities, production, consumption and logistics, including the entire transport system — all sectors are subject to the unstoppable advance of digitalization. Some areas have long been benefiting from this process, others have seen controversial debate, and still others are suddenly experiencing an unexpected boost. This is all driven by scientific and technological developments that are playing a decisive role around the globe in shaping our future.

In times of change and turbulence, such as we are currently experiencing, the value of science is immediately apparent. It provides new insights and thus the basis for practical solutions. This is the guiding principle of the non-profit Daimler and Benz Foundation: since it was established in 1986, it has been promoting science in the interplay between humanity, technology and the environment. Autonomous and interdisciplinary in its focus, this is the general theme for its various funding projects.

The foundation also contributes to the communication and dissemination of scientific knowledge and its implications, to make them publicly accessible. It promotes a dialogue between members of the public, scientists from various disciplines as well as experts from more practically oriented fields. Science is communicated in a wide variety of event formats, lecture series and publications, thereby generating added value for society.

In the research project AVENUE21 – Autonomous Traffic: Developments in Urban Europe in the 21st Century, scientists from TU Wien in Austria examined current problems and pressing future issues in connection with digitalized and automated road transport. The project was funded by the foundation over a period of four years. Since this topic transcends national borders, the scientists conducted their research in interdisciplinary and international teams. Their findings, which are of relevance to modern societies, are now published in this second report. While the first volume from 2020 (English edition published in 2022) dealt with the urban impact of automated vehicles, the focus here is on public space: what experience has been gained with automated shuttle buses in public transport? Are delivery robots the solution for the last mile in city centres? How can bicycle traffic be integrated into automated transport structures of the future? And how should politics, administration and society deal with all this?

Changes are opportunities — they create new scope for shaping the future. To this end, scientific research is indispensable. The Daimler and Benz Foundation supports selected projects such as "AVENUE21" to ensure that the insights gained are are not only gained, but are also put into effect in a targeted manner and are widely disseminated.

Prof. Dr. Julia Arlinghaus & Prof. Dr. Lutz H. Gade
Board of Directors of the Daimler and Benz Foundation

WHAT IS IN STORE FOR US?

FOREWORD BY THE AVENUE21 RESEARCH TEAM

Over a period of four years, the "AVENUE21" project gave us an opportunity to research the topic of connected and automated mobility at the interfaces of mobility and settlement development. With generous funding from the Daimler and Benz Foundation, we were able to examine this topic from various angles as an interdisciplinary team at the Institute of Spatial Planning at TU Wien and to reflect on it together with international experts. By adopting an open-ended approach at the outset of the project, we were able to develop various perspectives on the possible impact of connected and automated mobility in an iterative, interdisciplinary process, and on this basis to ask whether, where and under what circumstances connected and automated mobility can be used to the benefit of sustainable settlement and mobility development.

The increasing automation and networking of road transport will have an impact on existing planning tasks over several levels of magnitude and will also significantly influence forthcoming challenges: from European "connectivity" to urban, or urban-regional, development planning and the design and structure of roads as public space. A new diversity of mobility offers will emerge (vehicles, mobility services, information, data) that holds potential for sustainable and spatially compatible development, but may also call it into question. Generic solutions "from above", or mere implementation of technological innovations, are hardly sufficient for developing forward-looking strategies. Rather, the emerging trend requires all parties involved to recognize and take seriously the locally specific, differentiated challenges. Policy-makers and planning bodies are likewise called on to exercise consistent control, in order on the one hand to specifically define and constantly develop a framework oriented to the common good, and on the other hand to adopt a flexible approach towards new technologies.

Whereas in the first publication, *AVENUE21. Connected and Automated Driving: Prospects for Urban Europe*, we described the challenges posed by connected and automated transport on urban development and public space in a differentiated manner and outlined possible (ideal) developments on the basis of scenarios, the focus of the present volume is on the potential diversity of possible applications of connected and automated mobility – described for example on the basis of international examples – and the potential and necessity of their control. In-depth coverage by the research team is supplemented by input from international colleagues.

We would like to thank all the people and institutions who over the past four years have accompanied us, supported us and given us the opportunity to make a contribution with our research to other disciplines, also internationally. We are grateful for this highly constructive exchange and hope that with this volume we can make a further contribution to crucial discussions of the future of mobility.

The AVENUE21 research team

AUTHORS

Christof Abegg
University of Bern, Team Leader Urban and Regional Economy, EBP Schweiz AG
Bern/Zurich, Switzerland

Daniela Allmeier
TU Wien, future.lab Research Center and Research Unit of Local Planning (IFOER)
Vienna, Austria

Ian Banerjee
TU Wien, Research Unit Sociology (ISRA)
Vienna, Austria

Martin Berger
TU Wien, Research Unit Transport System Planning (MOVE)
Vienna, Austria

Stefan Bindreiter
TU Wien, simlab and Research Unit of Local Planning (IFOER)
Vienna, Austria

Emilia M. Bruck
TU Wien, future.lab Research Center and Research Unit of Local Planning (IFOER),
doctoral candidate in the AVENUE21 project
Vienna, Austria

Jens S. Dangschat
TU Wien, Research Unit Sociology (ISRA)
Vienna, Austria

Alexander Egoldt
TU Berlin, Department of Road Planning and Operation
Berlin, Germany

Lutz Eichholz
TU Kaiserslautern, Department of Spatial and Environmental Planning
Kaiserslautern, Germany

Steven Fleming
Cycle Space
Melbourne, Australia

Tomoyuki Furutani
Keio University, Faculty of Policy Management
Tokyo, Japan

Alexander Hamedinger
TU Wien, Research Unit Sociology (ISRA)
Vienna, Austria

Arne Holst
TU Berlin, Department of Road Planning and Operation
Berlin, Germany

Peraphan Jittrapirom
Nijmegen School of Management, Radboud University and Global Carbon Project (GCP),
National Institute for Environmental Studies (NIES)
Nijmegen, Netherlands and Tsukuba, Japan

AUTHORS

Detlef Kurth
  TU Kaiserslautern, Department of Spatial and Environmental Planning
  Kaiserslautern, Germany

Zoltán László
  Swiss Federal Railways (SBB), Lead On-Demand Mobility, New Mobility Services
  Bern, Switzerland

Bert Leerkamp
  University of Wuppertal, Institute of Freight Transport Planning and Logistics
  Wuppertal, Germany

Katharina Manderscheid
  University of Hamburg, Department of Socioeconomics, Subject Area Sociology
  Hamburg, Germany

Robert Martin
  JAJA Architects ApS and Aalborg University CPH, Department of Planning
  Copenhagen/Aalborg, Denmark

Mathias Mitteregger
  TU Wien, future.lab Research Center, AVENUE21 Project Lead
  Vienna, Austria

Lucia Paulhart
  TU Wien, future.lab Research Center
  Vienna, Austria

Fabienne Perret
  ETH, Head of Transportation Division, EBP Schweiz AG
  Zurich, Switzerland

Thomas Richter
  TU Berlin, Department of Road Planning and Operation
  Berlin, Germany

Rudolf Scheuvens
  TU Wien, future.lab Research Center and Research Unit of Local Planning (IFOER)
  Vienna, Austria

Aggelos Soteropoulos
  TU Wien, future.lab Research Center and Research Unit Transport System Planning (MOVE),
  doctoral candidate in the AVENUE21 project
  Vienna, Austria

Andrea Stickler
  TU Wien, future.lab Research Center and Research Unit Sociology (ISRA),
  doctoral candidate in the AVENUE21 project
  Vienna, Austria

FURTHER PARTICIPANTS

Bettina R. Algieri
German proofreading
Baden, Austria

Lukas Bast
Artwork, layout and typesetting
TU Wien, future.lab Research Center
Vienna, Austria

Lyam Bittar
English translation
Berlin, Germany

Jonathan Fetka
Content-related project assistance and publication management
TU Wien, future.lab Research Center and Research Unit Transport System Planning (MOVE)
Vienna, Austria

Michael Gidam
GIS-based analyses
TU Wien, Research Unit Transport System Planning (MOVE)
Vienna, Austria

Maximilian Kipke
Artwork
TU Wien, Research Unit Transport System Planning (MOVE)
Vienna, Austria

Melinda Klics
Project assistance
TU Wien, future.lab Research Center
Vienna, Austria

Andrew Leslie
English translation
Stuttgart, Germany

Rita Neulinger
Layout
Brandneulinger Grafik
Vienna, Austria

Nivene Raafat
English translation
Brighton, United Kingdom

Iain Reynolds
English translation
Lancaster, United Kingdom

Maria Slater
English translation and proofreading
Vienna, Austria

Andrea Wölfer
Organization, human resources and administration
TU Wien, future.lab Research Center
Vienna, Austria

LIST OF ABBREVIATIONS

ADRT	automated demand-responsive transport
ADS	automated driving system
AI	artificial intelligence
ANT	actor-network theory
API	application programming interface
CAT	connected and automated transport
CAV	connected and automated vehicle
CEP	services courier, express and parcel services
C-ITS	cooperative intelligent transport services
DRT	demand-responsive transport
EV	electric vehicle
FGSV	Forschungsgesellschaft für Straßen- und Verkehrswesen e. V.
FUA	functional urban area
FUR	functional urban region
GIP	Graph Integration Platform
GIS	geographic information system
IoT	Internet of Things
LPT	local public transport
LTS	large technical systems
MaaS	Mobility as a Service
MPT	motorized private transport
ODD	operational design domain
OSM	Open Street Map
ph	peak hour
PIT	public individual transport
PMV	personal mobility vehicle
PRT	personal rapid transport
PT	public transport
PT	policy transfer
SBB	Swiss Federal Railways
SMS	shared mobility services
STS	science and technology studies
TA-SWISS	Swiss Foundation for Technology Assessment
TNC	transport network company (e.g. Uber, Lyft, etc.)
UPM	urban policy mobilities
veh	vehicles
veh/ph	vehicles per peak hour
ZAMG	Austrian Central Institution for Meteorology and Geodynamics

CONTENTS

1 Connected and automated driving: The Long Level 4

Mathias Mitteregger, Ian Banerjee

Mathias Mitteregger
TU Wien, future.lab Research Center

Ian Banerjee
TU Wien, Research Unit Sociology (ISRA)

© The Author(s) 2023
M. Mitteregger et al. (eds.), *AVENUE21. Planning and Policy Considerations for an Age of Automated Mobility*, https://doi.org/10.1007/978-3-662-67004-0_1

1. INTRODUCTION

This volume comes at the conclusion of four years of research at TU Wien made possible by the Daimler and Benz Foundation. During this time, assessment of the potential and impact of connected and automated vehicles underwent a fundamental transformation. The findings of the core team and of the international authors of this volume have contributed to this process in various ways.

Four years of research were sufficient to expose as premature the often-repeated announcement that autonomous vehicles would enter series production in five years' time. This has long been a problem in this field: development is dictated by companies whose success narratives later often turn out to be mere declarations of intent (Matthaei et al. 2015: 1144): neither carmakers (Boudette 2016, Wang 2016) nor IT companies (Rogers 2015, Korosec 2015) were able to fulfil their five-year plan.

For research fields that are interested in the impact of a new technology or search for suitable applications for it, this extra time is certainly valuable, as is shown by a comparison: countless smart city projects and initiatives have been launched in recent years. While a very specific approach has been taken in all urban sectors with regard to data and connectivity, at best "initial general indications" of possible impact could be provided, or a consideration of impact could be mentioned under "research requirements" (Soike et al. 2019: 5).

2. AUTOMATION AND CONNECTIVITY OF TRANSPORT IN CONTEXT

At the beginning of this collaborative research project, practically unlimited expectations were placed in "autonomous driving robots" (those who had long been dealing with issues of automation and connectivity were more sceptical; cf. Shladover, Kornhauser and Miller in Simonite 2016). Connected and automated vehicles were to make traffic more efficient, safer and more sustainable, and were to respond more specifically to individual mobility needs and seamlessly integrate various services (cf. Dangschat/Stickler 2020). For a technology that even today is not yet in regular operation but is still only at the testing phase, all these attributions remain mere metaphors – linguistic images (in this specific case borrowed from the world of platform economies) that not only serve to describe things that elude immediate experience, but can also help influence their development (Kuhn 1979, Boyd 1979; on the role of narratives on connected and automated mobility in Europe, see Stickler 2022).

The unrealistic expectations placed on connected and automated vehicles are not entirely surprising. Various theoretical positions bear out the fact that people tend to overestimate their own accomplishments. Karl Marx spoke of commodities acquiring "theological niceties" through the intervention of their makers and thus being elevated to a quasi-religious sphere (Marx 1890/1962: 85–98) – an assessment that is remarkably shared by management consultants who speak of the hype cycle (Fenn/Raskino 2008).

However, the overall situation must be considered here. Firstly, the market in the mobility sector has been extremely turbulent since about 2010. It was around this time that new players from the IT sector started entering the mobility market; within a relatively short time, they had achieved considerable success through technological innovations and service-oriented business models (by the end of 2020, the stock of the Californian automobile manufacturer Tesla was worth as much as General Motors, Ford, Toyota, Honda, Fiat Chrysler and Volkswagen combined; Krisher 2020). At peak times, more than a quarter of all trips in parts of San Francisco are now carried out by ride-hailing services such as Uber or Lyft (SFCTA 2018).

Secondly, after years of inaction, there is now enormous pressure to radically reduce emissions from road transport. While all other sectors of the economy have managed to reduce greenhouse gas emissions effectively, the emissions from road transport within the EU have risen unabated and would continue to rise without resolute action (IEA 2020). This has led to announcements becoming hopes; cities throughout the world are commissioning simulation studies on the potential impact of a transport system consisting solely of automated vehicles (Soteropoulos et al. 2019). Even today, numerous policy and strategy papers attribute a key role to connectivity and automation in the transition to sustainable transport and mobility.

At the conclusion of the research project, both the technological developments and the possible positive contribution of connected and automated vehicles to the mobility transformation are being assessed more soberly. Rather than a revolution, we are experiencing a phase of transformation or transition lasting several decades, "during which CAVs will be deployed only in parts of the road network. During this transition period, conventional means of transport will continue to play an essential but increasingly specialized role" (Mitteregger et al. 2022: VIII). For this period, the term "Long Level 4" was coined by the core team in the first publication of this research project (Mitteregger et al. 2022).

3. WHAT CHARACTERIZES THE PERIOD OF TRANSITION IN THE MOBILITY SYSTEM?

Also with a view to transformation research, it is expected that a technology will initially be implemented in "niches" before being disseminated throughout society at large (see Chap. 19 by Dangschat in this volume). Since only parts of the road network could be suitable for the use of connected and automated vehicles, the impact must be fundamentally reassessed. Some aspects of this transition period are analysed here (on the role of interfaces, for example, see Chap. 8 by Bruck et al. in this volume; and on the interaction of bicycle traffic and automated vehicles see Chap. 11 by Eichholz and Kurth).

A fundamental change of course is already discernible here; this concerns the matter of how selective suitability will be treated in the future. In the long term, automated driving systems are likely to be overburdened in certain situations. Various factors play a role in this, such as prevailing weather conditions, the state of the road infrastructure or the desired speed of travel (see Chap. 5 by Soteropoulos in this volume). Two solutions are offered in an international comparison:

1. **Hybrid offers:** Ride service providers (e.g. Uber or Lyft) are already planning for the transition phase of Long Level 4. For the foreseeable future, it is expected that only some rides can be automated (with or without a safety driver). The prevailing environmental conditions are examined for each customer request. If analysis shows that an automated

trip is possible, a vehicle without a driver is dispatched; otherwise, a human driver will be at the wheel (Chaum 2019, Sheikh 2018).

2. **Demarcated operating areas:** Legislation in Germany, where the current federal government has set itself the goal of being the first country in the world to permit regular operation during the transition phase, takes a different approach: areas of operation are to be designated, each of which may be used by one specific automated driving system. Examples of possible operating areas are "public transport within municipalities", "service and supply trips in the municipal sector", "company shuttles for employee transport" and "trips between medical care centres and old people's or nursing homes" (cf. Kugoth 2020, Mitteregger et al. 2022).

Which of these two paths will ultimately prevail (locally) is likely to be one of the key decisions and will fundamentally shape the face of public streetscapes (Fig. 1).

Figure 1: Responses regarding selective suitability of the road network during the Long Level 4

	Dynamic response	Geographical response
Range of application	temporary, situation-dependent	fixed, spatially delimited
Space metaphor	experienced space	container space
Example	hybrid ride service providers	law in Germany

Source: the authors

The effects would be fundamentally different. Here too, we are initially dealing with (opposing) metaphors — in this case for space. Geographically demarcated operating areas allow for a very different legibility of space from what is possible with a situation-dependent assessment. From the perspective of other road users, entering an operating area would be a conscious act. The underlying container space, however, calls for abstractions that have come in for much criticism — above all concerning usage or living experience. In the other case, it would remain unclear why a vehicle is in the road space without a driver. The underlying analysis would not be accessible. The ability of authorities to carry out checks, the possibility of intervention and control, e.g. for the purpose of traffic management, forms of supply and working conditions — all this would be affected.

The two responses also fundamentally differ in terms of what media we can use today in order to understand possible impacts. A dynamic, mutable urban space is difficult to depict in pictures; videos or texts are better suited here. Waymo's promotional video is an example: a blind person can be seen in an automated ultracompact vehicle. The elderly man is happy that he has "regained an important part of his life" — a part that was "wrested from him by his loss of sight" (Waymo 2016). But whether such an experience is now open to this person during heavy rain or in places other than Phoenix, Arizona, remains unanswered.

Current national strategies almost exclusively emphasize or regulate geographical aspects of the transition period (see Chap. 13 by Banerjee and Furutani and Chap. 17 by Perret and Abegg in this volume). A significant task could be to account for situational aspects to a greater extent.

REFERENCES

Boudette, N. E. 2016. "Ford Promises Fleets of Driverless Cars Within Five Years", *The New York Times*, 16/8/2016. https://www.nytimes.com/2016/08/17/business/ford-promises-fleets-of-driverless-cars-within-five-years.html (1/9/2020).

Boyd, R. 1979. "Metaphor and theory change: What is 'metaphor' a metaphor for?", in *Metaphor and Thought*, ed. by A. Ortony. Cambridge: Cambridge University Press, 481–532.

Chaum, M. 2019. "Uber Video Keynote", lecture at the Swiss Mobility Arena, 16/9/2019.

Dangschat, J. S., and A. Stickler 2020. "Kritische Perspektiven auf eine automatisierte und vernetzte Mobilität", in *Jahrbuch StadtRegion 2019/2020*, ed. by C. Hannemann, F. Othengrafen, J. Pohlan, B. Schmidt-Lauber, R. Wehrhahn and S. Güntner. Wiesbaden: Springer VS, 53–74.

Fenn, J., and M. Raskino 2008. *Mastering the Hype Cycle: How to Choose the Right Innovation at the Right Time*. Brighton, MA: Harvard Business Press.

Geels, F. W. 2002. "Technological transitions as evolutionary reconfiguration processes: a multi-level perspective and a case-study", in *Research Policy* (31) 8/9, 1257–1274.

IEA (International Energy Agency) 2020. "Changes in transport behaviour during the Covid-19 crisis", 27/5/2020. Paris. https://tinyurl.com/y32cn6xa (1/9/2020).

Korosec, K. 2015. "Elon Musk Says Tesla Vehicles Will Drive Themselves in Two Years", *Fortune*, 21/12/2015. https://fortune.com/2015/12/21/elon-musk-interview/ (1/9/2020).

Krisher, T. 2020. "Tesla joins the S&P 500 today. It's already worth more than Toyota, Volkswagen, GM, Ford, Fiat Chrysler, Nissan and Daimler combined", *The Chicago Tribune*, 21/12/2020. https://www.chicagotribune.com/business/ct-biz-tesla-elon-musk-sp500-20201221-i3udv6vv7reg7pp2hr-zu5eyjjq-story.html (20/12/2020).

Kugoth, J. 2020. "Gesetzentwurf: Wo Roboshuttles rollen sollen", *Tagesspiegel Background*, 11/5/2020. https://tinyurl.com/y2h7lwx9 (1/9/2020).

Kuhn, T. 1979. "Metaphor in Science", in *Metaphor and Thought*, ed. by A. Ortony. Cambridge: Cambridge University Press, 533–542.

Matthaei, R., A. Reschka, J. Rieken, F. Dierkes, S. Ulbrich, T. Winkle and M. Maurer 2015. "Autonomes Fahren", in *Handbuch Fahrerassistenzsysteme*, ed. by H. Winner, S. Hakuli, F. Lotz and C. Singer. Wiesbaden: Springer Vieweg, 1139–1165.

Marx, K. 1890/1962. *Das Kapital. Kritik der politischen Ökonomie in 3 Bänden, Band 1: Der Produktionsprozeß des Kapitals*, ed. by F. Engels. Berlin: Dietz.

Mitteregger, M., E. M. Bruck, A. Soteropoulos, A. Stickler, M. Berger, J. S. Dangschat, R. Scheuvens and I. Banerjee 2022. *AVENUE21. Connected and Automated Driving: Prospects for Urban Europe*, trans. M. Slater and N. Raafat. Berlin: Springer Vieweg. DOI: 10.1007/978-3-662-64140-8.

Rogers, C. 2015. "Google Sees Self-Driving Car on Road Within Five Years", *The Wall Street Journal*, 14/1/2015. https://www.wsj.com/articles/google-sees-self-drive-car-on-road-within-five-years-1421267677 (1/9/2020).

SFCTA (San Francisco Country Transportation Authority) 2018. "TNCs & Congestion. Final Report". https://tinyurl.com/y6w76ta6 (15/9/2020).

Sheikh, N. 2018. "Applying a Hybrid Network Approach to Deployment of Self-Driving Mobility Services", Automated Vehicle Symposium. https://tinyurl.com/uhwtb3k (15/9/2020).

Simonite, T. 2016. "Prepare to be Underwhelmed by 2021's Autonomous Cars", *MIT Technological Review*. 23/8/2016. https://www.technologyreview.com/2016/08/23/157929/prepare-to-be-underwhelmed-by-2021s-autonomous-cars/ (1/9/2020).

Soike, R., J. Libbe, M. Konieczek-Woger and E. Plate 2019. "Räumliche Dimensionen der Digitalisierung. Handlungsbedarfe für die Stadtentwicklungsplanung. Ein Thesenpapier", special publication. Berlin: Deutsches Institut für Urbanistik.

Soteropoulos, A., M. Berger and F. Ciari 2019. "Impacts of automated vehicles in travel behaviour and land use: An international review of modelling studies", in *Transport Reviews* (39) 1, 29–49.

Stickler, A. 2022. "Negotiating a dominant narrative about connected and automated mobility in Europe", in *AVENUE21. Connected and Automated Driving: Prospects for Urban Europe, trans. M. Slater and N. Raafat*, ed. by M. Mitteregger, E. M. Bruck, A. Soteropoulos, A. Stickler, M. Berger, J. S. Dangschat, R. Scheuvens and I. Banerjee. Berlin: Springer Vieweg, 90–92. https://doi.org/10.1007/978-3-662-64140-8.

Wang, C. 2016. "Self-driving cars are coming, and the technology promises to save lives", *The Guardian*, 17/12/2015. https://www.theguardian.com/technology/2015/dec/17/self-driving-cars-safety-futureinteractive (15/12/2020).

Waymo 2016. "Say hello to Waymo", YouTube video, 13/12/2016. https://www.youtube.com/watch?v=uHbMt6WDhQ8 (1/9/2020).

2 Connected and automated driving: Consideration of the local, spatial context and spatial differentiation

Emilia M. Bruck, Aggelos Soteropoulos

Emilia M. Bruck
future.lab Research Center and Research Unit of Local Planning (IFOER)

Aggelos Soteropoulos
TU Wien, future.lab Research Center and Research Unit Transport System Planning (MOVE)

© The Author(s) 2023
M. Mitteregger et al. (eds.), *AVENUE21. Planning and Policy Considerations for an Age of Automated Mobility*, https://doi.org/10.1007/978-3-662-67004-0_2

1. INTRODUCTION

The articles in this volume deal with effects on spaces and with control and planning approaches at municipal and regional level. It is essential to distinguish here between the spatial conditions of rural communities and regions and those of urban centres and major cities. Neither rural nor urban areas are homogeneous, but are complex areas with a wide variety of settlement structures, natural spaces and infrastructures, functions and challenges (VCÖ 2019: 9). Their characteristics determine different basic conditions for connected and automated driving and consequently influence its spatial impact, which differs for individual locations.

Spatial structures and mobility behaviour (modal split, degree of motorization, CO_2 emissions), along with economic and infrastructural conditions, are just some of the factors that determine the prerequisite conditions for automated forms of application. In view of these differences, it must be borne in mind that approaches to mobility and settlement development can only be transferred to a limited extent from urban centres to regions and rural areas; this also applies to automated driving. A spatially differentiated view can be useful in examining opportunities for action against the background of local conditions, and adapting action plans accordingly.

The interdependency between transport media on the one hand and settlement development, streetscapes and urban forms on the other must be recognized here; this all gives shape to a space (Angerer/Hadler 2005; Mitteregger et al. 2022: 66). Just as the introduction of the railway and the automobile triggered a historic transformation in accessibility and settlement development, connected and automated driving likewise has the potential to fundamentally transform public spaces, townscapes, landscapes and settlement structures. Conversely, a change in land use or settlement structure can influence accessibility and thus lead to relatively rapid adaptation of mobility activities (Wegener/Fürst 1999; Bertolini 2012, 2017). The extent to which the impact of new transport modes and transport infrastructures is manifested in spatial terms will vary depending on the settlement structure: for example, there is a correlation between the effectiveness of investments in transport infrastructure, the degree of development of a settlement area and its current level of accessibility.

While saturation is reached in terms of improving accessibility in areas that are already highly accessible and/or have a highly developed settlement structure, in less developed areas more pronounced effects can be expected from new transport infrastructures or new modes of transport, since accessibility can thereby be noticeably increased (Kasraian et al. 2016, Mitteregger et al. 2022). For a differentiated understanding of spatial changes in interaction with new forms of mobility, settlement and infrastructure spaces must be specifically perceived both with a view to their functional and structural differences and in their historical context – i.e. the conditions under which they originated – in order to specifically address deficits and potentials. Finally, it is also important to mention the interdependence between processes of spatial development and policies of planning, mobility and settlement – see Part 4, "Governance", in this volume.

2. SPATIAL DIFFERENTIATION AND TYPES OF SPACE

For the purpose of analysis, a delimitation and categorization of the spatial system is required not only in urban and regional planning, but also in politics and industry. Since the early 20th century, a variety of different methods have been developed for a differentiation of types of space in order to define spatial categories on the basis of uniform criteria. More recent approaches include for example types of space as defined by the German Federal Institute for Research on Building, Urban Affairs and Spatial Development (BBSR 2010), types of transport space (Matthes/Gertz 2014) and an urban-rural typology (Statistik Austria 2017). The methods used mostly differ in their consideration of (a) functional characteristics, such as traffic flows, commuter flows and supply relationships, and (b) structural characteristics, such as the extent of built-up areas, morphology, population and job density, demographic-sociological population structure or employment figures in the various sectors of the economy.

While the classical differentiation of settlement areas distinguishes between cities, outlying areas and rural spaces, other models are based on an internal classification of the so-called urban-rural continuum (cf. Borsdorf/Bender 2010: 250). With the term *Stadtregion* (urban region), Boustedt (1953, 1970, 1975) in particular provided a significant basis in the German-speaking countries for differentiation of spatial units within urban regions. This model is founded on a combination of various functional (degree of interdependence) and structural characteristics (number of inhabitants, population density, agricultural quota) and served as a basis for the later concept of "densification areas", which were last renewed for Germany in 1993 (Borsdorf/Bender 2010).

In view of the continuing urban sprawl, however, a distinction between urban and rural areas is proving to be less and less productive. It would seem more appropriate to differentiate between various constellations of urban areas or various degrees of urban intensity (Kretz/Kueng 2016). Approaches that dispense with a traditional understanding of the urban-rural dichotomy include the "settlement-structural area types" introduced in 1986 (Borsdorf/Bender 2010). Here, seven types of region are defined on the basis of population density and are assigned to three basic types ("agglomeration areas", "urbanized areas" and "rural areas"). The "Spatial Categories 2010" developed by the BBSR also provide a comprehensive typification of spaces and refrain from delimitation according to administrative boundaries. Their classification is based on the three basic structural characteristics of settlement structure (population density and proportion of settlement area), location (accessibility and centrality) and economic strength (BBSR 2010).

This type of spatial differentiation has hardly been taken into account in previous considerations of spatial effects of connected and automated driving. The focus to date has largely been on macroscopic modelling of the spatial and transport-related effects of connected and automated driving, particularly in urban areas (Soteropoulos et al. 2019).

3. SPATIAL EFFECTS OF AUTOMATED DRIVING

In recent years, a number of studies have been published on the possible spatial effects of connected and automated driving. A much-cited basis is provided by the research of Milakis et al. (2017), according to which the impact of automated driving can be divided into primary, secondary and tertiary effects:

- **Primary effects:** Traffic volumes (road capacity/congestion and distances travelled), time benefits (perception of time and travel comfort) and choice of transport mode

- **Secondary effects:** Vehicle ownership, choice of location and land use, and transport infrastructure

- **Tertiary effects:** Energy consumption (fuel savings), environmental impact, road safety, social equity, economy and health.

The traffic-related and spatial effects of automated driving, in particular, have been investigated in numerous simulation studies (Soteropoulos et al. 2019). The focus here is often on scenaio -based modelling of possible changes in transport demand, settlement structure and land use due to various factors such as trends in traffic performance, the modal split, parking space requirements and the choice of location for households and businesses.

While earlier studies and models mostly assumed a state of full saturation with highly automated vehicles at SAE Level 5 (SAE 2018), this has been called into question to an increasing extent in subsequent studies (Mitteregger et al. 2022, Soteropoulos et al. 2020). Instead, a longer transition phase is now moving into the focus of attention, whereby different levels of automation will be permissible depending on surrounding conditions and will be used in mixed traffic scenarios (see Chap. 1 by Mitteregger and Banerjee in this volume). The spatial effects of connected and automated driving are dependent on the types of offers provided, settlement structures and streetscapes, and (applicable) planning policy objectives. The ongoing development of automation is also becoming part of a technological and organizational transformation of mobility systems that is paving the way for later usage modes. These include for example car and ride sharing, or changes in logistics and goods transport (Beckmann/Brügger 2013, Gerdes/Heinemann 2019, and Chap. 7 by Leerkamp et al. in this volume).

Accordingly, automated driving could contribute to spatial transformations in relation to (1) parking (e.g. street-based parking spaces, car parks, parking garages), (2) road space requirements (e.g. loading areas), (3) the necessary infrastructure (e.g. digital networking, lane markings) and (4) choice of location for businesses and households. Since these changes can trigger further developments both in urban planning and in settlement structure, the spatial consequences of connected and automated driving can be classified into primary and secondary effects (Fig. 1).

3.1 PRIMARY SPATIAL EFFECTS

One of the often-mentioned spatial consequences is a reduction in parking space requirements as a possible result of the lower number of vehicles on the roads. However, this effect in particular is dependent on a high number of ride-sharing trips, or at least widespread use of car-sharing offers (Zhang et al. 2015, Soteropoulos et al. 2019). The freeing up of parking spaces will thus result from the adoption of Mobility as a Service (MaaS) in general. If the shared ride services already available today prove not only to become increasingly popular but also result in widespread disuse of private cars, this could expedite reclamation of public space. However, automated driving is likely to lead to a relocation of parking garages and shared garages. As long as a sufficient number of loading areas are provided in selected urban spaces and streets (see Chap. 8 by Bruck et al. in this volume), parking spaces could be reduced and be replaced by decentralized shared garages. Depending on the type of provider and the space available, various locations would come into consideration here (Lewis/Anderson 2020: 104).

Changes to the streetscape also affect the competition for space between different modes and providers, which already today is increasing the pressure on public spaces in view of intensified delivery traffic, platform-based ride services and micromobility offers (see Chap. 8 by Bruck et al. and Chap. 9 by Martin et al. in this volume). Automated transport could also promote spatial separation effects, since an increase in mileage and stabilization of traffic flows could make it

Figure 1: Effects of automated driving systems (ADSs) on spatial development

Automated driving

refers to all forms of offers involving digitally connected and automated vehicles, which in some cases are electrically powered.

PRIMARY EFFECTS
of automated driving

Traffic volumes	Time advantages and travel comfort	Usage and operating costs	Choice of transport media
ADSs can open up new mobility options and reach new user groups. This can lead to an increase in transport demand, mileage and traffic volumes. If users switch to car-sharing and ride-sharing services, traffic volumes can be reduced.	Due to time benefits and increased convenience, ADSs can encourage longer commutes. Users could use the time for entertainment or work.	In the short term, the fixed costs of a vehicle purchase are higher than for a conventional vehicle. On the other hand, efficiency benefits can lead to reduced usage and operating costs. Low operating costs could facilitate the expansion of networks for public providers.	The Introduction of ADSs may bring about changes in travel behaviour. Users could increasingly use private car-sharing services instead of walking or cycling. There could also be a shift away from public transport.

PRIMARY EFFECTS
on settlement areas

Changed parking demand	Competition for space	Complementary infrastructure	Relocation
A shift to mobility services, micromobility and ADSs alters parking requirements. There would be a greater need for temporary stopping zones rather than on-street parking. This has an impact on the distribution of road space and surrounding building development.	The rise of mobility services, online commerce and micromobility is increasing the pressure on usage and distribution of public spaces. If ADSs also become established, the need for loading areas will increase; this will have an impact on mobility patterns and spatial separation effects.	In areas where ADSs are in operation, there can be a need for modifications to the infrastructure (e.g. digital infrastructure, loading bays, road markings). The need for e-charging stations and service options may also increase.	The cost benefits and convenience of ADSs could encourage relocation. This includes new locations for decentralized collective garages and distribution centres for goods traffic, migration of logistics and commercial areas, and choice of residential locations.

SECONDARY EFFECTS
on the settlement area

Conversion areas	Diversity in public space	New centralities	Dynamics of urban density
Relocations and reduced parking requirements free up spaces for development, above all in areas used by cars such as parking spaces, garages, petrol stations, etc., along with commercial and logistics areas. Spatial transformation and subsequent densification are supported.	ADSs increase the variety of demands on land use, which need to be reconciled. Along with the reduction of parking spaces and lane widths on roads, consideration must also be given to prioritization of modes, travel speeds and urban qualities, e.g. spatial permeability and spaces for rest and relaxation.	When ADSs are planned as part of the public transport network, centres of activity can arise at mobility hubs and new stations or bus stops. The transformation of suburban areas can also promote their development. Space demand for mobility hubs, collection and distribution centres can arise at the periphery of areas where ADSs are in use.	Longer commuting distances can help increase urban sprawl and average urban density. New densities could also emerge in areas with a high proportion of conversion space. Spatial corridors could develop along routes on which ADSs are in operation.

Source: Bruck, based on Larco/Tierney (2020), Streckler et al. (2019) and Mitteregger et al. (2022)

more difficult for pedestrians and cyclists to cross roads, with a resulting loss of permeability (Ghielmetti et al. 2017; see Chap. 5 by Soteropoulos in this volume).

In both respects, it will be crucial to reconcile the requirement for safe interaction between the transport modes with the need for spatial compatibility and quality. Complementary transport infrastructures include both digital systems (e.g. GPS, satellites, uniform roadmaps or "geo-fencing") and material installations such as loading areas in public spaces, at mobility hubs and stations (see Chap. 8 by Bruck et al. in this volume). In the quest for resource-saving spatial development, above all a reassignment of existing infrastructures (e.g. roadside parking spaces or park-and-ride facilities) has the potential to reduce space requirements in connection with shared automated mobility.

Finally, it is expected that within urban regions, industrial and commercial sites could be moved to locations mainly situated on large-capacity transport axes that would be accessed at an early stage by connected and automated driving (Lewis/Anderson 2020, Mitteregger et al. 2022). Spurred by current developments in online retail, small-scale and decentralized distribution centres in inner-city areas could become necessary in the future, while suburban logistics operations could continue to migrate. At the same time, the benefits of convenient travel, time savings and economic efficiency could lead to longer commuting distances and a migration of both workplaces and households (Milakis et al. 2017, Litman 2020).

3.2 SECONDARY SPATIAL EFFECTS

The secondary effects of connected and automated driving also include, for example, the freeing up of development and conversion areas. Automation could thus have a reinforcing effect on orientation towards service in the mobility and transport system ("shared mobility", Mobility as a Service) and help not only to free up individual parking spaces, but also to reduce space requirements for all car-related functions, especially in commercial areas and outer suburbs, where they are mostly located (see Chap. 15 by Mitteregger and Soteropoulos in this volume). The release of these land reserves, in particular, offers potential for spatial transformations (Bruck 2019). Planning policy measures, such as a revision of parking space regulations, reassignment of monofunctional structures or a focus on internal development, could trigger new developments in previously neglected areas. If these are utilized in coordination with areas of application of connected and automated driving, spatial densification could benefit, especially in the vicinity of transfer points.

Connected and automated driving also increases the urgency of preserving diversity in the public space and implementing appropriate controlling and formative approaches. Already today, increasingly diverse needs regarding space have to be reconciled (increasing variety of transport media, capacity for lingering, requirements in connection with climate adaptation, etc.), which is why the introduction of connected and automated driving is expected to further increase the pressure of utilization in the public space. Without planning policy measures, connected and automated driving would hinder the permeability of street spaces for pedestrians. However, if high occupancy rates and shared driving were to become widespread, there would be the potential to reduce the number of on-street parking spaces and to make traffic lanes narrower, in addition to fundamentally rethinking the prioritization of transport modes, travel speeds and forms of utilization.

If connected and automated driving applications are planned in this way as part of the public transport network, this could have an impact on the development of urban centres and on spatial concentration. Studies already carried out show positive impact trends here in terms of local urbanization processes and inner-city population growth (Soteropoulos et al. 2019, WEF 2018).

Centrality is expected to increase, especially at stopping points, mobility nodes and new mobility hubs, which would need to be promoted with appropriate focal points in planning. The above-mentioned reassignment of settlement structures that were previously monofunctional and were dominated by automobile-related functions can also trigger transformation for new urban centre developments. In the long term, a shift in spatial densities is to be expected; this could take the form of spatial corridors within the catchment areas of connected and automated driving networks (Larco/Tierney 2020: 123).

This process of displacement could also be manifested as locational dynamics and changes in settlement densities – especially where connected and automated vehicles (CAVs) are not integrated into the public transport system. If the benefits of automated private vehicles predominate, this could favour urban sprawl (Milakis et al. 2017, Litman 2020). It can thus be assumed that peripheral areas will grow, while the density of central areas could be reduced due to the migration of certain functions and operations (Larco/Tierney 2020: 122).

The uncertainty as to the direction to be taken by these developments is due not least to the expectation that the coming decades will be marked by a coexistence of different degrees of automation and forms of application. In addition, the spatial complexity of European streetscapes and the cost-intensive upgrading of existing infrastructure networks suggest that the introduction of CAVs will be limited to selected sections of the network and particularly suitable areas (Mitteregger et al. 2022, Soteropoulos et al. 2020).

4. SPATIALLY DIFFERENTIATED EFFECTS OF AUTOMATED DRIVING

Especially in view of a long phase of transition, it must be borne in mind that the effects described above will not be felt in all settlement structures and urban spaces, nor to the same extent. As the current use of services from transportation network companies (TNCs) bears out, this technology first found favour mainly in dense urban centres and among affluent users. This aspect is also instructive for automated driving, since ride-sharing models can be most efficiently operated in high-density areas and over short distances. The higher the user density, the more trips can be pooled with fewer vehicles (cf. Larco/Tierney 2020). On the other hand, the first pilot projects with self-driving minibuses and ride-hailing services are being carried out above all in suburban and rural areas, since extensive road networks, a minimum of environmental complexity and low traffic volumes offer the best conditions for technological trials. A feeder function for automated minibuses is also a preferred form of service from the perspective of urban transport operators and planning authorities.

Some of the studies on automated driving conducted to date emphasize a spatial differentiation in the impact of automated driving and the fact that this will be manifested in different ways. Zhang and Wang (2020) and Kondor et al. (2020), for example, examined the effects of CAV sharing on parking demand and space usage. These studies showed that for the urban regions considered (Atlanta and Singapore) the impact of the changes would not be uniform from the city centre through to the suburbs. Other studies (e.g. Legêne et al. 2020, Gelauff et al. 2017, Zhang 2017, Thakur et al. 2016) likewise showed an uneven distribution of the effects of CAVs on space usage in cities and urban regions.

To concretize the potentials and risks of the trends in spatial development, further studies are needed that take into account the spatially differentiated impact of automated driving and show the extent of potential effects in different urban structures and areas. This extends not only to the above-mentioned changes in space usage and parking demand, but also to potential changes in public space or transformations in certain types of area. Individual articles in this volume are dedicated to this task and deal with some of the aforementioned effects in different settlement structures and urban spaces. The article by Mitteregger and Soteropoulos (see Chap. 15 in this volume), for example, makes this clear for the effects of CAVs with regard to development and conversion areas that are becoming available. Bruck et al. (see Chap. 8) take a differentiated view of spatial interfaces in public space depending on settlement structures, while Soteropoulos (see Chap. 5) focuses on the compatibility and spatial separation effects of automated transport in different streetscapes.

REFERENCES

Angerer, F., and G. Hadler 2005. "Folgen und Wirkungen des Verkehrs – Städtebauliche Folgen", in *Stadtverkehrsplanung. Grundlagen, Methoden, Ziele* 2, 152–159.

BBSR (Federal Institute for Research on Building, Urban Affairs and Spatial Development) 2010. "Laufende Raumbeobachtung – Raumabgrenzungen. Raumtypen 2010". Bonn. https://tinyurl.com/yyp9cyea (28.12.2020).

Beckmann, J., & A. Brügger 2013. "Kollaborative Mobilität", in *Internationales Verkehrswesen 65*, 57–59.

Bertolini, L. 2012. "Integrating Mobility and Urban Development Agendas – A Manifesto", in *disP – The Planning Review* (48) 1, 16–26.

Bertolini, L. 2017. *Planning the Mobile Metropolis: Transport for People, Places and the Planet*. London: Palgrave/Red Grove Press.

Bormann, O. 2014. "Aktuelle Verkehrslage – Von der Rückgewinnung urbaner Infrastruktur", in *Architektur im Kontext,* K. von Keitz and S. Voggenreiter (eds.). Berlin: Jovis, 96–108.

Borsdorf, A., and O. Bender 2010. *Allgemeine Siedlungsgeographie*. Cologne/Weimar/Vienna: Böhlau.

Boustedt, O. 1953. "Die Stadtregion. Ein Beitrag zur Abgrenzung städtischer Agglomerationen", in *Allgemeines statistisches Archiv* 37, 13–26.

Boustedt, O. 1970. "Stadtregionen", in *Handwörterbuch der Raumforschung und Raumordnung, ed. by* Academy for Territorial Development, 2nd ed. Hanover: Gebrüder Jänecke, 3207–3237.

Boustedt, O. 1975. "Grundriß der empirischen Regionalforschung. Teil 1: Raumstrukturen", *Taschenbücher zur Raumplanung Band 4*. Hanover: Hermann Schroedel.

Bruck, E. M. 2019. "Automatisierte Mobilitätsdienste als Wandlungsimpuls für suburbane Räume?", in *Broadacre city 2.0 – postfossil. Ein urbanistisches Szenario für 2050*, ed. by J. Fiedler. Graz: Haus der Architektur.

Gelauff, G., I. Ossokina and C. Teulings 2017. "Spatial effects of automated driving: dispersion, concentration or both?", working paper, 18/9/2107. Download at https://tinyurl.com/2bsd2ttn (28/12/2020).

Gerdes, J., and G. Heinemann 2019. "Urbane Logistik der Zukunft – ganzheitlich, nachhaltig und effizient", in *Handel mit Mehrwert: Digitaler Wandel in Märkten, Geschäftsmodellen und Geschäftssystemen*, ed. by G. Heinemann, H. M. Gehrckens and T. Täuber. Wiesbaden: Gabler, 397–420.

Ghielmetti, M., R. Steiner, J. Leitner, M. Hackenfort, S. Diener and H. Topp 2017. "Forschungsprojekt SVI 2011/023 Flächiges Queren in Ortszentren: Langfristige Wirkung und Zweckmässigkeit". Bern: Verkehrsteiner AG. https://tinyurl.com/y8ys7453 (28/12/2020).

Kasraian, D., K. Maat, D. Stead and B. van Wee 2016. "Long-term impacts of transport infrastructure networks on land-use change: An international review of empirical studies", in *Transport Reviews* (36) 6, 772–792.

Kondor, D., P. Santi, D.-T. Le, X. Zhang, A. Millard-Ball and C. Ratti 2020. "Addressing the 'minimum parking' problem for on-demand mobility", in *Scientific Reports 10*, 15885. Download at https:// tinyurl.com/3yuvj25a (28/12/2020)..

Kretz, S., and L. Kueng (eds.) 2016. *Urbane Qualitäten – Ein Handbuch am Beispiel der Metropolitanregion Zürich*. Zurich: Edition Hochparterre.

Larco, N., and G. Tierney 2020. "Impacts on Urban Design", in *Multilevel Impacts of Emerging Technologies on City Form and Development*, ed. by A. Howell and K. Lewis Chamberlain. Eugene, OR: University of Oregon, 115–141. https://tinyurl.com/ybdh7udt (28/12/2020).

Legêne, M. F., W. L. Auping, G. Correia and B. van Arem 2020. "Spatial impact of automated driving in urban areas", in *Journal of Simulation 14*, 295–303. https://tinyurl.com/yaxqvh3y (28/12/2020).

Lewis, R., and M. Anderson 2020. "Impacts on Land Use", in *Multilevel Impacts of Emerging Technologies on City Form and Development*, ed. by A. Howell and K. Lewis Chamberlain. Eugene, OR: University of Oregon, 97–113. https://tinyurl.com/ybdh7udt (28/12/2020).

Litman, T. 2020. "Autonomous Vehicle Implementation Predictions: Implications for Transport Planning". Victoria, BC: Victoria Transport Policy Institute. https://www.vtpi.org/avip.pdf (4/6/2020).

Matthes, G., and C. Gertz 2014. "Raumtypen für Fragestellungen der handlungstheoretisch orientierten Personenverkehrsforschung", *ECTL Working Paper 45*. Hamburg: TU Hamburg, Institute for Transport Planning and Logistics.

Milakis, D., B. van Arem and B. van Wee 2017. "Policy and society related implications of automated driving: A review of literature and directions for future research", in *Journal of Intelligent Transportation Systems* (21) 4, 324–348.

Mitteregger, M., E. M. Bruck, A. Soteropoulos, A. Stickler, M. Berger, J. S. Dangschat, R. Scheuvens and I. Banerjee 2022. *AVENUE21. Connected and Automated Driving: Prospects for Urban Europe*, trans. M. Slater and N. Raafat. Berlin: Springer Vieweg. DOI: 10.1007/978-3-662-64140-8.

SAE International 2018. "Taxonomy and Definitions for Terms Related to Driving Automation Systems for On-Road Motor Vehicles – J3016", 15/6/2018. whttps://tinyurl.com/a5f53jus (20/4/2020).

Soteropoulos, A., M. Berger and F. Ciari 2019. "Impacts of automated vehicles on travel behaviour and land use: An international review of modelling studies", in *Transport Reviews* (39) 1, 29–49.

Soteropoulos, A., M. Mitteregger, M. Berger and J. Zwirchmayr 2020. "Automated drivability: Toward an assessment of the spatial deployment of level 4 automated vehicles", in *Transportation Research Part A: Policy and Practice* 136, 64–84.

Statistik Austria 2017. "Urban-Rural-Typologie". https://tinyurl.com/6y3yjkrc (4/6/2020).

Steckler, B. 2019. "Navigating New Mobility: Policy Approaches for Cities", *Urbanism Next Reports*, October 2019. Eugene, OR: Urbanism Next Center, University of Oregon. Download at https://scholarsbank.uoregon.edu/xmlui/handle/1794/25190 (1/12/2020).

Thakur, P., R. Kinghorn and R. Grace 2016. "Urban form and function in the autonomous era", *Australasian Transport Research Forum 2016 Proceedings*. Melbourne. https://tinyurl.com/yblrqyu4 (28/12/2020).

VCÖ (Verkehrsclub Österreich) 2019. "In Gemeinden und Regionen Mobilitätswende voranbringen", VCÖ Mobilität mit Zukunft. https://tinyurl.com/b2pcxdh5 (4/6/2020).

WEF (World Economic Forum) 2018. "Reshaping Urban Mobility with Autonomous Vehicles: Lessons from the City of Boston". https://tinyurl.com/2w6um58r (4/6/2020).

Wegener, M., and F. Fürst 1999. "Land-Use Transport Interaction – State of the Art", *Berichte aus dem Institut für Raumplanung 46*. Dortmund: TU Dortmund.

Weidmann, U., R. Dorbritz, H. Orth, M. Scherer and P. Spacek 2011. "Einsatzbereiche verschiedener Verkehrsmittel in Agglomeration". Swiss Confederation: Federal Department of the Environment, Transport, Energy and Communications UVEK. https://tinyurl.com/uncz38rk (4/6/2020).

Zhang, W. 2017. "The interaction between land use and transportation in the era of shared autonomous vehicles: A simulation model", dissertation, Georgia Institute of Technology, Atlanta, GA.

Zhang, W., and S. Guhathakurta 2018. "Residential Location Choice in the Era of Shared Autonomous Vehicles", in *Journal of Planning Education and Research*, 1–14.

Zhang, W., S. Guhathakurta, J. Fang and G. Zhang 2015. "Exploring the impact of shared autonomous vehicles on urban parking demand: An agent-based simulation approach", in *Sustainable Cities and Society 19*, 34–45.

Zhang, W., and K. Wang 2020. "Parking futures: Shared automated vehicles and parking demand reduction trajectories in Atlanta", in *Land Use Policy 91*, 103963. https://tinyurl.com/4h6er3dc.

3 Connected and automated driving in the context of a sustainable transport and mobility transformation

Andrea Stickler, Jens S. Dangschat, Ian Banerjee

Andrea Stickler
TU Vienna, future.lab Research Center and Research Unit Sociology (ISRA)

Jens S. Dangschat
TU Vienna, Research Unit Sociology (ISRA)

Ian Banerjee
TU Vienna, Research Unit Sociology (ISRA)

© The Author(s) 2023
M. Mitteregger et al. (eds.), *AVENUE21. Planning and Policy Considerations for an Age of Automated Mobility*, https://doi.org/10.1007/978-3-662-67004-0_3

1. INTRODUCTION

Under what conditions can connected and automated transport (CAT) help bring about sustainable mobility? This question has been at the focus of interest of the interdisciplinary research group AVENUE21 from the outset. In view of today's global challenges, no technologies or business models should be permitted that generate further traffic and exacerbate the urban, ecological and social problems faced by today's transport system (cf. Mitteregger et al. 2022). This volume highlights the opportunities and risks of connected and automated vehicles in the transition to sustainable mobility and derives action plans, above all for policymakers and planning bodies. The basic terms "transport", "mobility", "motility", "transition to sustainable transport" and "transition to sustainable mobility" are defined and the significance of CAT in the context of a transition to sustainable transport and mobility considered.

2. DEFINITIONS OF THE TERMS TRANSPORT, MOBILITY, MOTILITY, AND TRANSITION TO SUSTAINABLE TRANSPORT AND MOBILITY

Although the terms transport and mobility are often used synonymously, they have different meanings. The word transport is generally understood to mean the movement or change of location of people, goods and information (cf. Schopf 2001: 4–5; Canzler 2004: 342). "Transport" thus refers to the factual and physical movement of people and goods, including the resulting structural and infrastructural impact (cf. Wilde/Klinger 2017: 6). The term mobility (Latin "mobilitas"), on the other hand, refers to various phenomena such as processes of social advancement (vertical mobility) and value change (horizontal mobility), change of domicile, but also migration (cf. Dangschat 2021). In terms of overcoming distances, this is understood as the possibility of being mobile, i.e. the chance to be free of spatial restrictions to a certain extent. Unlike transport, the concept of mobility emphasizes the capabilities and needs associated with observable changes of location. Typical distinguishing features of these two terms and their use in research have been summarized by Wilde and Klinger (2017: 7; see Overview 1).

Overview 1: Terminological distinction between transport and mobility

Transport	Mobility
movement	capacity for movement
physical	physical, social, cultural
distances and routes as main parameter	activities and accessibility as main parameter
mainly aggregated	mainly individual
often constructional, infrastructural and planning-related problem areas	mainly social and psychological problem areas

Source: Wilde/Klinger (2017: 7)

Another term commonly used, in mobility research in the field of social sciences, is motility. This term stems from biology and refers to the mobility of individual units within a system; in mobility research, it is understood to mean the ability of people (but also of goods or information) to be mobile within a geographical space. Aspects of social inequality are also frequently addressed in this context (cf. Kaufmann et al. 2004: 750). Motility results from access to different forms of mobility and their range of operation, along with the competence to perceive, acquire and use these access options. In addition to financial or time restrictions and skills, relevance is ascribed to an unequal degree of access to digital media, end devices and apps (the "digital divide") in the analysis of innovative mobility offers.

In view of the global climate catastrophy, the urgency of a transition to sustainable transport and mobility has become increasingly clear in recent years. The transport sector is the only area in which CO_2 and NO_x emissions are not only not declining, but are continuing to rise (cf. Vieweg et al. 2018). According to the definition by Agora Verkehrswende (2017), the term "transition to sustainable transport" encompasses the demand for a turnaround in both mobility and energy. It is crucial that energy consumption be reduced and the remaining energy demand be met with climate-neutral energy. While the energy transition is above all a technical challenge, the mobility transition requires a new mobility culture. It is important here to extend the range of integrated transport to facilitate multimodal behaviour (cf. Agora Verkehrswende 2017).

A somewhat different distinction is given by Manderscheid (2020): she distinguishes between turnarounds in propulsion systems, transport and mobility. A transition to sustainable propulsion systems in the transport sector can be understood by analogy with the energy turnaround in electricity and heating that has been driven forward since the early 1990s, although changes in the transport sector have a much stronger impact on the everyday lives and perception of many people (cf. Wissen 2019: 231–232f.). Unlike the transition to sustainable propulsion systems, the transport turnaround comprises a reduction in car ownership or the replacement of private cars by other modes such as public transport, new mobility services or cycling and walking. The transition to sustainable mobility goes even further: it not only concentrates on the physical movements of people and goods in the streetscape, but also includes the virtual, symbolic and imagined dimensions of movement along with its associated meanings and experience horizons (cf. Manderscheid 2020 with reference to Cresswell 2006 and Urry 2007).

Overview 2: Conceptual distinction between the turnarounds in drive systems, transport and mobility

Drive turnaround	Transport turnaround	Mobility turnaround
Change in propulsion technology (electric or hybrid engine etc.)	Reduction in car ownership, or replacement of private cars by other modes (public transport, new mobility services, walking and cycling)	Change in physical movements of people and goods in space, and changed virtual, symbolic and imagined movements

Source: the authors, based on Manderscheid (2020)

The goals of the transition to sustainable (and also active) modes of transport can be described in keywords: better conditions for walking and cycling, more intermodal offers, i.e. being able to use various means of transport within a network, and electrification of the remaining motorized vehicles[1] (cf. Loske 2018). Taking sustainable development seriously in the transport and mobili-

---

1   The shift of emissions to the Global South, destruction of natural landscapes and biodiversity, and impairment of working and living spaces of the (often indigenous) population due to extraction of resources are often ignored in discussions of battery-powered electromobility. Moreover, automobility as a social context is negated within the scope of political strategies (cf. Dangschat/ Stickler 2020, Manderscheid 2012).

ty sector, with the goal of "zero-emission mobility: clean, connected, competitive" (STRIA 2019), means bringing about profound changes in mobility behaviour – above and beyond the technical challenges of the transition to sustainable energy and propulsion systems. This calls for both "push" and "pull" measures in integrated transport planning and policy. Over the past few years, there has therefore been increasing pressure on governments and industry, for example, to press ahead with radical initiatives in the mobility sector with a view to reducing CO_2 emissions. To organize mobility in a more sustainable way, three strategies must come into effect:

1. **avoiding** non-essential traffic,

2. **shifting** traffic to efficient, environmentally friendly modes of transport; and

3. **improving** the mobility offer and the quality of stay in public spaces.

A transition to sustainable transport also generates added value that goes beyond climate protection: air and noise pollution are to be reduced, and the health and quality of stay in public spaces should also be improved (cf. Agora Verkehrswende 2017: 91).

3. INNOVATIONS IN THE TRANSITION TO SUSTAINABLE TRANSPORT (SHARED MOBILITY, MOBILITY AS A SERVICE, CONNECTED AND AUTOMATED DRIVING)

Over the past few years, an increasing number of organizational and technological innovations have arisen in the mobility sector that are seen as beacons of hope for climate-friendly mobility. If CAT is to contribute to the transition to sustainable transport and mobility, (automated) vehicles must be emission-free, be embedded in an integrated Mobility-as-a-Service (MaaS) concept and be launched on the roads as shared mobility (cf. Lennert/Schönduwe 2017, Mitteregger et al. 2022). However, the far-reaching transformation of systems is a non-linear, complex process. Many factors must therefore interact in order for processes of innovation and implementation to be successful (cf. Kesselring 2020).

Geels (2012) adopts a multilevel perspective in describing socio-technical systems as a complex comprising three levels – "niches, regimes and landscapes" – thus reconciling structuralist and action-centred approaches. According to this perspective, technological or social innovations arise in protected niches (e.g. within funding programmes) before some are gradually adapted and become established within society. The process perspective makes it possible to systematically document innovations in transport, such as the introduction of MaaS offers, shared mobility or e-mobility, while taking into account both the logic of the agents' actions and the influence of the structural framework conditions (cf. Wilde/Klinger 2017: 18).

CAT entails technological innovations that will radically reorganize transport and our mobility. With increasing convenience through automation and connectivity, travel time will be freed up for other activities and will thus be differently assessed; this in turn influences accessibility and thus also transport demand (cf. Soteropoulos et al. 2018). Over the long term, it is therefore likely that CAT will also affect choices of location for households and businesses, thereby influencing settlement patterns. On the basis of various scenarios, it can be shown that automated driving will give rise in future to new modes of transport that may supplement the existing transport system, or even supersede or displace established modes of transport. There is a risk here that sustainable (transport policy-related) development goals will be counteracted by an increase in the number and length of journeys due to the new qualities of private vehicles (cf. Berger et al. 2020).

4. THE AMBIVALENT IMPACT OF CONNECTED AND AUTOMATED TRANSPORT ON THE TRANSPORT AND MOBILITY TRANSFORMATION

Continuing digitization and automation will have a significant impact on mobility – in a largely positive way, according to the general consensus of the mainstream: with greater road safety, higher energy efficiency, reduced environmental burdens and remobilization for persons with limited mobility (cf. STRIA 2019). App-controlled and platform-based connected mobility services could also pave the way for a less car-oriented multimodality, with different modes of transport being used for specific routes, and ultimately a reduction in motorized private transport. These "idealized" visions, however, contrast with scenarios which predict that connected and automated driving will create more traffic, urban sprawl will further increase, and public transport and active mobility in the form of walking and cycling will meet with a great deal of additional competition (cf. Dangschat 2017b, 2019; Soteropoulos et al. 2019; Milakis et al. 2017). Moreover, scepticism towards connected and automated driving is very widespread in Germany – especially among the older population and women (cf. Fraedrich/Lenz 2015).

If policymakers want to create incentives for a change in behaviour to support the new transportation revolution, they should become more aware of which potential target groups respond to which incentive schemes. Financial incentives are helpful if they are powerful, but are by no means tailor-made for different interest groups and motivational profiles among the population. In the mobility sector, there is also an urgent need to take greater account of differences among regions, but above all between urban and regional spaces, in the transport system when it comes to achieving multimodality that largely dispenses with cars.

5. THE CONTRIBUTION OF CONNECTED AND AUTOMATED TRANSPORT TO THE TRANSPORT AND MOBILITY TRANSFORMATION

CAT can make an important contribution to the transition to sustainable transport and mobility. However, this requires the technology to be controlled and supplemented by suitable political and planning measures. How and under what conditions this would be appropriate is discussed in detail by various authors in this volume (see e.g. Chap. 4 by Manderscheid, Chap. 6 by Soteropoulos et al., Chap. 11 by Eichholz and Kurth, Chap. 14 by Mitteregger et al. and Chap. 18 by Stickler).

For the time being, however, we can briefly point out some significant aspects: policymakers and planning bodies are usually somewhat perplexed by the question of how to make the mobility behaviour of the population more sustainable. Although "avoid", "shift" and "improve" are fitting keywords, they have hardly been consistently implemented to date. For the most part, the political and planning approach merely comprises blanket economic instruments in a complex landscape of lock-in effects – in the form of tax incentives such as the commuter allowance and home ownership subsidies, car-oriented settlement structures and an inappropriate settlement policy regarding specialist retail markets, shopping centres, etc. – and adherence to the image of the automobile as a symbol of social advancement and prestige, including its modernization through automation and networking (cf. Dangschat 2019). Technological developments such as CAT will not alone suffice to achieve the above-mentioned goal of a transition to sustainable transport and mobility. Rather, people's mobility behaviour must likewise adapt to these goals. This can be supported for example by clearly prioritizing alternative, post-fossil transport modes when it comes to policy and planning.

REFERENCES

Agora Verkehrswende 2017. "Mit der Verkehrswende die Mobilität von morgen sichern. 12 Thesen zur Verkehrswende". Berlin. https://tinyurl.com/ycjsvvzd (28/12/2020).

Bamberg, S., M. Hunecke and A. Blöbaum 2007. "Social context, personal norms and the use of public transportation: Two field studies", in *Journal of Environmental Psychology* (27) 3, 190–203.

Berger, M., V. Sodl, L. Dörrzapf, C. Kirchberger and A. Soteropoulos 2020. "Herausforderung Mobilitäts- und Verkehrswende – Stärkung einer integrierten Betrachtung von Raum und Verkehr sowie Wissenschaft und Praxis", in *50 Jahre Raumplanung an der TU Wien. Studieren – Lehren – Forschen. Jahrbuch Raumplanung 2020*, ed. by T. Dillinger, M. Getzner, A. Kanonier and S. Zech. Vienna: Neuer Wissenschaftlicher Verlag, 258–273.

Canzler, W. 2004. "Wege aus der 'verfahrenen' Verkehrspolitik?", in *Informationen zur Raumentwicklung 6*, 341–350.

Cresswell, T. 2006. *On the Move: Mobility in the Modern Western World*. New York/London: Routledge.

Dangschat, J. S. 2017a. "Wie bewegen sich die (Im-)Mobilen? Ein Beitrag zur Weiterentwicklung der Mobilitätsgenese", in *Verkehr und Mobilität zwischen Alltagspraxis und Planungstheorie. Ökologi-ogische und soziale Perspektiven*, ed. by M. Wilde, M. Gather, C. Neiberger and J. Scheiner Wiesbaden: Springer VS, 25–52.

Dangschat, J. S. 2017b. "Automatisierter Verkehr – was kommt da auf uns zu?", in *Zeitschrift für Politikwissenschaft* (ZPol) 27, 493–507.

Dangschat, J. S. 2019. "Gesellschaftlicher Wandel und Mobilitätsverhalten", in *Nachrichten der ARL* 01/2019, 8–11.

Dangschat, J. S. 2021. "Wohnen und Mobilität", in *Wohnen – interdisziplinäre Positionen und Perspektiven*, ed. by C. Hannemann, N. Hilti and C. Reutlinger. Stuttgart: Fraunhofer. In print.

Dangschat, J. S., and A. Stickler 2020. "Kritische Perspektiven auf eine automatisierte und vernetzte Mobilität", in: *Jahrbuch StadtRegion 2019/2020, Schwerpunkt: Digitale Transformation*, ed. by C. Hannemann, F. Othengrafen, J. Pohlan, B. Schmidt-Lauber, R. Wehrhahn and S. A. Güntner. Wiesbaden: Springer, 53–74.

European Commission 2018. "10. Clean, Connected and Competitive Mobility". Brussels. https://ec.europa.eu/clima/sites/clima/files/docs/pages/initiative_10_mobility_en.pdf (28/12/2020).

Fraedrich, H., and B. Lenz 2015. "Gesellschaftliche und individuelle Akzeptanz des autonomen Fahrens", in *Autonomes Fahren. Technische, rechtliche und gesellschaftliche Aspekte*, ed. by M. Maurer, J. C. Gerdes, B. Lenz and H. Winner Berlin/Heidelberg: Springer Vieweg, 638–660.

Geels, Frank W. 2012. "A socio-technical analysis of low-carbon transitions: introducing the multi-level perspective into transport studies", in *Journal of Transport Geography* 24, 471–482.

Götz, K. 2007. "Mobilitätsstile", in *Handbuch Verkehrspolitik*, ed. by O. Schöller, W. Canzler and A. Knie. Wiesbaden: VS Verlag für Sozialwissenschaften, 759–784.

Hunecke, M. 2015. *Mobilitätsverhalten verstehen und verändern. Psychologische Beiträge zur interdisziplinären Mobilitätsforschung.* Wiesbaden: VS Verlag für Sozialwissenschaften.

Kaufmann, V., M. M. Bergman and D. Joye 2004. "Motility: Mobility as Capital", in *International Journal of Urban and Regional Research* (28) 4, 745–756.

Kesselring, S. 2020. "Reflexive Mobilitäten", in: *Das Risiko – Gedanken über und ins Ungewisse. Interdisziplinäre Aushandlungen des Risikophänomens im Lichte der Reflexiven Moderne. Eine Festschrift für Wolfgang Bonß*, ed. by H. Pelizäus and L. Nieder Wiesbaden: Springer VS, 155–193.

Lennert, F., and R. Schönduwe 2017. "Disrupting Mobility: Decarbonising Transport?", in *Disrupting Mobility. Impacts of Sharing Economy and Innovative Transportation on Cities*, ed. by G. Meyer and S. Shaheen Basel: Springer International Publishing, 213–238.

Loske, R. 2018. "Klimafreundliche Mobilität für alle. Wo bleibt die Verkehrswende?", in *Neue Gesellschaft Frankfurter Hefte* 4, 25–28. https://www.frankfurter-hefte.de/artikel/klimafreundliche-mobilitaet-fuer-alle-2051/ (28/12/2020).

Manderscheid, K. 2012. "Automobilität als raumkonstituierendes Dispositiv der Moderne", in *Die Ordnung der Räume. Geographische Forschung im Anschluss an Michel Foucault*, ed. by H. Füller and B. Michel. Münster: Westfälisches Dampfboot, 145–178.

Manderscheid, K. 2020. "Antriebs-, Verkehrs- oder Mobilitätswende? Zur Elektrifizierung des Automobildispositivs", in *Baustelle Elektromobilität*, ed. by A. Brunnengräber and T. Haas. Berlin: transcript, 37–68.

Milakis, D., B. van Arem and B. van Wee 2017. "Policy and society related implications of automated driving: A review of literature and directions for future research", in *Journal of Intelligent Transportation Systems* (21) 4, 324–348.

Mitteregger, M., E. M. Bruck, A. Soteropoulos, A. Stickler, M. Berger, J. S. Dangschat, R. Scheuvens and I. Banerjee 2022. *AVENUE21. Connected and Automated Driving: Prospects for Urban Europe*, trans. M. Slater and N. Raafat. Berlin: Springer Vieweg. DOI: 10.1007/978-3-662-64140-8.

Schopf, J. M. 2001. "Mobilität & Verkehr – Begriffe im Wandel", in *Wissenschaft & Umwelt Interdisziplinär* 3, 3–11.

Soteropoulos, A., M. Berger and F. Ciari 2018. "Impacts of automated vehicles in travel behaviour and land use: An international review of modelling studies", in *Transport Reviews* (39) 1, 29–49.

Soteropoulos, A., A. Stickler, V. Sodl, M. Berger, J. S. Dangschat, P. Pfaffenbichler, G. Emberger, E. Frankus, R. Braun, F. Schneider, S. Kaiser, H. Wakolbinger and A. Mayerthaler 2019. "SAFiP – Systemszenarien Automatisiertes Fahren in der Personenmobilität. Endbericht". Vienna: Federal Ministry for Climate Action, Environment, Energy, Mobility, Innovation and Technology. https://projekte.ffg.at/anhang/5cee1b11a1eb7_SAFiP_Ergebnisbericht.pdf (2/2/2020).

STRIA (Strategic Transport Research and Innovation Agenda) 2019. *Roadmap on Connected and Automated Transport: Road, Rail and Waterborne*. Brussels: European Commission.

Urry, J. 2007. *Mobilities*. Cambridge: Polity Press.

Vieweg, M., D. Bongardt, C. Hochfeld, A. Jung, E. Scherer, R. Adib and F. Guerra 2018. "Towards Decarbonising Transport 2018. A 2018 Stocktake on Sectoral Ambition in the G20", Report on behalf of Agora Verkehrswende and the German Society for International Cooperation (GIZ). Download at https://tinyurl.com/yatbpjqw (28/12/2020).

Wilde, M., and T. Klinger 2017. "Integrierte Mobilitäts- und Verkehrsforschung: zwischen Lebenspraxis und Planungspraxis", in *Verkehr und Mobilität zwischen Alltagspraxis und Planungstheorie. Ökologische und soziale Perspektiven*, ed. by M. Wilde, M. Gather, C. Neiberger and J. Scheiner. Wiesbaden: Springer VS, 5–25.

Wissen, M. 2019. "Kommodifizierte Kollektivität? Die Transformation von Mobilität aus einer Polanyi'schen Perspektive", in *Große Transformation? Zur Zukunft moderner Gesellschaften, Sonderband des Berliner Journals für Soziologie*, ed. by K. Dörre, H. Rosa, K. Becker, S. Bose and B. Seyd. Wiesbaden: Springer VS, 231–244.

PART I
Mobility and transport

Aggelos Soteropulos, Martin Berger

Automated vehicles will fundamentally transform mobility as we know it today. But what areas of mobility, e.g. public transport or freight transport, will be influenced by automated driving, and in what ways? What developments are discernible here? How does transport and urban planning approach automated driving, and what contribution can it make in view of the climate crisis? The authors address these questions in their articles for this chapter, "Mobility and Transport". They provide specific, in-depth insights into these topics and issues, which they also discuss in connection with possible proposals for action from the perspective of transport and urban planning.

In her chapter *Self-driving turnaround or automotive continuity? Reflections on technology, innovation and social change*, Katharina Manderscheid discusses the possible roles of self-driving vehicles within the context of various manifestations of the current transformation in transport. She makes a distinction here between the drive and automation turnaround, the traffic turnaround and the mobility turnaround. She points out that the technology of self-driving vehicles alone does not define the transport and social mobility of the future, but merely represents one element within a complex process of negotiation among politics, economics and society. In future, it will therefore be important for scenarios of societal mobility futures with self-driving vehicles to take into account not only the interests and visions of individual major industries and players, but also the less vociferously propounded visions.

In the chapter *Automation, public transport and Mobility as a Service: Experience from tests with automated shuttle buses*, Aggelos Soteropoulos, Emilia Bruck and Martin Berger give an overview of what public transport with automated vehicles could look like. The authors focus here specifically on automated shuttle buses, which are being accorded priority in the course of tests in public transport projects with automated vehicles. Selected pilot projects with automated shuttle buses are presented in here the form of a comparative analysis, and two such projects are then examined in more detail: based on the example of the "autoNV OPR" project in Ostprignitz-Ruppin in Germany, Arne Holst and Alexander Egoldt highlight technical and legal aspects of tests with automated shuttle buses in public transport. Zoltán László then uses the example of the "MyShuttle" project in Zug, Switzerland, to present aspects of the operation and integration of such buses in public transport networks. These examples also show where the emphasis should be placed in future pilot projects involving automated shuttle buses in public transport.

Bert Leerkamp, Aggelos Soteropoulos and Martin Berger address the role of delivery robots in urban goods transport in their chapter *Delivery robots as a solution for the last mile in the city?*. The authors first describe the various types of delivery robot and give an overview of current tests. On the basis of previous experience in tests with delivery robots suitable for operation on pavements, they discuss their potential as a solution for the last mile along with the risks involved. The authors also derive implications for planning and describe why giving early strategic consideration to deploying delivery robots is particularly important for traffic and urban planning.

In the chapter *Automated drivability and streetscape compatibility in the urban-rural continuum using the example of Greater Vienna*, Aggelos Soteropoulos casts light on the role of road space in consideration of the effects of automated driving on traffic. Using case studies of municipalities in the Vienna city region, he first presents the technical-infrastructural suitability of road spaces for automated driving systems. The author then analyses the number of vehicles that are compatible for selected road spaces of certain municipalities. On the basis of these two analyses, he discusses which road spaces are or are not suitable for automated driving, both from a technical-infrastructural viewpoint and in terms of its compatibility with usage of the streetscape surroundings.

4   Self-driving turnaround or automotive continuity? Reflections on technology, innovation and social change

Katharina Manderscheid

Katharina Manderscheid
University of Hamburg, Department of Social Economics, Sociology

© The Author(s) 2023
M. Mitteregger et al. (eds.), *AVENUE21. Planning and Policy Considerations for an Age of Automated Mobility*, https://doi.org/10.1007/978-3-662-67004-0_4

1. INTRODUCTION

The automobile is one of the salient features of modern industrialized societies and symbolizes social progress and individual prosperity. In the countries of the Western world, the number of private cars per capita of the population and the distances travelled daily by car continue to increase. In 2018, there were 568 passenger cars per 1,000 inhabitants in Germany (Umweltbundesamt 2019b); prior to the Corona pandemic[1] each person covered an average of 29 out of 39 kilometres daily by car as a driver or passenger (Nobis/Kuhnimhof 2018: 46). There is clearly a close connection between economic growth and the growth of freight and passenger transport (Altvater 2007: 787; Verron et al. 2005: 7).

Motorized individual transport, especially traffic based on the so-called "driver car" (Dant 2004), is now also seen as a symbol of the ecological unsustainability of the modern lifestyle and is increasingly reaching its limits in cities. The negative implications of private motorized transport, and in particular its fossil fuel drive, have now moved into the political arena — not only since German courts started to give precedence to health protection over the interests of diesel car drivers (e.g. Stuttgart Administrative Court, 2017) or the Paris climate protection goals were resolved (cf. Umweltbundesamt 2019a). Indeed, in the European Union motorized road transport is the only sector (apart from aviation) in which CO_2 emissions have increased in relation to the reference year 1990 (European Commission 2017: 126, 134). For cities, the large number of private cars presents a further pressing problem in view of their high space requirements on roads and in parking areas. Especially in prospering metropolises, these requirements are increasingly competing with space demands on the part of slow and active traffic and for the purpose of recreation and leisure and for residential usage. In light of this, the need for a turnaround in mobility and transport is now generally acknowledged. Already today, various trends in development are discernible that can be understood as signs of such change. On the other hand, it remains to be seen where the change in this sector is heading, how far-reaching this transformation must or may be, and how it can be initiated.

With the aim of gaining a more differentiated view of current developments in the field of self-driving transport technology and mobility, the relationship between technological innovation and social change will first be examined in the following from a sociological perspective. Using the example of various autonomous vehicle prototypes, it will be shown that technologies are closely interwoven with specific notions of social structures. However, society is not simply being transformed by technological developments: not least the history of the automobile shows that technological innovations are dependent on politically generated framework conditions and on integration into social practices. An ideal-typical distinction is then made between three forms of change in transport: (1) a drive and automation turnaround, which above all comprises technological further development of the current dispositive of automobility (Manderscheid 2012), (2) a transport turnaround, in which other means of transport besides the private car play an increasingly significant role, and (3) a mobility turnaround, which focuses on the driving factors, dynamics and constraints involved in overcoming spatial distance and acceleration effects. In the following, the possible roles of self-driving vehicles are discussed for each type of turnaround. This makes it clear that the technology of self-driving vehicles does not in itself dictate the transport and social mobility of the future, but is merely one element of a complex process of negotiation in politics, economics and society at large.

---

1   The changes in transport behaviour during and immediately following the lockdown are examined in a study by Infas (infas/Motiontag 2020). It is doubtful whether transport behaviour will return to the former distribution patterns in the near future.

2. TECHNOLOGICAL INNOVATION AND SOCIAL TRANSFORMATION

Political and public discussion of the future of transport is focusing on technical solutions. New developments in the field of vehicle technology are likely to address the negative effects of the status quo or provide impulses for a more or less fundamental change within society. Technology and innovation are seen here as quasi extra-societal forces that more or less directly impact society from the outside and change the behaviour of individuals. An obstacle on the path to technological solutions is repeatedly seen in public acceptance of new technologies (Fraedrich/Lenz 2015). At times, attempts are made to generate acceptance by means of information and advertising strategies or socio-scientific accompanying measures.

Autonomous or self-driving cars are seen as a technological innovation that should make a significant contribution to solving today's traffic problems (e.g. Vaid 2018). Replacing the driver with automated control systems is expected to make the use of vehicles safer for passengers and other road users, ensure efficient driving and thus low fuel consumption, reduce space requirements and congestion by means of more precise driving and parking, and reduce the overall number of vehicles on the roads. The automation of vehicles and of traffic itself is widely accepted as a fact for the future. Socio-political and socio-scientific discussion concentrates above all on ethical issues, problems of public acceptance and political challenges (Dangschat 2017, Goodall 2019, Lenz/Fraedrich 2015a, Schreurs/Steuwer 2015, Thomopoulos/Givoni 2015).

A different view of the relationship between innovation and social change can be found in techno-sociology, which regards technology in its development, form and potential, its social significance and its fundamentally contingent use as a societal phenomenon (e.g. Blättel-Mink 2006, Rammert 2010). The development, dissemination and use of technological innovations are understood here as a multidimensional process within society that is by no means easy to plan or predict (e.g. Elias 1995). Regarding the question as to what autonomous vehicles mean for traffic and thus for society, it follows from these techno-sociological considerations that it is first necessary to broaden our field of vision and to illuminate the diverse interrelationships of social and technical development.

Certain views of problems, along with values and assumptions regarding target groups, are already incorporated into the development stage of technologies. From the point of view of sociological technology research, technological artefacts – and thus also vehicles – are seen not merely as material-technical objects, but also as an expression of specific social assumptions, values and perceptions of problems. The social scientist Madeleine Akrich describes this social dimension of technology with the concept of "script".

> "[W]hen technologists define the characteristics of their objects, they necessarily make hypotheses about the entities that make up the world into which the object is to be inserted. Designers thus define actors with specific tastes, competences, motives, aspirations, political prejudices, and the rest, and they assume that morality, technology, science, and economy will evolve in particular ways. A large part of the work of innovators is that of 'inscribing' this vision of (or prediction about) the world in the technical content of the new object. I will call the end product of this work a 'script' or a 'scenario'" (Akrich 1992: 207f.).

The diversified car market in particular reflects this diversity of differentiated target groups and assumed patterns of use: business limousines and sports cars contrast with family cars, panel vans and small and micro cars (Wajcman 1994). The more or less clearly defined target group described in this socio-technical script is accompanied by assumptions about their spatial environment and the broader social context. In addition to the development engineers, this script is also influenced by public discourse and media representations, and the perceptions that these condition among society.

Various differently accentuated narratives can be found in the scripts of autonomously driving cars, as is shown in a comparison of the marketing visualizations of a number of these vehicles (cf. Manderscheid 2018): the first widely known prototype of a self-driving car was the Google car Firefly (Google Self-Driving Car Project 2014); however, its development has now been discontinued. With this prototype, Google formulates the claim to make urban traffic more *socially inclusive and safer* – by reducing the scope of human agency. Social inclusion refers here to the mobilization of groups of people who did not previously use cars due to physical limitations, such as the blind or the elderly. In this narrative, social inclusion is technically mediated via individual auto-mobility, which makes the individual socially and temporally independent of their context. In this script, autonomous driving technology takes the place of other persons, who drive, or of timetables.

Figure 1: Google's Firefly

A second narrative regarding self-driving cars is the time saved for individuals by being relieved of the task of driving and thus having the opportunity to concentrate on the "essentials". The Mercedes F 015 concept car (Mercedes-Benz 2017), for example, visualizes this in the form of business meetings, while the Google subsidiary Waymo (2019) focuses on family, leisure and socializing while being driven. In addition, self-driving cars are to be able to autonomously take over everyday driving tasks, e.g. for shopping and errands, or taking children to and from their destinations (VW 2017, 2018).

In both of these cases, actively steering the vehicle, which to date has been an essential and identity-forming element of driving, is interpreted as a waste of time and a burden. Here, the new technology assumes the role of a chauffeur, available to reasonably affluent households. The promise is that this will create new time windows in the normally hectic everyday routine.

Figure 2: The Mercedes F 015 concept car (top) and the VW Sedric (bottom)

The technology of networked autonomous driving is not limited to use in passenger cars, but is also undergoing development for public transport. Self-driving buses are being tested or are already in use in urban contexts (e.g. in Helsinki, Hamburg, Berlin and Vienna), on company and airport premises (e.g. at Frankfurt Airport) and on university campuses. For rural areas, too, replacing the driver with technology is expected to open up new opportunities for transport connections. The technology of autonomous driving is thus by no means restricted to passenger cars; its use in larger vehicles such as buses, lorries or similar is also conceivable. In these applications, the focus is above all on economic efficiency due to the cost savings achieved by replacing the driver and on flexible, demand-driven use.

Figure 3: A Navya shuttle of the transport operator Wiener Linien

It is not possible to simply derive future traffic scenarios from these scripts of autonomous vehicles, which can be understood as current notions of future transport (cf. Grunwald 2009). The actual integration of new technologies into the social framework takes place in the course of multilayered interrelationships. In the tradition of cultural studies (cf. During 2000), various studies have shown that users repeatedly integrate technologies into everyday life in unpredictable ways and develop new practices. Examples are studies on the dissemination and development of idiosyncratic uses of the Walkman (Du Gay et al. 1997) or of the mobile phone (Rettie 2008, Thulin/Vilhelmson 2007, Wajcman 2008). These adaptations do not occur spontaneously, but are flanked by generated framework conditions. The macrostructural processes of integrating technical innovations into the social context are systematically elaborated in the approach of socio-technical transition (Geels/Schot 2007, Kanger et al. 2019). The history of the private car likewise impressively shows that the process of establishing new vehicle technologies is complex and has been significantly supported by political framework conditions and social attributions of meaning.

This sociopolitical basis, which founded the success of the car, is elaborated by various authors (Dennis/Urry 2009, Kuhm 1997, Norton 2008, Paterson 2007); this deconstructs the myth of a "natural spread" of the car that caters to pre-existing needs among the population. Rather, the social acceptance of the car was preceded by massive lobbying and political decisions on the direction to be taken, and by the largely publicly supported development of the corresponding infrastructure such as car-friendly roads and urban structures, along with the creation and adaptation of the corresponding legal framework and institutions. This development has been flanked by the decline of public rail vehicles, which were the means of mass transport in early industrial societies (Knie 2007: 51; cf. Norton 2008). Furthermore, the corresponding social practices had to emerge: motor sport, car travel, shopping and leisure, and social distinction through the car (Gerhard 2000, Miller 2001, Peters 2006); modes of social subjectivation (Bonham 2006, Laurier et al. 2008, Manderscheid 2013) and the concomitant structures of need and desire also had to become established. In other words, the specific form of automobility that appears "natural" and self-evident to us today came about in a complex interplay of political, economic, material-spatial and social power relations and cannot be derived from the potentiality of automotive technology alone. The way cars are integrated into contemporary societies and the everyday lives of individuals hardly corresponds any longer to the socio-technical script pursued by the developers of the first vehicles.

From a techno-sociological perspective, therefore, future integration into social practices and everyday organizations can be deduced neither from the potentiality of the technology nor from the intentions and ideas of the developers. Rather, the dissemination of new technologies is promoted, steered or impeded by a number of overriding conditions and is also subject to a societal obduracy in the attribution of significance to and the emergence of new socio-technical practices that cannot be anticipated in advance.

3. A TURNAROUND IN DRIVE AND AUTOMATION, TRAFFIC OR MOBILITY?

With regard to the depth and breadth of the turnaround in transport, an analytical distinction can be made between three concepts of varying depth: (1) a drive and automation turnaround, (2) a traffic turnaround and (3) a mobility turnaround. In the following, these three forms of transformation in road traffic will be examined in terms of the possible roles of self-driving vehicles and of the corresponding conditions for opening up opportunities for politics and society. In view of the above considerations, however, this typology should by no means be seen as a "blueprint" for generating corresponding futures, but rather as a demonstration of the socio-political preconditions that go beyond technological innovations.

In the field of individual mobility, politicians and companies focus above all on new drive systems such as electric and hybrid technologies. In this context, German Federal Minister of Transport Andreas Scheuer explicitly speaks of a "drive turnaround", i.e. the gradual replacement of combustion engines with units powered by hydrogen, fuel cells or batteries (Gathmann/ Traufetter 2018). In addition, autonomous vehicles and the gradual introduction of various automated driving assistance systems are expected to lead to more efficient driving and thus to fuel and emission savings and, above all, to greater road safety; this can be described as an *automation turnaround*. Automated safety technology such as turning assistants for lorries along with braking, emergency lane keeping, cruise control or reversing assistants will become the new standard in vehicle manufacture as a result of EU Regulation 2019/2144. These innovations can build on a long tradition of technical improvements for greater road safety – from uniform vehicle lighting, speed limits, seat belts for vehicle occupants, improved braking systems and airbags to automatic accident alerts. The introduction of new drive and assistance systems is currently being promoted by legal regulations, the public expansion of appropriate infrastructure and financial incentives for the purchase of corresponding vehicles. However, these political measures are not intended to bring about a fundamental transformation, but rather to further promote and develop hegemonic privatized automobility. It is quite conceivable that the partial automation already existing today will be gradually extended and that in a few years' time entirely self-driving passenger vehicles will also be seen on the roads. Such fully autonomous private vehicles would presumably first enable the economic elite to expand their radius of action in geographical space by spending their travel time with activities other than driving. This would increase the spatial and temporal independence of this social group, but in the long run would further accelerate social life overall while maintaining privatized automobility (Manderscheid 2012, Rosa 2005). The current pandemic situation could well also be a significant driver of this development.

A distinction can be made between a drive and automation turnaround and a *transport turnaround*, which sets out to supplement or reduce traffic in private cars both as an aggregate and in individual practice with other, more ecologically compatible modes, while retaining the overall level of mobility. Especially in large cities and metropolitan regions, the use of alternative modes of transport is increasingly being promoted – from the extension of public transport to the encouragement of so-called active transport (walking and cycling), the licensing of new micro-vehicles such as electric kick-scooters, and an offer of various mobility services (MaaS – Mobility as a Service). MaaS includes, for example, car-sharing offers – station-based and "one-way" (Lanzendorf/Hebsaker 2017: 137f.) – and app-based ride-hailing services, i.e. driving services that chauffeur individuals or groups with similar routes on demand, along individual or fixed routes. While public and private ride-hailing services are currently performed by a driver, experiments are already being carried out with self-driving vehicles in both of these areas (e.g. Lenz/Fraedrich 2015b). This politically intended extension of the urban mobility offer anticipates that users will flexibly and spontaneously assemble their own transport solutions from the various offers, depending on location, time of day, weather, occasion and destination (Lanzendorf/Hebsaker 2017: 145). The proponents of this initiative therefore believe that maintenance of current transport volumes in urban space could also be ensured beyond the sphere of private automobility. The focus of the transformation here is primarily on urban transport and its efficient organization. The driver of such a change is the increasing dysfunctionality of privatized car transport in cities, while alternative modes of transport are at the same time becoming socially and politically more attractive. In this connection, self-driving vehicles represent a technological option that – to the extent to which they are becoming embedded in information and communication technologies – facilitates a change from a vehicle approach to a system approach in road transport. The proposals of the National Platform Future of Mobility (NPM) in Germany, for example, which are formulated above all in the context of digitalization and autonomous vehicles, are directed towards systemically networked mobility concepts (NPM 2019: 46). Within such a paradigm shift in transport models, the electrically powered self-driving car would be understood as an element of multimodal transport concepts that is part and parcel of the energy transition and would no longer be seen as deficient in comparison with the hitherto predominant petrol-engined car (Lenz/ Fraedrich 2015b; Sauter-Servaes 2011: 37).

Thirdly, a distinction can also be made between these two transformation perspectives and the *mobility turnaround*, whereby such a clear differentiation between transport and mobility turnaround is often not made in social discourse. In the present context, the use of the term mobility rather than transport is intended to emphasize that the whole matter is to be thought of more holistically and that – in addition to the empirically observable physical movement of people and goods in the streetscape – virtual, symbolic and imagined movements are also intended, along with the associated implications and social horizons of meaning (cf. Cresswell 2006, Urry 2007). This extended understanding of movement originates from sociological mobility research and the so-called "new mobilities paradigm" (Sheller/Urry 2006). From this perspective, various authors argue that since the second half of the 20th century, automobility – as a system (Urry 2004), a regime (Böhm et al. 2006) or a dispositive (Manderscheid 2012) – has not been a purely technical-functional means of overcoming distance that plays a merely objective role in society, but on the contrary is a constitutive element of social and economic dynamics and of spatial organization (Kuhm 1997, Paterson 2007). To put it pointedly, the spatial, temporal, economic, social and symbolic order of contemporary societies can only be understood with and through the privately driven car as a hegemonic medium of mobility. Conversely, this also means that for a sociological understanding of automobility it is not sufficient to attribute car traffic and its growth to a single causal dimension such as needs of the individual, a superiority of technology or globalization. Accordingly, sociological research into mobility considers not only the routes and distances travelled and the means of transport used, but also the socio-economic, cultural and spatial dynamics and constraints that underlie social normalities and individual needs. The dynamics that lead to a sustained increase in these observable routes and distances, as well as the leeway and restraints that individuals face in dealing with the mobility expectations of society, can then be incorporated into this integral perspective (Cass/ Manderscheid 2019).

A mobility turnaround towards social and ecological sustainability would then aim at reducing the compulsion to be mobile – in other words, it would supplement the right to mobility with a right to immobility (cf. ibid., Rajan 2007). According to the basic tenet of the mobility paradigm, mobilities arise from social contexts such as employment, family, circles of friends, and the spatial organization of everyday life (Cass et al. 2005, Hammer/Scheiner 2006, Larsen et al. 2006, Shove 2002). Individuals and social groups differ significantly here in the extent to which they can autonomously shape their mobilities and immobilities. In particular, the demands of labour markets and social policy, and of the conditions of the housing market and the infrastructure of residential areas in terms of transport and supply, make mobility necessary in order to participate in society. The technology of automated vehicles can definitely be included in the development of ideas for a socially and ecologically sustainable future of mobility in order to ensure flexible transport connections above all in peripheral or rural regions. Self-driving cars would then chiefly be an element of integrated transport services in the sense of MaaS that can be flexibly ordered and used. However, it can be assumed that such an offer can only be provided by the private sector to a limited extent; this would require structures to be developed by public and civic bodies. The key drivers of such a mobility turnaround are therefore to be found outside the sphere of transport technology and presuppose fundamental changes in the social and economic framework conditions of everyday life.

4. ENVISIONING AND SHAPING THE FUTURE

The notions of possible futures are by no means inconsequential academic flights of fancy. Rather, they are potentially performative and have an influence on the production of future in society, in that they influence actions in the present. In terms of discourse theory, visions of the future are elements of discursive practices that "systematically form the objects of which they speak" (Foucault 1981: 71). As (visualized) elements of communication, social and technological visions of the future already exist in the present and become part of future knowledge by framing and limiting the discursive space of social production of the future, by defining and excluding what constitutes a problem that needs to be addressed. Although driverless cars and buses still only exist as prototypes, their incorporation into motorized traffic as part of current transport policy, planning and legislation is already being prepared materially and in terms of infrastructure. At the same time, this knowledge of the future legitimizes political and economic decisions such as pressing ahead with the expansion of the 5G network. Scientific research and discussion on driverless cars, their possible acceptance by society and their effects on the organization of traffic, along with ethical challenges, likewise contribute to the establishment of the driverless car as a knowledge-based traffic object.

These theoretical reflections draw attention to heterogeneous attempts to shape the future. In view of climate and health crises, the "peak car" hypothesis and outlines of a sharing economy, prior to the Corona pandemic the future of private automobility was perhaps more open than it had been for a long time. Nevertheless, numerous protagonists play a role here, among which the car industry is a powerful player with a strong interest in continued automobility for the future. The interests of IT companies and the automotive industry appear to coincide, at least in part, in the field of networked automated transport technology, without an underlying controlling centre being assumed. In this context, concepts of networked autonomous vehicles accentuate the need for the extension and use of ever-new networked and mutually communicating technologies and the extension of the necessary infrastructures such as the 5G network. Here, these interests can tie in with the predominant digitalization narrative, whereby there is also a relevant economic interest in the data produced during driving operation. The visions of driverless cars should therefore be seen as elements of the discursive approach to future mobility. Sociologically and socially significant here is not only what is said, but also what is excluded. Consequently, neither the interests associated with the technological innovations nor the less vociferously propounded ideas of societal mobility futures should be disregarded in the course of critical sociological research on self-driving cars.

REFERENCES

Akrich, M. 1992. "The De-Scription of Technical Objects", in *Shaping Technology/Building Society: Studies in Sociotechnical Change*, ed. by W. E. Bijker and J. Law. Cambridge, MA/London: MIT Press, 205–244.

Altvater, E. 2007. "Verkehrtes Wachstum", in *Handbuch Verkehrspolitik*, ed. by O. Schöller, W. Canzler and A. Knie. Wiesbaden: VS Verlag für Sozialwissenschaften, 787–802.

Blättel-Mink, B. 2006. "Veralltäglichung von Innovationen", in *Kompendium der Innovationsforschung*, B. Blättel-Mink and R. Menez (eds.). Wiesbaden: VS Verlag für Sozialwissenschaften, 77–92.

Böhm, S., C. Jones, C. Land and M. Paterson 2006. "Introduction: Impossibilities of automobility", in *Against Automobility*, ed. by S. Böhm, C. Jones, C. Land and M. Paterson. Malden, MA: Blackwell Publishing, 3–16.

Bonham, J. 2006. "Transport: disciplining the body that travels", in *Against Automobility*, ed. by S. Böhm, J. Campbell, C. Land and M. Paterson. Malden, MA/Oxford: Blackwell, 57–74.

Cass, N., and K. Manderscheid 2019. "The autonomobility system: Mobility justice and freedom under sustainability", in *Mobilities, Mobility Justice and Social Justice*, ed. by N. Cook and D. Butz. London/New York: Routledge, 101–115.

Cass, N., E. Shove and J. Urry 2005. "Social Exclusion, Mobility and Access", in Sociological Review (53) 3, 539–555.

Cresswell, T. 2006. *On the Move: Mobility in the Modern Western World*, New York: Routledge.

Dangschat, J. S. 2017. "Automatisierter Verkehr – was kommt da auf uns zu?", in *Zeitschrift für Politikwissenschaft* (27) 4, 493–507.

Dant, T. 2004. "The Driver-car", in Theory, *Culture & Society* (21) 4/5, 61–79.

Dennis, K., and J. Urry 2009. *After the Car*. Cambridge: Polity Press.

Du Gay, P., S. Hall, L. Janes, H. Mackay and K. Negus 1997. "Doing Cultural Studies: The Story of the Sony Walkman", *The British Journal of Sociology*, vol. 48. London/Thousand Oaks/New Delhi: Sage Publications in association with The Open University.

During, S. 2000. *The Cultural Studies Reader*. London/New York: Routledge.

Elias, N. 1995. "Technization and civilization", in *Theory, Culture & Society* (12) 3, 7–42.

European Commission 2017. "Transport in Figures. Statistical Pocketbook 2017", Luxembourg. https://ec.europa.eu/transport/facts-fundings/statistics/pocketbook-2017_en (10/8/2020).

Foucault, M. 1981. *Archäologie des Wissens*. Frankfurt am Main: Suhrkamp.

Fraedrich, E., and B. Lenz 2015. "Gesellschaftliche und individuelle Akzeptanz des autonomen Fahrens", in *Autonomes Fahren. Technische, rechtliche und gesellschaftliche Aspekte*, ed. by M. Maurer, J. C. Gerdes, B. Lenz and H. Winner. Berlin/Heidelberg: Springer Vieweg, 639–660.

Gathmann, F., and G. Traufetter 2018. "Verbote sind für mich kein Politikstil", interview with German Federal Minister of Transport Andreas Scheuer,, *Spiegel Online*, 26/4/2018. www.spiegel.de/politik/deutschland/andreas-scheuer-csu-verbote-sind-fuer-mich-kein-politikstil-a-1204886.html (12/8/2020).

Geels, F. W., and J. Schot 2007. "Typology of sociotechnical transition pathways", in *Research Policy* (36) 3, 399–417.

Gerhard, U. 2000. "'Nomaden'. Zur Geschichte eines rassistischen Stereotyps und seiner Applikation", in *Medien in Konflikten. Holocaust – Krieg – Ausgrenzung*, ed. by A. Grewenig and M. Jäger. Duisburg: DISS Duisburger Institut für Sprach- und Sozialforschung, 223–235.

Goodall, N. 2019. "More than Trolleys", in *Transfers* (9) 2, 45–58.

Google Self-Driving Car Project 2014. "A First Drive", 27/5/2014. https://youtu.be/CqSDWoAhvLU (17/4/2020).

Grunwald, A. 2009. "Wovon ist die Zukunftsforschung eine Wissenschaft?", in *Zukunftsforschung und Zukunftsgestaltung*, ed. by R. Popp and E. Schüll. Berlin/Heidelberg: Springer-Verlag Berlin Heidelberg, 25–35.

Hammer, A., and J. Scheiner 2006. "Lebensstile, Wohnmilieus, Raum und Mobilität – Der Untersuchungsansatz von StadtLeben" in *StadtLeben – Wohnen, Mobilität und Lebensstil*, ed. by K. J. Beckmann, M. Hesse, C. Holz-Rau and M. Hunecke. Wiesbaden: VS Verlag für Sozialwissenschaften, 15–30. DOI: 10.1007/978-3-531-90132-9_2.

infas/MotionTag 2020. "Unsere Alltagsmobilität in der Zeit von Ausgangsbeschränkung oder Quarantäne – alles anders oder nicht? Ergebnisse aus Beobachtungen per Mobilitätstracking", 9/4/2020. Bonn/Berlin. www.infas.de/fileadmin/user_upload/PDF/Tracking-Report_No1_infas-Motiontag_09042020.pdf (12/8/2020).

Kanger, L., F. W. Geels, B. Sovacool and J. Schot 2019. "Technological diffusion as a process of societal embedding: Lessons from historical automobile transitions for future electric mobility", in *Transportation Research Part D: Transport and Environment* 71, 47–66.

Knie, A. 2007. "Ergebnisse und Probleme sozialwissenschaftlicher Mobilitäts- und Verkehrsforschung", in *Handbuch Verkehrspolitik*, O. Schöller, W. Canzler and A. Knie (eds.). Wiesbaden: VS Verlag für Sozialwissenschaften, 43–60.

Kuhm, K. 1997. *Moderne und Asphalt. Die Automobilisierung als Prozeß technologischer Integration und sozialer Vernetzung*. Pfaffenweiler: Centaurus.

Lanzendorf, M., and J. Hebsaker 2017. "Mobilität 2.0 – Eine Systematisierung und sozial-räumliche Charakterisierung neuer Mobilitätsdienstleistungen", in *Verkehr und Mobilität zwischen Alltagspraxis und Planungstheorie*, M. Wilde, J. Scheiner, M. Gather and C. Neiberger (eds.). Wiesbaden: Springer Fachmedien Wiesbaden, 135–151.

Larsen, J., J. Urry and K. W. Axhausen 2006 *Mobilities, Networks, Geographies*. Hampshire: Ashgate.

Laurier, E., H. Lorimer, B. Brown, O. Jones, O. Juhlin, A. Noble, M. Perry, D. Pica, P. Sormani, I. Strebel, L. Swan, A. Taylor, L. Watts and A. Weilenmann 2008. "Driving and 'Passengering': Notes on the Ordinary Organization of Car Travel", in *Mobilities* (3) 1, 1–23.

Lenz, B., and E. Fraedrich 2015a. "Gesellschaftliche und individuelle Akzeptanz des autonomen Fahrens", in *Autonomes Fahren. Technische, rechtliche und gesellschaftliche Aspekte*, ed. by M. Maurer, J. C. Gerdes, B. Lenz and H. Winner. Berlin/Heidelberg: Springer Vieweg, 639–660.

Lenz, B., and E. Fraedrich 2015b."Neue Mobilitätskonzepte und autonomes Fahren: Potenziale der Veränderung", in *Autonomes Fahren. Technische, rechtliche und gesellschaftliche Aspekte*, ed. by M. Maurer, J. C. Gerdes, B. Lenz and H. Winner. Berlin/Heidelberg: Springer Vieweg, 175–195.

Manderscheid, K. 2012. "Automobilität als raumkonstituierendes Dispositiv der Moderne", in *Die Ordnung der Räume*, ed. by H. Füller and B. Michel. Münster: Westphälisches Dampfboot, 145–178.

Manderscheid, K. 2013. "Automobile Subjekte", in *Mobilitäten und Immobilitäten. Menschen – Ideen – Dinge – Kulturen – Kapital*, ed. by J. Scheiner, H.-H. Blotevogel, S. Frank, C. Holz-Rau and N. Schuster. Essen: Klartext, 105–120.

Manderscheid, K. 2018. "From the Auto-mobile to the Driven Subject?", in *Transfers* (8) 1, 24–43.

Manderscheid, K. 2020. "Antriebs-, Verkehrs- oder Mobilitätswende? Zur Elektrifizierung des Automobilitätsdispositivs", in *Baustelle Elektromobilität. Sozialwissenschaftliche Perspektiven auf die Transformation der (Auto-)Mobilität*, ed. by A. Brunnengräber and T. Haas. Bielefeld: transcript, 37–67.

Mercedes-Benz 2017. "Der Mercedes-Benz F 015 Luxury in Motion". www.mercedes-benz.com/de/mercedes-benz/innovation/forschungsfahrzeug-f-015-luxury-in-motion/ (24/4/2020).

Miller, D. (ed.) 2001. *Car Cultures: Materializing Culture*. Oxford: Berg Publishers.

NPM (Nationale Plattform Zukunft der Mobilität) 2019. "Wege zur Erreichung der Klimaziele 2030 im Verkehrssektor, Arbeitsgruppe 1: Klimaschutz m Verkehr", interim report 03/2019. Berlin: Federal Ministry for Digital and Transport, G20 paper. www.plattform-zukunft-mobilitaet.de/wp-content/uploads/2020/03/NPM-AG-1-Wege-zur-Erreichung-der-Klimaziele-2030-im-Verkehrssektor.pdf (12/8/2020).

Nobis, C., and T. Kuhnimhof 2018. "Mobilität in Deutschland – MiD Ergebnisbericht". Study by infas, DLR, IVT and infas 360 on commission from the Federal Ministry for Digital and Transport, Bonn/Berlin. www.mobilitaet-in-deutschland.de/pdf/MiD2017_Ergebnisbericht.pdf (10/8/2020).

Norton, P. D. 2008. *Fighting Traffic: The Dawn of the Motor Age in the American City*. Cambridge/London: MIT Press.

Paterson, M. 2007. *Automobile Politics. Ecology and Cultural Political Economy*. Cambridge: Cambridge University Press.

Peters, P. F. 2006. *Time, Innovation and Mobilities: Travel in technological cultures*. London/New York: Routledge.

Rajan, S. C. 2007. "Automobility, liberalism, and the ethics of driving", in *Environmental Ethics* (29) 1, 77–90.

Rammert, W. 2010. "Die Innovationen der Gesellschaft", in *Soziale Innovation: Auf dem Weg zu einem postindustriellen Innovationsparadigma*, ed. by J. Howaldt and H. Jacobsen. Wiesbaden: VS Verlag für Sozialwissenschaften, 21–51.

Rettie, R. 2008. "Mobile Phones as Network Capital: Facilitating Connections", in *Mobilities* (3) 2, 291–311.

Rosa, H. 2005. *Beschleunigung. Die Veränderung der Zeitstrukturen in der Moderne*. Frankfurt am Main: Suhrkamp.

Sauter-Servaes, T. 2011. "Technikgeneseleitbilder der Elektromobilität", in *Das Elektroauto. Bilder für eine zukünftige Mobilität*,ed. byS. Rammler and M. Weider. Berlin: Lit, 25–40.

Schreurs, M. A., and S. D. Steuwer 2015. "Autonomous Driving – Political, Legal, Social, and Sustainability Dimensions", in *Autonomes Fahren. Technische, rechtliche und gesellschaftliche Aspekte*, ed. by M. Maurer, J. C. Gerdes, B. Lenz and H. Winner. Berlin/Heidelberg: Springer Vieweg, 151–173.

Sheller, M., and J. Urry 2006. "The New Mobilities Paradigm", in *Environment and Planning* A (38) 2, 207–226.

Shove, E. 2002. "Rushing around: coordination, mobility and inequality – Draft paper for the Mobile Network meeting, October 2002". www.lancaster.ac.uk/staff/shove/choreography/rushingaround.pdf (12/8/2020).

Shu, C. 2015. "Google and Ford will reportedly team up to build Self-Driving Cars", in *TechCrunch*, 22/12/2015, https://tcrn.ch/34Eh1Z3 (14/10/2020).

Thomopoulos, N., and M. Givoni 2015. "The autonomous car – a blessing or a curse for the future of low carbon mobility? An exploration of likely vs. desirable outcomes", in *European Journal of Futures Research* (3) 1, 1–14.

Thulin, E., and B. Vilhelmson 2007. "Mobiles everywhere: Youth, the mobile phone, and changes in everyday practice", in *Young* (15) 3, 235–253. DOI: 10.1177/110330880701500302.

Umweltbundesamt 2019a. "Emissionen des Verkehrs". Dessau-Roßlau. www.umweltbundesamt.de/daten/verkehr/emissionen-des-verkehrs (10/8/2020).

Umweltbundesamt 2019b. "Mobilität privater Haushalte". Dessau-Roßlau. www.umweltbundesamt.de/daten/private-haushalte-konsum/mobilitaet-privater-haushalte#textpart-1 (17/7/2019).

Urry, J. 2004. "The 'System' of Automobility", in *Theory, Culture & Society* (21) 4/5, 25–39.

Urry, J. 2007. *Mobilities*. Cambridge: Polity.

Vaid, K. 2018. *Selbst ist das Auto – automatisiertes und autonomes Fahren. Die Zukunft der Mobilität*, Hamburg: Diplomica.

Verron, H., B. Huckestein, G. Penn-Bressel, P. Röthke, M. Bölke and W. Hülsmann 2005. *Determinanten der Verkehrsentstehung*, Texte 26/05. Dessau-Roßlau: German Environment Agency. www.umweltbundesamt.de/sites/default/files/medien/publikation/long/2967.pdf (10/8/2020).

VW 2017. "SEDRIC – Concept car – Film, Langfassung", 6/3/2017. www.volkswagen-newsroom.com/de/videos-und-footage/sedric-concept-car-film-langfassung-2791 (12/8/2020).

VW 2018. "SEDRIC: Das Auto der Zukunft zum Anfassen". www.volkswagenag.com/de/news/stories/2018/02/sedric-the-future.html (12/8/2020).

Wajcman, J. 1994. *Technik und Geschlecht. Die feministische Technikdebatte*. Frankfurt am Main: Campus.

Wajcman, J. 2008. "Life in the fast lane? Towards a sociology of technology and time", in *The British Journal of Sociology* (59) 1, 59–77.

Waymo 2019. "Waymo One". https://waymo.com/waymo-one/ (24/4/2020).

Wiener Linien (no date). "auto.Bus – Seestadt", "Über den Bus". https://www.wienerlinien.at/eportal3/ep/ channelView.do/pageTypeId/66528/channelId/-4400525 (15/12/2020).

5 Automated drivability and streetscape compatibility in the urban-rural continuum using the example of Greater Vienna

Aggelos Soteropoulos

Aggelos Soteropoulos
TU Wien, future.lab Research Center and Research Unit Transport System Planning (MOVE)

© The Author(s) 2023
M. Mitteregger et al. (eds.), *AVENUE21. Planning and Policy Considerations for an Age of Automated Mobility*, https://doi.org/10.1007/978-3-662-67004-0_5

1. INTRODUCTION

Automated vehicles will have a significant impact on our mobility and the way we travel. Already over the next few years, highly dynamic developments can be expected in the mobility sector, with fundamental changes and upheavals that will present both opportunities and risks. From the point of view of transport, infrastructure and urban planning, strategies are therefore required as to where and how automated vehicles can be best deployed.

Numerous studies have already been carried out on the traffic-related and spatial effects of automated vehicles; these can serve as a basis for developing strategies and measures for political decision makers. According to these studies, opportunities are opened up by automated vehicles for example as a result of higher-capacity utilization of the existing transport infrastructure, improved economic efficiency of public transport as long as personnel costs can be reduced, and extended mobility options for specific user groups such as persons with impaired mobility (cf. Milakis et al. 2017: 13; Soteropoulos et al. 2019a: 12). On the other hand, most of these studies also assume that automated vehicles will increase the attractiveness of vehicle use, and thus overall mileage, in the areas envisaged for automated vehicles due to their numerous convenience benefits and their attractiveness for new user groups (cf. Soteropoulos et al. 2019b: 40; Hörl et al. 2019: 26; Marsden et al. 2018: 31). Furthermore, the smoother traffic flow enabled by the shorter distances between automated vehicles, which enable higher capacity-utilization of the transport infrastructure, leads to denser traffic; as a result, it is more difficult for pedestrians, cyclists and turning vehicles to find gaps, thus increasing the barrier effect of roads and reducing the permeability of the streetscape for other road users (cf. Abegg et al. 2018: 26; Wissenschaftlicher Beirat beim Bundesminister für Verkehr und digitale Infrastruktur 2017: 23; Heinrichs 2015: 237).

To date, however, these studies have only dealt with the possible general trend of potential effects. The level of the streetscape has not been given sufficient consideration in this context. The streetscape is public space; i.e. it is fundamentally accessible to everyone at all times and is essentially publicly owned, but only represents a subset of all public space, which in addition to the streetscape also includes for example squares, parks and publicly accessible open spaces or ground-floor zones connected with settlements. These in turn are all strongly connected to the streetscape (cf. Stadt Wien 2018: 13). The level of the streetscape is particularly relevant in two aspects (Fig. 1):

1. Cities, but also rural settlement structures, have a wide variety of streetscapes, which differ in terms of their function (e.g. connection or access), traffic volumes, local characteristics (e.g. old village centre or new commercial area), peripheral usage and streetscape conditions (e.g. road margin, width and topography; cf. Marshall 2005: 54; Baier/FGSV 2007: 5). For the use of automated vehicles, these different streetscape contexts give rise to different requirements: increased complexity of the operational design domain (ODD), for example due to a large number of pedestrians or cyclists to be identified or intersections with complex shapes, has increased the demands placed on the automated driving system (cf. Czarnecki 2018; SAE International 2018: 14). For traffic, infrastructure and urban planning, analysis of streetscapes in terms of their suitability for automated vehicles from a technical-infrastructural point of view can thus provide information on where automated vehicles are likely to operate more efficiently, or with fewer adaptations to the infrastructure as a result of the surrounding streetscape and environmental conditions. This also makes it possible to identify streetscapes where the use of automated vehicles is possible with greater or less effort.

2. In principle, there is an abundance of demands for use, requirements and needs of people to seek out or to move within streetscapes. In addition to the function of roads as transport routes, the type and extent of further requirements on a streetscape result from its environment, i.e. the use of buildings, their orientation to the streetscape and other factors (cf. Bühlmann/Laube 2013: 9). These are at times contradictory and lead to conflicts of use (cf. Häfliger et al. 2015: 19). In particular, conflicts between the demands of motorized individual transport and other needs (other forms of movement such as walking and cycling) are only compatible up to a certain intensity or as are acceptable in a specific situation. However, once the traffic load exceeds this limit it reaches a predominance which affects the other usage demands to an extent that is no longer compatible (cf. Bühlmann/Laube 2013: 10). From a planning perspective, it must also be examined to what extent the possible use of automated vehicles is compatible with the other demands for usage of the streetscape, in light of their impact such as a rise in traffic volume and an increased barrier effect due to the denser spacing of these vehicles.

Figure 1: Overview of the relevance of streetscapes for the use of automated vehicles

Streetscapes

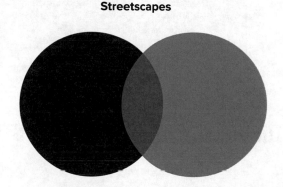

Suitability for automated vehicles from a technical-infrastructural point of view
(complexity of operational environment)

Compatibility with use of surroundings in view of the effects associated with automated vehicles
(increase in traffic volume, increased barrier effect, etc.)

Source: the authors

In this article, the suitability of streetscapes for automated vehicles in the urban-rural continuum is examined using the example of four different municipalities in Greater Vienna. Different types of criteria (e.g. intersections or speed limits), chosen on the basis of current problems of automated driving systems as described in test reports (cf. BMVIT 2018, Favarò et al. 2018), are combined here to assess the suitability of road sections for the use of automated driving systems. In an analysis of the compatibility of streetscapes, the conflicts arising from today's motorized traffic (traffic volumes, speed) for pedestrians and cyclists in parallel traffic, when stationary and when crossing roads are also evaluated (cf. Baier et al. 2011: 37). Here, individual sections of road in the four municipalities are investigated to determine whether and in what ways problems such as a high barrier effect or low permeability for pedestrians and cyclists are already evident in streetscapes today, by explicitly examining and taking into account criteria such as traffic volume, vehicle speed or the proportion of lorries (cf. Nørby/Meltofte 2012: VI; Litman 2009: 1). The assessment of streetscape compatibility is then linked to or considered together with an assessment of the suitability of the streetscape for automated driving systems. The aim is thus to determine the extent of correlation between the suitability of streetscapes for automated vehicles and the compatibility of these spaces (Fig. 2). Combining these two assessments is essential for determining where there will be a particular need for action in the near future from a traffic and urban planning perspective, and to develop suitable strategies for the best possible use of automated vehicles.

Figure 2: Correlation between technical-infrastructural suitability of streetscapes for automated vehicles and streetscape compatibility

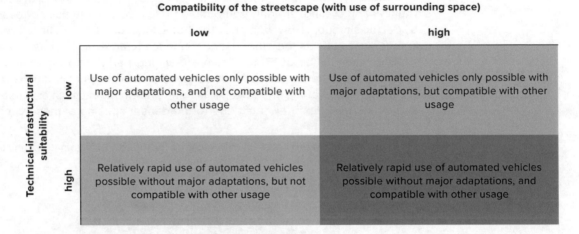

Source: the authors

2. METHODOLOGY OF THE PRESENT STUDY

To determine the suitability of streetscapes for the use of automated vehicles from a technical-infrastructural point of view and to assess their compatibility, extensive procedures were used, which are explained in the following.

2.1 DETERMINING THE SUITABILITY OF STREETSCAPES FOR THE USE OF AUTOMATED VEHICLES: AUTOMATED DRIVABILITY

The methodology for assessing the suitability of streetscapes for the use of automated vehicles is very much based on the concept of automated drivability as described in the work of Soteropoulos et al (2020). The methodology is only briefly described in the following; for a detailed description see Soteropoulos et al (2020).

2.1.1 POINT OF DEPARTURE, FRAMEWORK AND COMPONENTS OF AUTOMATED DRIVABILITY

The concept of automated drivability assumes that certain streetscape contexts and surrounding conditions (ODD) place increased requirements on automated driving systems to perform their driving task. Processes such as (1) recording and mapping the surroundings, (2) planning and making an appropriate driving decision, and (3) executing the appropriate driving decision (control), which are necessary for the automated vehicle to perform its task of driving, can be complicated by various factors and conditions. On the basis of this, a framework for the description of automated drivability was drawn up, with various components and sub-elements that increase complexity for automated driving systems. The framework comprises the following components:

1. number of objects in the streetscape

2. diversity of objects in the streetscape

3. condition and configuration of the road infrastructure

4. speed limit

5. stability of the operational design domain.

Figure 3: Principal tasks of an automated driving system and components of the framework for the assessment of automated drivability, including sub-elements

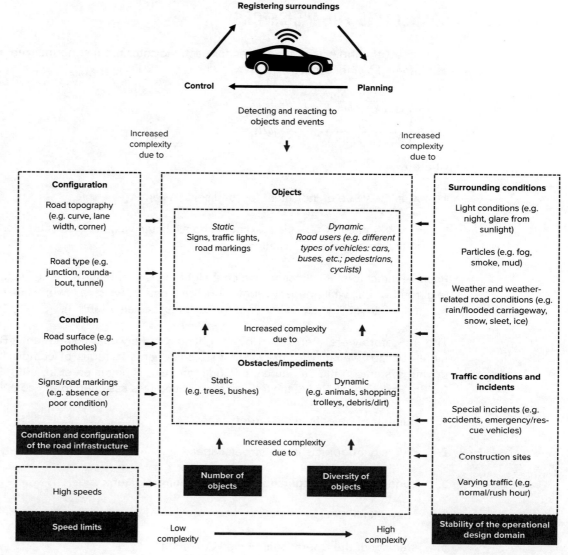

Figure 3 (on the previous page) gives an overview of the principal tasks of an automated driving system and the components of the framework for assessing automated drivability, including the sub-elements.

2.1.2 DERIVATION OF INDICATORS FOR THE VARIOUS COMPONENTS OF THE FRAMEWORK

For each component of the automated drivability framework described at the outset, suitable indicators are subsequently derived, taking into account (publicly) available data. To calculate of the indicators, data from the Austrian Graph Integration Platform (GIP), the Open Street Map (OSM) and the Austrian Central Institution for Meteorology and Geodynamics (ZAMG) were used in this study. The calculations were carried out using the geographic information system software ArcGIS. Table 1 (see next page) provides an overview of the derived indicators for the respective components and of the various data sources.

2.1.3 CONNECTING THE INDICATORS

To connect the indicators, the values for each indicator are first normalized and standardized using the following formula:

$$X'_{ij} = \begin{cases} \dfrac{X_{ij} - min\, X_{ij}}{max\, X_{ij} - min\, X_{ij}}, & positive \\[2ex] \dfrac{max\, X_{ij} - X_{ij}}{max\, X_{ij} - min\, X_{ij}}, & negative \end{cases}$$

whereby

- X_{ij} is the value of indicator i for road section j and

- $maxX_{ij}$ and $minX_{ij}$ are the maximum and minimum values of indicator i for all road sections in the area under examination.

Positive indicators are those for which a higher value represents higher automated drivability (e.g. road width), while negative indicators are those for which a lower value represents higher automated drivability (e.g. number of traffic lights; see also Table 1).

The indicators were then linked by weighting the individual components. Based on assessments of the relevance of the individual components in terms of complexity for automated driving systems in the literature (cf. Pendleton 2017; Brummelen et al. 2018; Shladover 2018a, 2018b; Favarò et al. 2018), the following weighting was assigned to the components:

1. number of objects in the streetscape: 3

2. diversity of objects in the streetscape: 7

3. condition and configuration of the road infrastructure: 2

4. speed limit: 3

5. stability of the operational design domain: 1.

Table 1: Derived indicators for the components of the framework

Indicator	Indicator type	Components of the framework					Data source	Effect on suitability for automated driving systems
		Number of objects in the streetscape	Diversity of objects in the streetscape	Condition and configuration of the road infrastructure	Speed limit	Stability of the operational design domain		
Road type	categorical	✓					GIP	positive
Number of traffic lights	numerical	✓		(✓)			OSM	negative
Number of public transport station	numerical	✓		(✓)			OSM	negative
Prohibition of pedestrians and cyclists in the streetscape	binary		✓				GIP	positive
Number of pedestrian crossings	numerical	(✓)	✓				OSM	negative
Pedestrian zone or shared space	binary	(✓)	✓				GIP	negative
Cycling infrastructure on the road	binary	(✓)	✓				GIP	negative
Road width	numerical	(✓)					GIP	positive
Type of crossing	categorical			✓			GIP	positive
Predominant type of area use	categorical			✓			OSM	positive
Speed limit	numerical				✓		GIP	negative
Average annual number of days with snow	numerical					✓	ZAMG	negative

✓ = complete representation of component by indicator; (✓) = partial representation of component by indicator. Source: the authors.

To establish a connection, the indicator values were subsequently summed for each section of road using the following formula, by adding the values of the indicator for each component of the framework and then dividing by the number of indicators per component and multiplying by the respective weighting, to obtain the aggregated automated drivability index (ADX) value for each section of road:

$$X_{ADXj} = \begin{aligned} &0.188 * ((Xc1_{i_1j} + Xc1_{i_2j} \dots)/Ic1_n) + 0.438 * ((Xc2_{i_1j} + Xc2_{i_2j} \dots)/Ic2_n) + \\ &0.125 * ((Xc3_{i_1j} + Xc3_{i_2j} \dots)/Ic3_n) + 0.188 * ((Xc4_{i_1j} + Xc4_{i_2j} \dots)/Ic4_n) + \\ &0.063 * ((Xc5_{i_1j} + Xc5_{i_2j} \dots)/Ic5_n) \end{aligned}$$

whereby

- X_{ADXj} is the value of the ADX for road section j,

- $Xc1_{i1j}$ is the value of indicator i1 of component 1,

- $Xc2_{i1j}$ is the value of indicator i1 of component 2 ...; and

- $Ic1_n$ is the total number of indicators for component 1 and

- $Ic2_n$ is the total number of indicators for component 2 ...

2.2 PROCEDURE FOR ASSESSING THE STREETSCAPE COMPATIBILITY OF THE ROAD SECTIONS

For assessment of streetscape compatibility, numerous procedures can be found in the literature, e.g. the M.A.R.S. model of autonomous and relative standards (cf. Baier 1992), the LADIR procedure (Müller et al. 1994) or the compensatory approach (Mörner et al. 1984). These methods originated in the 1980s and the 1990s, but have since been applied partly in adapted form in more recent years, for example in Bühlmann and Laube (2013), Frehn et al. (2013) or Baier et al. (2011).

The compensatory approach of Mörner et al. (1984) served as the basis for assessing streetscape compatibility in the context of this article and was adapted using the findings of the recent application of this approach by Bühlmann and Laube (2013). In the compensatory approach, a maximum traffic load in the dimension "number of motor vehicles per peak hour" (veh/ph) is determined for individual road sections, based on the degree of building development and usage of the surrounding area, and the function and significance of the locality. This maximum traffic load is corrected in line with various influencing factors, and the compatibility of the existing traffic volume with the road environment (measured traffic load in comparison with maximum compatible traffic load) is then assessed on this basis. Taking into account the effects described above in connection with automated driving with regard to an increase in barrier effect due to a possible denser sequence of vehicles, this barrier effect was also explicitly included in the framework of the compensatory approach applied here, or in some cases the approach was extended to include this aspect.

2.2.1 DETERMINING THE BASIC STILL COMPATIBLE TRAFFIC LOAD

The basic traffic load, i.e. the still compatible traffic load in terms of vehicles per peak hour (veh/ph), is determined by the degree of building development and usage of the surrounding area for the respective section of road. While Mörner et al. (1984) based their approach on urban areas, Bühlmann and Laube (2013) also mention a connection between traffic load that is still compatible and the function and significance of a locality, whereby traffic has a lower level of acceptance in rural areas than in urban centres (cf. Bühlmann/Laube 2013: 34). Since the sections of road examined are all located in relatively small towns within Greater Vienna (see Section 3), the following four categories have been assigned on the basis of Mörner et al. (1984) and Bühlmann/Laube (2013):

Category A Shopping area/centre, largely characterized by closed block development with more than two storeys and medium to intensive commercial use

Category B Mixed use, largely characterized by open to semi-open two- to four-storey buildings, or closed one- to two-storey buildings and medium commercial use

Category C Residential, largely characterized by open single- and two-family housing, with only isolated shops or other intensive public use

Category D Commercial and industrial area, with mainly low residual demands and no intensive public use such as residences or (pedestrian) shopping.

The basic traffic load (veh/ph) that is still compatible with the demands of residents on the road surroundings was defined on the basis of Mörner et al. (1984) and Bühlmann/Laube (2013) as follows for the four categories; as with Bühlmann/Laube (2013), it assumes that acceptance of traffic is lower in smaller towns and in rural areas than in larger towns (cf. ibid.: 17). In addition, it is assumed here that in city centres much higher demands are placed on quality of stay and thus on compatibility than in commercial and industrial areas:

Category A 150 vehicles per peak hour

Category B 250 vehicles per peak hour

Category C 400 vehicles per peak hour

Category D 1,000 vehicles per peak hour.

2.2.2 DETERMINING THE ACTUAL STILL COMPATIBLE TRAFFIC LOAD: ADAPTATION OF THE BASIC VALUE THROUGH USE AND DESIGN OF THE STREETSCAPE

The categories listed in the first step make only a rough distinction, and the use and design of a streetscape can differ significantly within the various categories. For this reason, the basic still compatible traffic load determined in the first step is subsequently adapted in line with the following evaluation criteria:

a. Use by pedestrians and cyclists

b. Area distribution

c. Greenery and design

d. Speed

e. Share of heavy-vehicle traffic/lorries and

f. Barrier effect: crossing aids/detours for pedestrians.

These criteria serve as compensatory aspects by means of which the compatible traffic load is increased or reduced in accordance with the other utilization demands placed on the streetscape. Based on Mörner et al. (1984) and with slight adaptation of these values, the compatibility levels of the respective evaluation criteria lead to a compensation or adaptation of the basic still compatible traffic load as shown in Table 2.

Table 2: Overview of the compatibility levels per assessment criterion and the corresponding adaptation values for the basic still compatible traffic load

Compatibility level of the respective assessment criterion	Adaptation of the basic still compatible traffic load by
++ (highly compatible)	+50 veh/ph
+ (compatible)	+25 veh/ph
o (just compatible)	±0 veh/ph
- (incompatible)	-25 veh/ph
- (entirely incompatible)	-50 veh/ph

Source: the authors, based on Mörner et al. (1984)

a) Use by pedestrians and cyclists

The assessment criterion "use by pedestrians and cyclists" is based on the assumption that motorized traffic disturbs or endangers pedestrians and cyclists in their activities in the streetscape. The assessment thus follows the logic that motorized traffic becomes less compatible as the number of pedestrians and cyclists on a road increases. For this purpose, the number of pedestrians and cyclists (i.e. persons inconvenienced or endangered by motorized traffic) during peak hour is used, with the resulting compatibility levels shown in Table 3.

Table 3: Compatibility levels for the criterion "use by pedestrians and cyclists"

Number of pedestrians and cyclists on the road in the peak hour	Compatibility levels (compatibility of motorized traffic with this number of pedestrians and cyclists)
< 100	++ (highly compatible)
over 100 to 200	+ (compatible)
over 200 to 400	o (just compatible)
over 400 to 600	- (incompatible)
over 600	- (entirely incompatible)

Source: the authors, based on Mörner et al. (1984)

b) Area distribution

The assessment criterion "area distribution" evaluates the width of the areas for pedestrians and cyclists (pavements, cycle paths and grass verges) in relation to the width of the area for motorized traffic (traffic area including multipurpose lanes and areas for parked motor vehicles). It is assumed that the lower the ratio of the width of the area for pedestrians and cyclists in the streetscape to the width of the area for motorized traffic, the less compatible is the motorized traffic. Accordingly, the compatibility levels shown in Table 4 are set on the basis of the ratio between the width of the area for pedestrians and cyclists and the width of the area for motorized traffic.

Table 4: Compatibility levels for the criterion "area distribution"

Ratio of the width of areas for pedestrians and cyclists to the width of areas for motorized traffic	Levels of compatibility (with demands of residents and the surrounding area)
> 1.00	++ (highly compatible)
0.75 to under 1.00	+ (compatible)
0.5 to under 0.75	o (just compatible)
0.25 to under 0.5	- (incompatible)
under 0.25	- (entirely incompatible)

Source: the authors, based on Mörner et al. (1984)

c) Greenery and design

The assessment criterion "greenery and design" is used to evaluate the design of a streetscape with green spaces and trees, along with accompanying design elements. It is assumed that a smaller number of green spaces, trees and accompanying design elements in the streetscape will diminish the quality of stay and thus make motorized traffic less compatible. The evaluation is carried out in qualitative terms (in lieu of quantifiable characteristics such as the number of trees or of greenery and design elements, etc.), whereby a distinction is made between the categories and compatibility levels described in Table 5.

Table 5: Compatibility levels for the criterion "greenery and design"

Greenery and design in the streetscape	Levels of compatibility (with demands of residents and the surrounding area)
Greenery and other design elements clearly define the streetscape and shape the character of the street. They lend the street a distinctive character with high experience value.	++ (highly compatible)
Greenery and other design elements are objectively perceptible. Overall, they still predominate over the technical traffic instalments in the appearance of the street.	+ (compatible)
There is an equal balance between greenery and design elements and undesigned areas; the street is not characterized by greenery and designed areas.	o (just compatible)
Greenery and other design elements are sporadically present. They do not shape the character of the street; a technical-functional image predominates.	- (incompatible)
Greenery and other design elements are entirely lacking; the streetscape is bare and naked.	- (entirely incompatible)

Source: the authors, based on Mörner et al. (1984)

d) Speed

The speeds travelled on the section of road are evaluated with the assessment criterion "speed". It is assumed that the speeds travelled have a significant influence on the usability of the streetscape for residents and on the safety of non-motorized road users (e.g. in terms of ease of crossing the road) and that higher speeds travelled on the section of road make motorized traffic less compatible. This assessment uses speeds travelled in the form of v85[1] and the compatibility levels shown in Table 6.

Table 6: Compatibility levels for the criterion "speed"

Speeds travelled by motor vehicles (v85)	Levels of compatibility (with demands of residents and the surrounding area)
< 30 km/h	++ (highly compatible)
over 30 to 35 km/h	+ (compatible)
over 35 to 40 km/h	o (just compatible)
over 40 to 50 km/h	- (incompatible)
over 50 km/h	- - (entirely incompatible)

Source: the authors, based on Mörner et al. (1984)

e) Heavy-vehicle traffic/lorry share

The assessment criterion "heavy-vehicle traffic/lorry share" evaluates the share of lorries in the motorized traffic volume during normal hours. It is based on the assumption that heavy-vehicle traffic with an overall low traffic volume is burdensome to residents particularly due to noise emissions, but also with regard to road safety. This assessment uses the share of heavy vehicles in the total volume of motorized traffic and the compatibility levels shown in Table 7.

Table 7: Compatibility levels for the criterion "heavy-vehicle traffic/lorry share"

Share of lorries in total motorized traffic volume	Levels of compatibility (with demands of residents and the surrounding area)
< 3%	++ (highly compatible)
over 3 to 6	+ (compatible)
over 6 to 9	o (just compatible)
over 9 to 12%	- (incompatible)
over 12%	- (entirely incompatible)

Source: the authors, based on Mörner et al. (1984)

f) Barrier effect: crossing aids/detours for pedestrians

The assessment criterion "barrier effect: crossing aids/detours for pedestrians" evaluates the number of crossing aids in relation to the length of the road section. Based on the development structure or the previously defined road environment categories, it is assumed that an appropriate number of crossing points is necessary in order to cater to the needs of pedestrians as well as possible and without lengthy detours. To obtain the final indicator value, the ratio of the number of crossing aids to the length of the road section is multiplied by 100: for example, a

---

1   v85 is the speed not exceeded by 85% of vehicles.

value of 1.0 indicates an average distance of 100 metres between two crossing aids (cf. Häfliger et al. 2015: 80f.). Table 8 gives an overview of the respective compatibility levels for different types of road environment or development structure of the area. Since a shared space can be crossed at any point, it is assumed in accordance with Häfliger et al. (2015) that this is always highly compatible with the surrounding area (cf. ibid.: 81).

Table 8: Compatibility levels for the criterion "barrier effect: crossing aids/detours for pedestrians"

Ratio of the number of crossing aids to road section length in metres (*100)				Levels of compatibility (with demands of residents and the surrounding area)
A Shopping/ centre	B Mixed use	C Residential	D Commercial/ industrial	
≥ 1.5	≥ 1.3	≥ 1.1	≥ 0.9	++ (highly compatible)
≥ 1.2 to under 1.5	≥ 1.0 to under 1.3	≥ 0.8 to under 1.1	≥ 0.6 to under 0.9	+ (compatible)
≥ 0.9 to under 1.2	≥ 0.7 to under 1.0	≥ 0.5 to under 0.8	≥ 0.3 to under 0.6	o (just compatible)
≥ 0.6 to under 0.9	≥ 0.4 to under 0.7	≥ 0.2 to under 0.5	≥ 0.1 to under 0.3	- (incompatible)
< 0.6	< 0.4	< 0.2	< 0.1	-- (entirely incompatible)

Source: the authors, based on Häfliger et al. (2015: 81)

2.2.3 COLLECTION OF DATA REQUIRED FOR ASSESSING STREETSCAPE COMPATIBILITY

For the procedure described above to assess the compatibility of streetscapes, the relevant criteria must be determined on site for the respective road sections. To record the relevant data for the assessment, the authors carried out surveys on all road sections examined. This involved traffic counts to determine the number of vehicles and of pedestrians and cyclists, and the proportion of lorries in peak hour. The traffic counts for all road sections were conducted on working days in November 2019 between 3 and 4 p.m., in accordance with the recommendations for traffic surveys – EVE (cf. FGSV 2012). The remaining data required for the respective criteria were (1) (approximately) determined from the Austrian Graphic Integration Platform (GIP) (criterion "speeds travelled (v85)") or (2) collected on site (criteria "area distribution", "greenery and design" and "barrier effect: crossing aids").

3. THE MUNICIPALITIES IN GREATER VIENNA AND ANALYSED ROAD SECTIONS

The investigation of automated drivability and streetscape compatibility was carried out in the four sample municipalities

1. Mödling,
2. Gumpoldskirchen,
3. Bad Vöslau und
4. Leobersdorf

in the south of Greater Vienna. These four municipalities are representative of suburban, rather rural municipalities in Greater Vienna. Figure 5 (on the next page) shows the location of the sample municipalities studied in Greater Vienna.

Figure 4: Overview of the road sections examined in the four municipalities of the study

Hauptstraße, Mödling

Wiener Straße, Gumpoldskirchen

Industriestraße, Bad Vöslau

Hauptstraße, Leobersdorf

Source: Aggelos Soteropoulos

The investigation of automated drivability was conducted over the entire road network approved for motorized traffic in the municipal areas of the respective four municipalities studied. The road network was divided into sections of up to 100 metres in length; these appeared to be the most suitable spatial reference units for the representation of automated drivability (cf. Su et al. 2019: 66).

Figure 5: Locations of the municipalities studied

Mödling

Gumpoldskirchen

Bad Vöslau

Leobersdorf

Vienna

0 km   15 km   30 km

Source: the authors

The assessment of streetscape compatibility was carried out by way of example for one road or road section in each of the four municipalities studied, including three through roads and one commercial road. These four roads were selected firstly on the basis of the significance of these types of road in a suburban, partly rural setting, and secondly in order to reflect and take into account the heterogeneous nature of streetscapes in these municipalities. Figure 4 shows an overview of the road sections examined.

4. RESULTS OF THE STUDY

In the following, the results of the study regarding the suitability of the streetscapes for the use of automated vehicles (automated drivability) and the compatibility of the streetscapes are discussed, and a synopsis of these two aspects is presented.

4.1 RESULTS FOR SUITABILITY OF STREETSCAPES FOR THE USE OF AUTOMATED VEHICLES (AUTOMATED DRIVABILITY)

Figure 6 provides an overview of the assessment of the suitability of streetscapes for automated vehicles, i.e. automated drivability, in the four sample municipalities:

1. Mödling,
2. Gumpoldskirchen,
3. Bad Vöslau and
4. Leobersdorf.

Figure 6: Automated drivability in the four municipalities examined

Mödling

Gumpoldskirchen

Bad Vöslau

Leobersdorf

Low suitability
(only with major adaptations)

High suitability
(without major adaptations)

Source: the authors

High values for automated drivability and thus a high level of suitability for automated vehicles can be seen in the municipalities studied, especially for motorways, which are located in the east of Bad Vöslau and west of the centre of Leobersdorf. Moreover, high values for automated drivability are mostly also observable in those parts of the municipalities located close to the motorway accesses/exits, such as north of the town centre in Bad Vöslau, or west of the motorway and of the town centre in Leobersdorf. In Mödling, high levels of automated drivability were recorded almost exclusively in the area around the railway axis, and in Gumpoldskirchen at best in the western part of the municipality.

By contrast, the town centres and some access/exit roads in the municipalities examined exhibit somewhat lower levels of automated drivability. In Bad Vöslau, low automated drivability values are evident for the road running parallel to the B212 from the town centre to the north, and in Mödling especially for the roads in the town centre and the road sections at the junction of the B11 (Spitalmühlgasse) and Neusiedler Straße. In Gumpoldskirchen, low automated drivability values can be seen especially for the road sections around the roundabout in the western part of the municipality; in Leobersdorf, this situation can be seen throughout the town centre and especially along the main road running through the town centre.

4.2 RESULTS FOR STREETSCAPE COMPATIBILITY

In the following, the streetscape compatibility is analysed for the four municipalities examined: Mödling, Gumpoldskirchen, Bad Vöslau and Leobersdorf.

4.2.1 BASIC VALUE OF THE STILL COMPATIBLE AND ACTUAL TRAFFIC LOADS

Table 9: Overview of integration into the road surroundings, and the basic value of the still compatible traffic load with compatibility levels for the road sections examined

Road section, municipality	Classification of street surroundings	Basic value of the still compatible traffic load and compatibility levels (veh/ph)
Hauptstraße, Mödling 	**Category A** Shopping/centre	< 75 (++ highly compatible) **75 to 150 (+ compatible)** > 150 to 250 (o just compatible) > 250 to 400 (- incompatible) > 400 (- - entirely incompatible)
Wiener Straße, Gumpoldskirchen 	**Category B** Mixed use	< 150 (++ highly compatible) **150 to 250 (+ compatible)** > 250 to 400 (o just compatible) > 400 to 1,000 (- incompatible) > 1,000 (- - entirely incompatible)

Industriestraße, Bad Vöslau

Category D

Commercial and industrial areat

< 400 (++ highly compatible)

400 to 1,000 (+ compatible)

> 1,000 to 1,200 (o just compatible)

> 1,200 to 1,500 (- incompatible)

> 1,550 (- - entirely incompatible)

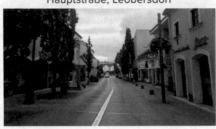

Hauptstraße, Leobersdorf

Category A

Shopping/centre

< 75 (++ highly compatible)

75 to 150 (+ compatible)

> 150 to 250 (o just compatible)

> 250 to 400 (- incompatible)

> 400 (- - entirely incompatible)

Source: the authors, based on Mörner et al. (1984) and Bühlmann/Laube (2013)

Table 9 shows the classification of the surroundings of the four examined road sections according to the categories described above, and the corresponding basic values of the still compatible traffic load. The road sections in the municipalities of Mödling and Leobersdorf were assigned to the category "A: Shopping/centre" with a basic value of the still compatible traffic load of 150 veh/ph. The road section in Gumpoldskirchen, on the other hand, can best be assigned to the category "B: Mixed use", with a basic value of the still compatible traffic load of 250 veh/ph. The road section in Bad Vöslau was assigned to category "D: Commercial and industrial area" and thus has a basic value of the still compatible traffic load of 1,000 vehicles in peak hour.

4.2.2 DETERMINING THE ACTUAL STILL COMPATIBLE TRAFFIC LOAD: ADAPTATION OF THE BASIC VALUE THROUGH USE AND DESIGN OF THE STREETSCAPE

a) Use by pedestrians and cyclists

In the criterion "use by pedestrians (pd) and cyclists (cy)", fewer than 100 pedestrians and cyclists were counted in peak hour for the road sections in both Gumpoldskirchen (94 pd and cy/ph) and Bad Vöslau (20 pd and cy/ph). This leads to an assessment of motorized traffic as being highly compatible with this number of pedestrians and cyclists, resulting in an adjustment of the basic value of the still compatible traffic load by +50 veh/ph.

On the road section in Leobersdorf, 196 pedestrians and cyclists per peak hour were recorded, so that the motorized traffic can be classified as compatible with this number of pedestrians and cyclists, with a resulting adjustment of the basic value of the still compatible traffic load by +25 veh/ph. On the other hand, motorized traffic is incompatible with the number of pedestrians and cyclists recorded on the road section in Mödling (504 pd and cy/ph), leading to an adjustment of the basic value of the still compatible traffic load by -25 veh/ph.

Table 10: Classification and adjustment of the basic value of the still compatible traffic load for the criterion "use by pedestrians and cyclists"

Road section, municipality	Number of pedestrians and cyclists on the road in the peak hour	Levels of compatibility (of motorized traffic with this number of pedestrians and cyclists)	Adjustment of the basic value of the still compatible traffic load (veh/ph) by
Hauptstraße, Mödling	504	- (incompatible)	-25
Wiener Straße, Gumpoldskirchen	94	++ (highly compatible)	+50
Industriestraße, Bad Vöslau	20	++ (highly compatible)	+50
Hauptstraße, Leobersdorf	196	+ (compatible)	+25

Source: the authors

b) Area distribution

In the criterion "area distribution", i.e. the ratio of the width of surfaces for pedestrians and cyclists to those for motorized traffic, significant differences can be seen in places between the various road sections. Figure 7 shows cross-sectional profiles of the four road sections studied.

Figure 7: Cross-sections of the four road sections in the municipalities examined

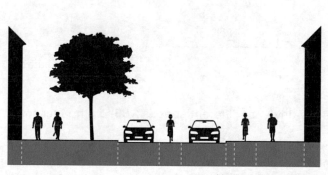

Hauptstraße, Mödling

Wiener Straße, Gumpoldskirchen

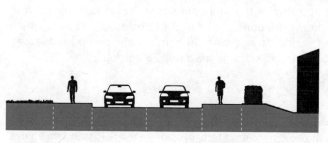

Industriestraße, Bad Vöslau

Hauptstraße, Leobersdorf

Source: the authors

For the road section in Gumpoldskirchen, the ratio of the width of surfaces for pedestrians and cyclists to those for motorized traffic areas is very low at 0.18, which leads to its classification as being entirely incompatible with the requirements of residents and the surroundings, and to an adjustment of the basic value of the still compatible traffic load by -50 veh/ph. The road section in Bad Vöslau likewise has a rather low value of 0.57, resulting in its classification as being just compatible with the residents' demands and with the surroundings, and an adjustment of the basic value of the still compatible traffic load by ±0 veh/ph.

With values of 1.17 and 0.75 respectively, the road sections in Mödling and Leobersdorf on the other hand show more positive ratios of the width of surfaces for pedestrians and cyclists to the those for motorized traffic; these are compatible and highly compatible respectively with the requirements of residents. This leads to an adjustment of the basic value of the still compatible traffic load by +50 and +25 veh/ph respectively.

Table 11: Classification and adjustment of the basic value of the still compatible traffic load for the criterion "area distribution"

Road section, municipality	Ratio of the width of surfaces for pedestrians and cyclists to those for motorized traffic	Levels of compatibility (with requirements of residents and the surrounding area)	Adjustment of the basic value of the still compatible traffic load (veh/ph) by
Hauptstraße, Mödling	1.17	++ (highly compatible)	+50
Wiener Straße, Gumpoldskirchen	0.18	- (completely incompatible)	-50
Industriestraße, Bad Vöslau	0.57	o (just compatible)	±0
Hauptstraße, Leobersdorf	0.75	+ (compatible)	+25

Source: the authors

c) Speed

In the criterion "speed", motor vehicles are driven at different speeds (v85) on the four road sections according to the GIP[2].

The road sections in Gumpoldskirchen and Leobersdorf registered relatively low speeds of 35 and 20 km/h respectively, which are highly compatible with the requirements of residents and with the surroundings and lead to an adjustment of the basic value of the still compatible traffic load by +25 and +50 veh/ph. The road section in Mödling registered a vehicle speed (v85) of 40 km/h, which is only just compatible with the requirements of residents and with the surroundings (adjustment of the basic value of the still compatible traffic load by ±0 veh/ph). The speed of motor vehicles (v85) on the section of road in Bad Vöslau is 50 km/h, which is incompatible with the requirements of residents and with the surroundings and leads to an adjustment of the basic value of the still compatible traffic load by -25 veh/ph.

---

2   The v85 speeds included in the GIP ideally originate from measurements. To the extent that no more reliable speeds were known in the course of data transmission to the GIP, standard values (from empirical studies) are specified by the GIP for entry, depending on the road categories and differentiated according to open countryside and urban areas (cf. GIP.at 2019: 123).

Table 12: Classification and adjustment of the basic value of the still compatible traffic load for the criterion "speed"

Road section, municipality	Vehicle speeds (v85)	Levels of compatibility (with requirements of residents and the surrounding area)	Adjustment of the basic value of the still compatible traffic load (veh/ph) by
Hauptstraße, Mödling	40 km/h	o (just compatible)	±0
Wiener Straße, Gumpoldskirchen	35 km/h	+ (compatible)	+25
Industriestraße, Bad Vöslau	50 km/h	- (incompatible)	-25
Hauptstraße, Leobersdorf	20 km/h	++ (highly compatible)	+50

Source: the authors

d) Share of heavy-vehicle traffic/lorries

In the criterion "share of heavy-vehicle traffic/lorries", the road section in Bad Vöslau has the highest share of lorries in the total motorized traffic volume with 5%, which is compatible with the requirements of residents and the surroundings and leads to an adjustment of the basic value of the still compatible traffic load by +25 veh/ph. The same classification applies for the road section in Gumpoldskirchen, for which a lorry share of 4% was identified. The road sections in both Mödling and Leobersdorf have very low lorry shares of 1 and 2% respectively, which is highly compatible with the demands of residents and the surrounding area and leads to an djustment of the basic value of the still compatible traffic load by +50 veh/ph.

Table 13: Classification and adjustment of the basic value of the still compatible traffic load for the criterion "share of heavy vehicle traffic/lorries"

Road section, municipality	Share of lorries in total motorized traffic volume	Levels of compatibility (with requirements of residents and the surrounding area)	Adjustment of the basic value of the still compatible traffic load (veh/ph) by
Hauptstraße, Mödling	1%	++ (highly compatible)	+50
Wiener Straße, Gumpoldskirchen	4%	+ (compatible)	+25
Industriestraße, Bad Vöslau	5%	+ (compatible)	+25
Hauptstraße, Leobersdorf	2%	++ (highly compatible)	+50

Source: the authors

e) Barrier effect: crossing aids/detours for pedestrians

In the criterion "barrier effect: crossing aids/detours for pedestrians", the road sections in Mödling and Gumpoldskirchen have a low number of crossing aids compared to the length of the road section and the road environment; this is incompatible with the requirements of residents and the surroundings and leads to an adjustment of the basic value of the still compatible traffic load by -25 veh/ph in each case.

On the road section in Bad Vöslau, the number of crossing aids in comparison to the length of the road section is just compatible with the demands of residents and the surroundings (adjustment of the basic value of the still compatible traffic load by ±0 veh/ph). The road section in Leobersdorf, on the other hand, is highly compatible with the requirements of residents and the surroundings with regard to crossing aids due to the existing shared space, which leads to an adjustment of the basic value of the still compatible traffic load by +50 veh/ph.

Table 14: Classification and adjustment of the basic value of the still compatible traffic load for the criterion "barrier effect: crossing aids/detours for pedestrians"

Road section, municipality	Road surroundings category	Ratio of the number of crossing aids to road section length (*100)	Levels of compatibility (with requirements of residents and the surrounding area)	Adjustment of the basic value of the still compatible traffic load (veh/ph) by
Hauptstraße, Mödling	A	0.87	- (incompatible)	-25
Wiener Straße, Gumpoldskirchen	B	0.48	- (incompatible)	-25
Industriestraße, Bad Vöslau	D	0.58	o (just compatible)	±0
Hauptstraße, Leobersdorf	A	shared space	++ (highly compatible)	+50

Source: the authors

f) Greenery and design

In the criterion "greenery and design", the road sections in Mödling and Leobersdorf demonstrate that greenery and other design elements clearly characterize the streetscape and are objectively perceptible, and that these can be respectively classified as highly compatible and compatible with the requirements of residents.

This yields an adjustment of the basic value of the still compatible traffic load by +25 and +50 respectively. For the road section in Gumpoldskirchen, on the other hand, it is apparent that the greenery and design elements tend to be offset in their effectiveness by untreated areas; this is only just compatible with the requirements of residents and with the surroundings (adjustment of the basic value of the still compatible traffic load by ±0 veh/ph). The section of road in Bad Vöslau is completely lacking greenery and other design elements; this is entirely incompatible with the requirements of residents and with the surroundings and leads to an adjustment of the basic value of the still compatible traffic load by -50 veh/ph.

Table 15: Classification and adjustment of the basic value of the still compatible traffic load for the criterion "greenery and design"

Road section, municipality	Greenery and design of the streetscape	Levels of compatibility (with requirements of residents and the surrounding area)	Adjustment of the basic value of the still compatible traffic load (veh/ph) by
Hauptstraße, Mödling	Greenery and other design elements are objectively perceptible. Overall, they still predominate over the technical traffic instalments in the appearance of the street.	+ (compatible)	+25
Wiener Straße, Gumpoldskirchen	Greenery and design elements are in balance with undesigned areas; the street is not characterized by greenery and design.	o (just compatible)	±0
Industriestraße, Bad Vöslau	Greenery and other design elements are entirely lacking; the streetscape is bare and naked.	- (entirely incompatible)	-50
Hauptstraße, Leobersdorf	Greenery and other design elements clearly define the streetscape and shape its character. They lend the street a distinctive character with high experience value.	++ (highly compatible)	+50

Source: the authors

4.2.3 OVERALL ASSESSMENT OF STREETSCAPE COMPATIBILITY

Table 16 (see next page) provides an overview of the

a. basic values of the still compatible traffic load

b. adapted total values of the compatible traffic load and compatibility levels

c. actual traffic load and

d. classification of streetscape compatibility for all four road sections studied.

In contrast with the road section in Bad Vöslau (where the adjustment does not lead to any change in the basic values of the respective compatibility levels; the basic value of the compatible traffic load remains at 1,000 veh/ph following adjustment), for the other three municipalities the adjustment of the basic value of the still compatible traffic load is positive.

For the road section in Mödling, the basic values of the respective compatibility levels are increased by 75 veh/ph, and the adapted basic value of the compatible traffic load is thus 225 veh/ph; for Gumpoldskirchen, the basic values of the respective compatibility levels are increased by 25 veh/ph, and the adapted basic value of the compatible traffic load is thus 275 veh/ph. For Leobersdorf, the basic values of the respective compatibility levels for the road section are increased by 250 veh/ph, yielding an adapted basic value of the compatible traffic load of 400 veh/ph.

Table 16: Overall assessment of streetscape compatibility

Road section, municipality	Basic value of compatible traffic load (veh/ph)	Adjustment of the basic values (veh/ph)							Adapted total values (veh/ph)	Actual traffic load (veh/ph)	Compatibility classification
		Use by pedestrians and cyclists	Surface allocation	Greenery and design	Speed	Share of heavy vehicle traffic/lorries	Barrier effect: crossing aids	Total adjustment			
Hauptstraße, Mödling	150	-25	+50	+25	±0	+50	-25	**+75**	<150 (++) 150 to **225** (+) >225 to 325 (o) >325 to 475 (-) >475 (--)	454	incompatible -
Wiener Straße, Gumpoldskirchen	250	+50	-50	±0	+25	+25	-25	**+25**	<175 (++) 175 to **275** (+) >275 to 425 (o) >425 to 1.025 (-) >1.025 (--)	158	highly compatible ++
Industriestraße, Bad Vöslau	1,000	+50	±0	-50	-25	+25	±0	**±0**	<400 (++) 400 to **1.000** (+) >1.000 to 1.200 (o) >1.200 to 1.500 (-) >1.500 (--)	1,032	just compatible o
Hauptstraße, Leobersdorf	150	+25	+25	+50	+50	+50	+50	**+250**	<325 (++) 325 to **400** (+) >400 to 500 (o) >500 to 650 (-) >650 (--)	512	incompatible -

Source: the authors

On the basis of the actual traffic load (veh/ph) and the adapted compatibility levels for the four road sections, it can be seen that only for the road section in Gumpoldskirchen is the actual motorized traffic load highly compatible with the requirements of the road surroundings. For the road section in Bad Vöslau, the actual motorized traffic load is just compatible with the requirements of the road surroundings. By contrast, the actual motorized traffic load on the road sections in Mödling and Leobersdorf is incompatible with the requirements of the road surroundings.

Table 17: Comparison of the assessment of automated drivability and streetscape compatibility for all road sections studied

Road section, municipality	Classification of automated drivability	Classification of compatibility
Hauptstraße, Mödling 	rather low (0.641)	- incompatible
Wiener Straße, Gumpoldskirchen 	low to medium (0.665)	++ highly compatible
Industriestraße, Bad Vöslau 	rather high (0.762)	o just compatible
Hauptstraße, Leobersdorf 	low (0.592)	- incompatible

Source: the authors

4.3 SUMMARY OF THE RESULTS FOR AUTOMATED DRIVABILITY AND STREETS-CAPE COMPATIBILITY

Table 17 (on the previous page) shows a comparison of the assessment of the road sections in terms of streetscape compatibility with that of automated drivability, i.e. suitability for use by automated vehicles; the tendency is as follows. For the road sections in Mödling and Leobersdorf, which have a rather low rating for automated drivability, the actual motorized traffic load is incompatible with the requirements of the surroundings. For the road section in Bad Vöslau, on the other hand, which has a rather high level of suitability for automated vehicles, the actual motorized traffic load is just compatible with the requirements of the road surroundings. For the road section in Gumpoldskirchen, with a low to medium level of suitability for automated vehicles, the actual vehicle traffic load is in fact highly compatible with the requirements of the road surroundings.

5. DISCUSSION AND CONCLUSION

This chapter examined the suitability of streetscapes for automated vehicles (automated drivability) in the urban-rural continuum, using the example of four different municipalities in Greater Vienna. An analysis of streetscape compatibility, i.e. the extent to which the current actual motorized traffic load on the road sections is compatible with the requirements of the road surroundings, was also carried out for example road sections in the four municipalities under examination. These two assessment criteria were compared for the four road sections.

The results show that for the historic town centres often encountered in the municipalities under examination, the values for automated drivability are rather low and these streets are therefore poorly suited to the use of automated vehicles. They are typically narrow, with shops and a high volume of pedestrians and cyclists; these present difficult conditions for automated vehicles, which could only be used at very low speeds or following corresponding (structural or digital) modifications to the infrastructure. On the other hand, greater suitability for automated vehicles is discernible especially on the sections of the motorways and in some parts of the municipalities outside the town centre – and here above all in commercial or residential areas with low speed profiles. Automated vehicles could be deployed here more easily and with less need for modification.

With regard to streetscape compatibility, very different results were registered for the three selected sections of local through roads and for the selected road section in the industrial area. The current actual motorized traffic load is only compatible with the requirements of the road surroundings for the through road in Gumpoldskirchen. In the case of the other through roads (Mödling and Leobersdorf), on the other hand, the actual motorized traffic load is incompatible with the requirements of the road surroundings. For the road section in the industrial area in Bad Vöslau, the actual motorized traffic load is just compatible with the requirements of the road surroundings.

A combined examination of these two assessments suggests that for the road sections with a rather low automated drivability rating, i.e. rather low suitability for automated vehicles (Mödling and Leobersdorf), the actual motorized traffic load is incompatible with the requirements of the road surroundings. However, this does not apply to the same extent in the case of the road section in Gumpoldskirchen, with low to medium suitability for automated vehicles; here, the actual motorized traffic load is compatible with the requirements of the road surroundings. For the road section in Bad Vöslau, which has a rather high level of suitability for automated vehicles, the actual motorized traffic load is just compatible with the requirements of the road surroundings.

For the purpose of planning, this shows that an assessment of suitability for automated vehicles can help in identifying the areas in which the use of automated vehicles is more practicable from a technological-infrastructural point of view, or with fewer modifications to the streetscape. Regarding potential areas of use of automated vehicles, however, it should also be closely examined whether the current motorized traffic loads are compatible with the requirements of the road surroundings; the analysis shows that already today, the current situation is problematical on some road sections. Where the use of automated vehicles – which leads to increased or additional traffic volume and a denser vehicle sequence and thus further reduces the permeability of the streetscape for other road users – is in fact not compatible with the requirements of the road surroundings, it should possibly be avoided in these areas.

Table 18 (see next page) once more gives an overview of the assessment of streetscape compatibility of the various road sections, taking into account an increase in traffic performance with automated vehicles. In terms of the effects on traffic performance, the meta-study by Soteropoulos et al. (2019b) mentions ranges of between +1% and +59% for private automated vehicles and between +8% and +80 % for automated carsharing, as a result of shifts from other traffic modes and empty trips. A reduction in transport volume in the range of -25% to -10% could be achieved merely with a very high share of ride-sharing and thus a high vehicle occupancy rate (cf. Soteropoulos et al. 2019b: 40). To accurately depict the increase in traffic performance brought about by automated vehicles in the respective road sections, a transport demand model would be required in order to specifically model the respective use cases of automated vehicles, such as those in private use or as deployed in car or ride-sharing (together with the associated assumptions), for the road sections in the municipalities examined. However, since such modelling is not possible within the scope of this study, increases in traffic performance of +5% to +30% serve as an exemplary basis for the table and for the respective classifications of compatibility.

Even a small increase in traffic load (+5%) due to automated vehicles on the road section in Mödling has been shown to be entirely incompatible with the requirements of the road surroundings. This is likewise the case for the road section in Leobersdorf, with a higher increase in traffic load (+30 %). Also in the case of the road section in Bad Vöslau, a 20% increase in traffic load due to automated vehicles is already incompatible with the requirements of the road surroundings. Only for the road section in Gumpoldskirchen – in view of the low initial traffic load – is an increase in traffic load of up to 30% due to automated vehicles shown to be compatible with the requirements of the road surroundings.

In cases where the motorized traffic load of a streetscape is not compatible with the requirements of the road surroundings, it can thus be assumed – depending on the respective increase in traffic load due to automated vehicles – that the use of these vehicles will make the streetscape even more incompatible with the requirements of the road surroundings. In these cases, the streetscape should instead be designed to be more compatible with the requirements of the road surroundings (e.g. with reduced speeds, additional direct crossing aids for pedestrians, conversion of parking spaces), or the use of automated vehicles should be coupled to these measures (cf. Anciaes/Jones 2016: 4). In order to also reduce the increase in traffic performance through automated vehicles, measures such as dynamic "road pricing" or "mobility pricing" – i.e. imposing a quota of vehicles or of distances covered on certain road sections – or measures to increase vehicle occupancy rates (e.g. bans on or charges for empty

Table 18: Assessment of streetscape compatibility, taking into account an increase in traffic load due to automated vehicles

Road section, municipality	Adapted total values (veh/ph)	Increase in traffic load due to automated vehicles	Traffic load (veh/ph)	Classification of compatibility
Hauptstraße, Mödling				
	< 150 (++)	±0%	454	- incompatible
	150 to 225 (+)	+5%	477	- entirely incompatible
	> 225 to 325 (o)	+10%	499	- entirely incompatible
	> 325 to 475 (-)	+20%	545	- entirely incompatible
	> 475 (- -)	+30%	590	- entirely incompatible
Wiener Straße, Gumpoldskirchen				
	< 175 (++)	±0%	158	++ highly compatible
	175 to 275 (+)	+5%	166	++ highly compatible
	> 275 to 425 (o)	+10%	174	++ highly compatible
	> 425 to 1025 (-)	+20%	190	+ compatible
	> 1025 (- -)	+30%	205	+ compatible
Industriestraße, Bad Vöslau				
	< 400 (++)	±0%	1,032	o just compatible
	400 to 1000 (+)	+5%	1,084	o just compatible
	> 1000 to 1200 (o)	+10%	1,135	o just compatible
	> 1200 to 1500 (-)	+20%	1,238	- incompatible
	> 1500 (- -)	+30%	1,342	- incompatible
Hauptstraße, Leobersdorf				
	< 325 (++)	±0%	512	- incompatible
	325 to 400 (+)	+5%	538	- incompatible
	> 400 to 500 (o)	+10%	563	- incompatible
	> 500 to 650 (-)	+20%	614	- incompatible
	> 650 (- -)	+30%	666	- entirely incompatible

Source: the authors

runs) are also appropriate (cf. Soteropoulos et al. 2019a: 133). The use of automated vehicles as public transport media can also be of benefit, but careful consideration should be given to whether this would be compatible with the requirements of the road surroundings – especially if it would involve structural or infrastructural modifications. Further extensive studies would be necessary to investigate the connection between the technological-infrastructural suitability of streetscapes for the use of automated vehicles and the compatibility of these streetscapes.

Additional research is also needed to establish the extent to which modifications for improved streetscape compatibility, or measures to enable certain applications of automated driving (e.g. hop-on/hop-off areas for automated vehicle fleets; cf. Chap. 8 by Bruck et al. in this volume), affect the suitability of these streetscapes for automated driving systems and vice versa. However, a number of aspects should be taken into account regarding the results of the streetscape compatibility assessment, and further research is necessary:

a. The counts of motorized vehicles and of pedestrians and cyclists on the respective road sections would have to be carried out more comprehensively in order to further refine the results.[3] Firstly, the traffic counts were conducted in November, a month in which significantly fewer cyclists and pedestrians are on the respective sections of road than at other times of the year. Secondly, the traffic counts did not take place for all road users at the respective peak hours (e.g. pedestrians 12 noon–2 p.m. and 4–6 p.m., cyclists 12 noon–2 p.m., motorized traffic 7–11 a.m. and 3–7 p.m.; cf. FGSV 2012: 28), but were carried out over the period from 3 to 4 p.m. The results of the traffic count may therefore deviate from the actual volumes in the respective peak hours. It can thus be assumed that the assessment of streetscape compatibility tends to be somewhat worse for some of the road sections examined, since more road users are presumably on the move during the respective peak hours than were registered in the current assessment. As well as more extensive manual traffic counts, automatic traffic counts – e.g. using radar devices at the side of the road – could also be used here to give a much more accurate picture of the actual traffic volumes on the respective road sections (cf. FGSV 2012: 35). In addition, the values derived from the GIP for the v85 speeds of vehicles on the road sections would also have to be verified. Extensive speed measurements should also be made in order to account for certain variances in the speeds travelled (e.g. different times of day: higher vehicle densities and lower speeds at rush hour, higher speeds at night), so as to depict the v85 speed of vehicles in more detail.

b. Determining the basic value of the compatible traffic load is a highly significant factor for assessing streetscape compatibility. The basic values used in the studies by Mörner et al. (1984) and by Bühlmann and Laube (2013) were used and slightly adapted for this purpose, but it would be advisable to carry out a sensitivity analysis with regard to the basic values on the basis of further exemplary road sections.

c. The compensatory approach used for assessing of streetscape compatibility does not sufficiently take into account numerous aspects which however would be relevant to this assessment. Firstly, relevant factors such as accessibility for pedestrians with disabilities are entirely lacking. Although there is invariably some tension between the applicability of the approach in terms of data available and its level of detail or the number of aspects taken into account with regard to streetscape compatibility, consideration of some additional factors would appear to be relevant for depicting streetscape compatibility more comprehensively. To take further aspects into account, the quality criteria for public space described by Gehl Architects (2009) can be used, or these criteria can be applied in a two-stage process following an assessment of streetscape compatibility (cf. Gehl Architects 2009: 43).

d. Secondly, due to the logic of the compensatory approach, aspects such as (1) allocation of space between motor vehicles and pedestrians or cyclists, or (2) green spaces and design

---

3 The traffic counts to determine the number of vehicles during peak hour, the number of pedestrians and cyclists, and the share of lorries were carried out by Michael Haudum in the course of his diploma thesis.

elements in the streetscape, are each only taken into account as one of several criteria on the basis of which the compatible traffic load is merely compensated for. However, the results for the road section in Gumpoldskirchen – where the pavement widths only barely comply with minimum standards (cf. FSV 2015), but good compatibility is achieved – show that topics such as sufficient pavement width for pedestrians are not sufficiently taken into account in this approach. An adaptation of the compensatory approach with a higher weighting (higher compensation values) would appear important in terms of the distribution of space between motor vehicles and pedestrians or cyclists. This also applies to green spaces, which should be accorded greater significance in the assessment of streetscape compatibility in view of climate change and their potential to counteract the formation of heat islands (cf. Sandholz/Sett 2019: 11).

e. For an assessment of streetscape compatibility in connection with automated driving, the procedure should be more strongly linked to the effects of automated vehicles or be extended to include such aspects. An example of this is the barrier effect in the streetscape resulting from the more compact spacing of automated vehicles. In the context of this article, the procedure was already extended to take crossing aids into account. However, further consideration should also be given for example to an assessment of the need to cross roads. A detailed examination of the crossing behaviour of pedestrians and cyclists is necessary here, in order to determine whether from the perspective of these road users there is a need to cross at specific points that can be catered to with individual crossing points, or whether linear crossing (e.g. in shopping streets) or planar crossing (e.g. at railway station forecourts) is required (cf. Häfliger et al. 2015: 80). It should also be taken into consideration that according to FGSV (2002), on roads up to 8.50 metres wide with two lanes, crossing by pedestrians is hardly possible (1) with traffic volumes greater than 1,000 veh/ph and a vehicle speed of 50 km/h or (2) with traffic volumes greater than 500 veh/ph and a vehicle speed of more than 50 km/h. In these two cases, crossing aids are always required (cf. FGSV 2002). An evaluation of the waiting time at crossing aids (e.g. traffic lights and pedestrian crossings) would also be relevant in this context.

f. Greater consideration of these criteria would be particularly important in order to assess not only how much additional traffic due to automated vehicles is compatible with the streetscape, but also to what extent the denser spacing of automated vehicles leading to an increase in capacity is compatible with the road surroundings, and what measures and adaptations (e.g. a possible channelling of pedestrians crossing the carriageway) would be necessary to balance the effects of such a capacity increase and barrier effect in the best possible way.

g. Finally, even more consideration should be given to how the effects of automated vehicles other than their intended function as transport media could likewise influence their compatibility with use of the surrounding area; the evaluation of streetscape compatibility should be extended to include these aspects. Already today, for example, the sensors of vehicles equipped with technology solely for the purpose of automated driving (e.g. Tesla Autopilot) monitor their surroundings, and thus also inevitably register such use along with people and their activities in the streetscape. This "continuous" recording of data or monitoring of persons by automated vehicles may seem unproblematic for some applications and activities, but for other sensitive applications, for example in the context of political rallies or similar (cf. Heger 2008: 93), in view of privacy considerations it must be clarified whether and to what extent this data can be collected. The question is therefore to what extent automated vehicles are "compatible" with such applications (cf. Chap. 10 by Mitteregger in this volume). In the context of automated vehicles, it thus seems relevant to extend streetscape compatibility by additional aspects; this should be more closely investigated.

REFERENCES

Abegg, C., C. Girod, K. Fischer, N. Pahud, L. Raymann and F. Perret 2018. "Einsatz automatisierter Fahrzeuge im Alltag – Denkbare Anwendungen und Effekte in der Schweiz. Schlussbericht Modul 3d 'Städte und Agglomerationen'", version of 30/8/2018. www.ebp.ch/sites/default/files/project/uploads/2018-08-30%20aFn_3d%20St%C3%A4dte-Agglomerationen%20Schlussbericht_1.pdf (4/5/2020).

Anciaes, P., and P. Jones 2016. "How do pedestrians balance safety, walking time, and the utility of crossing the road? A stated preference study", *Street Mobility and Network Accessibility Series*, Working Paper 8. London: UCL. https://pdfs.semanticscholar.org/bac7/2aa3228d282fe2cd-f260ac4388be21b23dd5.pdf (6/5/2020).

Baier, R. 1992. "Verträglichkeit des Kraftfahrzeugverkehrs in Straßenräumen und Straßennetzen – Praxisorientiertes Verfahren in der Verkehrsentwicklungsplanung", in *Internationales Verkehrswesen* 10, 395–399.

Baier, R., and FGSV (Forschungsgesellschaft für Straßen- und Verkehrswesen) 2007. *Richtlinien für die Anlage von Stadtstraßen – RASt 06*, FGSV – Forschungsgesellschaft für Straßen- und Verkehrswesen (ed.). Cologne: FGSV Verlag.

Baier, R., C. Hebel, Y. Jachtmann, A. Reinartz, K.-H. Schäfer and A. Warnecke 2011. "Stadt Mönchengladbach. Untersuchungen zur Verkehrsentwicklungsplan", BSV – Büro für Stadt- und Verkehrsplanung Reinhold Baier, Aachen. https://tinyurl.com/yaqvgpwk (4/5/2020).

BMVIT (Federal Ministry of Transport, Innovation Stadt Wien) 2018. "Testberichte. Tests auf Straßen mit öffentlichem Verkehr in Österreich gemäß der Automatisiertes Fahren Verordnung. Zeitraum: 2016–2018", Vienna. https://bit.ly/3ereyGW (4/5/2020).

Brummelen, J. van, M. O'Brien, D. Gruyer and H. Najjaran 2018. "Autonomous vehicle perception: The technology of today and tomorrow", in Transportation Research Part C: Emerging Technologies 89, 384–406.

Bühlmann, F., and M. Laube 2013. "Verträglichkeit Straßenraum. Methodik und Ergebnisse", Canton of Zurich (ed.). Zurich: Traffic Office. https://tinyurl.com/y7nou6qc (4/5/2020).

Czarnecki, K. 2018. "Operational Design Domain for Automated Driving Systems. Taxonomy of Basic Terms", Waterloo Intelligent Systems Engineering (WISE) Lab, University of Waterloo, Canada.

Favarò, F. M., S. Eurich and N. Nader 2018. "Autonomous vehicles' disengagements: Trends, triggers, and regulatory limitations", in *Accident Analysis and Prevention* 110, 136–148.

FGSV (Road and Transportation Research Association) 2002. Empfehlungen für Fußgängerverkehrsanlagen (EFA). Cologne: FGSV Verlag.

FGSV (Road and Transportation Research Association) 2012. *EVE – Empfehlungen für Verkehrserhebungen*. Cologne: FGSV Verlag.

Frehn, M., G. Steinberg and S. Schröder 2013. "Methodik und Ergebnisse der Straßenraumverträglichkeit. Verkehrsentwicklungsplan Bremen 2025". www.bau.bremen.de/sixcms/media.php/13/130228_E03_Straßenraumvertraeglichkeit_Methodik_Ergebnisse.pdf (5/5/2020).

FSV (Forschungsgesellschaft Straße – Schiene – Verkehr) 2015. "RVS 03.02.12 Fußgängerverkehr". Vienna.

Gehl Architects 2009. "Downtown Seattle: Public Spaces & Public Life", ed. by City of Seattle. www.seattle.gov/Documents/Departments/SDCI/Codes/PublicSpacesLifeIntro.pdf (7/5/2020).

GIP.at 2019. "Intermodaler Verkehrsgraph Österreich. Standardbeschreibung der Graphenintegrationsplattform (GIP), Version 2.3". www.gip.gv.at/assets/downloads/GIP_Standardbeschreibung_2.3.pdf (5/5/2020).

Häfliger, R., J. Bubenhofer, C. Hagedorn, K. Zweibrücken, S. Condrau and R. Baier 2015. "Verträglichkeitskriterien für den Straßenraum innerorts. Forschungsprojekt SVI 2004/058 on commission from the Swiss Association of Transport Engineers and Experts (SVI)". Zurich: Schweizerischer Verband der Strassen- und Verkehrsfachleute (VSS).

Heger, Nora 2008. "Entgrenzte Räume. Kontrolle des öffentlichen Raums am Beispiel der Videoüberwachung am Wiener Schwedenplatz", diploma thesis, University of Vienna. http://othes.univie.ac.at/2690/1/2008-10-21_0002326.pdf (7/5/2020).

Heinrichs, D. 2015. "Autonomes Fahren und Stadtstruktur", in *Autonomes Fahren. Technische, rechtliche und gesellschaftliche Aspekte*, ed. by M. Maurer, J. C. Gerdes, B. Lenz and H. Winner. Berlin/Heidelberg: Springer Vieweg, 219–239.

Hörl, S., F. Becker, T. Dubernet and K. W. Axhausen 2019. "Induzierter Verkehr durch autonome Fahrzeuge: Eine Abschätzung". Bern: Federal Roads Office. https://ethz.ch/content/dam/ethz/special-interest/baug/ivt/ivt-dam/vpl/reports/1401-1500/ab1433.pdf (4/5/2020).

Litman, T. A. 2009. "Barrier Effect", in *Transportation Cost and Benefit Analysis. Techniques, Estimates and Implication*, 2nd ed., ed. by T. A. Litman. Victoria: Victoria Transport Policy Institute, Chap. 5.13. www. vtpi.org/tca/tca0513.pdf (4/5/2020).

Marsden, G., J. Dales, P. Jones, E. Seagriff and N. Spurling 2018. "All change? The future of travel demand and the implications for policy and planning. The First Report of the Commission on Travel Demand". www.demand.ac.uk/wp-content/uploads/2018/04/FutureTravel_report_final.pdf (4/5/2020).

Marshall, S. 2005. *Street & Patterns*. London: Spon Press.

Milakis, D., B. van Arem and B. van Wee 2017. "Policy and society related implications of automated driving: A review of literature and directions for future research", in *Journal of Intelligent Transportation Systems. Technology. Planning and Operations* (21) 4, 324–348.

Mörner, J. von, P. Müller and H. Topp 1984. "Forschung Straßenbau und Straßenverkehrstechnik, Entwurf und Gestaltung innerörtlicher Straßen", Report 425 from the series *Forschung Straßenbau und Straßenverkehrstechnik*. Bonn: Federal Ministry of Transport.

Müller, P., H.-J. Collin, A. Ratschow and W. Rüthrich 1994. "Das LADIR-Verfahren zur Bestimmung stadtverträglicher Belastungen durch Autoverkehr", final report on the research project of the research field "Urban Development and Transport" in the Federal Ministry of Regional Planning, Building and Urban Development, Darmstadt/Braunschweig.

Nørby, L. E., and K. R. Meltofte 2012. "Over Vejen. Vejen som trafikal barriere for fodgængere". Aalborg: Aalborg University. https://projekter.aau.dk/projekter/files/63452678/Over_vejen_hovedrapport_.pdf (4/5/2020).

OECD 2020. "Functional urban areas by country". www.oecd.org/cfe/regional-policy/functionalurbanareasbycountry.htm (5/5/2020).

Pendleton, S. D., H. Andersen, X. Du, X. Shen, M. Meghjani, Y. H. Eng, D. Rus and M. H. Ang Jr. 2017. "Perception, Planning, Control, and Coordination for Autonomous Vehicles", in *Machines* (5) 6, 1–54.

SAE International 2018. "Taxonomy and Definitions for Terms Related to Driving Automation Systems for On-Road Motor Vehicles – J3016", June 2018. www.sae.org/standards/content/j3016_201806/ (4/5/2020).

Sandholz, S., and D. Sett 2019. "Erfahrungen und Bedarfe von Akteuren der Stadtplanung im Hinblick auf Vulnerabilität gegenüber Hitzestress. Ergebnisse einer Haushalts-Umfrage zum Hitzeempfinden in Bonn", *ZURES Working Paper 2*, 08/2019. https://collections.unu.edu/eserv/UNU:7510/ZURES_workingpaper2_ErgebnisseHHUmfrage_UNU-EHS-1.pdf (7/5/2020).

Shladover, S. E. 2018a. "Connected and automated vehicle systems: Introduction and overview", in *Journal of Intelligent Transportation Systems* (22) 3, 190–200.

Shladover, S. E. 2018b. "Practical Challenges to Deploying Highly Automated Vehicles", presentation at Drive Sweden, Gothenburg.

Soteropoulos, A., A. Stickler, V. Sodl, M. Berger, J. Dangschat, P. Pfaffenbichler, G. Emberger, E. Frankus, R. Braun, F. Schneider, S. Kaiser, H. Walkobinger and A. Mayerthaler 2019a. "SAFiP – Systemszenarien Automatisiertes Fahren in der Personenmobilität. Endbericht". Vienna: Federal Ministry of Transport, Innovation and Technology. https://projekte.ffg.at/anhang/5cee1b11a1eb7_SAFiP_Ergebnisbericht.pdf (4/5/2020).

Soteropoulos, A., M. Berger and F. Ciari 2019b. "Impacts of automated vehicles in travel behaviour and land use: An international review of modelling studies", in *Transport Reviews* (39) 1, 29–49.

Soteropoulos, A., M. Mitteregger, M. Berger and J. Zwirchmayr 2020. "Automated drivability: Toward an assessment of the spatial deployment of level 4 automated vehicles", in *Transportation Research Part A: Policy and Practice* 136, 64–84.

Stadt Wien 2018. "STEP 2025. Fachkonzept Öffentlicher Raum", Stadtentwicklung Wien, Magistratsabteilung 18 – Stadtentwicklung und Stadtplanung (ed.). www.wien.gv.at/stadtentwicklung/studien/pdf/ b008522.pdf (4/5/2020).

Su, S., H. Zhou, M. Xu, H. Ru, W. Wang and M. Wenig 2019. "Auditing street walkability and associated social inequalities for planning implications", in *Journal of Transport Geography* 74, 62–76.

Wissenschaftlicher Beirat beim Bundesminister für Verkehr und digitale Infrastruktur 2017. "Automatisiertes Fahren im Straßenverkehr. Herausforderungen für die zukünftige Verkehrspolitik". Berlin: Federal Ministry for Digital and Transport. www.bmvi.de/SharedDocs/DE/Anlage/G/ wissenschaftlicher-beirat-gutachten-2017-1.pdf?__blob=publicationFile (4/5/2020).

6   Automation, public transport and Mobility as a Service: Experience from tests with automated shuttle buses

Aggelos Soteropoulos, Emilia M. Bruck, Martin Berger, Alexander Egoldt, Arne Holst, Thomas Richter, Zoltán László

Aggelos Soteropoulos
TU Wien, future.lab Research Center and Research Unit Transportation System Planning (MOVE)

Emilia M. Bruck
TU Wien, future.lab Research Center and Research Unit of Local Planning (IFOER)

Martin Berger
TU Wien, Research Unit Transportation System Planning (MOVE)

Alexander Egoldt
TU Berlin, Department of Road Planning and Operation

© The Author(s) 2023
M. Mitteregger et al. (eds.), *AVENUE21. Planning and Policy Considerations for an Age of Automated Mobility*, https://doi.org/10.1007/978-3-662-67004-0_6

Arne Holst
TU Berlin, Department of Road Planning and Operation

Thomas Richter
TU Berlin, Department of Road Planning and Operation

Zoltán László
Swiss Federal Railways (SBB), Lead On-Demand Mobility, New Mobility Services

1. INTRODUCTION

Automated driving will fundamentally transform future mobility and will also affect public transport. In this context, there is often talk of a further shift of the boundaries between classic public transport and motorized individual transport, with an area of transition in public individual transport with automated vehicles, or an individualization of public transport (cf. Lenz/Fraedrich 2015: 189; Röhrleef 2017: 15; Bruns et al. 2018: 12; Barillère-Scholz et al. 2020: 16): already today, mobility is becoming differentiated through new forms of services such as car sharing and ride hailing.

Especially in cities, new mobility service providers are offering demand-oriented, individualized transport options – so-called on-demand mobility – thereby extending the mobility offer, which will continue to expand in view of the advance of digitalization (cf. Barillère-Scholz et al. 2020: 15; Buffat et al. 2018: 90; Lenz/Fraedrich 2015: 183). The technological development of automated driving provides opportunities to develop unprecedented business models that will open up the market to further providers: it is conceivable that automation will bring about disruptive developments in the mobility sector and a further transformation of the forms of service offered today (cf. Gertz/Dörnemann 2016: 5). Mobility as a Service (MaaS) is also becoming increasingly significant: this entails combining public and private transport offers, along with different modes of transport, by means of a uniform digital access portal (platform or app), thus offering customised mobility solutions that cater to individual requirements (cf. EPOMM 2017; Jittrapirom et al. 2017: 14).

Automated driving on public roads is also expected to offer potential for improving the economic efficiency of public transport, if personnel costs can be reduced: if a driver is no longer needed and new supplementary forms of service in the form of smaller, more versatile units are provided, and if vehicle concepts are increasingly matched to current demand, there will be greater scope for more economical, efficient and demand-oriented use of mobility services (cf. Hörl 2020: 2; Hörl et al. 2019: 60; Bösch et al. 2018: 7; Gertz/Dörnemann 2016: 22) – even if new or additional costs arise in some cases, for example for scheduling systems or for additional personnel to repair and clean the vehicles (cf. Bruns et al. 2018: 5). Furthermore, unlike the often long-term licensing arrangements in use today, which normally do not allow for any significant modifications or adjustments, the new forms of service make it possible to direct offers more towards individual personal needs, to make specific adjustments and thus to make public transport more attractive and stronger (Barillère-Scholz et al. 2020: 15).

Figure 1: Shifting the boundaries between public transport and motorized individual transport

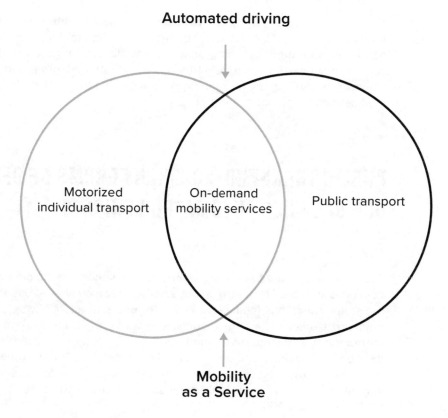

Source: the authors

Numerous pilot projects are currently being conducted for automated driving on public roads, in the course of which automated technologies are being put into practice at an early stage in selected application cases, and new transport services are being created and holistic vehicle concepts developed. These projects ensure that the operation of on-demand mobility services, data-based traffic analysis and platform solutions are ready today for the automated mobility of tomorrow. The aim here is to provide system modules as a perspective to enable customers such as transport providers, municipalities and cities to operate new forms of mobility in public transport (cf. Barillère-Scholz et al. 2020: 18).

Especially in German-speaking countries, the pilot projects for automated driving on public roads currently focus on the test operation of automated shuttle buses. These run on specially approved, fixed routes and are for the most part still accompanied on board by safety drivers or operators. In most of these projects involving automated shuttle buses, the tests focus on aspects of technological, organizational, operational and economic feasibility (cf. Jürgens 2020).

On the basis of experience gained in a number of pilot projects with automated shuttle buses, this chapter specifically addresses the aspects of (1) technological feasibility and the possibility of operating in streetscapes, and (2) integration of automated shuttle buses into existing public transport systems.

For this purpose, the relevance of automation for public transport is first presented and various use cases of automated driving in public transport are shown. The chapter then specifically examines the use case of automated shuttle buses and provides an overview of the various tests with automated shuttle buses being conducted in Europe, above all in German-speaking countries.

The technical and legal aspects of testing automated shuttle buses will then be illustrated using the example of the "autoNV OPR" project in Ostprignitz-Ruppin in Germany, and the experience gained in operating and integrating the shuttle bus into the existing public transport system will be examined in more detail using the "MyShuttle" project in Zug, Switzerland. Finally, a summary is given of the main findings from the two example projects, along with a derivation of implications for planning and policy.

2. PUBLIC TRANSPORT: CURRENT FORMS OF OFFER AND FUTURE USE CASES OF AUTOMATED VEHICLES

Public transport fundamentally covers all passenger transport offers provided on a regular, commercial basis. These are largely characterized by shared use that is accessible to all under the same conditions (cf. Hörold 2016: 38; Bruns et al. 2018: 15). Since the provision of public transport services is a largely public task within the framework of services of general interest for mobility (cf. Rollinger/Amtmann 2009: 6), public transport services usually do not fully cover their operating costs and are thus publicly subsidized. Public transport services also have the following characteristics, which as a rule are legally founded and thus include the familiar, now largely fixed elements of public transport services such as departure times, stops and routes (cf. Bruns et al. 2018):

- Operation by authorized transport companies on licensed lines or routes

- Obligatory timetable: devising and publishing a timetable

- Obligation to operate: carrying out the published offer, independently of external conditions or momentary demand

- Obligatory tariffs: fixing and publishing conditions of carriage and fares.

In addition to the more classic scheduled transport services, however, some forms of service are more strongly oriented towards the individual needs of passengers through flexibilization of departure times (on-demand transport) and of routes/lines and variable stops (without fixed stopping points) or a combination of these elements (cf. Bruns et al. 2018: 15). A distinction can be made here between flexible forms of service, i.e. micro public transport systems or on-demand services, and alternative forms of service such as car or ride sharing. Sommer (2018: 3f.) speaks in this context of the publicly accessible offers of an "extended" public transport service. Table 1 gives an overview of the various current forms of public transport services. However, elements of the flexible forms of offer are also often combined with those of the alternative forms (e.g. registration of use, etc.).

Table 1: Overview of the characteristics of different forms of public transport services

	Classic regular service	Flexible forms of service	Alternative forms of service
Charac-teristics	Fixed service, timetable and clearly defined route	For use in times and areas of low demand	Usually permit-free and no guarantee of carriage, one-time registration or application required (except for social driving service)
Examples	• Superordinate basic rail-bus network • Local scheduled transport for area coverage	**With timetable:** • Linear service area (line service on demand) • Corridor service area (usually two fixed stops): travels to requested stops (e.g. call bus) • Sector service area (one connection point): sector operation (e.g. ring-and-ride, feeder service) **Without timetable:** • Zonal service area and with trip grouping (zonal operation or zonal service)	**Passenger as ride-sharer:** • Private ride provider • Ride-pooling: provider = transport, taxi or rental car company with obligation to operate, obligation to carry, compulsory travel area and tariff requirements • Ride sharing (public transport offer): journey takes place even if no third person rides or if only persons ride who did not order via a platform; without obligation to operate, obligation to carry, compulsory travel area or tariff requirements (e.g. BlaBlaCar) • Ride-selling and ride-hailing: provider = commercial platform provider such as Uber, Moia etc. without obligation to operate, obligation to carry, compulsory travel area or tariff requirements • Social institution/association as ride provider: social ride service, passenger as self-driver • Car sharing (station-based/"free-floating"): public car
Means of transport	Underground, commuter/regional train, tram, articulated bus, standard bus, minibus (citizens' bus), shuttle bus	Standard bus, call bus, minibus, shuttle bus, van, call-sharing taxi	Shared taxi/van/car, single taxi/car

Source: the authors, based on BMVI (2016) and Sommer (2018)

The extension and combination of classic scheduled transport with flexible and alternative, and in this case also private, forms of service is currently under discussion above all in the context of Mobility as a Service (MaaS). This combination is implemented via a uniform digital access portal (e.g. platform, app). The mobility offer allows interaction between different modes in such a way as to best meet various personal needs – i.e. individually tailored mobility solutions are offered (cf. EPOMM 2017; Jittrapirom et al., 2017: 14). According to this idea, MaaS should improve the efficiency both of existing mobility systems and of public resources (cf. Hoadley 2017: 5ff.).

In view of the existing forms of service in extended public transport, automated driving provides options for further differentiation of the service, along with opportunities for redesigning inter-mobility, for further flexibilization and individualization, and for temporal and spatial densification of the service (cf. Lenz/Fraedrich 2015: 189). This is made possible by the elimination of personnel costs, which account for a large portion of the overall costs in public transport (cf. Hell 2006: 169). By eliminating or reducing personnel costs, the services could be operated more economically (cf. Hörl 2020: 2; Hörl et al. 2019: 60; Bösch et al. 2018: 7; Gertz/Dörnemann 2016: 22). However, it remains to be clarified to what extent accompanying personnel is still needed to ensure the safety of passengers (cf. Salonen/Haavisto 2019: 13; Mitteregger et al. 2019: 610).

The advance of automation could lead to new forms of service or transport media (cf. Soteropoulos et al. 2019: 104). This includes classic scheduled transport (e.g. automated standard or articulated bus), flexible forms of service (e.g. automated minibus/shuttle bus, automated van, automated ride sharing) and alternative forms of service (e.g. automated shared taxi/ride sharing, automated single taxi/car sharing; cf. Bruns et al. 2018: 21). In classic scheduled services, automation will also extend to large vehicles that will provide a minimum public service, while the flexible and alternative forms of service will tend to be provided by small vehicles. Their market niches result from additional offers and from offers for the first and last mile in combination with regular services, especially in times and areas of low demand (cf. Ohnemus/Perl 2016: 591).

Table 2 (on the next page) gives an overview of the characteristics of various forms of service and transport media in extended public transport, taking into account automation in road traffic. The automated shuttle bus occupies a special position here: its size makes it suitable both for regular service on fixed routes with a fixed timetable and fixed stops, and for flexible use according to demand with request stops.

Particularly in the case of the possible flexible and alternative forms of service with automation, these could come into consideration for different forms of service (see Table 3). The (automated) minibus or shuttle bus is most suited to scheduled service or as a call bus that also travels to requested destinations within a specified area. The (automated) van (ride sharing) and the (automated) shared taxi (ride sharing), on the other hand, are most suitable as ring-and-ride taxis, as a feeder, i.e. for sector operation, or for use in zonal operation or zonal service. The (automated) single taxi (car sharing) is likewise most suited to sector and zonal operation.

All in all, a wide range of applications are therefore possible for automated vehicles in public transport. These different vehicle concepts must be ideally implemented in future. To sound out their best possible use already today, numerous pilot projects are being conducted for automated driving in public transport. The majority of these projects focus on automated shuttle buses (Barillère-Scholz et al. 2020: 18), which are mostly used here in regular service. Relevant aspects are now already emerging that will retain their relevance in the future for the use of automated shuttle buses or of other automated vehicle concepts in the context of flexible or alternative forms of service. In the next section, the use case of the automated shuttle bus will therefore be examined in more detail, and an exemplary overview of the various pilot projects will be given.

Table 2: Characteristics of different forms of service and transport media in extended public transport, and possible characteristics taking into account automation in road traffic

Characteristics/elements	Commuter/regional train	Underground	Tram	(AT) articulated bus	(AT) standard bus	(AT) minibus/shuttle bus	(AT) van (ride sharing)	(AT) shared taxi (ride sharing)	(AT) single taxi (car sharing)
Temporal availability									
Fixed timetable	✓	✓	✓	✓	✓	✓			
Transport on demand						✓	✓	✓	✓
Frequency (mins.)	15–60	2–15	3–30	3–60	3–120	5–30/–	–	–	–
Request time (mins.)	–	–	–	–	–	–/15–45	10–45	5–30	4–20
Spatial or local availability									
Fixed stops	✓	✓	✓	✓	✓	✓			
Stop on demand						✓	✓	✓	✓
Distance between stops (m)	2500–3500	500–900	300–700	300–700	300–700	250–500 / –	–	–	–
Travel speed (km/h)	30–60	30	15–40	15–30	15–30	20–40	25–50	25–50	25–60
Convenience	medium	medium	medium	low	low	medium	high	high	very high
Payment system									
Frequent-use tariff	✓	✓	✓	✓	✓	✓	✓	✓	
Payment per trip	✓	✓	✓	✓	✓	✓	✓	✓	✓
Distance-/time-dependent	✓				✓			✓	✓
Vehicle capacity (persons)	1000	750	140–200	75–140	25–75	8–20	5–8	2–5	1–2
Operator	mostly public	mostly public	mostly public	mostly public	mostly public	public/private	public/private	public/private	public/private
Offer type	scheduled	scheduled	scheduled	scheduled	scheduled/flexible	scheduled/flexible	flexible/alternative	flexible/alternative	flexible/alternative
Demand	high	very high	high	high	medium	medium	low	low	low

Source: the authors, based on Wolf-Eberl et al. (2011: 27ff.), Weidmann et al. (2011: 89ff.), BMVI (2016: 23ff.), Bruns et al. (2018: 20) and Sommer (2018: 10)

Table 3: Flexible and alternative forms of service with automation

	Designation	Service principle	Fixed timetable	Registration required	Departure from	Destination
(AT) minibus/ shuttle	schedule	●—●—●—●—●	yes	no	(H)	(H)
	call bus (to requested stops)	⬡ network	yes	yes	(H)	(H)
(AT) van (ride sharing)	call-sharing taxi (sector operation)	●→ ○○○	yes	yes	(H)	⌂
(AT) shared taxi (ride sharing)	feeder (sector operation)	●←← ○○○	yes	yes	⌂	(H)
(AT) single taxi (car sharing)	area operation (area service)	○○○	yes	yes	⌂	⌂

●—	Stop served according to timetable	(H)	Journey from/ to a stop
○—	Stop served on request	⌂	Journey from/ to front door
○○○	Service zone, with boarding and alighting possible at any place		

Source: the authors, based on Wolf-Eberl et al. (2011: 27), BMVI (2016: 23), Mörner (2018: 11) and Sommer (2018: 6)

3. AUTOMATED SHUTTLE BUSES AS A PRIORITY USE CASE IN PUBLIC TRANSPORT PILOT PROJECTS

Automated shuttle buses, or driverless electric minibuses, represent a special use case for automated vehicles as described in Section 2, and are currently undergoing testing by public transport companies. The automated shuttle buses that are currently on the market and are undergoing field tests, such as *NAVYA Arma* or *EasyMile EZ10*, correspond to automation level 2 and drive on a specially approved and prepared fixed route on which they assume both longitudinal and lateral control (cf. Rentschler et al. 2020: 320). In most cases, a safety driver or operator is still on board. For an increasing number of test drives, however, vehicles are only monitored from a control centre.

Due to the low capacity of the shuttle buses, which can usually carry 8 to 12 passengers, these models can function as an appropriate and demand-based supplement to the public transport system. As described above, several fields of application are now emerging for automated shuttle buses in public transport; due to technological limitations, however, these have so far only been tested in scheduled service operation (cf. Derer/Geis 2020: 7; Földes/Csiszár 2018: 2).

In the future, automated shuttle buses could cover a wide range of applications, be used flexibly and thus efficiently replace classic scheduled services in accordance with demand. Especially in cases of low transport demand, in expansive residential and commercial areas, in large-scale hospital centres or on university campuses and at research locations, automated shuttle buses could help to better serve and connect such areas (cf. Derer/Geis 2020: 7). In view of the above-mentioned gaps in the public transport network, tests with automated shuttles are now also being given more attention by municipalities, which are eager to promote a "transit-first" approach to automation (cf. Heinrichs et al. 2019: 248). Figure 2 gives an overview of possible fields of application for automated shuttle buses.

Figure 2: Possible fields of application for automated shuttle buses

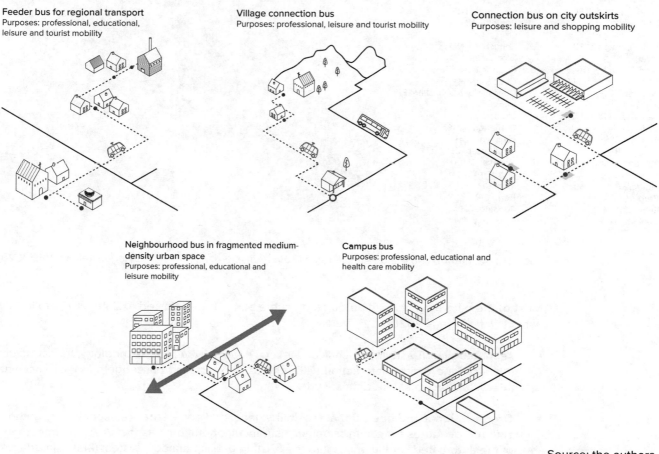

Feeder bus for regional transport
Purposes: professional, educational, leisure and tourist mobility

Village connection bus
Purposes: professional, leisure and tourist mobility

Connection bus on city outskirts
Purposes: leisure and shopping mobility

Neighbourhood bus in fragmented medium-density urban space
Purposes: professional, educational and leisure mobility

Campus bus
Purposes: professional, educational and health care mobility

Source: the authors

In Europe, the first tests with automated shuttle buses on public roads were carried out between 2012 and 2016 as part of the "CityMobil2" project (cf. Alessandrini et al. 2015). Prior to this, most of the tests carried out were for demonstration projects. Numerous further pilot projects with automated shuttle buses have since been initiated in Europe as a result of research funding and adapted legal requirements. Figure 3 gives an overview of pilot projects with automated shuttle buses in Europe.

Figure 3: Locations of pilot projects and demonstrations with automated shuttle buses on public roads and private premises in Europe since 2008 (without claim to completeness)

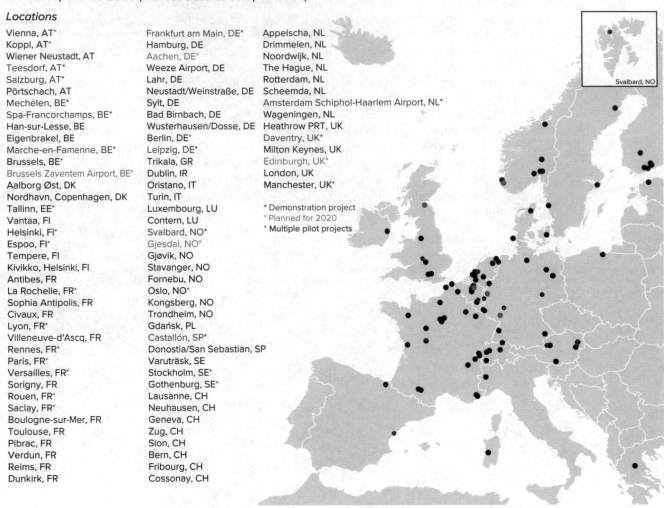

Locations

Vienna, AT⁺
Koppl, AT⁺
Wiener Neustadt, AT
Teesdorf, AT*
Salzburg, AT*
Pörtschach, AT
Mechelen, BE*
Spa-Francorchamps, BE*
Han-sur-Lesse, BE
Eigenbrakel, BE
Marche-en-Famenne, BE*
Brussels, BE⁺
Brussels Zaventem Airport, BE°
Aalborg Øst, DK
Nordhavn, Copenhagen, DK
Tallinn, EE⁺
Vantaa, FI
Helsinki, FI⁺
Espoo, FI⁺
Tempere, FI
Kivikko, Helsinki, FI
Antibes, FR
La Rochelle, FR⁺
Sophia Antipolis, FR
Civaux, FR
Lyon, FR⁺
Villeneuve-d'Ascq, FR
Rennes, FR⁺
Paris, FR⁺
Versailles, FR⁺
Sorigny, FR
Rouen, FR⁺
Saclay, FR⁺
Boulogne-sur-Mer, FR
Toulouse, FR
Pibrac, FR
Verdun, FR
Reims, FR
Dunkirk, FR

Frankfurt am Main, DE*
Hamburg, DE
Aachen, DE°
Weeze Airport, DE
Lahr, DE
Neustadt/Weinstraße, DE
Sylt, DE
Bad Birnbach, DE
Wusterhausen/Dosse, DE
Berlin, DE⁺
Leipzig, DE*
Trikala, GR
Dublin, IR
Oristano, IT
Turin, IT
Luxembourg, LU
Contern, LU
Svalbard, NO*
Gjesdal, NO°
Gjøvik, NO
Stavanger, NO
Fornebu, NO
Oslo, NO⁺
Kongsberg, NO
Trondheim, NO
Gdańsk, PL
Castallón, SP*
Donostia/San Sebastian, SP
Varuträsk, SE
Stockholm, SE⁺
Gothenburg, SE⁺
Lausanne, CH
Neuhausen, CH
Geneva, CH
Zug, CH
Sion, CH
Bern, CH
Fribourg, CH
Cossonay, CH

Appelscha, NL
Drimmelen, NL
Noordwijk, NL
The Hague, NL
Rotterdam, NL
Scheemda, NL
Amsterdam Schiphol-Haarlem Airport, NL*
Wageningen, NL
Heathrow PRT, UK
Daventry, UK*
Milton Keynes, UK
Edinburgh, UK°
London, UK
Manchester, UK⁺

* Demonstration project
° Planned for 2020
⁺ Multiple pilot projects

Svalbard, NO

Source: the authors, based on Alessandrini (2016), Ainsalu et al. (2018) and Hagenzieker et al. (2020)

Most of the pilot projects with automated shuttle buses have focused to date on the following aspects (cf. Jürgens 2020):

- **Technological feasibility:** technical aspects, especially vehicle technology, infrastructure (physical and digital), interaction with other road users and passengers (human-machine interaction), road safety, data security

- **Organizational and operational feasibility:** legal and administrative aspects (e.g. authorization procedures for commissioning, insurance), operational aspects/service, integration of the automated shuttle buses into the existing public transport system (MaaS, interfaces, data infrastructure)

- **Economic feasibility:** economic aspects and financing (e.g. operating costs), user acceptance

- **Social dimension:** inclusion.

However, the individual pilot projects differ in their intensity of focus. They can also be differentiated in terms of their spatial location and of their integration into the transport system and their operating concepts.

Using the example of selected pilot projects in Europe with automated shuttle buses, the possible points of focus are shown in the following, along with the various spatial deployment environments and operating concepts in which automated shuttle buses are currently undergoing testing. To cover as broad a spectrum as possible, pilot projects in Vienna and Koppl in Austria, in Aalborg Øst in Denmark, in Wusterhausen/Dosse in Germany, and in Zug in Switzerland were selected for this comparison (see also Table 4 on the following pages).

The pilot projects are all similar in terms of technological and formal criteria. The shuttle buses, all of which are used in scheduled services, generally operate on routes 2 to 3.5 kilometres in length and serve up to ten predefined stops. In this sense, the test operations hardly fulfil the expectations of flexible booking or routing, but they do demonstrate both the long-term potential and the limitations of supplementary use in the public transport network. To evaluate these aspects, the forms of operation tested should be seen above all in the context of local conditions and requirements. In the course of these projects, the tests mostly involve connections for the first or last mile with automated shuttles, which however are differently suited in terms of space, topography and demand:

- In the "auto.Bus – Seestadt" project in Vienna, shuttle buses are in operation on the residential streets of a densely inhabited new neighbourhood development to test its connection to the underground terminus in the form of an access bus. In view of the high population density, neither does the vehicle size seem economically viable in the long term, nor does the competition with active forms of mobility and new forms of micromobility appear expedient.

- On the other hand, in Koppl, a rural village in Austria, the connection of the various districts to a regional bus route was tested as part of the "Digibus 2017" project in order to facilitate commuting to the supraregional centre in the future (feeder bus for regional transport). The potential here is seen in providing demand-oriented access to dispersed settlement areas in the future, thereby creating an affordable alternative to travel by private car for residents, tourists and the transport of goods, and thus stabilizing the location.

- With the "smartbusaalborg" project in Aalborg Øst in Denmark, the gap in the public transport network is being filled by an automated shuttle in the form of a neighbourhood bus that operates on a now central foot and cycle path to connect the various residential areas and facilities in this highly fragmented suburb. The focus here is not only on providing a tangential transport connection, but above all on mobilizing the population and socially integrating a suburban district of Aalborg that has to date been characterized by functionalistic planning principles.

- In Wusterhausen/Dosse, a small town in the northwest of Brandenburg in Germany, mobilization of the residents, and especially of the ageing population, is likewise at the focus of the pilot project. The automated shuttle bus provides access to this small town, connects it to the regional rail network, and in a second phase will also serve more remote parts of the town (village connection bus). In view of the exodus of residents, high commuter numbers and an ageing population, the extension and flexibilization of the local public transport system is seen as an opportunity to make the region more attractive for locals and tourists and to stabilize dispersed settlement areas as residential locations.

- With the "MyShuttle" project in Zug, a small town in Switzerland, the automated shuttle bus was used to better connect the site of the company V-Zug, in the north of the town, to the railway station and thus create a possible service for the company's commuters (*feeder bus for regional transport*). A particular focal point was the integration of the automated shuttle into the existing public transport information systems.

Table 4 gives an overview of the pilot projects examined involving automated shuttle buses.

Table 4: Overview of selected pilot projects with automated shuttle buses

	Vienna, AT	Aalborg Øst, DK	Wusterhausen/Dosse, DE	Koppl, AT	Zug, CH
Project	auto.Bus Seestadt	smartbusaalborg	autoNV OPR	Digibus© 2017	MyShuttle
Duration of test operation	06/2019–07/2020	03/2020 –06/2021	10/2019–06/2020	04–11/2017	01/2019–12/2019
Regional context or type of area	• Vienna city fringe or city fringe centre • High-density urban development area with mixed use and restricted use of private vehicles • Positive population development	• Suburban or fringe area of urban agglomeration • City fringe area from the 1970s with urban revitalization projects • Positive population development	• Leisure and tourism resort in rural region far from urban agglomeration • Historic village centre, dispersed settlement structure • Ageing and dwindling population • Leisure/tourist traffic	• Rural community east of Salzburg • Dispersed settlement structure • Positive population development • Commuter traffic	• Small town south of Zurich • Historic city centre and dense settlement structure • Positive population development • Commuter traffic
Test environment and route	• Mixed traffic on roads in urban residential area • Length approx. 2 km, 10 stops	• Mixed traffic on a foot and cycle path • Length 2.1 km, 10 stops	• Mixed traffic on village and country roads • 8 km, 18 stops	• Mixed traffic on a country road and village access road • Length 1.4 km, 6 stops	• Mixed traffic on urban local road • Length 1.5 km, 3 stops
Transport connections and urban integration, operating concept	• Linking the underground terminus to neighbouring dense residential areas	• Linking a new district centre with social facilities, fragmented residential areas and a university campus	• Linking the town or city centre with the railway station, shopping centre and supermarket, and an outlying residential area	• Linking the village centre with a regional (public transport) corridor	• Linking the railway station and Metalli shopping centre with the V-Zug company site (first and last mile)
Municipal participation	• Embedded in the public transport company Wiener Linien • Municipal administration involved in coordination processes	• Embedded in the urban and communal administration of Aalborg • Extensive participation of municipal administration	• Rural district of Ostprignitz-Ruppin • Operations of the public transport company ORP-Busse as associated partners	• Provision of framework conditions by the municipality of Koppl	• City of Zug as project partner

Source: the authors

To examine specific aspects of the selected pilot projects with automated shuttle buses in more detail, the following two sections each provide a detailed description of two such tests and give insights into different aspects that are relevant to both the testing of automated shuttle buses and their operation. The first in-depth study, dealing with the "autoNV OPR" project in Ostprignitz-Ruppin, Germany, gives insights into the technical and legal aspects of the use of automated shuttle buses in public transport. The second in-depth study, using the example of the "MyShuttle" project in Zug, Switzerland, sheds light on the operation of automated shuttles and their integration into existing public transport systems.

Table 5: Overview of selected pilot projects with automated shuttle buses

	Priority topics in the project				
Project	auto.Bus Seestadt	smartbusaalborg	autoNV OPR	Digibus© 2017	MyShuttle
Technological feasibility					
Vehicle technology	✓	✓	✓	✓	✓
Infrastructure (physical and digital)	✓	✓	✓	✓	✓
Interaction with other road users (and passengers)	✓			✓	
Road safety	✓			✓	
Data security	✓	✓		✓	
Organizational and operational feasibility					
Legal and administrative aspects			✓	✓	✓
Operational aspects	✓	✓	✓		✓
Integration into the existing public transport system	✓	✓	✓	✓	✓
Economic feasibility					
Business aspects/ financing			✓		✓
User acceptance	✓		✓	✓	✓
Social dimension					
Social inclusion		✓			

Source: the authors

4. TECHNICAL AND LEGAL ASPECTS OF TESTING AUTOMATED SHUTTLE BUSES IN PUBLIC TRANSPORT

based on the example of the "autoNV OPR" project in Ostprignitz-Ruppin, Germany
Arne Holst, Alexander Egoldt, Thomas Richter

In a rural area, specifically in the district of Ostprignitz-Ruppin in the north-west of Brandenburg in Germany, the cooperative project "autoNV OPR" is researching the use and effects of automated minibuses in public streetscapes. The project consortium comprises the universities TU Dresden and TU Berlin, the development company Regionalentwicklungsgesellschaft Nordwestbrandenburg, the public transport operator Ostprignitz-Ruppiner Personennahverkehrsgesellschaft and the subcontractors IGES Institut and Büro autoBus. In this project, the general conditions for the use of automated forms of operation in traffic are investigated, along with acceptance on the part of users and stakeholders. Scenarios and effects of automated forms of operation are also being investigated in terms of financing routine for public transport, and statements on transferability are derived. The project was launched in autumn 2017, and the automated shuttle bus line developed in this connection has been in operation since July 2019. In the following, technical and legal aspects of the operation of the automated shuttle bus are examined in more detail.

4.1 THE SHUTTLE BUS

This project uses the second-generation *EZ10* shuttle from the French manufacturer EasyMile (Figure 4). The shuttle is classified by the manufacturer as an automation level 4 vehicle – however, as described above, it actually only corresponds to automation level 2. The dimensions of the shuttle are 4.02 x 2.00 x 2.87 metres (L x W x H). It has six passenger seats, is electrically powered and can be manually remote-controlled. The shuttle has a technical maximum speed of 45 km/h, although it is only approved for this project at speeds of up to 20 km/h. In operation, however, it travels at a maximum of 15 km/h (cf. EasyMile 2019a). By means of GPS, a correction

Figure 4: The *EZ10* shuttle in operation

Photo: www.autonv.de

signal, inertial sensors, odometry and a pre-stored map, the shuttle bus can localize itself and adhere to the specified trajectory (cf. EasyMile 2019b). The map, including trajectory and route environment, is generated in the course of numerous test drives prior to operation. For this purpose, the shuttle bus scans its route environment while in manual operation mode (cf. Rutanen/ Arffman 2017). The bus uses roof-mounted lidar sensors for localization and stores all scanned geometric features in the corresponding positions on the map (cf. Regional Transportation District 2019). By comparing the stored geometric features with the data detected in real time, the shuttle bus can locate itself by recognizing these features, known as landmarks. The vehicle detects obstacles at a height of 35 centimetres by means of its four safety lidar sensors, which enable 360-degree obstacle detection (cf. ibid.).

4.2 CHOICE OF ROUTE, INFRASTRUCTURE ADAPTATION AND ORIENTATION OF THE SHUTTLE BUS

With regard to the area used for testing the automated shuttle bus, an initial delineation in the project outline was already made by defining the rural district of Ostprignitz-Ruppin as the test area. To further narrow down possible areas of application, all superordinate traffic nodes (PlusBus and rail traffic) in the district were subsequently recorded in order to provide appropriate connections for the automated bus line. This line thus became part of the public transport network of Ostprignitz-Ruppin, and through this integration can make a contribution to securing provision of public services.

In view of the technical requirements, to date only the last mile has come into consideration for operation of the bus. For this purpose, the next step was to assess the demand on the basis of population figures for the localities and structural data. With these two steps, 25 potential routes throughout the district of Ostprignitz-Ruppin were examined and then assessed in accordance with further criteria such as traffic density, route length, road category, maximum speed, mobile network and the need for adaptation. This process yielded two suitable routes, of which that in the municipality of Wusterhausen/Dosse was finally selected.

The route selected for operation of the automated bus line has a total length of 8 kilometres and connects the railway station of the municipality of Wusterhausen/Dosse, which is located outside the town to the east and also has a PlusBus connection (local public transport service), to the town centre, the supermarkets located there, the nursing home, a housing estate outside the town centre, and the lake landscape on the northern edge of the town. Taking into account the technical capabilities of the vehicle, the introduction of the route was realized in three stages or sections (yellow, red, green), on the basis of which the length of the route can be 2, 4 or 8 kilometres (Figure 5 on the next page).

As a preparatory measure for the respective sections, the vehicle manufacturer (EasyMile) provides an assessment, including recommendations for adaptations of the infrastructure. Using the example of the second route section (shown in red in Figure 5), a description of these adaptations with regard to infrastructure and the orientation of the shuttle bus is given in the following.

According to the Road and Transportation Research Association (FGSV; FGSV 2006), the second section, Berliner Straße, is a local access road in the town of Wusterhausen with an average daily traffic volume of 646 motor vehicles per 24 hours (2018 traffic census) and many property access roads. The carriageway is 8.7 metres in width, and parking is permitted on both sides of the road. The speed limit is 50 km/h (Figure 6, left: "Initial situation"). Since the trajectory of the shuttle bus cannot be dynamically changed, it must be ensured that it is not blocked by

Figure 5: Overview of the automated shuttle bus route in the municipality of Wusterhausen/Dosse

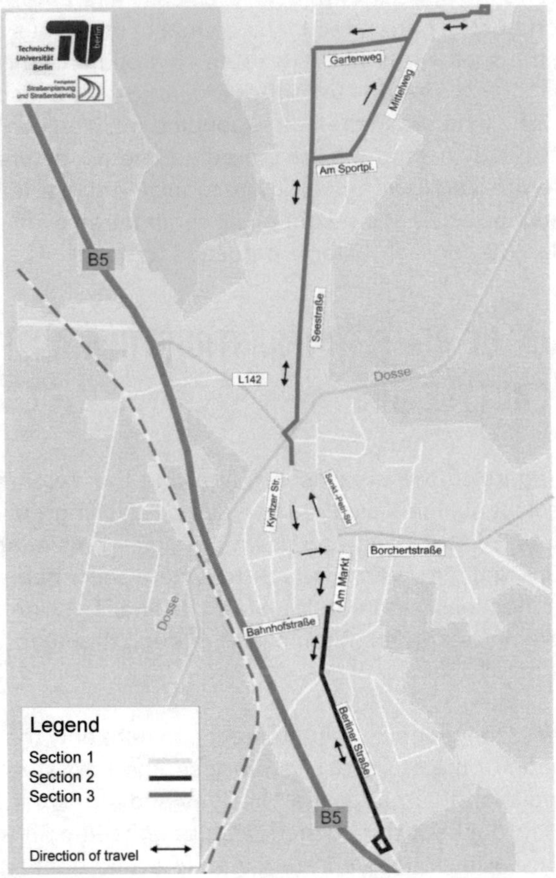

Source: www.autonv.de/fahrplan-strecke

parked vehicles. However, it is not possible to shift the trajectory to the centre of the carriageway, due to the German requirement to drive on the right of the road. In order not to violate this regulation, a one-sided stopping ban was introduced for the western side of the road during operation and longitudinal parking spaces were marked on the eastern side. The southbound trajectory can thus run along the edge of the road and the northbound trajectory adjacent to the parking strip. This prevents conflict between oncoming vehicles in the streetscape of Berliner Straße, since the shuttle bus no longer has to anticipate the behaviour of oncoming vehicles. In addition, the speed limit was lowered to 30 km/h due to the problematical road surface conditions, at times resulting from damage and cobblestones. The plan for the no-stopping zone and parking space markings (red line) is shown in the middle of Figure 6 and the situation following implementation at the right.

The cross sections in Figure 6 show the safety spaces specified by FGSV (2006) and those for the shuttle bus (cf. Rutanen/Arffman 2017). In the cross section on the right, it can be seen that the specifications lead to overlapping of the safety spaces, which in turn necessitates a speed reduction for the shuttle bus at such points. Introduction of a no-stopping zone on both sides was not feasible due to the high requirement for parking space.

For the orientation of the shuttle bus, static geometric features in the surroundings of the route are essential in view of the localization process based on landmark recognition. In view of the position of the localization lidar sensor on the roof of the shuttle bus and its small aperture angle, geometric objects at a height of approx. 3 metres were required (cf. EasyMile 2019b). Build-

Figure 6: Overview of the infrastructural adaptation for route section 2, Berliner Straße in Wusterhausen/Dosse, in the three phases initial situation, plan and implementation

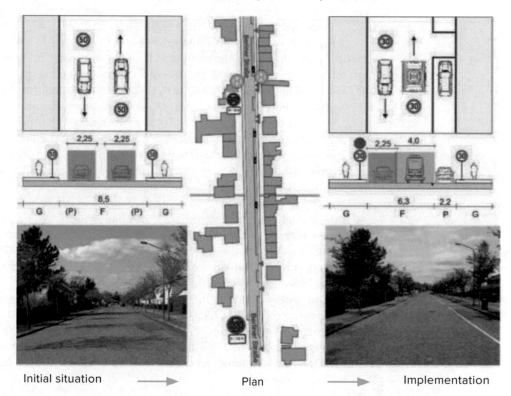

Initial situation ⟶ Plan ⟶ Implementation

Source: the authors; left photo 17/4/2018, right photo 4/7/2019; photos: Holst/Egoldt

ings or corners of buildings, which are normally to be found on inner-city streets, can serve this purpose. The southern part of Berliner Straße, however, included an area devoid of buildings where no landmarks were present. After consultation with the manufacturer EasyMile, three artificial landmarks in the form of signs were set up in this area. In order to interfere as little as possible with the streetscape, these were attached to existing lampposts. The initial situation, plan with dimensions, and implementation are shown in Figure 7. The third road section also includes areas without buildings; here, however, trees at the edge of the roadway served as landmarks, since they fulfilled the requirements.

Figure 7: Overview of the road surroundings adaptation for orientation of the shuttle on route section 2, Berliner Straße in Wusterhausen/Dosse, in the three phases Initial situation, plan and implementation

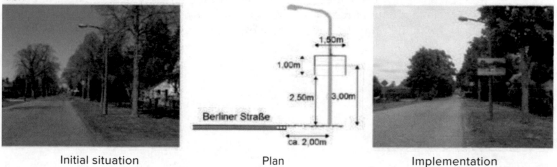

Initial situation   Plan   Implementation

Source: the authors; 17/4/2018, photo at right 4/7/2019; photos: Holst/Egoldt

4.3   REGISTRATION AND OPERATION

As with any vehicle used on public roads, operation of the automated shuttle bus required registration, For this project, exemption from this requirement was granted in the form of a special permit. To obtain such a permit in Germany, an expert assessment from an official testing centre is required. Since the route and the time span of the project must also be specified in this assessment, both the special permit and the registration of the shuttle bus are restricted to the nominated route and time span. Once approval was granted, the route concession was applied for. For this purpose numerous documents had to be submitted, such as the exact route of the line including the positions and names of all bus stops, along with registration documents for the vehicle. According to German law, it was only possible to put the line into operation once the route concession was granted. In addition, however, approval was also required from the vehicle manufacturer, which only gave its consent to operation once the infrastructural modifications it had previously specified had been implemented.

For operation of the shuttle bus, test runs without passengers were carried out in the first few days following the release of one section, so that all road users and especially the accompanying personnel could get used to the driving characteristics of the bus. The accompanying persons are necessary not only for legal reasons; they also have to carry out driving tasks. At complex intersections, driving must be specifically activated by the accompanying person so that the shuttle bus can proceed. Due to the safety space for the shuttle bus, on roads with standard cross sections according to RASt 06 (cf. FGSV 2006) braking is always initiated when oncoming traffic is encountered, as the oncoming vehicle then enters the safety space of the shuttle bus. Dynamic overtaking manoeuvres by other road users, followed by cutting in directly in front of the shuttle bus, likewise lead to delays that affect the operation of the bus and were perceived as negative by the passengers.

4.4   INSIGHTS GAINED

Described above are the approval procedure, selection and adaptation of the route, and an assessment of operations after around six months of experience gained. The choice of manufacturer is important for the project, as this factor plays a major role both in the choice of route and in the specification of infrastructural adaptations. The needs of the users are also highly significant. The wish for an extension of the route was already expressed at the time of release of the first section, above all on the part of the elderly population. However, it became clear that a local transport system in a small town must likewise cater to the needs of the passengers, especially in terms of time: firstly, an adjustment of travel speed and an increase in reliability is necessary; and secondly, the timetable must be coordinated with user demand. The provision of a demand-oriented timetable makes it all the more difficult to establish a traditional bus route for a small town in a rural area and requires alternatives to be examined. To ensure smooth operation, infrastructural measures are also currently required, which however could become obsolete with the ongoing development of automated driving functions and should therefore be avoided

5. OPERATION AND INTEGRATION OF AUTOMATED SHUTTLE BUSES IN PUBLIC TRANSPORT SYSTEMS

based on the example of the "MyShuttle" project in Zug, Switzerland
Zoltán László

As part of the "MyShuttle" project, various possible applications of automated vehicles and associated service concepts were analysed, developed and subjected to extensive field tests in the city of Zug (around 30,000 inhabitants) in Switzerland using an automated shuttle bus. The point of departure for this project was the radical transformation of passenger transport triggered by automated driving. In particular, the opportunities opened up by automated driving for a more customer-friendly, efficient, environmentally friendly and cost-effective overall transport system – as a convergence of public and individual transport, with a focus on car- and ride sharing services – motivated the Swiss Federal Railways (SBB) to prepare for these changes so as to be in a position to offer customers a suitable range of services in the future.

With these considerations in mind, the "MyShuttle" project was launched by SBB, Switzerland's largest passenger transport company, together with the Swiss company Mobility Carsharing, the transport operator Zugerland Verkehrsbetriebe, the City of Zug as the local partner, and the Tech Cluster Zug. The operational testing focused on three main areas: (1) the ability of automated vehicles to be integrated into the existing public transport customer information system, (2) specific experience regarding the technical maturity of the available software and hardware, and (3) customer acceptance of automated shuttle services. The one-year pilot operation was launched in January 2019, after a year and a half of planning and preparation. "MyShuttle" was

Figure 8: The *EZ10* shuttle in operation in Zug

Source: SBB (2020: 1)

thus the first automated shuttle bus in Switzerland to operate in mixed traffic on a public road with a constantly high traffic volume. In the following, the operation and integration of the automated shuttle bus into the existing public transport system is examined in more detail.

5.1 THE SHUTTLE BUS AND ITS OPERATION

Prior to the project launch, none of the leading technology providers or vehicle manufacturers in the field of automated vehicles had agreed to test vehicles in Switzerland or to participate as partners in a pilot project. For the "MyShuttle" project, the responsible parties therefore took a pragmatic approach and procured the most suitable automated shuttle bus available for purchase from the point of view of this project: the *EZ10* model from the French start-up EasyMile (Figure 8, previous page).

Figure 9: Overview of the automated shuttle routes in the city of Zug

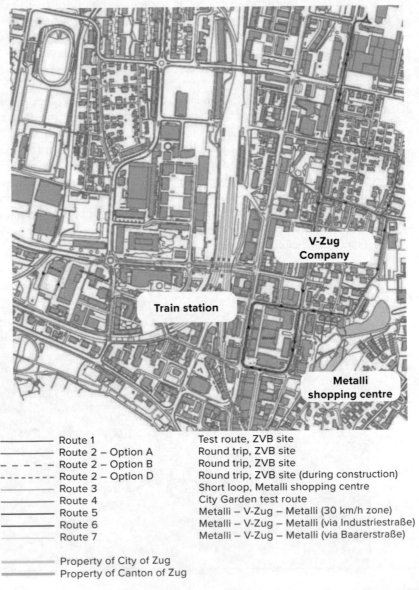

——————	Route 1	Test route, ZVB site
——————	Route 2 – Option A	Round trip, ZVB site
– – – – –	Route 2 – Option B	Round trip, ZVB site
– – – – – –	Route 2 – Option D	Round trip, ZVB site (during construction)
——————	Route 3	Short loop, Metalli shopping centre
——————	Route 4	City Garden test route
——————	Route 5	Metalli – V-Zug – Metalli (30 km/h zone)
——————	Route 6	Metalli – V-Zug – Metalli (via Industriestraße)
——————	Route 7	Metalli – V-Zug – Metalli (via Baarerstraße)

————	Property of City of Zug
————	Property of Canton of Zug

In this project, the automated shuttle bus was mainly used to serve the first and last mile within the city of Zug, especially between the railway station and the V-Zug company site. Figure 9 gives an overview of the routes for the shuttle bus in Zug. While Routes 1 to 4 were only used for functional tests of the vehicle or for brief demonstrations, the regular, public operation of the automated shuttle bus took place on Route 5 (pink) between the railway station or the Metalli shopping centre and the V-Zug location, and on Routes 6 and 7 (purple and green), each of which is an adapted, partly shortened form of Route 5 (Figure 9).

EasyMile, the manufacturer of the automated shuttle bus, worked hard to meet the requirements of this project. As a result, it was finally possible for the bus to run in mixed traffic between the station and the V-Zug company site. However, operation of the automated shuttle repeatedly met with challenges: reliably recognizing traffic lights, anticipating the flow of traffic, avoiding obstacles, driving in heavy rain and autonomously "learning" from new traffic situations could not yet be mastered by the vehicle model in the course of operation. Sensor malfunctions due to pollen, turning left at intersections, problems caused by vegetation growth and construction sites necessitated the presence of safety drivers on board. In view of the technical complexity, in the course of this project it was also not possible for EasyMile to offer a demand-driven zonal service (on-demand service within a selected area, where passengers can board or alight at any location). Passengers could only board or alight during intermediate stops at fixed bus stops along the route.

5.2   INTEGRATION INTO PUBLIC TRANSPORT CUSTOMER INFORMATION SYSTEMS

One main focus of the project was to investigate possible integration of the shuttle bus into the customer information systems of the SBB – and therefore into the existing public transport landscape in Switzerland. For this purpose, the project cooperated with Bestmile, a company that provides scheduling and fleet management systems for automated vehicles. The aim was to integrate the automated shuttle bus into the various public transport customer information systems, e.g. displays of (1) departure times on monitors at bus stops, (2) ongoing connections on board trains or (3) departure times in the SBB apps.

Customer information systems inform passengers about the availability of and changes to transport services. Switzerland is a pioneer in this field for public transport: on behalf of the Federal Office of Transport (FOT), SBB bundles information on the majority of public transport services and makes it available to the industry and to customers.

In very simplified terms, these systems are oriented towards current forms of public transport services, i.e. on predefined routes with fixed stops and long-term timetables. As long as services with automated vehicles – such as the line-bound shuttle bus with fixed stops as implemented in this project – adhere to this logic, they can be easily integrated into the current system landscape and communicated to customers. The decisive factor here is not the degree of automation of the vehicle, but whether it runs according to a timetable or is offered on demand. In the course of this project, the subsequent public transport connections were displayed on a monitor in the shuttle bus (Figure 10, left). The scheduled departure times of the shuttle bus from Zug main station were also shown on the departure displays at the station (Figure 10, right). For this purpose the shuttle bus, running as a classic scheduled service, was integrated into the public transport timetable and customer information system as Line 17 and a timetable for the shuttle bus was drawn up.

As mentioned at the outset, automated vehicles open up great opportunities for new forms of offer, especially on-demand transport services. Thanks to their automation, these vehicles can be ordered to arrive at a defined location at a certain time, whereby it is possible to be brought to one's destination without changing. Such an offer would make ownership of a private car

Figure 10: Display of ongoing connections on board the shuttle bus (left) and indication of the shuttle bus on the departure screen at Zug station (right)

Source: SBB (2020: 32)

superfluous. If a number of journeys are then also bundled to increase efficiency, this new form of offer combines the advantages of public transport with those of owning a car. The result is known as public individual transport.

To enrich the strong Swiss public transport system with these new forms of service, initial attempts were made in the "MyShuttle" pilot project to integrate such a public individual transport offer, e.g. a shuttle bus with flexible stop requests, into the customer information system. The required chain of information between the existing SBB customer information system, the scheduling or fleet management system of Bestmile and the vehicle is shown in Figure 11. The timetable from the public transport systems is transmitted to the control or scheduling system, which divides the time schedule into individual trips and conveys these in sequence to the vehicle (mission management). The vehicle reports all positional, speed and further data back to the Bestmile scheduling system. The control and scheduling system then calculates the discrepancy between the real position of the vehicle and that indicated on the timetable and reports this back to the customer information system in order to inform customers accordingly.

Figure 11: Schematic depiction of the information chain between customer information system, scheduling system and vehicle

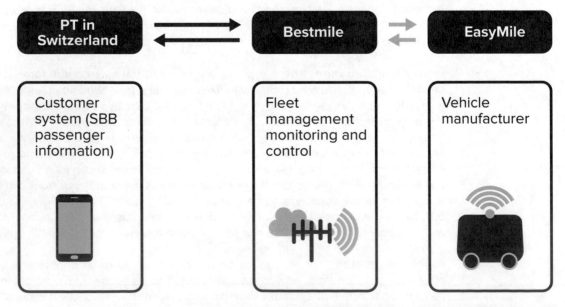

Source: the authors, based on SBB (2020: 32)

A significant aspect is therefore that the scheduling system sends driving orders, so-called missions, to the vehicle, which it then carries out. The scheduling system must also be able to implement changes to the mission, e.g. in the form of a requested stop of the shuttle bus, during a journey. Due to the technical complexity involved, in the course of the pilot operation of the shuttle bus EasyMile was not able to provide an interface that could process a mission and handle such changes, or provide an on-demand zonal service that enables customers to board and alight at any place. According to EasyMile, as things stand, externally accessing a vehicle in the form of a mission is critical for security and safety reasons and is subject to abuse.

To investigate the integration of such forms of service into the existing customer information system, (1) fictitious vehicles were then "generated", i.e. simulated, in Bestmile's scheduling system for proof of concept, and (2) the functionality of flexible stops for the shuttle bus was evaluated by the safety drivers using an app developed by Bestmile (City of Zug on-demand app) for end customers.

For (1) the proof of concept, the experts set about establishing a link between the SBB customer information system and Bestmile's scheduling system. This was necessary because the formats of the customer information system are currently based on fixed stops and predefined routes and timetables, whereas the scheduling system is designed for a demand-based zonal service. These tests were carried out on test systems; SBB's actual productive systems and customer information systems were not available for this purpose.

For the tests, the Bestmile system and SBB's customer information system (CUS – customer system) were connected by an adapter, which unites the logics of these two systems, and subsequently by a router (MetaRouter) and by the information channels of SBB's MIKU (mobile information tool for customer contacts) or KIB (customer information at the station, i.e. loudspeakers and screens; see Figure 12).

Figure 12: Information flow for integration of public transport and proof of concept, and thus for the adapter between SBB's customer system (CUS) and Bestmile

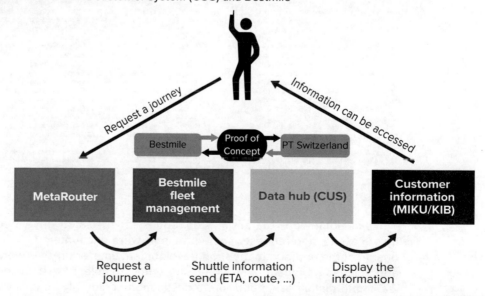

Source: the authors, based on SBB (2020: 26)

However, CUS reached its limits here. As long as operations remained within the framework of scheduled services, the vehicles simulated by Bestmile could be successfully displayed in the core information systems for public transport (see also the successful implementation of scheduled shuttle bus services into departure displays at the station, mentioned above); an upstream logic recognized and even appropriately displayed bundled journeys (grouping together of several bookings from the MetaRouter for assignment to a single vehicle). However, as soon as operations went beyond scheduled service to include zonal services without timetables, or attempted to simulate this feature (trips and stops on demand), the system could no longer process and display the information.

To test the functionality of flexible stop requests (2), Bestmile developed an app for the City of Zug (Zug On-Demand), whereby stop requests were transmitted to the safety driver via the app alone. These then appeared in the app of the safety driver, who could manually accept or reject the displayed requests (Figure 13).

Figure 13: "Train On Demand" app (top) and driver app interface for journey and stop requests (bottom)

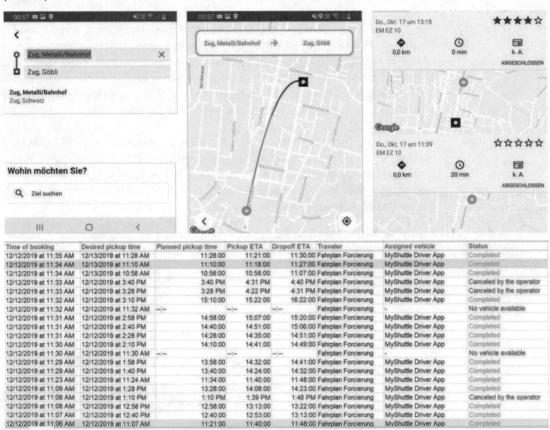

Time of booking	Desired pickup time	Planned pickup time	Pickup ETA	Dropoff ETA	Traveler	Assigned vehicle	Status
12/12/2019 at 11:35 AM	12/13/2019 at 11:28 AM		11:28:00	11:21:00	11:30:00 Fahrplan Forcierung	MyShuttle Driver App	Completed
12/12/2019 at 11:34 AM	12/13/2019 at 11:10 AM		11:10:00	11:18:00	11:27:00 Fahrplan Forcierung	MyShuttle Driver App	Completed
12/12/2019 at 11:34 AM	12/13/2019 at 10:58 AM		10:58:00	10:58:00	11:07:00 Fahrplan Forcierung	MyShuttle Driver App	Completed
12/12/2019 at 11:33 AM	12/12/2019 at 3:40 PM		3:40 PM	4:31 PM	4:40 PM Fahrplan Forcierung	MyShuttle Driver App	Canceled by the operator
12/12/2019 at 11:33 AM	12/12/2019 at 3:28 PM		3:28 PM	4:22 PM	4:31 PM Fahrplan Forcierung	MyShuttle Driver App	Canceled by the operator
12/12/2019 at 11:32 AM	12/12/2019 at 3:10 PM		15:10:00	15:22:00	16:22:00 Fahrplan Forcierung	MyShuttle Driver App	Completed
12/12/2019 at 11:32 AM	12/12/2019 at 11:32 AM	--:--	--:--	--:--	Fahrplan Forcierung	-	No vehicle available
12/12/2019 at 11:31 AM	12/12/2019 at 2:58 PM		14:58:00	15:07:00	15:20:00 Fahrplan Forcierung	MyShuttle Driver App	Completed
12/12/2019 at 11:31 AM	12/12/2019 at 2:40 PM		14:40:00	14:51:00	15:06:00 Fahrplan Forcierung	MyShuttle Driver App	Completed
12/12/2019 at 11:31 AM	12/12/2019 at 2:28 PM		14:28:00	14:35:00	14:51:00 Fahrplan Forcierung	MyShuttle Driver App	Completed
12/12/2019 at 11:30 AM	12/12/2019 at 2:10 PM		14:10:00	14:41:00	14:49:00 Fahrplan Forcierung	MyShuttle Driver App	Completed
12/12/2019 at 11:30 AM	12/12/2019 at 11:30 AM	--:--	--:--	--:--	Fahrplan Forcierung	-	No vehicle available
12/12/2019 at 11:29 AM	12/12/2019 at 1:58 PM		13:58:00	14:32:00	14:41:00 Fahrplan Forcierung	MyShuttle Driver App	Completed
12/12/2019 at 11:29 AM	12/12/2019 at 1:40 PM		13:40:00	14:24:00	14:32:00 Fahrplan Forcierung	MyShuttle Driver App	Completed
12/12/2019 at 11:23 AM	12/12/2019 at 11:24 AM		11:34:00	11:40:00	11:48:00 Fahrplan Forcierung	MyShuttle Driver App	Completed
12/12/2019 at 11:09 AM	12/12/2019 at 1:28 PM		13:28:00	14:08:00	14:23:00 Fahrplan Forcierung	MyShuttle Driver App	Completed
12/12/2019 at 11:08 AM	12/12/2019 at 1:10 PM		1:10 PM	1:39 PM	1:48 PM Fahrplan Forcierung	MyShuttle Driver App	Canceled by the operator
12/12/2019 at 11:08 AM	12/12/2019 at 12:58 PM		12:58:00	13:13:00	13:22:00 Fahrplan Forcierung	MyShuttle Driver App	Completed
12/12/2019 at 11:07 AM	12/12/2019 at 12:40 PM		12:40:00	12:53:00	13:13:00 Fahrplan Forcierung	MyShuttle Driver App	Completed
12/12/2019 at 11:06 AM	12/12/2019 at 11:07 AM		11:21:00	11:40:00	11:48:00 Fahrplan Forcierung	MyShuttle Driver App	Completed

Source: SBB (2020: 32)

While use of the app to transmit stop requests to the safety driver helped to test the functionality of flexible stop requests, some problems were encountered – especially in view of the lean design of the app due to the short development time: addresses were not correctly recognized and at times assigned to the wrong stops, for example, so that customers unnecessarily had to walk long distances. In addition, a destination could only be changed when the bus was stationary. It was thus possible that customers were assigned to a trip but the vehicle, which had already set out on its journey, failed to stop to pick them up. All in all, rudimentary on-demand functionalities could be accommodated in the project, but the software provided by Bestmile did not have the necessary degree of maturity at that time.

5.3   INSIGHTS GAINED

While the project did not achieve all the originally formulated goals, the overarching learning objectives were nevertheless exceeded. Within SBB, essential know-how was built up in this field and customer acceptance of automated driving was reinforced, especially in terms of its integration into the public transport system. To establish automated shuttle buses as a new form of service in public transport, they should be available for ordering, along with the appropriate customer information, via classic public transport channels (e.g. the SBB app). The use of flexible shuttle buses in line with demand must also be planned to match the arrival and departure times of superordinate public transport media. However, the tests showed that today's core public transport systems, which are geared to fixed stops, lines and timetables, cannot yet deal with on-demand transport (without fixed timetables), route changes or flexible stops, but can only handle shuttle buses that operate in scheduled service. This calls for new systems for the integration and development of standard application programming interfaces (APIs) that can be processed by the core public transport systems.

Overall, however, the project was able to provide important insights into new types of services, so that an economically optimized, customer-oriented service for Switzerland can be put in place once the automation technology reaches market maturity. In particular, the findings from the project with regard to the integration of automated shuttle buses or of automated on-demand services into the customer information systems of SBB and other public transport services are now being used to make the above-mentioned systems fit for this new type of mobility.

6.   CONCLUSION

Automated driving enables a wide range of possible applications for new forms of service in public transport that could be integrated into the existing public transport system. By merging traditional public transport with more individual forms of service that focus on car- and ride sharing offers, public transport can be made more personalized and flexible in future, and a temporal and spatial densification of the overall service can be achieved. This could ultimately meet the needs of customers more broadly and effectively.

To evaluate the requirements for the best possible use of new, automated forms of service, extensive trial operations are needed – not only with regard to the technological functioning and ongoing development of automated vehicles, but also in order to gain experience in integrating these vehicles into the existing public transport system and in their interaction with the local surroundings. At the focus of current debate regarding automated vehicles in the sphere of public transport are pilot projects for the operation and integration of automated shuttle buses. These projects are located in various types of area or spatial setting (e.g. on the outskirts of a large city, or in small towns and villages), make use of different operating concepts and also place greater or lesser emphasis on specific aspects (e.g. technological, organizational-operational or economic feasibility). Many of the pilot projects also serve to demonstrate power of innovation and to provide a future-oriented image for the actors involved, such as public transport companies or municipalities (cf. Perkins et al. 2018: 10). However, these example in-depth studies demonstrated that participating in research into future mobility options in public transport and testing them in practice also offer further advantages and potential for those involved. For example, pilot projects with automated vehicles, and in particular with automated shuttle buses in public transport, can serve to

- test possible operating concepts in practice,

- enable various stakeholders to gain experience,

- involve users in the operation and to determine their acceptance,

- acquire know-how with regard to operation, repair and service, and

- create appropriate data infrastructure (e.g. operational and analytical data sets, definition of requirements for data transfer, interfaces etc.) that will be relevant for the evaluation of test operations carried out to date and for future projects.

The pilot projects "autoNV OPR" and "MyShuttle" have shown that the automated shuttle buses used still have limitations in terms of operation, above all regarding aspects of technological and organizational-operational feasibility.

Technological feasibility

- Trouble-free use and operation of the shuttle buses would at present be challenging without modifications to the infrastructure (e.g. installation of artificial landmarks in the form of signs, introduction of no-stopping zones, etc.).

- At complex intersections (especially when turning left), driving must be specifically activated by the operators so that the shuttle bus can proceed. Problems are also encountered in reliably recognizing traffic lights, anticipating the traffic flow, avoiding obstacles, dealing with (sensor) malfunctions due to pollen, vegetation growth or construction sites, and when driving in heavy precipitation.

- Overtaking manoeuvres by other road users, followed by cutting in directly in front of the shuttle bus, result in deceleration that passengers perceive as negative.

Organizational and operational feasibility

- Integrating automated shuttle buses into existing core public transport systems is currently only possible if they operate in scheduled service (with fixed timetables and stops). Demand-based operation of automated shuttle buses, with flexible stops, is hardly feasible as a part of today's core public transport systems.

Overall, it is worthwhile for public transport operators and urban stakeholders to invest financial resources in the testing of automated vehicles or in pilot projects with automated shuttle buses, as new insights and important know-how can be gained by this means. However, actually being able to use automated shuttle buses as a fully-fledged mobility option to supplement existing public transport services, e.g. in areas of low demand, necessitates the following framework conditions and measures, taking into account the results from the example projects:

1. **Further development of the vehicle technology to enable stable operation and higher driving speeds:** More intelligent and more anticipatory vehicle technology is necessary for the operation of automated shuttle buses in order to reduce possible delays and the need for intervention by the operator, and to ensure stable operation. The driving speed of the automated shuttle bus is an important aspect for the use, attractiveness and operation of the vehicle and should be increased in line with further development of the vehicle technology. This aspect is not as important in the finely detailed development of smaller residential areas, where the automated shuttle buses do not necessarily have to

travel at high speeds, but above all in cases where larger areas are to be developed or where facilities or locations have to be connected over greater distances and the automated shuttle buses tend to travel on larger-capacity roads.

2. **No extensive adaptation of the infrastructure, or only where other modes of transport (e.g. cycling, walking) would also benefit:** Adaptation of the infrastructure, which is already being carried out to a small extent in the course of tests with automated shuttle buses, should also be kept to a minimum in the course of further tests. Extensive infrastructural modifications (whether structural or digital, e.g. "vehicle-to-infrastructure") should be avoided, since the infrastructural requirements of the vehicles can rapidly change, and the costs for such modifications are relatively high. Moreover, the public space is already characterized by conflicts of use, and further land consumption or obstruction of the public space by infrastructure for automated shuttle buses should be avoided. Extensive modification of the infrastructure should therefore only take place where this would be of benefit not only to the automated shuttle buses, but also to pedestrian and bicycle traffic.

3. **Giving in-depth consideration to linking automated forms of service and the environmental network (public transport, cycling, walking):** It is essential to link automated (and networked) forms of service with other modes of transport, especially the existing public transport network. To integrate on-demand automated shuttle buses with flexible stops into existing public transport customer information and booking systems, new systems and standard interfaces or APIs are required that can be processed by the core public transport systems; these must also be adhered to on the part of the providers. With a view to customer convenience, booking rides should be as simple as possible and largely digital. Particular attention must also be given to ensuring compatible connections between the superordinate public transport media and the automated shuttle bus (orientation of demand-based, flexible shuttle bus operation to the arrival and departure times of the superordinate media; cf. SBB 2020: 62ff.).

Further framework conditions and measures are also important:

1. **Orientation of the operating concepts to the transport demand and settlement structure of the respective area:** The use of automated shuttle buses should always be oriented to the transport demand and settlement structure of the respective area in terms of operating time, frequency, route, area coverage etc. It can be advisable to link the use of these buses to the demands and goals described in the respective transport development plans and to integrate them accordingly.

2. **Spatial integration into existing public transport services:** Integration into or linking with existing public transport services is necessary not only from a technological point of view, but also spatially: the focus here should be on flexible and alternative forms of service as feeders to the classic scheduled services and on so-called "transit-oriented developments" (TODs), i.e. above all the functional enrichment and development of a settlement in the vicinity of the bus station should be promoted. The use of automated shuttle buses as feeders to rail services is necessary to prevent automated private vehicles from assuming this function (cf. Sinner 2019) – on the one hand since the space requirements at stations and transfer points are much lower in the case of on-demand buses and feeder services than for automated private vehicles or car sharing services (cf. Sinner et al. 2018), and on the other hand because a spatially facilitated transfer between transport modes can shorten overall travel times and intermodal travel can become more attractive for users (see Chap. 8 by Bruck et al. in this volume).

3. **Involvement and participation of relevant stakeholders:** In the context of further pilot projects with automated shuttle buses, the relevant stakeholders and agents of the municipalities and related administrative units should be involved. Furthermore, opportunities should be created for the local population to participate, in order to align the research and development process with local requirements and goals.

There is also a need for further research into the use of automated vehicles, and especially of automated shuttle buses, in public transport with regard to the following aspects, which should be investigated in the course of further pilot projects:

- **Costs:** Investigation of possible cost reductions through the use of automated vehicles, and of possible newly arising costs (e.g. scheduling systems, accompanying personnel, protection against vandalism, more frequent cleaning).

- **Infrastructure:** This aspect deals with infrastructure necessary not only for the actual operation of the vehicles, but for many further factors in connection with their use, e.g. charging infrastructure or bus stops (traffic areas).

- **User acceptance:** Acceptance by users is relevant not only with regard to the technology itself (e.g. travelling on the shuttle bus, comfort, quality), but also specifically at night (feeling of safety) and regarding the question of whether accompanying personnel are needed at certain times.

- **Dimensioning of fleets:** Investigations must be carried out regarding the profile of the required automated vehicle fleet, e.g. in terms of bundling potential, routes or scheduling.

REFERENCES

Ainsalu, J., V. Arffman, M. Bellone, M. Ellner, T. Haapamäki, N. Haavisto, E. Josefson, A. Ismailogullari, B. Lee, O. Madland, R. Madžulis, J. Müür, S. Mäkinen, V. Nousiainen, E. Pilli-Sihvola, E. Rutanen, S. Sahala, B. Schønfeldt, P. M. Smolnicki, R.-M. Soe, J. Sääski, M. Szymańska, I. Vaskinn and M. Åman 2018. "State of the Art of Automated Buses", in *Sustainability* (10) 3118, 1–34.

Alessandrini, A. 2016. "Final Report Summary – CITYMOBIL2 (Cities demonstrating cybernetic mobility)". Download at https://cordis.europa.eu/project/id/314190/reporting (24/7/2020).

Alessandrini, A., A. Campagna, P. Delle Site, F. Filippi and L. Persia 2015. "Automated Vehicles and the Rethinking of Mobility and Cities", in *Transport Research Procedia* 5, 145–160.

Barillère-Scholz, M., C. Büttner and A. Becker 2020. "Mobilität 4.0: Deutschlands erste autonome Buslinie in Bad Birnbach als Pionierleistung für neue Verkehrskonzepte", in *Autonome Shuttlebusse im ÖPNV. Analysen und Bewertungen zum Fallbeispiel Bad Birnbach aus technischer, gesellschaftlicher und planerischer Sicht*, ed. by A. Riener, A. Appel, W. Dorner, T. Huber, J. C. Kolb and H. Wagner. Wiesbaden: Springer Vieweg, 15–22.

BMVI (Federal Ministry for Digital and Transport) 2016. "Mobilitäts- und Angebotsstrategien in ländlichen Räumen. Planungsleitfaden für Handlungsmöglichkeiten von ÖPNV-Aufgabenträgern und Verkehrsunternehmen unter besonderer Berücksichtigung wirtschaftlicher Aspekte flexibler Bedienungsformen", Berlin. www.bmvi.de/SharedDocs/DE/Publikationen/G/mobilitaets-und-angebotsstrategien-in-laendlichen-raeumen-neu.pdf?__blob=publicationFile (19/8/2020).

bmvit (Federal Ministry for Transport, Innovation and Technology) 2019. "Nahverkehr. Recht". Download at: www.bmvit.gv.at/verkehr/nahverkehr/recht/index.html (20/1/2020).

Bösch, P. M., F. Becker, H. Becker and K. W. Axhausen 2018. "Cost-based analysis of autonomous mobility services", in *Transport Policy* 64, 76–91. DOI: /10.1016/j.tranpol.2017.09.005.

Bruns, F., M. Rothenfluh, M. Neuenschwander, M. Sutter, B. Belart and M. Egger 2018. "Einsatz automatisierter Fahrzeuge im Alltag – Denkbare Anwendungen und Effekte in der Schweiz. Schlussbericht Modul 3c 'Mögliche Angebotsformen im kollektiven Verkehr (ÖV und ÖIV)'", final version of 19/4/2018. www.ebp.ch/sites/default/files/project/uploads/2018-04-19%20aFn_3c%20Mögliche%20Angebotsformen%20im%20kollektiven%20Verkehr_Schlussbericht_0.pdf (18/8/2020).

Buffat, M., H. Sommer, M. Amacher, R. Mohagheghi, J. Beckmann and A. Brügger 2018. "Individualisierung des ÖV-Angebots. Analyse der Auswirkungen der Individualisierung und weiterer angebots- und nachfragerelevanten Trends auf die zukünftige Ausgestaltung des ÖV-Angebots", research project SVI 2014/004 on commission from the Swiss Association of Transport Engineers and Experts (SVI). Bern: Federal Department of the Environment, Transport, Energy and Communications UVEK.

Derer, M., and F. Geis 2020. "Entwicklungen im ÖPNV", in: *Autonome Shuttlebusse im ÖPNV. Analysen und Bewertungen zum Fallbeispiel Bad Birnbach aus technischer, gesellschaftlicher und planerischer Sicht*, ed. by A. Riener, A. Appel, W. Dorner, T. Huber, J. C. Kolb and H. Wagner. Wiesbaden: Springer Vieweg, 7–14.

EasyMile 2019a. "EZ 10". https://easymile.com/solutions-easymile/ez10-autonomous-shuttle-easymile/ (2/11/2019).

EasyMile 2019b. "Autonomous Technology Thanks To A Unique & Versatile Software Package". https://easymile.com/driverless-technology-easymile-how-does-it-work/ (9/3/2020).

EPOMM – European Platform on Mobility Management 2017. "Die Rolle von Mobilität als Dienstleistung für Mobilitätsmanagement", *E-update December* 2017. www.epomm.eu/newsletter/v2/content/2017/1217_2/doc/eupdate_de.pdf (7/1/2020).

FGSV (Road and Transportation Research Association) 2006. "Richtlinien für die Anlage von Stadtstraßen – RASt 06", 2006 edition. Download at www.forschungsinformationssystem.de/servlet/is/232185/ (19/8/2020).

Földes, D., and C. Csiszár 2018. "Framework for planning the mobility service based on autonomous vehicles", conference paper, Smart City Symposium Prague (SCSP) 2018. Prague.

Gertz, C., and M. Dörnemann 2016. "Wirkungen des autonomen/fahrerlosen Fahrens in der Stadt – Entwicklung von Szenarien und Ableitung der Wirkungsketten". Bremen: Senator for Environment, Construction and Transport.

Hagenzieker, M., R. Boersma, P. Nuñez Velasco, M. Ozturker, I. Zubin and D. Heikop 2020. "Automated Buses in Europe. An Inventory of Pilots", version 0.5. TU Delft.

Heinrichs, D., S. Rupprecht and S. Smith 2019. "Making Automation Work for Cities: Impacts and Policy Responses", in: *Road Vehicle Automation 5*, ed. by G. Meyer and S. Beiker. Cham: Springer International Publishing, 243–252.

Hell, W. (ed.) 2006. *Öffentlicher Personennahverkehr. Herausforderungen und Chancen*. Berlin/Heidelberg: Springer-Verlag.

Hoadley, S. (ed.) 2017. "Mobility as a service: Implications for urban and regional transport. Discussion paper offering the perspective of Polis member cities and regions on Mobility as a Service (MaaS)", Polis Traffic Efficiency & Mobility Working Group, September 2017. Brussels. www.polisnetwork.eu/wp-content/uploads/2017/12/polis-maas-discussion-paper-2017-final_.pdf (19/8/2020).

Hörl, S. 2020. "Dynamic Demand Simulation for Automated Mobility on Demand", Dissertation, ETH Zurich.

Hörl, S., F. Becker, T. Dubernet and K. W. Axhausen 2019. "Induzierter Verkehr durch autonome Fahrzeuge: Eine Abschätzung", research project SVI 2016/001 on commission from the Swiss Association of Transport Engineers and Experts (SVI). Bern: Federal Department of the Environment, Transport, Energy and Communications UVEK. https://ethz.ch/content/dam/ethz/special-interest/baug/ivt/ivt-dam/vpl/reports/1401-1500/ab1433.pdf (18/8/2020).

Hörold, S. 2016. *Instrumentarium zur Qualitätsevaluation von Mobilitätsinformation. Schriften zur Medienproduktion*. Wiesbaden: Springer Vieweg.

Jittrapirom, P., V. Caiati, A.-M. Feneri, S. Ebrahimigharehbaghi, M. J. Alonso González and J. Narayan 2017. "Mobility as a Service: A Critical Review of Definitions, Assessments of Schemes, and Key Challenges", in *Urban Planning* (2) 2, 13–25.

Jonuschat, H., A. Knie and L. Ruhrort 2016. "Zukunftsfenster in eine disruptive Mobilität. Teil 1: Mobilität in einer vernetzten Welt". Berlin: Innovation Centre for Mobility and Social Change (InnoZ). https://docplayer.org/57659946-Zukunftsfenster-in-eine-disruptive-mobilitaet.html (25/8/2020).

Jürgens, L. 2020. "Konnektivitätsveränderungen im ÖPNV-Netz durch die Einführung eines autonomen Shuttlebusses", in: *Autonome Shuttlebusse im ÖPNV. Analysen und Bewertungen zum Fallbeispiel Bad Birnbach aus technischer, gesellschaftlicher und planerischer Sicht*, ed. by A. Riener, A. Appel, W. Dorner, T. Huber, J. C. Kolb and H. Wagner. Wiesbaden: Springer Vieweg, 39–54.

Lenz, B., and E. Fraedrich 2015. "Neue Mobilitätskonzepte und autonomes Fahren: Potenziale der Veränderung", in *Autonomes Fahren. Technische, rechtliche und gesellschaftliche Aspekte*, ed. by M. Maurer, J. C. Gerdes, B. Lenz and H. Winner. Berlin/Heidelberg: Springer Vieweg, 175–196.

Mitteregger, M., A. Soteropoulos and M. Berger 2019. "A Framework for Assessing Use Cases of high and full Driving Automation based on transport-related Experiences", in *Transportation Research Procedia* 41, 609–613. DOI: 10.1016/j.trpro.2019.09.108.

Mörner, M. von 2018. "Sammelverkehr mit autonomen Fahrzeugen im ländlichen Raum", dissertation, TU Darmstadt.

Ohnemus, M., and A. Perl 2016. "Shared Autonomous Vehicles: Catalyst of New Mobility for the Last Mile?", in *Built Environment* (42) 4, 589–602.

Perkins, L., N. Dupuis and B. Rainwater 2018. "Autonomous Vehicle Pilots Across America – Municipal Action Guide". Washington D.C.: National League of Cities – Center for City Solutions. www.nlc.org/sites/default/files/2018-10/AV%20MAG%20Web.pdf (19/8/2020).

Regional Transportation District 2019. "University of Denver Autonomous Vehicle Shuttle – Automated Driving Demonstration Grants", 21/3/2019. Washington D.C. www.transportation.gov/ sites/dot.gov/files/docs/policy-initiatives/automated-vehicles/351416/69-university-denver.pdf (19/8/2020).

Rentschler, C., L. Herrmann, D. Kurth, W. Manz and M. Rumberg 2020. "Technische und rechtliche Systemgrenzen in der Routenplanung autonomer Shuttlebusse", in: *Neue Dimensionen der Mobilität. Technische und betriebswirtschaftliche Aspekte*, ed. by H. Proff. Wiesbaden: Gabler, 319–331.

Röhrleef, M. 2017. "Autonomes Fahren: Himmel oder Hölle für den ÖPNV?". https://bildungsservice.vcd.org/fileadmin/user_upload/Weiterbildung/Fachtagung_2016/Roehrleff_VCD_2050_autonomes_Fahren.pdf (7/1/2020).

Rollinger, W., and G. Amtmann 2009. *Handbuch Öffentlicher Verkehr. Schwerpunkt Österreich*, ed. by ÖVG Österreichische Verkehrswissenschaftliche Gesellschaft, Arbeitskreis Öffentlicher Verkehr. Vienna: Bohmann.

Rutanen, E., and V. Arffman 2017. "Autonomous robot bus experiments on public roads – SOHJOA Project report autumn 2017", 11/12/2017. Download at https://tinyurl.com/y567zhgb (19/8/2020).

Salonen, A. O., and N. Haavisto 2019. "Towards Autonomous Transportation. Passengers' Experiences, Perceptions and Feelings in a Driverless Shuttle Bus in Finland", in *Sustainability* (11) 588, 1–19.

SBB (Schweizer Bundesbahnen) 2020. "MyShuttle Abschlussbericht", 30/4/2020. Download at www.astra.admin.ch/dam/astra/de/dokumente/abteilung_straßennetzeallgemein/sbb-myshuttle-abschlussbericht.pdf.download.pdf/SBB_MyShuttle_Abschlussbericht.pdf (7/6/2020).

Sinner, M. 2019. "Effects of the Autonomous Bus on the Railway System", thesis, Institute for Transport Planning and Systems (IVT), ETH Zurich.

Sinner, M., P. Khaligh and U. Weidmann 2018. "Consequences of automated transport systems as feeder services to rail: SBB fund for research into management in the field of transport", *IVT Schriftenreihe Band 184*. ETH Zurich. Download at www.research-collection.ethz.ch/handle/20.500.11850/266025.

Sommer, C. 2018. "Neue Angebote für den ländlichen Raum", presentation at the conference "Zentralitäten 4.0 – Mittelzentren im Zeitalter der Digitalisierung?", 22/11/2018, Kassel.

Soteropoulos, A., A. Stickler, V. Sodl, M. Berger, J. Dangschat, P. Pfaffenbichler, G. Emberger, E. Frankus, R. Braun, F. Schneider, S. Kaiser, H. Walkobinger and A. Mayerthaler 2019. "SAFiP – Systemszenarien Automatisiertes Fahren in der Personenmobilität. Endbericht", ed. by Federal Ministry for Transport, Innovation and Technology, Vienna. https://projekte.ffg.at/anhang/5cee-1b11a1eb7_SAFiP_Ergebnisbericht.pdf (11/11/2020).

Weidmann, U., R. Dorbitz, H. Orth, M. Scherer and P. Spacek 2011. "Einsatzbereiche verschiedener Verkehrsmittel in Agglomerationen", research project SVI 2004/039 on commission from the Swiss Association of Transport Engineers and Experts (SVI). Bern: Federal Department of the Environment, Transport, Energy and Communications UVEK.

Wolf-Eberl, S., H. Koch, G. Estermann and A. Fürdös 2011. "Ohne eigenes Auto mobil – Ein Handbuch für Planung, Errichtung und Betrieb von Mikro-ÖV Systemen im ländlichen Raum", Blue Globe Manual Mobilität 10/2011. In cooperation with the Federal Ministry for Transport, Innovation and Technology (bmvit) and on commission from the Climate and Energy Fund. Vienna. http://rdc.co.at/wp-content/uploads/2017/11/Mikro_ÖV_Handbuch_publiziert_2011.pdf (19/8/2020).

7 Delivery robots as a solution for the last mile in the city?

Bert Leerkamp, Aggelos Soteropoulos, Martin Berger

Bert Leerkamp
University of Wuppertal, Department of Freight Transport Planning and Logistics

Aggelos Soteropoulos
TU Wien, future.lab Research Center and Research Unit Transportation System Planning (MOVE)

Martin Berger
TU Wien, Research Unit Transportation System Planning (MOVE)

© The Author(s) 2023
M. Mitteregger et al. (eds.), *AVENUE21. Planning and Policy Considerations for an Age of Automated Mobility*, https://doi.org/10.1007/978-3-662-67004-0_7

1. INTRODUCTION

Urban freight transport is assuming an increasingly important role in the field of urban mobility as well as in urban transport policy, as e-commerce is booming: with the advance of digitalization, goods are increasingly being ordered via the internet, with a resulting sharp increase in delivery transport volumes (cf. Muschkiet/Schückhaus 2019: 358; German Federal Government 2019: 44f.). New sustainable solutions and concepts are therefore needed in urban logistics, with the last mile in particular posing a major challenge due to a lack of bundling and the great effort involved (cf. Gerdes/Heinemann 2019: 399; Buthe et al. 2018: 30; Lierow/Wisotzky 2019). At the same time, new delivery media are emerging, such as delivery robots (cf. Baum et al. 2019: 2455; Jennings/Figliozzi 2019: 317), which are seen as having great potential for the last mile.

To date, the development and use of delivery robots have been driven above all by logistics companies and technology developers, without consideration of the municipal perspective. This raises the question of whether and to what extent delivery robots are at all compatible with urban public spaces. This is all the more important since public space is already under increasing strain due to new forms of mobility, adaptation to climate change, rising population figures in cities and the resulting pressure of use. Conflicts of use and interest are inevitable in particular when automated delivery robots are on the move on pavements or in pedestrian zones (cf. Buthe et al. 2018: 121). In view of the necessary transformation of public space from being a transit space to an area with quality of stay, the diverse implications of delivery robots are therefore discussed and options for (transport) policy and planning identified.

2. E-COMMERCE AND DELIVERY TRAFFIC ON THE RISE

Online trade – also known as "e-commerce" or "distance trade" – is booming. But delivery offers in stationary trade have likewise led to steady market growth for courier, express and parcel (CEP) services in recent years. In Austria, for example, the number of parcels delivered increased by around 69% between 2009 and 2018, and the corresponding global figure more than doubled between 2014 and 2018 alone. Similar developments can also be seen in Germany (cf. BIEK 2020: 11). Forecasts are predicting a further increase in the number of parcels delivered both in Austria and worldwide (cf. Umundum 2020: 151; Buchholz 2019). E-commerce sales volumes in Austria also recorded a significant increase of 21% in recent years, between 2015 and 2019, with an online share of total retail sales in Austria of currently already more than 5% (Fig. 1; cf. WIFO 2019: 15).

As a result of increasing delivery traffic, the need for action by cities and municipalities to proactively develop strategies and concepts for municipal freight transport, but also to plan and implement appropriate measures, is likewise increasing (cf. Schönberg et al. 2018: 4): since very many end users are served in urban freight transport, this results in a high number of small individual deliveries, which in turn leads to high mileages (cf. Vienna Business Agency 2016: 5). The result is an increase in particulate pollution and in CO_2 and noise emissions (cf. Muschkiet/Schückhaus 2019: 366), but above all conflicts in public space that are manifested in many ways as competition for space, personal endangerment, but also "commercialization of public space".

In view of the predicted increase in the volume of consignments – in Austria, for example, the number of parcels delivered is expected to increase by around 14% in 2020 over the figure for 2018 – the cost-intensive last mile will continue to gain significance, with all the negative effects on public space (cf. Leerkamp 2017: 12; Umundum 2020: 151).

Figure 1: Development of the Austrian parcel market in millions of units, 2009 to 2018

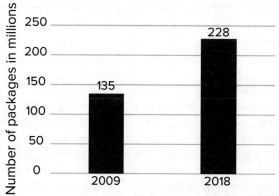

Source: Umundum (2020: 151)

Figure 2: Development of e-commerce sales in Austria, 2015 to 2019

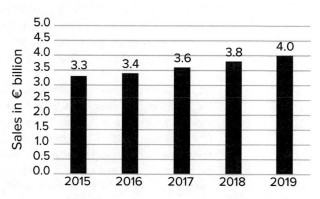

Source: WIFO (2019: 15)

3. NEW DELIVERY CONCEPTS FOR THE LAST MILE

The last mile, i.e. the final stage of delivery of goods to the customer's premises, is still one of the most pressing problems in urban freight transport: the degree of utilization of transport carriers in supply and disposal decreases with proximity to the destination, and bundling becomes increasingly difficult over the last link of the supply chain (cf. Just 2018: 5; open4innovation 2019). More than 50% of costs in parcel delivery are incurred in the last mile (cf. Schnedlitz et al. 2013: 251; Schocke 2019). Particularly outside the effective delivery window, i.e. when the probability of the recipient being at home decreases, efficiency is even lower due to the need for multiple trips.

New logistics concepts are being implemented in the area of conflict between commercial efficiency on the one hand and the demands by municipalities for traffic avoidance and displacement and environmentally compatible delivery on the other: these concepts should help to achieve bundling effects, increase the "stop factor" in end-customer business and reduce transport requirements (Buthe et al. 2018: 30). A promising logistics concept is delivery to collection points – so-called city hubs – in the urban core zone by a small number of large lorries from the periphery. From there, the parcels are delivered over the last mile either directly to the customers or to micro-depots and parcel boxes. Various vehicle and drive concepts or delivery by (e-)cargo bike are suitable for covering the last mile (cf. Wittenbrink et al. 2016: 79f.; Leerkamp 2019: 21; Gerdes/Heinemann 2019: 406).

Automation and digitalization, and the delivery concepts based on these, are driving innovation (cf. Umundum 2020: 157). In addition to delivery drones, tests have recently been carried out with electric delivery robots – so-called automated "delivery bots" – in the USA for example, but also in Europe. Last-mile delivery is often seen as one of the first areas of application for automated driving, as these robots travel at low speeds and in a perhaps simple operational design domain (ODD), for example on pavements in a residential area on the outskirts of a city (cf. Soteropoulos et al. 2020; Mitteregger et al. 2022; Leitner et al. 2018: 22).

4. OPERATING CONCEPTS OF AUTOMATED VEHICLES IN LOGISTICS

Automated vehicles are by no means new to the field of logistics: they have already been used for a long time, especially in internal logistics. These vehicles have been used to transport goods in production and logistics systems since the 1950s, mainly for transport without a driver (1) indoors or within the demarcations of buildings, (2) in private outdoor areas, e.g. on company premises or at container terminals, and (3) in hazardous or barely accessible areas (cf. Flämig 2015: 378; Hörl et al. 2019: 35; Paddeu et al. 2019: 9ff.; Hofer et al. 2018: 11ff.).

Today transport within company premises is still the typical domain of automated driving in logistics, e.g. in the autonomous yard logistics of Austrian Post, and is subject to specific framework conditions in terms of infrastructure and processes (cf. Clausen 2017: 16; Muschkiet/ Schückhaus 2019: 374; Umundum 2020: 156). As a result of progress in automation and digitalization, increasing attention is now also being given to applications in distribution logistics. In addition to automated lorries in long-distance transport (e.g. platooning – although some tests in this field have been discontinued, e.g. by Daimler; cf. Daimler 2019) and automated delivery concepts with goods delivered by drones, the use of delivery robots in the public spaces of cities and municipalities is now also being tested (cf. Baum et al. 2019: 2457; Howell et al. 2020: 36; Schröder et al. 2018: 7; Hofer et al. 2018: 14ff.). Figure 3 gives an overview of the operational concepts of automated vehicles in the field of logistics.

Figure 3: Overview of different delivery robots

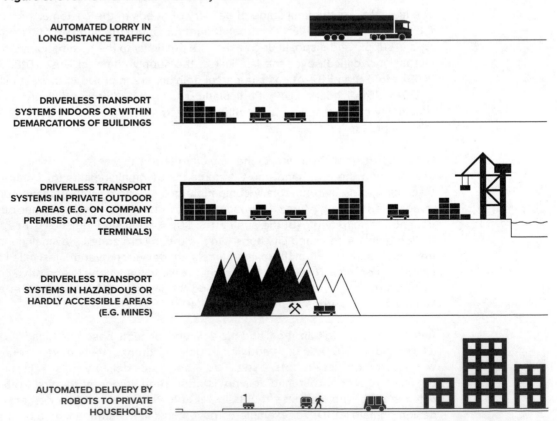

AUTOMATED LORRY IN
LONG-DISTANCE TRAFFIC

DRIVERLESS TRANSPORT
SYSTEMS INDOORS OR WITHIN
DEMARCATIONS OF BUILDINGS

DRIVERLESS TRANSPORT
SYSTEMS IN PRIVATE OUTDOOR
AREAS (E.G. ON COMPANY
PREMISES OR AT CONTAINER
TERMINALS)

DRIVERLESS TRANSPORT
SYSTEMS IN HAZARDOUS OR
HARDLY ACCESSIBLE AREAS
(E.G. MINES)

AUTOMATED DELIVERY BY
ROBOTS TO PRIVATE
HOUSEHOLDS

Source: the authors

5. DELIVERY ROBOTS

Delivery robots are driverless, often electric transport vehicles that assume the last mile of delivery from an inner-city warehouse or stationary retailer to customers within a defined permissible area (cf. Vogler et al. 2018: 152; Leerkamp 2017: 17). Their applications include special shipments that need to be delivered flexibly, rapidly and cheaply in a local environment, same-day or same-hour delivery, food consignments and home deliveries of medical products (cf. Hofer et al. 2018: 17). Some logistics concepts also involve lorries taking delivery robots to a large delivery area, where they carry out final delivery to customers (cf. Jennings/Figliozzi 2019: 321; DHL 2014: 32).

Various different types of delivery robot are now being developed by numerous companies (cf. Baum et al. 2019: 2457; Steer 2020: XVIII; and Fig. 4 below).

Figure 4: Overview of different delivery robots

Automated delivery robots for operation on pavements

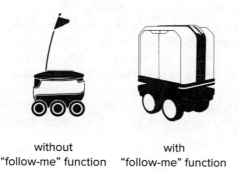

without "follow-me" function   with "follow-me" function

Automated delivery robots for operation on public roads

Source: the authors

5.1 DELIVERY ROBOTS FOR OPERATION ON PAVEMENTS

This type of delivery robot is not suitable for operation on the carriageways of public roads, but is used on pavements and in pedestrian zones, where the speed is limited to 6 km/h. These robots are currently used in office parks (e.g. in Mountain View, USA) or other special areas: the requirements on the automated driving system are not as high in such places as for operation on public roads (cf. Hern 2018). In addition, these robots usually only have a small loading volume (cf. Leerkamp 2017: 17).

a) Delivery robots for autonomous delivery

These robots are suitable for the autonomous delivery of individual items within a short timeframe (20–30 minutes), particularly in dense city districts. The manufacturers of these delivery robots include Starship Technologies, Kiwi and Marble. The robots usually have a small container in which parcels can be placed (cf. Marks 2019: 13ff.). In its current version, the delivery robot from Starship Technologies for example can transport a maximum of two parcels, each measuring 35 x 34 x 15 cm (cf. Ninnemann et al. 2017: 86). Once the delivery robot arrives at its destination, the customer can open the lid of the container at their front door using a one-time PIN that is sent to them via smartphone (cf. Vogler et al. 2018: 152).

b) Delivery robots with "follow-me" function
Delivery robots that drive on pavements can also support distribution logistics by means of a follow-me function; i.e. loaded with the parcel, they follow the recipient or orderer of the delivery, who is thus relieved of the burden. An example is the *PostBOT* delivery robot from the company Effidence S.A.S., developed together with Deutsche Post. Unlike delivery robots, which carry out their deliveries autonomously, the follow-me models are usually somewhat larger. The *PostBOT*, for example, has six package containers and can transport consignments weighing up to 150 kg (cf. Gerdes/Heinemann 2019: 411).

5.2   DELIVERY ROBOTS FOR OPERATION ON PUBLIC ROADS

Delivery robots that drive on public roads travel at speeds of up to 40 or 50 km/h and due to their size have a larger load volume than the models described above. They are typically used for autonomous delivery of individual items within a short timeframe (20–30 minutes), especially in areas with high customer density. Examples of this type of delivery robot are the *Nuro R1* and its successor *Nuro R2*, or *Robomart* and *Udelv* (Baum et al. 2019: 2458; Marks 2019: 22). The *Nuro R2* recently received the first nationwide temporary exemption from the U.S. Department of Transportation for testing on public roads without a driver (USDOT 2020: 5). These vehicles are not only suitable for last-mile delivery: Waymo, for example, recently announced that in the course of its ongoing test operations in Chandler, Arizona, its vehicles will also be used to deliver packages from local UPS shops to a UPS package sorting facility; however, delivery of packages directly to customers is not yet planned (UPS 2020). In the typical suburban areas of the USA where these vehicles are on the roads, with very wide streets, single-family homes and only few pedestrians and cyclists (e.g. *Nuro R2* in Scottsdale, Arizona, or Houston, Texas; cf. Nuro 2020), it is much easier for customers to pick up their goods from the vehicle at the kerb or in special areas where the robots are permitted to stop (so-called "self-driving pick-ups") than in corresponding neighbourhoods in Europe, which often have narrower streets and higher densities.

6.   SELECTED EXAMPLES OF TESTS WITH DELIVERY ROBOTS

Numerous countries are testing delivery robots in pilot trials. While most of the robots being tested in Germany, Austria and Switzerland are not intended for use on public roads but only operate on pavements and in pedestrian zones (cf. Baum et al. 2019: 2459; Hofer et al. 2018: 17), tests and pilot trials with delivery robots have already been carried out on public roads in the USA. Table 1 on the next page gives an overview of selected examples.

With the exception of Switzerland, the test areas were mostly in the centre or business district of a large city. The follow-me delivery robot was tested in the small town of Bad Hersfeld, but here too in the central, commercial district. In the USA, the road-going model was tested on public roads in residential and commercial areas on the outskirts of cities and in suburbs.

The tests served to determine how safe, practicable and economical the operation of delivery robots can be from the perspective of the logistics companies. In addition to the postal service,

Table 1: Overview of selected tests and pilot trials carried out to date with delivery robots

Location	Period	Type of delivery robot	Company	Neighbourhood type	Priority issues					Evaluation method		
					Commercial exploitability (business cases)	Safety and reliability	Technical feasibility	Customer acceptance	Interaction with other road users	Recording of vehicle data	Focus group discussions with stakeholders	Recording feedback from passers-by
Hamburg (DE)[1]	12/2016–3/2017, 5/2017 and 1/2018	Starship	Hermes, Domino's Pizza, Foodora	Mixed	✓	✓	✓	✓		✓	✓	
Bad Hersfeld (DE)[2]	10/2017–11/2018	PostBOT	Deutsche Post	Centre, commercial	✓	✓				✓		
Graz (AT)[3]	autumn 2017	Jetflyer	Austrian Post	Centre, commercial		✓	✓			✓		
Zurich, Bern, Köniz, Biberist, Dübendorf, Zuchwil (CH)[4]	8/2016–12/2016 and 9/2017–1/2018	Starship	Schweizer Post, Jelmoli	Centre, commercial, residential	✓	✓	✓		✓	✓		✓
Scottsdale, AZ / Houston, TX (USA)[5]	8/2018–3/2019 and 3/2019	Nuro R1, R2	Kroger, Fry's Food, Walmart, Domino's Pizza	Residential, commercial	✓	✓	✓	✓		✓		

Source: 1 – Brandt et al. (2018: 8); Ninnemann et al. (2017: 85f.); Hermes (2017a, b); Leitschuh (2018) | 2 – Gerdes/Heinemann (2019: 411f.) | 3 – APA 2017, Eigner 2017 | 4 – Marazzo/Mischler (2018) | 5 – Nuro (2018); Shaheen/Cohen (2020: 249); Wiles (2019)

the test users include CEP services and retail, restaurant and supermarket chains. However, no comprehensive evaluation or scientific monitoring took place aside from the companies' own interest in gaining knowledge. The tests did not address important issues such as conflicts of goals and use in the public streetscape, or especially interaction with other road users. By way of exception, however, the tests in Switzerland took more detailed account of these aspects, although they were not subjected to comprehensive scientific evaluation.

7. "PAVEMENT-COMPATIBLE" DELIVERY ROBOTS AS A SOLUTION FOR THE LAST MILE?

As the analysis shows, developments worldwide are mostly concerned with robots that travel on pavements and/or in pedestrian zones (cf. Baum et al. 2019: 2459); the following remarks therefore focus on this use case (see also Fig. 5).

As mentioned, much is expected of "pavement-compatible" delivery robots in high-density urban areas for delivery of individual items within short timeframes; however, significant economic potential can only be exploited once the robots no longer have to be accompanied by humans, but merely be remotely monitored by a human operator due to legal requirements (cf. Hermes 2017b). On the other hand, it remains to be clarified what requirements the legislators will place on this monitoring and what tensions will arise regarding the precision and type of monitoring and the resulting personnel costs – e.g. with or without simultaneous monitoring of several vehicles. Delivery robots with a follow-me function, on the other hand, could reduce the physical burden on delivery personnel, with the additional advantage that these persons could intervene in the event of malfunction or conflict. If the vehicles are electrically powered, this could bring about a reduction in CO_2 and noise emissions, although well-founded impact analyses are still lacking here. Only Jennings and Figliozzi have established in comparative simulations that the combination of delivery robots and conventional delivery vehicles makes for reduced delivery times, mileage and costs, especially in areas with high customer density, as compared with the use of delivery vehicles alone (cf. Jennings/Figliozzi 2019: 324). For delivery companies, the costs for acquisition and operation of a delivery robot are more than offset by the transport revenue that can be generated over the last mile (cf. Hofer et al. 2018: 48). Delivery robots are currently still too expensive, which is why only pilot trials have been implemented so far in German-speaking countries (cf. Hermes 2017b, Marazzo/Mischler 2018: 1; Wittenhorst 2019). Even in the long term, it remains to be seen to what extent aspects such as customer density, settlement density or purpose of use would yield a positive cost-benefit ratio, as the delivery robots only have a relatively small load volume (cf. Hofer et al. 2018: 48). In addition, security aspects (vehicle theft) and vandalism would have to be taken into account in the operation of these vehicles (cf. Paddeu et al. 2019: 32; Kunze 2016: 292; Hofer et al. 2018: 18).

The pilot tests also reveal specific technical problems such as limited battery power or user-friendliness of the user interface. The delivery robots also require a powerful LTE mobile network, which is not always available in all areas (Hermes 2017a; Marrazzo/Mischler 2018: 2). However, these problems are expected to be solved in the near future.

The situation is different when it comes to the challenges encountered in the interaction between humans and delivery robots in public spaces. These involve lack of acceptance, specific disruptive effects and hazards:

- Delivery robots restrict the freedom of movement of all pedestrians, but especially of people with limited mobility – above all those with walking aids (cf. Lenthang 2019; Hofer et al. 2018: 48), and of children and the elderly (cf. Marks 2019: 14).

- They can pose a hazard in public spaces if pedestrians cannot avoid them in time to prevent a collision due to impaired reactions, mobility, etc., or if they unexpectedly change direction.

- This effect is amplified on narrow pavements or pedestrian crossings with a high number of pedestrians walking at different speeds and in various directions (cf. Leerkamp 2017: 17; Marazzo/Mischler 2018: 4; Hsu 2019).

Conflicts are inevitable especially in the following driving situations that delivery robots constantly have to master (cf. Keesmaat 2020: 9; Groot 2019: 64):

- driving around obstacles,

- crossing lanes at a pedestrian crossing or traffic lights; Marazzo and Mischler (2018: 4) report for example that the green phase was too short for the delivery robot and accompanying person to cross,

- overtaking slow-moving pedestrians,

- contact with a playing child, a group of children or a large crowd of people, and

- driving onto/off a kerb or ramp (cf. Leerkamp 2017: 17).

Automated driving of the delivery robot in dense mixed traffic, which involves interaction with numerous and diverse road users, is a relatively complex driving task that only allows travel at low speeds (cf. Hofer et al. 2018: 49; cf. Fig. 6). It also remains largely unclear what technical requirements the pavements must fulfil in view of the above-mentioned conflicts in driving situations, especially with regard to kerbstones, pedestrian crossings at intersections or walking and cycling paths designated with different colours.

There are also numerous barriers to be overcome in the delivery process. For example, the logistics concept must determine how consignments are to reach customers who are not at home – or how deliveries are to be made in multistorey buildings, since the robot cannot climb stairs. Drop-off or parcel boxes that are accessible to delivery robots at ground level currently only exist in a few places.

At present, the legal framework for operation of delivery robots on public roads is relatively rigid and restrictive in Germany and Switzerland as compared to some states of the USA (cf. Jennings/Figliozzi 2019). No automated delivery robot may be operated in public spaces without an accompanying person. In Germany, an exemption is usually granted for this purpose in accordance with the Road Traffic Code (StVO) and the Road Traffic Licensing Regulations (StVZO), which include specific conditions and requirements for the operation of delivery robots (cf. Brandt et al. 2018: 7). Data protection also plays a role here, as delivery robots use image-based sensors to record their surroundings and also collect "critical" personal data of other road users in order to recognize objects. The issue here is compliance with national data protection laws and the European General Data Protection Regulation (GDPR), especially Article 25 (data protection through technology design). It must be ensured that in the course of recognition of other road users by the delivery robots, only personal data are processed that are necessary for this purpose (cf. Brandt et al. 2018: 8; Hoffmann/Prause 2018: 11). This also applies to possible remote video-based monitoring of the delivery robots by the operators. In particular, the requirement that the delivery robot be accompanied by an attendant at all times in public spaces makes the use of such robots unprofitable for the operating companies. Rather, the tests with automated delivery robots serve above all to gain practical experience with new technologies as a basis for exploring scope for action from the perspective of the companies (cf. Ninnemann et al. 2017: 138), so that they can advocate for their interests more specifically, for example to gain authorization for operation without an accompanying person. However, these interests stand in contrast to the major challenges experienced in the public space.

Figure 5: Overview of strengths, weaknesses, opportunities and risks of automated delivery robots driving on pavements for the last mile

Strengths	Weaknesses
• optimized delivery of individual consignments in short timeframes • increased efficiency by supporting the delivery agent (parallel execution of other tasks) • reduction of CO_2 and noise emissions through electric drive	• technical problems: battery power, user interface, flexibility of the system, complex mixed traffic • poor economic efficiency due to the need for an accompanying person, high purchase price, low payload • inability to climb stairs • special drop-off boxes needed for the recipient if no one is at home
Opportunities	**Risks**
• remote monitoring by a human operator, enabling exploitation of economic potential • reduced physical burden on delivery personnel due to delivery robots with follow-me function • in combination with delivery vehicles, lower delivery times and costs compared to delivery with conventional vans alone (especially with high customer density) and possibly also reduced mileage (but more journeys/mileage may be necessary due to low loading capacity)	• impaired freedom of movement for all pedestrians • danger to pedestrians, especially persons with limited mobility, children, the elderly, etc. • potential for conflict when crossing a carriageway at pedestrian crossings or traffic lights, when overtaking slow-moving pedestrians or when encountering a large crowd of people • security aspects (e.g. vehicle theft) and vandalism • collection of personal data

Source: the authors

8.   IMPLICATIONS FOR PLANNING

Initial experience from tests with automated delivery robots gives rise to hopes of economic potential for distribution logistics over the last mile. In the future, however, further detailed and spatially differentiated analyses will be needed to determine what areas offer what potential for covering the last mile. The experience gained in the tests also demonstrates the problems and risks posed by delivery robots in operation on public spaces, especially on pavements. The aspects that need to be considered for planning and (transport) policy, along with the existing scope for action, are therefore briefly outlined in the following.

In addition to the distribution of goods, already today there are a number of further demands on the public streetscape in urban areas that imply conflicts of use and interest (cf. Buthe et al. 2018: 121). Delivery robots – together with scooters, loading areas, etc. – additionally increase the already high pressure of use on public space particularly in dense urban neighbourhoods, above all on pavements. Public space is not only traffic space, but also a place to stay and meet people (cf. Stadt Wien 2018: 11), especially on pavements in the area of transition between buildings and the streetscape, where people talk to each other, look into shop windows, etc. The competing space requirements on the part of delivery robots to use pavements for driving and parking gives rise to conflict, especially in densely populated urban neighbourhoods (cf. Peters 2019: 76). This is all the more problematical since (1) it is precisely here that delivery concepts with robots are better suited by their very nature, due to the high customer density and thus economic efficiency; and (2) delivery robots operating on pavements will always have a low loading capacity (the vehicle width can hardly be greater than the scope of movement of a human, and requirements on the automated transfer of items from the robot to the parcel box, e.g. involving sorting the parcels in reverse order of delivery, do not allow optimal use to be made of the robot's storage space). A greater number of vehicles are therefore expected to be required than with the use of conventional delivery vehicles, and also in comparison with cargo bikes.

This also contrasts with the desire to "reclaim public space" by reducing traffic areas and increasing recreational areas for better quality of life and an attractive living environment. Delivery robots from commercial providers also restrict the free use of public space by all and contribute to a "privatization of public spaces" (cf. Marks 2019: 14; Wong 2017): the pavement clearly belongs to pedestrians.

If delivery robots become established, an additional need for adaptation of the infrastructure can be expected, along with further costs. Delivery concepts using robots require parcel boxes for households, for example, since personal delivery on the doorstep is hardly feasible due to the robots' inability to climb stairs. The green traffic light phases for pedestrians would also have to be extended if the delivery robots move more slowly than pedestrians. This raises the question as to who would bear the costs and ultimately also the responsibility for implementation (cf. Hofer et al. 2018: 48).

Even though the technological development is not yet sufficiently advanced to allow delivery robots to operate in public spaces on pavements at all times and under all conditions, it is necessary to give thought now to delivery robots in terms of planning. Firstly, we are living in an age of transformation of public spaces – away from traffic space, towards open space for all – and secondly, the qualities of public spaces must be secured at a very early stage. If this does not happen now, a blend of rapid technological advances in delivery robots and a marked increase in deliveries as a result of e-commerce could raise the political pressure to act so quickly that the regulatory measures outlined below would take effect either too late or not at all.

At the strategic level, data protection law and road traffic law – as a part of national (or EU) legislation – influence whether and in what ways delivery robots can be deployed in public spaces. Cities and municipalities can influence national legislation through their associations or by means of the countervailing principle in terms of spatial planning, but they also have planning levers of their own. To this end, it is necessary to address the topic of delivery robots in strategic concepts related to public space (e.g. the specialized concepts "Public Space" and "Mobility in the Planning Context of Vienna"). Furthermore, the required responsibilities, competences and resources must be established within the administrative sphere. Real experiments initiated by municipalities, which take into account conflicts of use and the effects of delivery robots, appear to be a first important step towards assessing possible applications in the urban space. It is important that the cities and municipalities take the initiative here.

Specific traffic planning measures that cities and municipalities could implement as regulatory instruments include "geofencing" – the spatially and temporally differentiated regulation of restricted zones – wherever the compatibility of delivery robots in public space cannot be ensured. A further measure would be the licensing of delivery robots in urban areas, in order to limit their number and to impose conditions on their operation. The measure of real-time pricing is a market-based instrument that levies dynamic charges based on the spatial and temporal compatibility of the delivery robots in public spaces. How these measures can be specifically devised and adapted to spatial situations, and how they can optimally complement each other, is currently still open and requires further research.

9. CONCLUSION

E-commerce is gaining an ever-greater share of the market, with a rapid increase in the number of deliveries to households. Logistics companies benefit from automation and digitalization whenever they succeed in optimizing the processes of the "last mile" – especially in terms of costs, but also of delivery times. High expectations are placed here in delivery robots. Even though numerous technological matters are yet to be resolved, delivery robots are already on the roads today in some cities and will be in seen more cities in the future. Low speeds for the delivery robots, combined with a simple operational design domain (ODD), favour early deployment. To what extent pavements can provide simple ODDs is the subject of much speculation. While unambiguous traffic regulations apply on carriageways, with defined directions of motion for road users, pavements are characterized by dynamic rules of distancing and walking behaviour: pedestrians arbitrarily change their speed and direction of motion. As pedestrian density increases, so too do the demands on the delivery robots' ability to navigate, with the result that simple design domains on the pavement become highly complex (cf. Keesmaat 2020: 16).

What deployment scenarios are conceivable for delivery robots? A possible, and in fact probable scenario is that they will not encounter much conflict in loosely built-up suburban areas or on very wide carriageways with little traffic (since issues such as passing stopped vehicles play a less significant role here), although the economic viability is doubtful. Real-time pricing based on pedestrian density and/or regulation by means of licensing would be possible as control measures in traffic planning.

The other scenario of using delivery robots in densely built-up urban areas entails much higher risks and negative impact. Strong regulation by means of exclusion zones, licensing, etc. seems necessary here. Pavements should really not be used; the delivery robots should only travel on the carriageway or in the vicinity of parking spaces (alternative use of parking lanes).

However, current pilot tests show that developments are focusing on delivery robots travelling on pavements. This trend is diametrically opposed to the use of public space for more stay and less transit. Pavements are highly important parts of the public arena that fulfil a variety of functions as linear open spaces: meeting people briefly, resting, waiting, playing, walking, observing, looking and sitting are part and parcel of everyday urban life. These qualities are often underestimated today, as frequent parking on pavements has shown.

Delivery robots – whether travelling on the pavement or parked at pick-up stations, charging stations or on the pavement – give rise to more pressure of usage, intensify competition for space and endanger passers-by. They are therefore especially critical in crowded spaces

Figure 6: Overview of the various usage demands and possible conflicts in the public space when using of delivery robots

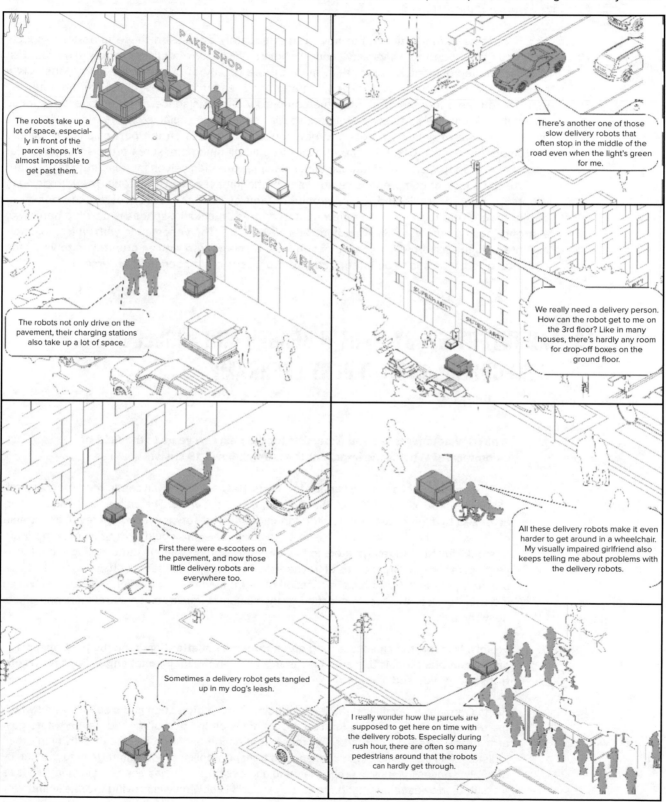

Source: the authors

where many people are on the move, and inconvenience above all people with limited mobility, children and the elderly.

The pilot tests carried out to date have not adequately addressed these important aspects and focus solely on the "drivability" and economic efficiency of delivery robots. The real test of whether deployment is appropriate must be the quality of public space. Here, there is a great need for interdisciplinary experimentation and research in these "real-life laboratories" in order to take adequate account of the complexity of public space. The key questions focus on what and how the streetscape-related, spatial, situational, social, etc. conditions determine the acceptance of delivery robots and new logistics concepts on the part of passers-by and customers (cf. Groot 2019: 64). The effects of traffic planning measures are also of particular interest here, so that cities and municipalities can prepare themselves for this future task. There is currently a lack of comprehensive analyses of the impact of logistics concepts with delivery robots in terms of traffic and the environment, and of their consequences for the quality of public spaces and road safety; a differentiated approach to social spaces is needed here, with a strong focus on the user (Soteropoulos et al. 2019: 163). The vehemence with which commercial interests are asserted in this connection must be countered with an orientation towards the common good, which is to be integrated into processes of democratic discourse.

10. AN INTERVIEW BY MARTIN BERGER AND AGGELOS SOTEROPOULOS WITH BERT LEERKAMP

1.   What developments are you expecting for the next few years in the field of e-commerce? What is the impact of the current Covid-19 crisis?

Bert Leerkamp: According to various forecasts, B2C deliveries will continue to grow at a relatively high rate of 5% to 10% annually over the next ten years. The online food trade is characterized by high growth rates, although currently at a very low absolute level. In extreme cases, where the costs are not passed on to consumers, this could partly eliminate the "buffer stock" function of refrigerators in favour of widespread on-demand ordering – but this is yet to come about. Other areas of the non-food sector also still seem to have above-average demand potential. However, a distinction must be made here between online sales, which includes "click and collect" (collection by the customer from a stationary retailer), and parcel delivery in B2C.

2.   Freight transport in cities was long given rather scant attention in municipal transport and urban planning. Is this still the case? Where do the major challenges lie in terms of transport policy and planning?

BL: In Germany, the public debate surrounding climate protection and clean air – and with it the establishment of various funding programmes on the part of the federal and state governments – has led to a marked increase in attention and a great deal of activity in urban logistics. As I see it, the major challenge in transport policy consists in regaining the quality of public spaces in terms of urban planning and design, which has been lost in many cities as a result of increased orientation towards cars and is now only improving gradually and very laboriously. It will be challenging to reach a social consensus for this urban redevelopment that goes beyond the elimination of these problems and requirements for action that are the subject of current debate on air pollution, and which is described as a comprehensive transport transformation. The climate protection argument is perhaps neither forceful nor persistent enough to bring about this transformation. But other European metropolises may well lead the way here and thereby gain highly effective competitive advantages in attracting technology-oriented companies, which will compete for highly qualified workers. This could give a boost to urban redevelopment, which must go hand in hand with changes in mobility behaviour.

In my opinion, the logistics of urban supply and disposal do not face any really major challenges as a result of such a transport transformation. Logistics is accustomed to finding optimal solutions under the given conditions and constantly optimizing itself. Conversely, this means that cities must define these conditions unambiguously, clearly and reliably. For example, changes to the accessibility of inner-city areas for delivery traffic, as we outlined in the current guideline "Liefern ohne Lasten" (delivering without burden; publisher: Agora Verkehrswende 2020), must be announced in advance and implemented in a binding manner and with sufficient lead time. On the other hand, in the process of exchange between municipal planning, trade, services and the transport industry it must be ensured that no counterproductive measures are planned.

3. **Why is the "last mile" of delivery so cost-intensive for logistics companies? Will this cost factor change over the years? What are the trends that influence cost dynamics? And how great is the motivation of logistics companies to save costs?**

BL: To name just a few examples, the following factors have contributed to rises in costs or will do so in the future:

1. the constantly increasing distance between the last transfer points (forwarding and CEP depots) and the delivery destinations, which necessitate longer journeys and therefore also increased deployment of personnel and vehicles (outsourcing of logistics locations from the city centres) and

2. increasing requirements on service, especially deadline deliveries, which reduce the scope of bundling consignments.

3. The diversification of the range of goods in the consumer sector and, at the same time, emerging competition from online retail have forced retailers to keep increasing their responsiveness; this involves more frequent deliveries from a greater number of senders with smaller consignment sizes, so here too there are negative effects on bundling capability.

4. In online retail, deadline deliveries also have a cost-driving effect: existing, well-founded delivery timeframes in the cities mean that the CEP service providers have to drive into city centres with several delivery vehicles at once during the short morning delivery timeframe – which is further shortened due to the trend towards later shop opening times – in order to successfully deliver all their consignments. The remaining consignments are then distributed in the wider city area; this reduces area-based bundling capability.

5. Further CEP volume growth in B2C overburdens the capacities of the logistics companies, leading to negative economies of scale: additional volume generates disproportionately increasing handling costs, with no increase in revenue per consignment.

6. The cost reduction potential has been exhausted – also as a result of wage dumping and outsourcing to subcontractors – and can no longer absorb the cost increases. Wages

will rise in future due to a shortage of personnel. The procurement of e-vehicles following the introduction of zero-emission zones will likewise lead to higher costs.

4. **How do you assess the technological development of delivery robots? What is the current focus of research? Is much progress being made? What are the biggest technological hurdles facing economical operation in practice? Which development path is likely to be pursued – towards operation on roads or on pavements?**

BL: The projects I know of often seem to me to be demonstration projects with a generous share of marketing intent – companies like to come across as being innovative and show that they are part of the solution to existing problems. I still fail to see any independent, comprehensive (holistic) assessment of the technological impact. In my opinion, the regulatory frameworks are kept more or less in the background; it is all about technical feasibility, and the impression to be conveyed is that autonomous vehicles could assume a large part of the delivery operations. Positive environmental effects are sometimes wrongly attributed, or it is assumed that they are a specific feature of autonomous vehicles. In fact, however, these effects are due to the electric drive, and a comparison with other solutions such as the cargo bike is lacking.

The autonomous or automated technical systems must function reliably under "chaotic" conditions, since highly diverse combinations of individual disturbances are encountered in practical operation. Humans quickly find acceptable solutions in such situations; technical systems, on the other hand, must be programmed to come up with solutions to all sorts of malfunctions occurring individually or in combination. In addition, functionable systems must include redundancies in their safety-relevant features. This all reduces the economic efficiency of the systems and makes them more susceptible to technical failure (e.g. of sensors, mechanics or energy supply). The monitoring and maintenance, and interventions in case of malfunction, give rise to additional costs.

Conditions of the system environment – in this case public space – that can be created in the laboratory are often not able to be transferred to reality. For example, I think it would be difficult to navigate a delivery robot only on a pavement, and not on an adjacent cycle path if this is not very clearly marked – what sensor system could do that? Demands for designing public spaces to suit the technology are to be viewed with scepticism – who should bear the costs, and who would benefit? This would not be practicable on a widespread basis.

In "mass transport", i.e. parcel delivery in densely populated areas, deliveries will still largely be carried out by humans because this is more economical. In very sparsely populated areas, I could envisage (ground-based or airborne) automated or autonomous systems in special situations, for example delivery of urgent goods to islands by drone, or delivery to individual farmsteads in mountainous regions; as I've mentioned, niche operations will be carried out on the carriageways rather than on pavements, because there are simply no pavements in these areas.

5. **Conflicts between delivery robots and pedestrians in the streetscape are inevitable. What are the most critical issues here? What problems of acceptance are encountered? In your view, should the pavement be taboo for delivery robots? Do you see any areas that are more suitable for delivery robots, and any that are particularly problematical? What criteria should apply here?**

BL: People will not accept obstruction by technical systems, as this would be seen as an unfair distribution of individual benefits in favour of the recipients and the logistics companies, and as a collective burden. Already today, the pavements of typical main roads are too narrow and force people to come into close contact with each other. If technical systems are to interact, presumably the only solution can be that they must evade humans, but this would impair the operation of the autonomous systems.

In practice, pavements are taboo for the above reasons, besides being unsuitable from a technical point of view. Criteria for use are: unconditional compatibility with the current environment (no requirement on the part of the environment to adapt to the system), and no obstruction to pedestrians, bicycles, or to persons with impaired mobility, vision or hearing, economic advantages under the overall conditions mentioned above, no negative effects on road safety.

6. **As e-commerce gains increasing market share and brick-and-mortar retail is on the decline, pedestrian shopping is also decreasing in volume. How do you assess this shift from physical to virtual mobility? Are there now fewer pedestrians on the pavements who can "hinder" delivery robots, resulting in their further proliferation?**

BL: From mobility surveys, it is possible to estimate what proportion of pedestrian traffic serves the purpose of shopping. Although this is concentrated in commercial areas, only a small part of pedestrian traffic can be transferred to online retail. In any case, people are likely to spend the time they save by not having to go shopping with a visit to a café – and will thus still travel on foot. So all in all, the effect is marginal and rather theoretical. And do we want streets devoid of people?

7. **For many years, efforts have been undertaken to give more room to pedestrians and cyclists in public space – but as yet with only moderate success. However, in the course of the Covid-19 crisis carriageways and parking spaces have been temporarily reallocated to pedestrians and cyclists in numerous cities such as Berlin, Vienna or Brussels. Can this crisis be seen as a tipping point – as an opportunity to actually bring about a redistribution of public space?**

BL: In my experience, municipalities have been very hesitant to take up this opportunity, although Berlin is an exception here. People often point out existing concepts and bemoan a lack of work assignments from city councils, thus implying that there is no basis for action. At the same time, I have noticed that the matter of redistribution is being raised more and more frequently and emphatically by a larger number of population groups. This should hopefully become sustainable even without Corona and influence political decision-making.

8. **Municipalities and cities are important stakeholders when it comes to locally implementing new mobility solutions such as delivery robots. What framework conditions are required at the other political levels? Is the problem being perceived and discussed at the level of local politics and administration? What protagonists are pushing this issue? What are their interests and motives? How do you perceive the vehemence with which commercial actors are asserting their interests?**

BL: The Road Traffic Act, which is administered by the federal government in consultation with the states, is decisive in Germany. The Federal Ordinance on Very Small Electric Vehicles (eKDV), which was only recently introduced, would have to be modified. In local and national politics, I have noticed that the concepts discussed here (delivery robots) are often seen in an undifferentiated way as innovative solutions with a fundamentally positive connotation. At times I gain the impression that the focus on and the undifferentiated welcoming of autonomous delivery systems serve to distract from a need for action that does not align with people's own objectives and would be more inconvenient to implement (primacy of technology versus an integral approach to the transport transformation). Commercial protagonists in logistics are approaching this topic very gingerly and, in my estimation, see little potential in this regard. Recently, in a working group on urban logistics in Düsseldorf, none of the CEP companies in attendance mentioned delivery robots as a possible solution.

9. **Does it make sense for municipalities and cities to act now and regulate the use of public spaces by delivery robots, or are the conditions still far too uncertain? Is the development of regulatory provisions keeping pace with technological progress? What regulatory innovations are being discussed, how do you rate the chances of these being implemented, and what risks are involved? In San Francisco, for example, licences are issued to individual providers for the operation of delivery robots. Would this model also be conceivable for municipalities and cities in German-speaking countries?**

BL: An early signal from the municipalities regarding their preferences for the use of pavements may be helpful for those who take a purely technical view of the whole issue and ignore the problematical framework conditions. I don't have an overview of where which regulatory innovations are being discussed. Municipalities in Germany are currently having ambivalent experience with licensed sharing systems (e-scooters). There is a great need for regulation and subsequent adjustment (e.g. prohibited areas), and the contributions to sustainable transport that are of benefit to municipalities are in conflict with observable negative effects (e.g. widespread abandonment of e-scooters on pavements, use for fun at night with disturbance of the peace, etc.). The municipalities would have to venture into new legal territory here. The relationship between the basic right to general and unrestricted use of public space on the one hand and licensing (i.e. restriction of use) on the other would have to be fundamentally clarified.

10. **What alternatives are there for covering the last mile apart from the use of delivery robots? Are there any other technological developments with potential, such as drones? Or would this necessitate social and organizational innovations? Do customers need to change their behaviour, or are technical or infrastructural measures sufficient to ensure the quality of public spaces? Could "nudging" – influencing people's behaviour (e.g. by displaying CO_2 consumption for the various different delivery options) – prove useful here?**

BL: I would give priority to two options: (1) CEP service providers could work with micro-hubs, from where they deliver consignments by cargo bike, and (2) forwarding agents working for the receivers could bundle consignments by area: a sender would write the address of such an agent on the package, which then serves as the delivery address for the CEP service provider. The receiving agent then bundles consignments for delivery on the basis of the recipients' addresses. This system is being successfully used by the company ABC-Logistik in Düsseldorf with around 150 retailers; it bundles B2B package consignments for trade operators and large office locations. A cargo bike is also used for this purpose.

11. **Where do you see the need for research at the interface of public space and last-mile logistics, with or without delivery robots? Should the public sector actively regulate innovations such as delivery robots, or rely on the market?**

BL: At present, I see a need for transformative research that would support or enable change and would test, evaluate and then disseminate good solutions. I don't see any need for the public sector to initiate research in order to promote the use of delivery robots, since I don't think the economic potential for their use in cities, or their positive effects for society, are sufficient.

REFERENCES

APA 2017. "Österreich-Premiere: Autonomes E-Fahrzeug stellt Pakete im Alleingang zu!". www.ots.at/presseaussendung/OTS_20171018_OTS0033/oesterreich-premiere-autonomes-e-fahrzeug-stellt-pakete-im-alleingang-zu (4/9/2020).

Baum, L., T. Assmann and H. Strubelt 2019. "State of the art – Automated micro-vehicles for urban logistics", in *IFAC-PapersOnLine* (52) 13, 2455–2462.

BIEK (Bundesverband Paket und Expresslogistik) 2020. "KEP-Studie 2020 – Analyse des Marktes in Deutschland". Cologne. Download at www.biek.de/download.html?getfile=2623 (26/8/2020).

Brandt, C., B. Böker, A. Bullinger, M. Conrads, A. Duisberg and S. Stahl-Rolf 2018. "Fallstudie: Delivery Robot Hamburg für KEP-Zustellung", on commission from the Federal Ministry for Economic Affairs and Climate Action (BMWi), Berlin/Düsseldorf: VDI Technologiezentrum. www.bmwi.de/Redaktion/DE/Downloads/C-D/delivery-robot-hamburg.pdf?__blob=publicationFile&v=4 (27/8/2020).

Buchholz, K. 2019. "87 Billion Parcels Were Shipped in 2018", *Statista*, 8/11/2019. www.statista.com/chart/10922/parcel-shipping-volume-and-parcel-spend-in-selected-countries/ (26/8/2020).

Buthe, B., J. Modes, B. Richter, H.-P. Kienzler, S. Altenburg, K. Esser, J. Kurter, D. Wittowsky, K. Konrad, A.-L. van der Vlugt and S. Groth 2018. "Verkehrlich-Städtebauliche Auswirkungen des Online-Handels", final report, 16/8/2018. Bonn: Federal Institute for Research on Building, Urban Affairs and Spatial Development. www.prognos.com/uploads/tx_atwpubdb/180927_BBSR_Endbericht_final_out.pdf (26/8/2020).

Clausen, U. 2017. "Was tut sich auf der letzten Meile?", in *Handelsblatt Journal 5, Sonderveröffentlichung zum Thema "Die Zukunft der Automobilindustrie"*, 16f.

Daimler 2019. "Daimler Trucks investiert eine halbe Milliarde Euro in hochautomatisierte Lkw". https://media.daimler.com/marsMediaSite/de/instance/ko/Daimler-Trucks-investiert-eine-halbe-Milliarde-Euro-in-hochautomatisierte-Lkw.xhtml?oid=42188247 (19/8/2020).

Deutsche Bundesregierung 2019. "Projektionsbericht 2019 für Deutschland gemäß Verordnung (EU) Nr. 525/2013". Berlin. https://cdr.eionet.europa.eu/de/eu/mmr/art04-13-14_lcds_pams_projections/projections/envxnw7wq/Projektionsbericht-der-Bundesregierung-2019.pdf (19/8/2020).

DHL 2014. "Self-driving vehicles in Logistics. A DHL Persperctive on implication and use cases for the logistics industry". Troisdorf. https://discover.dhl.com/content/dam/dhl/downloads/interim/full/dhl-self-driving-vehicles.pdf (27/8/2020).

Eigner, S. 2017. "Post und TU Graz erproben autonome Transportlogistik auf der 'letzten Meile'". www.tugraz.at/tu-graz/services/news-stories/medienservice/einzelansicht/article/post-ag-und-tu-graz-erproben-autonome-transportlogistik-auf-der-letzten-meile/ (4/9/2020).

ERTRAC (European Road Transport Research Advisory Council) 2014. "Urban Freight research roadmap". Brussels. www.ertrac.org/uploads/documentsearch/id36/ERTRAC_Alice_Urban_Freight.pdf (7/1/2020).

Flämig, H. 2015. "Autonome Fahrzeuge und autonomes Fahren im Bereich des Gütertransportes", in *Autonomes Fahren. Technische, rechtliche und gesellschaftliche Aspekte*, ed. by M. Maurer, J. C. Gerdes, B. Lenz and H. Winner. Berlin/Heidelberg: Springer Vieweg, 377–398.

Gerdes, J., and G. Heinemann 2019. "Urbane Logistik der Zukunft – ganzheitlich, nachhaltig und effizient", in *Handel mit Mehrwert. Digitaler Wandel in Märkten, Geschäftsmodellen und Geschäftssystemen*, ed. byG. Heinemann, H. M. Gehrckens and T. Täuber, Wiesbaden: Gabler, 397–420.

Groot, S. de 2019. "Pedestrian Acceptance of Delivery Robots: Appearance, interaction and intelligence design", master's thesis, TU Delft.

Heinemann, G., H. M. Gehrckens and T. Täuber 2019. *Handel mit Mehrwert. Digitaler Wandel in Märkten, Geschäftsmodellen und Geschäftssystemen*. Wiesbaden: Gabler.

Hermes 2017a. "'Wir haben etwas Neues auf den Fußweg gebracht': Erste Learnings aus dem Pilottest mit unserem Zustellroboter", *Hermes bloggt*, 31/3/2017. https://blog.myhermes.de/2017/03/wir-ha-ben-etwas-neues-auf-den-fussweg-gebracht-erste-learnings-aus-dem-pilottest-mit-unserem-zus-tellroboter/ (8/1/2020).

Hermes 2017b. "Starship-Roboter in Hamburg: 'Innovation passiert nicht am Schreibtisch'", *Hermes Newsroom*, 29/3/2017. https://newsroom.hermesworld.com/starship-roboter-in-hamburg-innova-tion-passiert-nicht-am-schreibtisch-12146/ (7/1/2020).

Hern, A. 2018. "First robot delivery drivers start work at Silicon Valley campus", *The Guardian*, 30/4/2018. www.theguardian.com/cities/2018/apr/30/robot-delivery-drivers-coming-to-a-campus-near-you-starship-technologies (7/1/2020).

Hörl, S., F. Becker, T. Dubernet and K. W. Axhausen 2019. "Induzierter Verkehr durch autonome Fahrze-uge: Eine Abschätzung", research project SVI 2016/001, on commission from the Swiss Association of Transport Engineers and Experts (SVI). Bern: Federal Department of the Environment, Transport, Energy and Communications UVEK. https://ethz.ch/content/dam/ethz/special-interest/baug/ivt/ivt-dam/vpl/reports/1401-1500/ab1433.pdf (18/8/2020).

Hofer, M., L. Raymann and F. Perret 2018. "Einsatz automatisierter Fahrzeuge im Alltag – Denkbare Anwendungen und Effekte in der Schweiz. Schlussbericht Modul 3f 'Güterverkehr/City Logistik (Straße)'", final version of 28/3/2018. Zurich: EBP Schweiz. www.ebp.ch/sites/default/files/project/uploads/2018-03-28%20aFn_3f%20Güterverkehr%20und%20Citylogistik_Schlussbericht_0.pdf (27/8/2020).

Hoffmann, T., and G. Prause 2018. "On the Regulatory Framework for Last-Mile Delivery Robots", in *Machines* (6) 3, 1–16.

Howell, A., H. Tan, A. Brown, M. Schlossberg, J. Karlin-Resnick, R. Lewis, M. Anderson, N. Larco, G. Tier-ney, J. Carlton, J. Kim and B. Steckler 2020. "Multilevel Impacts of Emerging Technologies on City Form and Development". Eugene, OR: Urbanism Next Center, University of Oregon. Download at https://scholarsbank.uoregon.edu/xmlui/bitstream/handle/1794/25191/R_NSF_Multilevellm-pacts.pdf?sequence=1&isAllowed=y (27/8/2020).

Hsu, J. 2019. "Out of the Way, Human! Delivery Robots Want a Share of Your Sidewalk", *Scientific American*, 19/2/2019. www.scientificamerican.com/article/out-of-the-way-human-delivery-robots-want-a-share-of-your-sidewalk/ (27/8/2020).

Jennings, D., and M. A. Figliozzi 2019. "Study of Sidewalk Autonomous Delivery Robots and Their Poten-tial Impacts on Freight Efficiency and Travel", in *Transportation Research Record* (2673) 6, 317–326.

Just, M. 2018. "Lieferservice und Güterverkehr in der Stadt – Historie, zukünftige Entwicklungen und Lösungsmöglichkeiten", lecture, 25/1/2018. Munich. https://docplayer.org/73063319-Lieferser-vice-und-gueterverkehr-in-der-stadt-historie-zukuenftige-entwicklungen-und-loesungsmoeglich-keiten.html (26/8/2020).

Keesmaat, P. 2020. "Designing socially adaptive behavior for mobile robots", master's thesis, TU Delft.

Knoppe, M., and M. Wild 2018. *Digitalisierung im Handel. Geschäftsmodelle, Trends und Best Practice*. Wiesbaden: Gabler.

Kunze, O. 2016. "Replicators, Ground Drones and Crowd Logistics: A Vision of Urban Logistics in the Year 2030", in *Transportation Research Procedia* 19, 286–299.

Leerkamp, B. 2017. "Städtische Güterverkehrskonzepte – die erste und die letzte Meile", lecture, Zukun-ftsfähige Mobilität in Wuppertal – Handel und Verkehr, 23/5/2017. https://docplayer.org/61079802-Staedtische-gueterverkehrskonzepte-die-erste-und-die-letzte-meile.html (26/8/2020).

Leerkamp, B. 2019. "Beiträge der Raumplanung zum Klimaschutz im Güterverkehr", in *Nachrichten der ARL* 1, 20–23. https://shop.arl-net.de/media/direct/pdf/nachrichten/2019-1/nr_1-19_leerkamp.pdf (26/8/2020).

Leitner, K.-H., T. Bacher, S. Humpl, A. Kasztler, A. Millonig, W. Rhomber and P. Wagner 2018. "Berufs-bilder und Chancen für die Beschäftigung in einem automatisierten und digitalisierten Mobil-itätssektor 2040". Vienna: Federal Ministry of Transport, Innovation and Technology. https://mobil-itaetderzukunft.at/resources/pdf/projektberichte/Mob_2040_Endbericht_2018_Septemberfinal.pdf (12/3/2020).

Leitschuh, V. 2018. "Zusteller der Zukunft: Lieferroboter in neuem Auftrag unterwegs", *Eimsbütteler Na-chrichten*, 10/1/2018. www.eimsbuetteler-nachrichten.de/lieferroboter-wieder-in-eimsbuettel-un-terwegs/ (7/1/2020).

Lenthang, M. 2019. "Autonomous food delivery robots that are just knee-high and travel on the sidewalk on college campuses are branded a menace for disabled people", *Daily Mail online*, 20/11/2019. www.dailymail.co.uk/news/article-7706813/Self-driving-delivery-robots-college-campuses-menace-disabled-people.html (9/1/2020).

Lierow, M., and D. Wisotzky 2019. "Letzte Meile macht E-Food zu schaffen. Zustellung fordert Lebensmittel-Onlinehandel heraus – Alternative Lieferkonzepte gefragt". www.oliverwyman.de/our-expertise/insights/2019/may/Letzte-Meile-macht-E-Food-zu-schaffen.html (19/8/2020).

Marks, M. 2019. "Robots in Space: Sharing our World with Autonomous Delivery Vehicles". https://robots.law.miami.edu/2019/wp-content/uploads/2019/04/Marks_Robots-in-Space.pdf (7/1/2020).

Marazzo, A. and J. Mischler 2018. "Abschlussbericht Lieferroboter – Testphase 2". Bern: Post CH AG. Download at: www.astra.admin.ch/dam/astra/fr/dokumente/abteilung_straßennetzallgemein/abschlussbericht_lieferrobotter.pdf.download.pdf/2018-08-23_%20Post%20_%20Abschlussbericht%20-%20Lieferroboter%20II%20.pdf (4/9/2020).

Maurer, M., J. C. Gerdes, B. Lenz and H. Winner 2015. *Autonomes Fahren. Technische, rechtliche und gesellschaftliche Aspekte.* Berlin/Heidelberg: Springer Vieweg.

Mitteregger, M., E. M. Bruck, A. Soteropoulos, A. Stickler, M. Berger, J. S. Dangschat, R. Scheuvens and I. Banerjee 2022. *AVENUE21. Connected and Automated Driving: Prospects for Urban Europe, trans. M. Slater and N. Raafat.* Berlin: Springer Vieweg DOI: 10.1007/978-3-662-64140-8.

Muschkiet, M., and U. Schückhaus 2019. "Anforderungen an die Handelslogistik der Zukunft", in *Handel mit Mehrwert. Digitaler Wandel in Märkten, Geschäftsmodellen und Geschäftssystemen*, ed. by G. Heinemann, H. M. Gehrckens and T. Täuber. Wiesbaden: Gabler, 357–378.

Ninnemann, J., A.-K. Hölter, W. Beecken, R. Thyssen and T. Tesch 2017. "Last-Mile-Logistics Hamburg – Innerstädtische Zustelllogistik. Studie im Auftrag der Behörde für Wirtschaft, Verkehr und Innovation der Freien und Hansestadt Hamburg". Hamburg: HSBA Hamburg School of Business Administration. www.hsba.de/fileadmin/user_upload/bereiche/forschung/Forschungsprojekte/Abschlussbericht_Last_Mile_Logistics.pdf (7/1/2019).

Nuro 2018. "Scottsdale, meet Nuro", 16/8/2018. https://medium.com/nuro/az-pilot-launch-33cceb55c871 (27/8/2020).

Nuro 2020. "Nuro. FAQ. How will I retrieve my order? Who can participate in Nuro's service?". https://nuro.al/faq (28/8/2020).

open4innovation 2019. "Nachhaltige Lösungen für die First/Last-Mile: Innovative Logistiklösungen für Zustellung, Auslieferung und Abholung". Vienna: Federal Ministry of Transport, Innovation and Technology (BMK). https://mobilitaetderzukunft.at/de/highlights/first-last-mile.php (4/9/2020).

Paddeu, D., T. Calvert, B. Clark and G. Parkhurst 2019. "New Technology and Automation in Freight Transport and Handling Systems". Download at https://uwe-repository.worktribe.com/output/851875/new-technology-and-automation-in-freight-transport-and-handling-systems (27/8/2020).

Peters, N. 2019. "Die Letzte Meile im urbanen Güterverkehr und ihre Auswirkungen auf öffentliche Räume", diploma thesis, TU Wien.

Schnedlitz, P., E. Lienbacher, B. Waldegg-Lindl and M. Waldegg-Lindl 2013. "Last Mile: Die letzten – und teuersten – Meter zum Kunden im B2C E-Commerce", in *Handel in Theorie und Praxis. Festschrift zum 60. Geburtstag von Prof. Dr. Dirk Möhlenbruch*, ed. by G. Crockford, F. Ritschel and U.-M. Schmieder. Wiesbaden: Gabler, 249–273.

Schocke, K.-O. 2019. "Boom mit Problemen: Paketbranche feilt an 'letzter Meile'", *Die Zeit*, 6/3/2019. www.zeit.de/news/2019-03/06/boom-mit-problemen-paketbranche-feilt-an-letztermeile-190306-99-264826 (26/8/2020).

Schönberg, T., T. Wunder and M. S. Huster 2018. "Urbane Logistik 2030 in Deutschland – Gemeinsam gegen den Wilden Westen", ed. by Roland Berger. Munich. Download at https://www.bvl.de/schriften/schriften/urbane-logistik-2030 (26/8/2020).

Schröder, J., B. Heid, F. Neuhaus, M. Kässer, C. Klink and S. Tatomir 2018. "Fast forwarding last-mile delivery – implications for the ecosystem: Travel, Transport, and Logistics and Advanced Industries", ed. by McKinsey & Company. www.mckinsey.com/de/~/media/mckinsey/locations/europe%20and%20middle%20east/deutschland/publikationen/fast%20forwarding%20last%20mile/180712-fast-forwarding-last-mile-delivery.ashx (27/8/2020).

Shaheen, S., and A. Cohen 2020. "Mobility on Demand in the United States. From Operational Concepts and Definitions to Early Pilot Projects and Future Automation", in *Analytics for the Sharing Economy: Mathematics, Engineering and Business Perspectives*, ed. by E. Crisostomi, B. Ghaddar, F. Häusler, J. Naoum-Sawaya, G. Russo and R. Shorten. Cham: Springer International Publishing, 227–254.

Soteropoulos, A., A. Stickler, V. Sodl, M. Berger, J. Dangschat, P. Pfaffenbichler, G. Emberger, E. Frankus, R. Braun, F. Schneider, S. Kaiser, H. Walkobinger and A. Mayerthaler 2019. "SAFiP – Systemszenarien Automatisiertes Fahren in der Personenmobilität". Vienna: Federal Ministry of Transport, Innovation and Technology. https://projekte.ffg.at/anhang/5cee1b11a1eb7_SAFiP_Ergebnisbericht.pdf (27/8/2020).

Soteropoulos, A., M. Mitteregger, M. Berger and J. Zwirchmayr 2020. "Automated drivability: Toward an assessment of the spatial deployment of level 4 automated vehicles", in *Transportation Research Part A: Policy and Practice* 136, 64–84.

Stadt Wien 2018. "STEP 2025. Fachkonzept: Öffentlicher Raum", Magistratsabteilung 18 – Stadtentwicklung und Stadtplanung (ed.). https://tinyurl.com/d6ppydrj (28/8/2020).

Steer 2020. "Economic Impacts of Automous Delivery Services in the US." Final report. New York. In https://www.steergroup.com/sites/default/files/2020-09/200910_%20Nuro_Final_Report_Public.pdf (18/9/2020).

Umundum, P. 2020. "Die letzte Meile – Königsdisziplin der Logistik", in *Logistik – die unterschätzte Zukunftsindustrie. Strategien und Lösungen entlang der Supply Chain 4.0*, P. H. Voß (ed.), 149–162.

UPS 2020. "UPS And Waymo Partner To Begin Self-Driving Package Pickup In Arizona", *UPS Pressroom*, 29/1/2020. https://pressroom.ups.com/pressroom/ContentDetailsViewer.page?ConceptType=PressReleases&id=1580327674120-833 (19/2/2020).

USDOT (U.S. Department of Transportation) 2020. "Nuro, Inc.; Receipt of Petition for Temporary Exemption for an Electric Vehicle with an Automated Driving System". Washington DC. www.nhtsa.gov/sites/nhtsa.dot.gov/files/documents/nuro_notice_of_reciept_unofficial.pdf (19/2/2020).

Vogler, T., J.-P. Labus and O. Specht 2018. "Mögliche Auswirkungen von Digitalisierung auf die Organisation von Handelsunternehmen", in *Digitalisierung im Handel. Geschäftsmodelle, Trends und Best Practice*, ed. by M. Knoppe and M. Wild. Wiesbaden: Gabler, 149–172.

Voß, P. H. 2020. *Logistik – die unterschätzte Zukunftsindustrie. Strategien und Lösungen entlang der Supply Chain 4.0*. Wiesbaden: Gabler.

WIFO (Österreichisches Institut für Wirtschaftsforschung) 2019. "Wie wird das Weihnachtsgeschäft 2019", press conference. Vienna. https://tinyurl.com/e5frybts (16/3/2020).

Wiles, R. 2019. "Kroger ends its unmanned-vehicle grocery delivery pilot program in Arizona", *USA Today*, 14/12/2019. https://tinyurl.com/m4syhubk (27/8/2020).

Wirtschaftsagentur Wien 2016. "City Logistik. Technologie Report". Vienna. https://wirtschaftsagentur.at/fileadmin/user_upload/Technologie/Factsheets_T-Reports/DE_CityLogistik_Technologie_Report.pdf (4/9/2020).

Wittenbrink, P., B. Leerkamp and T. Holthaus 2016. "Städtisches Güterverkehrskonzept Basel. Schlussbericht". Basel: Department of Building and Transport, Canton of Basel-City. Download at www.mobilitaet.bs.ch/gesamtverkehr/verkehrskonzepte/gueterverkehrskonzept.html (26/8/2020).

Wittenhorst, T. 2019. "Deutsche Post stoppt Paketkästen für Privathäuser und Zustellroboter Postbot", *heise online*, 3/2/2019. www.heise.de/newsticker/meldung/Deutsche-Post-stoppt-Paketkaesten-fuer-Privathaeuser-und-Zustellroboter-Postbot-4296767.html (9/1/2020).

Wong, J. C. 2017. "Delivery robots: a revolutionary step or sidewalk-clogging nightmare?", *The Guardian*, 12/4/2017. www.theguardian.com/technology/2017/apr/12/delivery-robots-doordash-yelp-sidewalk-problems (28/8/2020).

PART II
Public space

Mathias Mitteregger, Emilia M. Bruck, Andrea Stickler

With or without connected and automated transport – the streetscape is the scene of a collision between two fundamental needs that characterize a functioning and habitable city. On the one hand, a steady stream of goods and people must find its way into, through and out of the dense conurbation, while on the other hand, the streetscape plays a decisive role in determining the residential quality of a city. Where no space is available for greenery and recreation, and where no sense of security can evolve, the streetscape loses its function as a public space and a city loses its much-cited human quality. For this reason, every intervention in the streetscape is rightly viewed with a critical eye – whether it be the right of free expression in protests or a new means of transport. Connected and automated vehicles must therefore prove themselves on two counts: new transport options encounter city-dwellers in the limited space of the streetscape. The contributions selected for this section, "Public space", provide a multifaceted overview of the opportunities and risks involved, along with possibilities for design. The contributions all point out how important it is to critically evaluate this new technology from the perspective of public space: the qualities of the streetscape have invariably determined the intrinsic character of a city. There is no reason to assume that this will be any different in the future.

In their chapter *Control and design of spatial mobility interfaces: Considerations in the light of automated driving*, Emilia M. Bruck, Martin Berger and Rudolf Scheuvens discuss the possibilities of shaping interfaces in a partly connected and automated mobility system. The authors demonstrate that the question of where and how transfer between (automated) modes of transport will take place in the future is likely to hinge on three key factors: technological maturity, the forms of service in operation and (local) control measures. The authors speak in favour of an integrated view as a point of departure for this essential planning task of the future. They take on this view themselves in their article, in order to draw up an initial catalogue of requirements for the design of interchanges and stops. Differentiating according to urban, suburban and rural contexts, they point out the demands that are already starting to be placed on these new typologies already today in terms of urban planning, and outline what can be done to enable connected and automated vehicles to be used as an extension of the public transport system.

In their chapter *Transformations of European public spaces with AVs*, Robert Martin, Emilia M. Bruck and Aggelos Soteropoulos present concepts devised by the Danish firm JaJa Architects especially for this volume. JaJa Architects deliberately position their designs in contrast to existing, modular design maxims, to which they attest a "twist" in the direction of "total design". With the metropolitan region of Copenhagen, the firm has chosen an icon of sustainable mobility. The privately owned car no longer plays a role in their vision of the future: it is replaced by even more active mobility, along with a public transport system extended by automated mobility services. JaJa Architects illustrate their concept of the public space of the streetscape with three examples: a residential street in the suburban area of the metropolitan region (in the little finger of Copenhagen's well-known "Finger Plan"); a railway station forecourt (also in the suburban area); and finally a busy inner-city street in the urban centre. The possible future of Copenhagen is impressively illustrated in visualizations and diagrams.

Lutz Eichholz and Detlef Kurth discuss in their chapter *Integration of cycling into future urban transport structures with connected and automated vehicles* the role and opportunities of incorporating cycling into a future transport system. They give some initial insights into "RAD-

AUTO-NOM", a research project at TU Kaiserslautern. Taking the existing (legal) regulations as a point of departure, the authors first present a traffic-based perspective on the streetscape. They demonstrate that conflict-free coexistence of bicycles and automated vehicles could trigger pressure for action in terms of infrastructure. Cyclists are fast, agile and nimble when moving through urban space – behaviour that is likely to remain a difficult obstacle for the algorithms of automated vehicles to overcome in the long term. On the basis of selected traffic situations – overtaking manoeuvres, intersections and "shared spaces" – they point out the problems and opportunities of a streetscape that allocates appropriate space to both modes of transport.

Steven Fleming takes a different approach: in his essay *Against the driverless city*, he extends the vision of a bicycle city from his publication *Velotopia*. Fleming opposes urban planning that he sees as being "intoxicated with the glamour of the new". He contrasts current plans for and with automated vehicles with a radically different city: everything in Velotopia is tailored to the bicycle – from the streets to the living space. The ground undulates in places where cyclists are to brake or accelerate. The network of cycle paths is canopied throughout to keep out inclement weather, and instead of staircases there are ramps. Here too, the reader is guided through Velotopia with illustrations and diagrams. The author almost positions his image of the future among the icons of architectural history: the designs of Le Corbusier and Frank Lloyd Wright – authorities to whom we still look (critically) today when it comes to the city and the car – or the visions of the British urban planner Ebenezer Howard, who developed his garden city on the basis of rail transport; this is still a fixed element of curricula in architecture and in urban and transport planning. Time will tell whether this comparison holds.

In his chapter *At the end of the road: Total safety – how the safety concept of connected and automated driving systems is changing the streetscape*, Mathias Mitteregger identifies a problem area generated by the quest for increased road safety through automation and networking of road traffic: constant surveillance and subtle influence could fundamentally change the public space. Road safety is now the main argument used in promoting the development of automated driving systems and digital infrastructure. The author demonstrates that the intended effect is likely to remain very limited in local terms: more automation and networking will only enhance road safety on a small fraction of the road network, and only in wealthy countries. The author sees the decisive moment along the path already taken in a transition from passive to active safety systems. This would transform not only the safety concept of a vehicle, but also a fundamental principle of public life: a space generated by mutual respect becomes an environment with top-down control. Mitteregger points out the potential impact of such a transformation and emphasizes the far-reaching consequences for the cities of the future.

8 Control and design of spatial mobility interfaces

Considerations in the light of automated driving

Emilia M. Bruck, Rudolf Scheuvens, Martin Berger

Emilia M. Bruck
TU Wien, future.lab Research Center and Research Unit of Local Planning (IFOER)

Rudolf Scheuvens
TU Wien, future.lab Research Center and Research Unit of Local Planning (IFOER)

Martin Berger
TU Wien, Research Unit Transportation System Planning (MOVE)

© The Author(s) 2023
M. Mitteregger et al. (eds.), *AVENUE21. Planning and Policy Considerations for an Age of Automated Mobility*, https://doi.org/10.1007/978-3-662-67004-0_8

1. INTRODUCTION

This chapter deals with the changes in public space that accompany digital networking and automation in the mobility sector. The focus here is specifically on spatial mobility interfaces. This refers to locations characterized by interaction, by the interplay of different transport modes and by transitions in public space. Due to the increasing diversity of mobility offers and new usage requirements, the design of spatial interfaces is already acquiring importance in urban planning, especially since the locations of access to platform-based mobility offers often turn out to be focal points in the public space. In view of the possible introduction of self-driving vehicles, spatial interfaces – as thresholds between traffic space and pavements, and between motion and stationary usage – are expected to acquire intensified relevance as formative elements in a townscape: firstly because the introduction of automated mobility could contribute to a decline in private car ownership and a proliferation of sharing offers (Zhang et al. 2015), and secondly because is it assumed that automated vehicles will not necessarily park at their initial destinations, but can then continue autonomously to remote parking garages (Zhang/Wang 2020). In both cases, the requirement for parking places in the public space would decrease, while there would be increased demand for options for boarding and alighting, for changing to other modes and for delivering goods.

Against this background, it is important to think beyond the purely functional solution of spatial-physical networking and to take into account the interactions with the specific context – urban profile, usage neighbourhood, traffic volume and accessibility – in holistic approaches to planning. Possible approaches in design would for example involve designating stopping zones and gathering points in the public space, extending existing tram or bus stops and stations, or enhancing mobility points with the addition of supply functions so that they could gain significance as local centres.

Although it is currently unclear as to whether and under what circumstances automated vehicles will be widely used, it is crucial from an urban development perspective that future regulations for automated transport are drawn up on the basis of spatial control and usage patterns from the transition phase (Marsden et al. 2020). Rather than allowing traditional planning approaches – in which the automobile dominates the mobility system and public space – to be merely replicated for automated systems, it is important to determine at an early stage to what extent these can be overcome with the introduction of new forms of mobility. Consideration of small-scale modifications to accommodate automated mobility thus raises questions not only regarding the future linking of and interaction between different transport modes, but also regarding diversity of use in public space, a smooth transition to surrounding functions and the renewal of urban qualities. Last but not least, thought must be given to what consequences the application of new mobility solutions would have for the objective of implementing high-quality public spaces, which are invariably characterized by unplanned and unforeseen new developments. The authors' intention is to make a practical contribution to the discussion of appropriate forms of urban and mobility planning in the context of automated mobility. To this end, the ambivalent mechanisms of new mobility technologies are first discussed and eight considerations relevant to planning are then presented, which are concerned with two levels of action:

- strategic framework conditions of spatial interfaces and

- small-scale design of multimodal public spaces.

This chapter should also be understood as a plea: since public spaces are acquiring the status of increasingly contested terrain in view of automated driving, planning approaches must be finely adjusted in order to preserve the urban qualities of a location (Schmid 2016). To this end, and in view of new mobility offers and self-driving vehicles, these analyses emphasize the relevance of integrated approaches in order to promote intermodal routes and active mobility – especially since spatial design is a significant external factor influencing the choice of transport mode (Konrad/Groth 2019).

2. THE IMPACT OF AUTOMATED DRIVING ON PUBLIC SPACES

2.1 THE SITUATION

As a result of networking and automation, urban and regional mobility systems are undergoing a gradual transformation. Station-independent ride-sharing services (e.g. Uber or Lyft) in particular, which have also spread to large and medium-sized European cities in recent years, are seen as harbingers of automated transport (Soike et al. 2019, Erhardt et al. 2019). In view of the superior convenience of door-to-door transport, these services trump conventional local public transport in areas with high densities of population and demand. Factors relevant to planning, such as travel time or the spatial efficiency of the system, are receding into the background, while the flexibility, availability and individual adaptability of these services are gaining significance.

With highly automated vehicles (SAE Level 4 according to SAE International 2018), as the industry promises, travel experience could be further enhanced in terms of comfort, safety and efficiency: travelling in automated vehicles would benefit those who cannot drive a conventional private vehicle themselves, i.e. persons with impaired mobility, children or the elderly (Gavanas 2019); furthermore, the automation of public offers and private mobility services is expected to extend the range of transport media available to those who live in car-dependent areas or cannot afford to own a car. Accordingly, especially for suburban and rural or peripheral areas, it is hoped that cost-efficient coverage can be ensured through savings in operating and personnel costs brought about by self-driving feeder services, local buses and village connections (cf. Jürgens 2020: 40).

In the meantime, however, with a widespread switch to ride-sharing services, traffic volumes in North American inner cities have drastically increased and unregulated stopping along the kerb has led to conflicts of use with public transport media and cyclists (Schaller 2017, Erhardt et al. 2019). The accompanying increase in boarding and alighting manoeuvres exemplifies the intensification and diversification of usage demand in public space (Marsden et al. 2020) that go hand in hand with the emergence of delivery services, micromobility offers and platform-based ride-sharing services. In addition, at times conflicting interests are arising, since the above-mentioned technological solutions foster the establishment of new companies, while awareness for an environmentally compatible mobility transformation is on the rise and there are increasing calls for the deconstruction of streetscapes in favour of soft mobility (see Fig. 1).

In addition to the above-mentioned risks regarding interaction in traffic, a widespread increase in boarding and alighting in public space can have a negative impact. Double parking and the spatial spread of traffic reduce permeability and accessibility. With regard to automated driving, it can be assumed that this effect will be exacerbated. High penetration rates of automated vehicles may give rise to additional trips and congestion due to boarding and alighting, which in turn could significantly reduce opportunities for crossing certain streetscapes. This would clearly disadvantage weaker road users, and public spaces could lose their quality of stay. In addition, a dense spacing of stopping points could lead to intensified demand for automated vehicles on short routes due to the enhanced comfort they provide, with a resulting further increase in traffic volume (González-González et al. 2020, Kondor et al. 2020, Cavoli et al. 2017).

A widespread increase in boarding and alighting situations must also be viewed critically insofar as the desired spatial benefits of automated driving can only be realized if a large proportion of vehicle-kilometres is covered in shared rides. In this context, it has been shown that the bundling of journeys and higher occupation rates are enhanced by driving to dedicated collection points, with an overall decrease in mileage (Stiglic et al. 2015).

Against this background, it is clear that with automated driving, urban planning is faced with the challenge of reconciling increasingly diverse forms of usage and at times conflicting interests within the public space. The increasing need for temporary stopping points must be taken into account here along with the demands of road safety, efficient traffic flow, climate compatibility and quality of urban space.

Figure 1: The transformation in usage within the public space

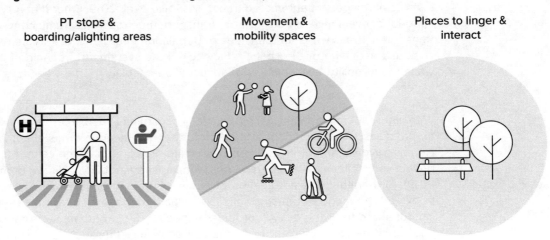

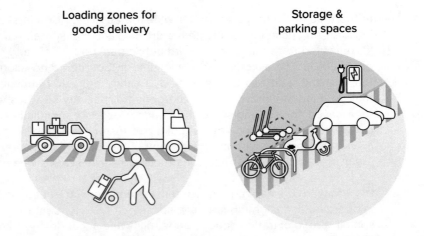

Source: the authors

2.2   POINTS OF FOCUS

In view of the problem of increased competition for room in public spaces and the need for integrated planning approaches, the following questions will now be considered:

- Where do spatial and modal interfaces arise? What should be taken into account when planning areas of operation and station networks?

- What "rules" are needed to make interfaces visible in public space and to design them with a view to heterogeneous forms of mobility and usage demands?

- To what extent can mobility interfaces provide positive impulses for urban development and serve as pivot points in the urban fabric?

Furthermore, the effects and the approaches for action must be seen in a differentiated light depending on the settlement structure, urban space and adjacent uses, since the potential and limitations of automated driving differ between inner-city, suburban and rural areas. In this context, not only must existing factors such as the settlement structure, the cross section of the streetscape or the intensity of use of the surrounding ground-floor zones be taken into account, but also the potential for urban space that can only become effective with a redistribution of mobility-related spaces, involving for example upgrading the perceptibility and quality of stay of public space. In the following analysis, a differentiation is made between urban, suburban and rural areas.

3.   STRATEGIC FRAMEWORK CONDITIONS OF SPATIAL INTERFACES

One space-related opportunity for connected and automated mobility concepts is based on the bundling of trips (Greenblatt/Shaheen 2015, Zhang et al. 2015): this can take the form of car-pooling among colleagues, pooling services or the integration of passenger and goods traffic. However, in order to realize associated spatial benefits (e.g. reduced parking demand, freeing up of existing space or transformation of existing structures), modes of control need to be strategically defined. In the context of automated driving, measures such as priority for vehicles with high occupancy rates, mileage-based pricing or restrictive parking management are typically mentioned as means of counteracting the individual use of vehicles (Sousa et al. 2018, Zhang 2017). However, spatial strategies can also be used to promote the bundling of transport flows – for example by providing appropriate mobility infrastructure or by concentrating offers and activities, thereby combining routes in terms of space and time. In both cases, spatial interfaces are created in which different requirements need to be reconciled and existing elements rendered compatible with future systems.

The locations where new spatial interfaces emerge and the extent to which they will shape structures of settlement and urban space is influenced at a superordinate level by the strategic guiding principles and underlying modes of control of automated driving. This primarily concerns the areas of application of automated fleets, whereby the suitability of the various forms of service in particular requires spatially differentiated consideration. Furthermore, integration of automated services into a network is a prerequisite for planning the locations of stops and transfer points. As mentioned, while modelling studies have shown that the establishment of collection points can facilitate the bundling of trips in automated mobility services (Stiglic et al.

2015), on an international scale – apart from individual pilot projects – regulatory approaches for the strategic planning of dedicated stopping zones and collection points are lacking (ITF 2018: 58; Schaller 2019). Last but not least, the development of centres in the vicinity of new hubs and the long-term adaptability of infrastructure and urban spaces must also be considered at the strategic level. These aspects are analysed in more detail below.

3.1 INTEGRATING ASSESSMENTS: AREAS OF OPERATION, MODES AND PERMISSIBLE SPEEDS

In view of a possibly long transition phase during which automated vehicles will only be used in appropriate areas, a spatial assessment and prioritization of areas of operation, modes and permissible speeds is necessary (see also Chap. 5 by Soteropoulos in this volume). In addition to the classic criteria for the use of different means of transport, the spatial effects of digital networking and automation must also be taken into account here; these are dealt with in existing studies in approaches for inner-city areas, whereby no systematic differentiation is made between urban space types or road typologies. González-González et al. (2020: 9), for example, describe a future vision in which so-called "Core Attractive Mixed-use Spaces (CAMS)" are designated in urban core areas. Within these zones, automated driving would only be permitted to a limited extent and public space would be reclaimed instead. The definition of spatially delineated zones of this kind has a fundamental influence on the location of spatial interfaces. Both within the area of operation and on its fringes, a need arises for transfer points and transition zones where various modes of transport are coordinated spatially, temporally and in terms of charges.

To identify possible areas of application under consideration of urban spatial effects, the suitability of new mobility concepts must be understood in terms of spatial structure. In transport planning, the following criteria are normally taken into account when deciding on the use of transport modes: cost-efficiency, functionality (travel speed and capacity), the hierarchy of services and suitability for use in settlements (Weidmann et al. 2011). To evaluate this last factor, population density and density of usage are named as the most important parameters, in addition to settlement sizes and core usage densities. According to Weidmann et al. (2011), these parameters can be used to determine which types of spatial structure favour the use of certain transport systems. For example, high-capacity transport systems (e.g. underground or commuter trains) are suitable in settlement areas with a high population density and differentiated corridors. On the other hand, dispersed areas do not normally justify rail transport in view of low demand or difficult topography, but instead require smaller vehicle units for connection and distribution systems along with feeder services linking to higher-ranking nodes (Weidmann et al. 2011). Here in particular, automated forms of service are seen as a solution, since higher economic efficiency and flexible forms of service can be assumed. Furthermore, in the integrated development of urban centres and mobility, the degree of functional blend is given high priority. In this connection, for example, potential for development within the catchment areas of public transport services can be sounded out, and the relationship between the surrounding area of a station and transport demand can be investigated (Bertolini 1999). The use of automated services and the planning of interfaces should therefore be evaluated in relation to structural factors, such as population and workplace densities or the diversity of surrounding usage, but also factors such as the availability of public transport and alternative modes.

In view of the above-mentioned suitability criteria, two major transformational trends should be taken into account in connection with automated driving that make an assessment of its appropriateness more complex. Firstly, the appearance of new players on the mobility market is increasingly blurring the distinction between public and private services (Lenz/Fraedrich 2015; Mitteregger et al. 2022: 44). A competitive situation is now arising between scheduled public services and flexible on-demand services from private providers, which in certain areas could lead to a decrease in demand density that is necessary for public transport operations. To maintain the public transport network as the backbone of mobility, areas of operation for automated mobility services should be designated on the basis of the existing public transport network, and settlement areas should be prioritized in which existing gaps can be closed or weak points strengthened in accordance with demand. Especially for areas of medium to low density, this can mean that the catchment areas of interurban transport hubs should be expanded to include the range of automated feeder services, with newly designated routes accorded priority as transformational and development areas. This also implies the need to spatially integrate new interfaces as transitional points and nodes, in order to facilitate the use of intermodal routes for users.

Secondly, the appropriateness of automated services should be assessed with a view to their possible impact on the public space, which can vary depending on the type of area and streetscape. For this purpose, streetscape compatibility can be analysed at the neighbourhood or district level (see Chap. 5 by Soteropoulos in this volume). It should be examined, for example, to what extent the approval of individual modes and vehicle speeds can lead to barrier effects in the form of greater difficultly for pedestrians and cyclists in crossing the road in certain street typologies (e.g. residential streets, district streets, access roads/main roads, connecting roads, etc.); the space requirements that could result from an increase in short stops and parking by vehicle fleets also need to be identified. Both of these factors should be assessed in relation to the existing demands for space usage on the part of surrounding functions. For example, an inner-city neighbourhood street with a concentration of retail and gastronomy services gives rise to different usage densities and spatial requirements (e.g. road-crossing opportunities, pavement cafés, lingering spaces, etc.) to a purely residential street or a main axis on the outskirts of a city. On this basis, it will have to be examined whether certain applications, such as automated delivery traffic, should be excluded either entirely from certain categories of road due to conflicts of use or only at certain times (see Chap. 7 by Leerkamp et al. in this volume). In this connection, it should be considered to what extent it could be expedient to allow platform-based ride-sharing services to operate in designated bus and taxi lanes. While this can serve as an incentive to use mobility services rather than private vehicles during the transition phase, in certain streetscapes it can cause obstruction to public transport vehicles and make crossing more difficult for pedestrians and cyclists.

Table 1: Integrating assessments: areas of application, modes and permissible speeds

	Spatially differentiated consideration
Urban space	In high-density areas it will be important to prioritize existing PT services, and only to supplement these with automated on-demand buses in selected peripheral areas and at off-peak times.
Suburban space	In suburban interurban areas, flexibly deployed forms of automated service could help to cover detailed accessibility throughout the area and create an alternative to the car (Mitteregger et al. 2022). There are greater opportunities for spatially and temporally bundling routes here than in rural areas.
Rural areas	Since transport routes in low-density settlement areas are spatially and temporally diverse, flexible forms of service with automated shuttle buses provide an opportunity to close existing gaps in the PT network and secure public connections (Jürgens 2020).

Source: the authors

3.2 EXTENDING THE STATION NETWORKS: ADDING STOPPING ZONES AND COLLECTION POINTS

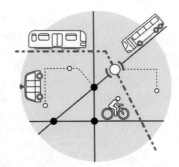

In view of the increasing mobility options, cities are called upon to coordinate various forms of service and to integrate collection points into their strategic network planning. A differentiation must be made here between classic creation of a network for local public transport, with individual lines integrated into a mostly hierarchical route network, and homogeneous network coverage with flexible (demand-oriented) forms of service (Schnieder 2018). The latter has the objective of offering direct connections between origins and destinations and accordingly requires a finely detailed distribution of collection points throughout the area. A possible approach has been investigated in a modelling study on the impact of flexible sharing services on the streetscape (ITF 2018). For an economic and commercial centre in an inner-city area of Lisbon it was established that in order to meet the demand, an even distribution of stopping zones would be required at places of origin and destinations throughout the road network. The levels and duration of congestion during the day were taken as indicators of transport demand (ITF 2018: 70). Accordingly, in this model stopping zones for boarding and alighting were located at the beginning and end of each block, in the middle of very long streets and at places of high transport demand (e.g. shopping centres or schools) in order to achieve the highest possible network coverage (Fig. 2).

Figure 2: Locations of pick-up and drop-off points for shared mobility services in Lisbon

Legend

☐ Area investigated
■ Pick-up and drop-off points
— Road network

0   200   400   600 m

Source: ITF (2018: 67)

Integration into existing station networks
Similarly to the areas of operation, the density of collection points must also be planned in accordance with the specific context. The various settlement structures and road typologies

must be taken into account here, along with utilization densities and availability of the public transport network. In dense settlement areas with well-developed public transport networks, integration into the existing network hierarchy is crucial, in addition to orientation towards source and destination locations and places of high demand (e.g. airports, business districts or shopping centres). The distance between stops influences not only the hierarchical allocation of functions within the overall transport system, but also potential usage rates. Dense station networks, such as that proposed for Lisbon with a spacing of 100 to 200 metres, thus tend to generate higher demand. When it comes to prioritizing environmentally compatible forms of mobility, a network concept with comparatively short distances between stops can definitely be seen in a critical light. In inner-city areas, this could lead to a weakening of local public transport and especially of active mobility, whereas in fringe locations and areas with low public transport coverage, it is advisable to seek ways of filling gaps in the network by means of densely planned collection points.

In inner cities, existing station networks in particular should therefore be extended to include stopping zones for new mobility services. These usually consist of higher-ranking stations (railway and underground/commuter rail stations), medium-sized and smaller nodes (tram stops and line intersections) and individual stops. At these locations, various interfaces are created that require integration in terms of urban development and settlement structure and can be fundamentally diverse in terms of design (Fig. 3 on the next page). In contrast to the development of rail transport in the 19th century and the construction of railway stations, this creates not so much a new "type" of infrastructure buildings, but rather the need to spatially and digitally connect the wide variety of mobility offers. Nevertheless, it cannot be ruled out that locations such as parking garages, which to date have been given little attention, will be transformed into new nodes through enriched usage.

The space requirements for the stopping zones and collection points of automated services can vary drastically depending on the type of use and demand. A study of Swiss regional railway stations found for example that the stopping zones for automated bus systems require a significantly smaller overall area than if automated private vehicles or car-sharing were to become established as feeders (Sinner et al. 2018). This is due on the one hand to the smaller number of vehicles that an automated bus system would require, and on the other hand to the possibility that scheduled arrival and departure times would allow stopping zones to be dimensioned in accordance with actual demand (ibid.). However, the amount of stopping and parking space identified for a system based on automated car-sharing hardly appears sustainable over the long term. The enormous space requirements would ultimately restrict the spatial opportunities for upgrading higher-ranking intersections as centres that can be reached on foot.

Extending networks by means of automated services
In suburban and rural or peripheral areas, which are often difficult to access due to low network coverage, automated forms of service enable existing station networks to be extended, whereby existing centre structures must be given special consideration in route and network planning. Rather than bypassing old town centres in favour of efficiency, a balance between provision of service and avoidance of barrier effects can help stabilize and upgrade existing structures (cf. Angélil et al. 2012). However, since in many places the demand for mobility fluctuates strongly depending on the time of day, it must be examined to what extent certain route sections can be served at specific times, and access points provided on a dynamic basis. For example, dedicated collection points can be served mainly during the day to facilitate bundling of trips, while convenient door-to-door service is offered at off-peak times. Wherever the public transport network is supplemented by means of demand-responsive automated shuttle buses or extended operating times, safe boarding and alighting as well as convenient transfer to a bicycle or pedelec must also be ensured. Integrated mobility management and cooperation between the public sector and private mobility service providers are essential prerequisites here.

Figure 3: Integration of spatial interfaces at various types of station, categorized according to traffic space type

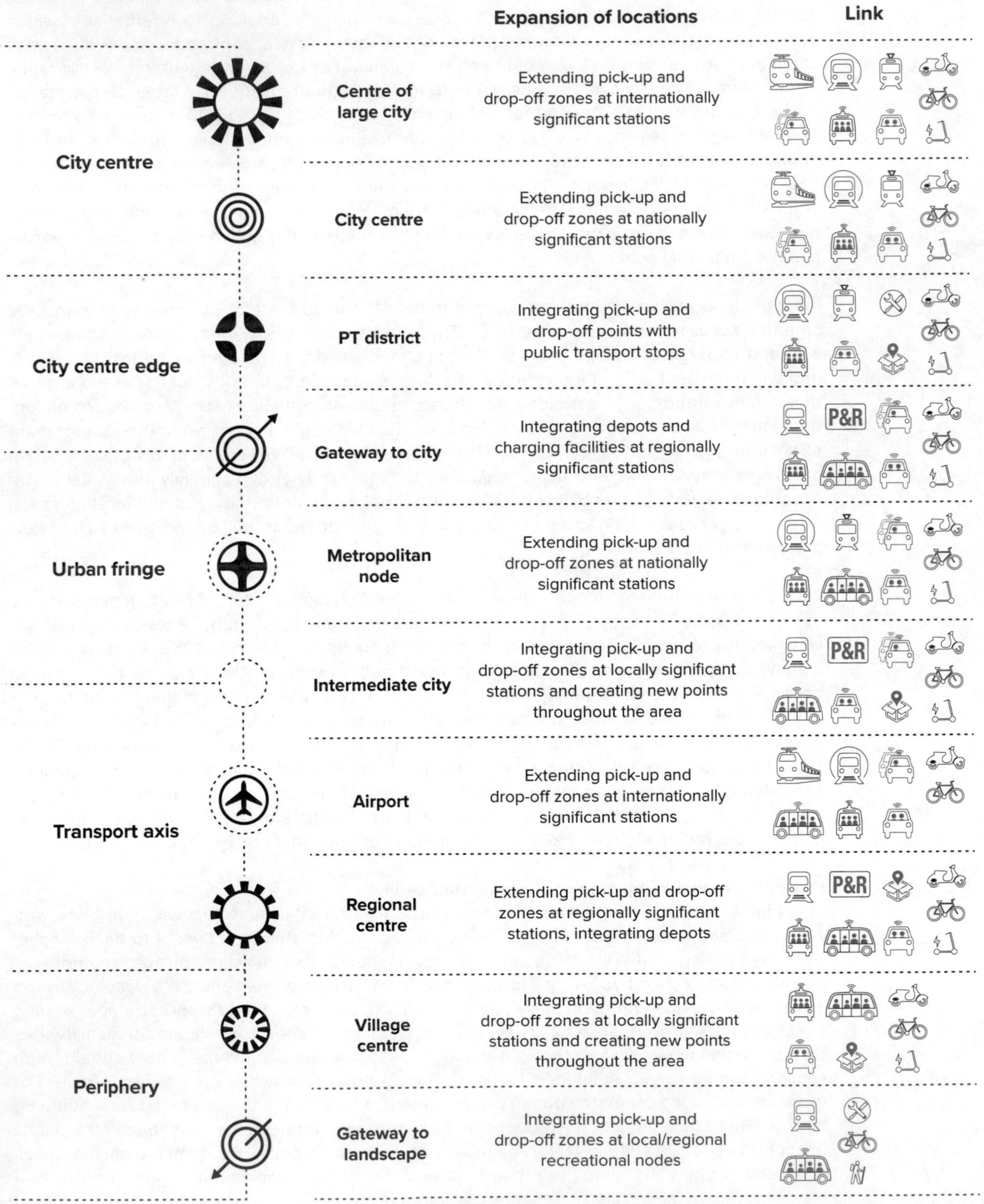

		Expansion of locations	**Link**
City centre	Centre of large city	Extending pick-up and drop-off zones at internationally significant stations	
	City centre	Extending pick-up and drop-off zones at nationally significant stations	
City centre edge	PT district	Integrating pick-up and drop-off points with public transport stops	
	Gateway to city	Integrating depots and charging facilities at regionally significant stations	
Urban fringe	Metropolitan node	Extending pick-up and drop-off zones at nationally significant stations	
	Intermediate city	Integrating pick-up and drop-off zones at locally significant stations and creating new points throughout the area	
Transport axis	Airport	Extending pick-up and drop-off zones at internationally significant stations	
	Regional centre	Extending pick-up and drop-off zones at regionally significant stations, integrating depots	
Periphery	Village centre	Integrating pick-up and drop-off zones at locally significant stations and creating new points throughout the area	
	Gateway to landscape	Integrating pick-up and drop-off zones at local/regional recreational nodes	

Source: the authors, based on Ram et al. (2013: 100) and Matthes/Gertz (2014: 61)

Table 2: Extending station networks: adding stopping zones and collection points

	Spatially differentiated consideration
Urban space	In dense settlement areas, stopping zones for automated forms of service should be integrated into the existing PT station network and new collection points only added where necessary.
Suburban space	The often inadequate connections to suburban areas calls for an attractive densification of the station network throughout an area. With automated neighbourhood buses, gaps between axes can be served tangentially on the basis of a finely meshed station network.
Rural areas	In rural or peripheral areas, remote settlement areas in particular should be made accessible by demand-oriented services. For an attractive design of automated services, stops and collection points should be planned in residential areas.

Source: the authors

3.3 STRENGTHENING CENTRALITIES: LINKING MODES OF USAGE, ROUTES AND DESTINATIONS

With a view to creating a "city and region of short distances" (Beckmann et al. 2011: 63), station networks of automated services should also be planned in an integrated manner with the development of mixed centres – especially since automated driving bears the risk of promoting land consumption through increasing urban sprawl due to the above-mentioned advantages of convenience. Risks of induced traffic (e.g. due to shifts in mobility patterns or newly acquired user groups) must be counteracted by traffic avoidance strategies (Hörl et al. 2019). Similarly to the extension of commuter train lines, which in the past often gave impetus to the inner development of areas close to stations (ARE 2004, Weidmann et al. 2011), the extent to which newly developed structures can be functionally enriched and developed as centres must also be considered in view of automated mobility networks. Functional densification in the vicinity of transport hubs can contribute to the upgrading of a location and provide an incentive for using sharing services. At the scale of a neighbourhood or district, it should be examined to what extent the potential for infrastructure to divide a neighbourhood can be mitigated and new spatial connections created. Particularly in suburban and rural-peripheral areas, where automated mobility is expected to bring about the greatest gains in accessibility, the planning of mobility nodes at the interfaces of settlement areas could provide impetus for the development of a settlement (cf. Kretz/Kueng 2016, Bruck 2019).

The development of centralities is also gaining significance in the form of attractive interchanges and acceptance of intermodal routes. In addition to short walking distances, an agglomeration of everyday usage in the vicinity of mobility nodes should also be taken into account. Time is saved when the purposes of individual trips, for example buying groceries and going to work, can be combined in an uncomplicated way, and the location also gains significance in the everyday networks of individual people (Kretz/Kueng 2016: 46).

By integrating a variety of different usages, a mobility hub can be enlivened and made experienceable at various different times of day (cf. ibid.). Opportunities for synergies also arise in view of the increasing trend towards using digital networking to transfer urban functional spaces and processes to outlying areas. The establishment of co-working spaces and of hybrid office and service offerings, for example, is gaining importance in the context of polycentric decentralization and increasingly flexible lifestyles; in view of automated mobility, this should be given consideration as an element of the integrated planning of new station networks, vibrant centres and strong settlement fringes.

Table 3: Strengthening centralities: linking modes of usage, routes and destinations

	Spatially differentiated consideration
Urban space	As a rule, areas of high settlement density already have a large number of centralities. In inner cities, it may therefore be necessary to relieve the burden on existing centres, or to interconnect them more effectively, by relocating collection points.
Suburban space	In suburban areas, the freeing up of parking spaces (see Chap. 15 by Mitteregger and Soteropoulos in this volume) can create spaces of opportunity to functionally remix the areas adjacent to mobility nodes, redensify them and enhance them with open spaces.
Rural areas	In addition to the challenge of separation by long distances, rural areas are often beset by a trend towards exodus and vacancy. Here too, the strategic upgrading of mobility nodes can provide an important impulse for small-scale centres.

Source: the authors

3.4 ENSURING ADAPTIVITY: OPENNESS TO FUTURE CHANGES IN USAGE

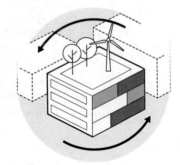

In view of the uncertainties surrounding future developments and the consequences of automated driving, adaptivity is an essential planning principle. This entails an unreserved openness to future developments and to the mutability of urban spaces and their modes of use (Taut 1977, Kretz/Kueng 2016). Since it is currently impossible to determine which scenarios and forms of automated mobility will prevail, the challenge consists in planning for various different future scenarios.

With regard to spatial interfaces of future mobility networks, this means that a high degree of adaptability or even deconstructability of mobility infrastructures, public spaces and building structures must be provided. It must be ensured, for example, that today's park-and-ride facilities and elevated garages can be converted for other uses in the future. Should a widespread shift in mobility to ride-sharing services and environment-friendly transport result in a reduction in parking space requirements, charging stations for electric fleets or decentralized logistics centres could be provided in these places (cf. Lewis/Anderson 2020: 104; Bienzeisler et al. 2019) – also because local distribution centres are an important prerequisite for delivery of goods over the last mile using environmentally friendly forms of mobility (González-González et al. 2020). In the long term, however, entirely new functions could also give rise to space requirements that cannot be anticipated from a current perspective. This requires space reserves in the vicinity of these charging stations, and leeway must be secured for future usage requirements and densification processes.

At the small-scale level of public space, structural and usage-related adaptability can be regulated by modular design or by ordinances based on time of day. By avoiding excessive determination of public spaces in terms of design, a flexible framework for designation of usage can be provided, with an interplay of permanent and temporary activities. For example, a streetscape can serve as a spatial extension of a schoolyard at certain times of day when driving is prohibited, or it can be made available for markets and cultural events on certain days of the week (Bendiks/Degros 2019). If opportunities for digital networking are then also integrated into a space, for example, stopping zones could be made available on a dynamic basis in line with demand; vehicles could be informed of this digitally, which could enable flexible routing during off-peak hours or facilitate alternative use of space at times of low demand.

Table 4: Securing adaptivity: open to future changes in use

	Spatially differentiated consideration
Urban space	Adaptable and open design of public spaces plays a significant role in view of the high pressure of use and low availability of space in inner-city areas. The variety of different requirements can only be accommodated by largely avoiding specific functional designation.
Suburban space	In view of the enormous consumption of resources in suburban areas, particular attention must be given to the adaptability of structures for multiple usage. This includes, for example, the future transformation of park-and-ride facilities, extensive parking areas and former railway installations.
Rural areas	In rural areas, there is potential for structural, functional and landscape reactivation. Vacant structures can be revitalized through improved accessibility and conversion and be made experienceable as elements of cultural landscapes.

Source: the authors

4. SMALL-SCALE DESIGN OF MULTIMODAL PUBLIC SPACES

Changes to public space require not only strategic framework conditions, but also the definition of clear-cut design approaches and goals. The ongoing transformation of mobility gives us cause to rethink the current distribution of public space usage with a view to more sustainable mobility. Rather than adapting urban space to the technological characteristics of automated vehicles, as is now under discussion with a view to installing uniform lane markings and sensors for example, municipalities can play a role in influencing the development of this technology by establishing spatial requirements and objectives. Traffic installations, open spaces and surrounding buildings are therefore seen as integral elements of public space, so that an integrated view can be taken of both the functionality of automated forms of service and approaches to their implementation.

In the short term, the changes with the most significant spatial impact mainly concern the organization of brief stopping and of transfer options in public spaces – firstly, because the already increasing demand for boarding and alighting options will probably continue to rise with the introduction of automated door-to-door mobility. Secondly, it is particularly significant from a municipal perspective that automated mobility services can make the greatest contribution to fortifying the public transport network when the use of feeder services reduces overall travel time in public transport (Sinner/Weidmann 2019). Quality of connection security is crucial not only for increasing the efficiency of the system, but also in order to make transfers more attractive to users (verkehrsplus 2015).

Since mobility options are expected to become increasingly diverse as a result of new technologies and business models, it will be advisable in future to differentiate between various forms of temporary stopping and to prioritize spaces accordingly. In light of this, a monomodal orientation of streets and public spaces no longer seems appropriate, and it is therefore up to the municipalities and planning departments to initiate the change from long-term parking to diverse usage options and flexible allocation of space.

In view of the increasing pressure of use, various methods of control are being discussed for the regulation of temporary parking in public spaces. Pricing strategies are repeatedly mentioned here as an obvious measure for prioritizing different modes of usage and imposing time limits (ITF 2018, NACTO 2019, Marsden et al. 2020). This raises the fundamental question of the usability of public space: this firstly relates to the question of who may charge fees for what purpose (cf. Marsden et al. 2020: 8); secondly, it must be fundamentally clarified which user groups would benefit from modifications to the streetscape and to what extent. For example, simply converting parking spaces on the street into stopping zones could lead to a significant loss of revenue for the public sector, while free access would bestow above-average benefits on private mobility service providers (ibid., Mitteregger et al. 2019).

In the following, we shall mainly focus on design measures, which along with fiscal instruments provide considerable scope of action for municipalities and city authorities to both regulate public spaces and make them available for a wide variety of activities. The focus is therefore on the qualification of spatial interfaces, i.e. places where different modes are interconnected and the spatial demands of new mobility services become apparent, and which – as spatial thresholds – influence the accessibility of surrounding neighbourhoods.

4.1 MULTIMODAL PUBLIC SPACES: UNBUNDLING AND BUNDLING OF MODES

The proliferation of mobility offers in the streetscape confronts planning with the familiar challenge of reconciling safety demands with those of quality of stay and ecological sustainability (green spaces, provision of shade, desealing of surfaces). Understanding this as a municipal design task is becoming increasingly urgent in view of the advance of digitization and automation in the sphere of mobility: firstly, since the automation of vehicles is being developed on the basis of existing road traffic regulations and the prevailing road traffic culture. Largely established rules of behaviour, such as the priority of modes, generally accepted speeds or individual vehicle use, will therefore be incorporated into future systems and reproduced by them. Secondly, municipalities have a direct influence on urban design and can therefore control the speed, flow and direction of traffic at the level of the streetscape by means of spatial measures (NACTO 2019). In terms of multimodal public spaces, the most important consideration is to ensure the safe coexistence of different modes. Streetscapes hitherto dominated by parked cars should be reassigned in favour of pavements, cycle paths and green spaces, and be extended with the addition of designated loading areas for short-term stops.

Depending on street typology and surrounding usage, the designation of loading areas can serve various purposes (e.g. public transport stops, boarding and alighting points for mobility services and private vehicles or delivery of goods) and they can take various forms. Coordination with surrounding usages is necessary, particularly since these place differing demands on public space, often depending on the time of day. When planning specific locations, or restrictions on use at certain times of day, consideration must therefore be given to local peaks of activity such as the start of school in the morning, in addition to mobility needs. On narrow roads and in traffic-calmed areas such as shared spaces, brief stops for the purpose of boarding and alighting may even be permissible on the carriageway itself.

While studies have shown that this may impede the flow of traffic and result in longer travel times for automated mobility services, the reduced attractiveness of motorized transport may favour active mobility (Elvarsson 2017, ITF 2018). Alternatively, stopping zones can be located at the side of the road and be identified by lane markings or a distinctive road surface. Such designation promotes the spatial concentration of short-term stops and safe interaction with pedestrian and bicycle traffic. On roads with wider road cross sections, dedicated stopping bays can also be provided, which would require vehicles to depart from the lane of traffic and would have the least impact on traffic flow (ITF 2018: 70). Depending on the user environment, lay-bys can be planned as an element of multifunctional roadside strips; this would facilitate connection to the pedestrian and cycle network and the integration of supplementary mobility infrastructure, green spaces and amenity areas.

In the planning of stopping zones, attention should also be given to the existing public transport station network. Making intermodal transfer and access to other modes as low-threshold as possible calls for an efficient configuration of spaces, with short distances between stopping zones for automated mobility services, public transport and micro-mobility offers (e.g. e-scooters or cargo bikes). To this end, a number of functional requirements must be considered (verkehrplus 2015):

- simple orientation,

- clarity,

- a logical sequence of facilities such as ticket machines, waiting areas and information services,

- several lines should use the same stops or stopping zones,

- several providers should use the same stops or stopping zones, and

- transfer points should be close together.

Table 5: Multimodal public spaces: unbundling and bundling of modes

Spatially differentiated consideration	
Urban space	The pressure on public spaces is highest in dense settlement areas. It is important here to replace street-bound parking spaces with stopping zones, encounter spaces and green areas. The different usage requirements must be coordinated with the surroundings and activities, and stopping zones should be integrated into existing public transport stations where possible.
Suburban space	In suburban areas, extension of the cycling infrastructure should also be considered when planning stopping zones for automated mobility services. Both modes of transport can fulfil important feeder functions for higher-ranking public transport media and require corresponding infrastructure facilities and enhanced visibility.
Rural areas	In rural areas, stopping points for automated mobility services should be planned as part of interconnected pavement and cycle path networks and be made highly visible as places with quality of stay.

Source: the authors

4.2 FLEXIBLE DIMENSIONING OF SPACES: BY DEMAND OR BY TIME OF DAY

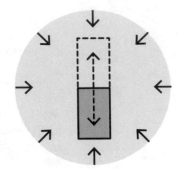

Determining the appropriate amount of space for stopping zones requires demand modelling; space can be designated for example on the basis of activity patterns, i.e. the traffic generation rates of surrounding usage and the expected mobility behaviour in relation to structural parameters (e.g. population and employment density, accessibility of public transport or modal split) of the respective catchment area (Schwartz 2017, Bruns et al. 2018). However, it can be assumed that automation of vehicles will encourage the mobilization of previously mobility-restricted groups of people and can thus trigger a latent increase in traffic volume. Basic assumptions, e.g. that a stopping area can serve up to 80 boarding and alighting operations per hour, each lasting 45 seconds (ITF 2018: 66), must be verified and adjusted on the basis of local demand surveys. The dimensioning of spaces can also be approached differently, depending on their position along the course of a road. In the case of lay-bys, for example, additional areas for boarding and alighting must be provided, although these require less space if they are suitably located on the near or far side of an intersection (Fehr & Peers 2019: 19).

As studies on the requirements of automated services for parking space have shown, the space demands for short waiting times between individual trips must also be taken into account in addition to stopping zones for boarding and alighting (Kondor et al. 2020: 17). Especially as an increased number of empty trips are to be avoided, these can constitute a decisive share of the future parking space requirements of automated fleets (ibid.). It must be taken into account that the required space can increase with settlement and demand density and the maximum waiting time for passengers, since the number of vehicles waiting at the same time could then also increase. Among other factors, the location density of stopping zones and the catchment area of mobility fleets are decisive here (ibid.). Both of these factors can be regulated by means of various measures on the part of the public sector, such as the definition of service areas, usage fees or a maximum number of usable parking spaces.

It must also be taken into consideration that space requirements can vary depending on the time of day and network capacity. With an appropriate database and real-time data collection, stopping zones could be designated dynamically in future and the public space could be used outside of peak times for leisure and active mobility. This would particularly benefit places with a high usage density such as schools, office buildings or cultural venues, where temporary extension areas could provide relief at times of peak activity (Fehr & Peers 2019). Stopping areas for delivery transport, mobile services or outdoor catering (food trucks) could also be restricted to specific times of day or night in order to make the required areas available in the most resource-efficient way (NACTO 2019: 120). Access to stopping zones can be regulated by means of usage fees that can vary depending on demand, type of use, occupancy rate and propulsion system.

The potential reclamation of public spaces ultimately depends on the prioritization of space requirements. Cities and municipalities will have to clarify which mobility types and usage groups are to be given priority in future and to what extent priority should be given to unhindered stopping opportunities for ambulances or ride services for mobility-impaired persons in addition to public transport stops. With regard to a comprehensive reallocation of parking spaces, public acceptance will be decisive. Future planning processes will have to give particular consideration to negotiation of usage claims between public and private mobility providers, residents and local businesses.

Table 6: Flexible dimensioning of spaces: dependent on demand or on time of day

	Spatially differentiated consideration
Urban space	Particularly in inner-city areas, dynamic stopping zones designated on the basis of real-time data could provide relief in places of high usage density or before/after major events as a supplement to static spaces.
Suburban space	In view of the lower usage densities, the provision of static stopping zones is expected to be sufficient in suburban areas. Here too, however, prioritization according to time of day must be coordinated with the respective surrounding usage.
Rural areas	Due to the low residential density of rural areas, low space requirements for boarding and alighting can be expected, so more space can be allocated to supplementary functions (e.g. passenger information, parking facilities or parcel delivery) as standard.

Source: the authors

4.3 SMOOTH TRANSITIONS: PERMEABLE, WITH QUALITY OF STAY

Since automated vehicles could give rise to increased space requirements and spatial barrier effects (see Chap. 5 by Soteropoulos in this volume), specific design approaches are needed to ensure that public spaces retain their permeability and quality of stay in future. The design of "smooth transitions" (Gehl 1971: 183), with a view to a relationship between order (in the sense of traffic safety) and freedom (in the sense of unregulated lingering and subjective appropriation) that is appropriate to the situation, is a key task of urban planning (cf. Bormann 2014).

On a small scale, this first of all concerns the need for pedestrians to cross the road, which is proportionate to the activity density of the surrounding area – the concentration of trade, gastronomy and public facilities. To enable safe interaction with pedestrian and bicycle traffic, low speeds for motorized vehicles should be specified, depending on the contextual conditions and the location within the transport network. In addition, measures appropriate to the specific road typology, such as narrower lane widths, flexible traffic islands and multipurpose lanes or different road surface materials, can help to ensure that streetscapes can be crossed and visibility spaces are created (Larco/Tierney 2020, NACTO 2019). Consequently, stopping zones and lay-bys should also be designed with barrier-free accessibility from several sides. Last but not least, large-area crossing opportunities as shared spaces enhance the permeability of a public space; they are particularly suitable in inner-city zones and in areas with high centre quality, which normally have an above-average need for crossing opportunities (Ghielmetti et al. 2017). Other factors that favour the designation of shared spaces include narrow road cross sections and low traffic volumes. When the permeability of a public space is ensured for pedestrians, this can have many positive effects: shorter distances for pedestrians, increased pedestrian frequency along surrounding ground-floor installations, higher continuity of traffic and the absence of negative effects on road safety (cf. Ghielmetti et al. 2017: 14).

In order not to reduce spatial interfaces to mere functional transit points, they should also be seen as spaces to linger and interact. In addition to spatial accessibility in the form of direct access and exit routes, this also calls for the provision of attractive rest areas where opportunities for encounter can arise. According to the narrative of the IT and automotive industries, the digital networking and automation of mobility systems serve among other things to establish "seamless mobility". This describes a future scenario of integrated processes that facilitate operation, along with door-to-door mobility for maximum convenience.

If a public space is oriented purely towards fast coming and going, however, activities of longer duration that significantly contribute to its lively character would disappear (Gehl 1971). Smooth transitions in this sense also mean reduced logic of efficiency and more deceleration, so that different activities can take place simultaneously in the vicinity of stopping zones and gathering points: waiting, sitting, borrowing, booking, picking up/dispatching, repairing, informing, learning, exchanging, meeting, refreshing. The stopping zone or collection point thus serves not only as a node in the transport network, but also as a space of social encounter (Soike et al. 2019).

The extent to which the social demands of stay and communication are taken into account ultimately influences quality of life in public spaces (cf. Weidmann et al. 2011). In view of the trend towards de-spatialization associated with digitalization, this aspect is becoming increasingly important. Contrary to some fears that the liveliness of public spaces could lose significance due to the internet, it is precisely phases of technological acceleration – such as the exceptional situation triggered by the Covid-19 pandemic in early 2020 – which make it clear that social interaction and movement within built space are an elementary counterpoint to the navigation of digital spheres (Banerjee 2001). This ultimately lends increased importance to design quality, which encourages lingering, safe play and outdoor encounter.

Table 7: Smooth transitions: permeable, with quality of stay

Spatially differentiated consideration	
Urban space	In dense settlement structures, the most important factor is the permeability of public spaces. Maintaining neighbourly relations and a high quality of stay for public spaces should be prioritized in view of increasing human-machine interaction.
Suburban space	In suburban areas too, public spaces should be upgraded as shared spaces to facilitate crossing by pedestrians. Diversity of usage and quality of stay can be enhanced with low speeds for automated vehicles and ground-level design.
Rural areas	Especially in rural areas, the design quality of public spaces is a key element in the revitalization of town centres. Smooth transitions between the road and surrounding usage, stopping zones flush with the carriageway and integrated opportunities to linger can significantly enhance the character of a town centre.

Source: the authors

4.4 AMENITIES AND ADDRESS GENERATION: INTERPLAY WITH THE SURROUNDINGS

Depending on location, usage requirements and the space available, various amenities and items of equipment need to be integrated at spatial interfaces. Stopping zones can also be supplemented with a number of structural fixtures such as seats, roofs, light and information columns, parking facilities for bicycles and scooters, and charging stations for electric vehicles. A modular design of these elements can promote long-term adaptability (Lehmann 2011). It is important that the bundling of mobility offers can contribute to the orderliness of stationary traffic and improve quality of stay. To enhance the public visibility and acceptance of mobility offers, existing concepts for multimodal mobility points also rely on a uniform system of identification, with distinctive colouring and signage for the various stations (Schlump et al. 2014: 81, Stadt Wien 2018). When establishing dedicated stopping zones and collection points for automated services, attention will likewise have to be given to attractive equipment and address generation in order to promote public perception and user acceptance.

For an assessment of the suitability and structural integration of "function-supporting infrastructure" (Schlump et al. 2014: 10), coordination with the immediate surroundings is crucial: adjacent public spaces (e.g. parks, squares or markets) and ground-floor usage of the surrounding buildings (e.g. residential, education, commercial or tourism) must be taken into account in a strategic selection of equipment to be installed. Especially in inner-city areas, pavements not only enable active mobility and access to other modes, but also function as public spaces connecting adjacent ground-floor zones. As Jane Jacobs stated (1961), public space is enlivened by the public character of accompanying usage (e.g. businesses, service facilities, cafés and restaurants or cultural venues). A positive relationship between indoor and outdoor areas can influence both the sense of security and the diversity of activity in public spaces. Structural fixtures should therefore be located in public spaces in such a way as to contribute to their clarity and promote permeability for pedestrians and cyclists (Ghielmetti et al. 2017).

To relieve the burden on public spaces, it should also be examined to what extent adjacent ground-floor zones can be brought into play as elements of mobility points. Especially garage areas facing the streetscape or vacant premises would lend themselves to mobility-promoting facilities such as charging stations, repair workshops or rental services. Direct access allows activities to flow out into and enrich the public space, which could thus be kept free from the spatial requirements of new mobility and delivery concepts and instead be designed so as to be open to a wide variety of uses.

Table 8: Amenities and address generation: interplay with the surroundings

	Spatially differentiated consideration
Urban space	Especially inner-city areas are faced with the challenge of integrating heterogeneous usage demands into the public space. In view of the high settlement density, installations such as e-charging stations or parcel stations should ideally be relocated to adjacent ground-floor areas to relieve the burden on streetscapes.
Suburban space	Suburban areas offer potential for creating new places of encounter by enriching the usage of stopping zones and collection points (e.g. rental stations or local supply depots), as long as these are made sufficiently experienceable by means of spatial design.
Rural areas	Rural areas can also benefit from enhanced visibility of gathering points; architectural-cultural accents in particular can serve here to generate a sense of local identity (Kaltenbach 2013). These not only have a distinctive visual impact on the landscape, but can also promote tourist activity and thus have a stabilising effect.

Source: the authors

5. SUMMARY AND OUTLOOK

Even with automated driving, the quality of life of public spaces is the outcome of a process of design in which the areas of application and objectives of new mobility concepts are determined (cf. Weidmann et al. 2011). The ongoing heterogenization of mobility requirements and the urgency to respond to climate and health crises make it clear that appropriate consideration of the partly conflicting demands for usage in public space is a key challenge for urban planning. The phase of transition towards automated mobility adds complexity to this formative remit: firstly because the course to be taken in actual technological introduction currently remains unclear, and secondly because the spatial effects of the transition phase (e.g. changes in parking space requirements) have hardly been researched on the basis of scenarios of mixed traffic, in which the streetscape is shared by pedestrians, cyclists and (automated) vehicles (Zhang/Wang 2020). Nevertheless, addressing the potential impact of automated driving at an early stage offers the opportunity to call into question and re-evaluate the planning principles and space requirements that underlie the current development of public spaces and are about to shape the future course of automated mobility.

The observations presented here make it clear that urban planning will be even more challenged in the future to coordinate the control approaches of traffic management with the control and design approaches of urban development, and to take their interaction into consideration. It is essential in this connection to evaluate the developments in mobility from the perspective of public space, as the design quality of spatial interfaces has an impact on mobility behaviour, urban profile and settlement development – especially since the extent of spatial change can vary depending on urban space, density, traffic volumes and accessibility (Larco/Tierney 2020). This calls for spatial differentiation in the course of further debate and research, for example on the basis of types of urban space and street typologies, with specific spatial case studies in urban to rural areas.

With a view to the future of automated driving, a strategic definition of areas of application is required, along with spatially differentiated approval of transport modes and travel speeds. Urban planning therefore has the important task of refining possible approaches by evaluating urban compatibility and spatial added value. There is also a need for public debate on the matter of how to achieve flexible and temporary use of public spaces. To what extent can the provision of infrastructure for shared mobility services and future automated transport already be tested today in pilot projects? What types of usage should be given priority in view of the increasing diversity of space requirements? Under what conditions can a dynamic distribution of usage be advantageous? And to what extent can this issue be integrated into existing planning strategies and concepts so that medium-term projects can be aligned with existing development initiatives? The task at hand for municipalities consists in coordinating the various time horizons and arriving at a suitable balance between openness and strategic control, with a view to reinforcing the character of public spaces that incorporate automated driving as versatile and transformable living spaces.

REFERENCES

Angélil, M., K. Christiaanse, V. M. Lampugnani, C. Schmid and G. Vogt 2012. "Urbane Potenziale und Strategien in metropolitanen Territorien. Am Beispiel des Metropolitanraums Zürich", National Research Programme NRP65 – New Urban Quality, ETH Zurich: Department of Architecture. https://www.christiaanse.arch.ethz.ch/upload/up.pdf (14/5/2020).

ARE (Federal Office for Spatial Development) 2004. "Räumliche Auswirkungen der Zürcher S-Bahn – eine ex-post Analyse. Zusammenfassung". Bern: Federal Department of the Environment, Transport, Energy and Communications DETEC. Download at https://www.are.admin.ch/are/de/home/mobilitaet/programmes-and-projects/raeumliche-auswirkungen-der-verkehrsinfrastrukturen/raeumliche-auswirkungen-der-zuercher-s-bahn.html (20/10/2020).

Arndt, W.-H., and F. Drews 2019. "Mobilität nachhaltig planen. Erfolge und Hindernisse in deutschen Städten – Ergebnisse einer Umfrage zu kommunalen Verkehrsentwicklungsplänen". Special publications. German Institute of Urban Affairs. https://difu.de/publikationen/2019/mobilitaet-nachhaltig-planen (14/5/2020).

Backhaus, W., S. Rupprecht and D. Franco 2019. "Practitioner Briefing: Road vehicle automation in sustainable urban mobility planning", ed. by Rupprecht Consult – Research & Consulting. https://www.eltis.org/sites/default/files/road_vehicle_automation_in_sustainable_urban_mobility_planning_0.pdf (14/5/2020).

Banerjee, T. 2001. "The future of public space: Beyond invented streets and reinvented places", in *Journal of the American Planning Association* (67) 1, 9–24. DOI: 10.1080/01944360108976352.

Banister, D. 2008. "The sustainable mobility paradigm", in *Transport Policy* (15) 2, 73–80.

Beckmann, K. D., J. Gies, J. Thiemann-Linden and T. Preuß 2011. "Leitkonzept – Stadt und Region der kurzen Wege. Gutachten im Kontext der Biodiversitätsstrategie", Texte 48, Sachverständigengutachten. Dessau-Roßlau: Federal Ministry for the Environment, Nature Conservation and Nuclear Safety. www.umweltbundesamt.de/sites/default/files/medien/461/publikationen/4151.pdf (14/5/2020).

Bendiks, S., and A. Degros 2019. *Traffic Space is Public Space. A Handbook for Transformation*, 1st ed. Zurich: Park Books.

Bertolini, L. 1999. "Spatial Development Patterns and Public Transport: The Application of an Analytical Model in the Netherlands", in *Planning Practice and Research* 14, 199–210. DOI: 10.1080/02697459915724.

Bienzeisler, B., S. Bengel, M. Handrich and S. Martinetz 2019. "Die digitale Transformation des städtischen Parkens. Eine Analyse der Veränderung des kommunalen Parkraummanagements vor dem Hintergrund der Herausforderungen einer Verkehrswende". Stuttgart: Fraunhofer IAO – Institute for Industrial Engineering. http://publica.fraunhofer.de/eprints/urn_nbn_de_0011-n-5381331.pdf (14/5/2020).

Bormann, O. 2014. "Aktuelle Verkehrslage – Von der Rückgewinnung urbaner Infrastruktur", in *Architecture in Context*, ed. by K. von Keitz and S. Voggenreiter. Berlin: jovis, 96–109.

Bremer, S. 2017. "Kommunale Mobilitätspläne und ihre Umsetzung", presentation, Zukunftsnetzwerk Mobilität NRW. 16/1/2017. https://mobilitaetsmanagement.nrw.de/sites/default/files/downloads/05_bremer_mobilitaetsplaene.pdf (14/5/2020).

Bruck, E. M. 2019. "Automatisierte Mobilitätsdienste als Wandlungsimpuls für suburbane Räume?", in *Broadacre City 2.0 – postfossil. An urbanistic scenario for 2050*, ed. by J. Fiedler. Graz: Haus der Architektur.

Bruns, F., B. Tasnady, N. de Vries, N. Frischknecht, E. Selz, S. Grössl and M. Berger 2018. "Verfahren und Kennwerte zur Abschätzung von Verkehrswirkungen", research project SVI 2014/005 on commission from the Swiss Association of Transport Engineers and Experts (SVI). Bern: Federal Department of the Environment, Transport, Energy and Communications DETEC. http://www.yverkehrsplanung.at/images/Projektauswahl/SVI_2014_005_Schlussbericht_2018-09-18.pdf (22/10/2020).

Cavoli, C., B. Phillips, T. Cohen and P. Jones 2017. "Social and behavioural questions associated with Automated Vehicles. A Literature Review." London: UCL – University College, Department for Transport. https://www.ucl.ac.uk/transport/sites/transport/files/social-and-behavioural-literature-review.pdf (14/5/2020).

Cohen, T., and C. Cavoli 2019. "Automated vehicles: Exploring possible consequences of government (non)intervention for congestion and accessibility", in *Transport Reviews* (39) 1, 129–151.

Elvarsson, A. B. 2017. "Modelling Urban Driving and Stopping Behavior for Automated Vehicles, Semester Project". Zurich: ETH, IVT – Institute of Transport Planning and Systems. https://ethz.ch/content/dam/ethz/special-interest/baug/ivt/ivt-dam/publications/students/501-600/sa597.pdf (15/10/2020).

Erhardt, G. D., S. Roy, D. Cooper, B. Sana, M. Chen and J. Castiglione 2019. "Do transportation network companies decrease or increase congestion?", in *Science Advances* (5) 5, eaau2670. DOI: 10.1126/sciadv.aau2670.

Fehr & Peers 2019. "Cincinnati Curb Study". www.fehrandpeers.com/curbs-of-the-future/ (14/5/2020).

Gavanas, N. 2019. "Autonomous Road Vehicles: Challenges for Urban Planning in European Cities", in *Urban Science* (3) 2, 61. DOI: 10.3390/urbansci3020061.

Gehl, J. 2012. *Living between houses. Concepts for public space.* Berlin: jovis.

Ghielmetti, M., R. Steiner, J. Leitner, M. Hackenfort, S. Diener and H. Topp 2017. "Flächiges Queren in Ortszentren – Langfristige Wirkung und Zweckmässigkeit", research project SVI 2011/023 on commission from the Swiss Association of Traffic Engineers and Experts (SVI). Bern: Federal Department of the Environment, Transport, Energy and Communications DETEC. Download at duct/24100/?q=-fl%C3%A4chiges%20queren&tx_solr%5Bfilter%5D%5B0%5D=facet_212_stringM%253AAktiv&tx_solr%5Bpage%5D=0&cHash=6b09fe1968d32b1011e3865d-5564cc4a (14/5/2020).

González-González, E., S. Nogués and D. Stead 2020. "Parking futures: Preparing European cities for the advent of automated vehicles", in *Land Use Policy* 91, 104010. DOI: 10.1016/j.landusepol.2019.05.029.

Graehler, M., R. A. Mucci, and G. D. Erhardt 2019. "Understanding the Recent Transit Ridership Decline in Major US Cities: Service Cuts or Emerging Modes?" Conference: Transportation Research Board Annual Meeting, Washington, DC. https://www.researchgate.net/publication/330599129_Understanding_the_Recent_Transit_Ridership_Decline_in_Major_US_Cities_Service_Cuts_or_Emerging_Modes (5/14/2020).

Greenblatt, J. B., and S. Shaheen 2015. "Automated Vehicles, On-Demand Mobility, and Environmental Impacts," in *Current Sustainable/Renewable Energy Reports* (2) 3, 74–81. Download at https://link.springer.com/article/10.1007/s40518-015-0038-5 (14/10/2020).

Groth, S. 2019. "Multioptionalität: Ein neuer ('alter') Terminus in der Alltagsmobilität der modernen Gesellschaft?", in *Raumforschung und Raumordnung/Spatial Research and Planning* (77) 1, 17–34.

Herget, M., F. Hunsicker, J. Koch, B. Chlond, C. Minster and T. Soylu 2019. "Ökologische und ökonomische Potenziale von Mobilitätskonzepten in Klein- und Mittelzentren sowie dem ländlichen Raum vor dem Hintergrund des demographischen Wandels", Texte 14. Dessau-Roßlau: Umweltbundesamt. https://www.umweltbundesamt.de/en/publikationen/oekologische-oekonomische-potenziale-von (14/5/2020).

Hörl, S., F. Becker, T. Dubernet and K. W. Axhausen 2019. "Induzierter Verkehr durch autonome Fahrzeuge: Eine Abschätzung", research project SVI 2016/001 on commission from the Swiss Association of Traffic Engineers and Experts (SVI). Bern: Federal Department of the Environment, Transport, Energy and Communications DETEC. https://ethz.ch/content/dam/ethz/special-interest/baug/ivt/ivt-dam/vpl/reports/1401-1500/ab1433.pdf (18/8/2020).

ITF (International Transport Forum) 2018. "The Shared-Use City: Managing the Curb", Corporate Partnership Board Report. https://www.itf-oecd.org/sites/default/files/docs/shared-use-city-managing-curb_3.pdf (14/5/2020).

Jacobs, J. 1961. *The Death and Life of Great American Cities.* New York: Random House.

Jin, S. T., H. Kong, R. Wu and D. Z. Sui 2018. "Ridesourcing, the sharing economy, and the future of cities", in *Cities* 76, 96–104.

Jürgens, L. 2020. "Konnektivitätsveränderungen im ÖPNV-Netz durch die Einführung eines autonomen *Shuttlebusses*", in *Autonome Shuttlebusse im ÖPNV: Analysen und Bewertungen zum Fallbeispiel Bad Birnbach aus technischen, gesellschaftlicher und planerischer Sicht*, ed. by A. Riener, A. Appel, W. Dorner, T. Huber, J. C. Kolb and H. Wagner. Berlin/Heidelberg: Springer Vieweg, 39–54.

Kaltenbach, F. 2013. "Typologie: Vom Verkehrsknoten zur Drehscheibe – Haltestellen für integrierte Mobilitätskonzepte", DETAIL 9. Download at https://inspiration.detail.de/typologie-vom-verkehrsknoten-zur-drehscheibe-haltestellen-fuer-integrierte-mobilitaetskonzepte-107171.html?lang=en (14/5/2020).

Kondor, D., P. Santi, D.-T. Le, X. Zhang, A. Millard-Ball and C. Ratti 2020. "Addressing the 'minimum parking' problem for on-demand mobility". https://arxiv.org/pdf/1808.05935.pdf (14/10/2020).

Konrad, K., and S. Groth 2019. "Consistency or contradiction? Mobility-Related Attitudes and Travel Mode Use of the Young New Generation", in *Raumforschung und Raumordnung | Spatial Research and Planning* (77) 6, 1–17.

Kretz, S., and L. Kueng (eds.). 2016. *Urbane Qualitäten – Ein Handbuch am Beispiel der Metropolitanregion Zürich*. Zurich: Edition Hochparterre.

Larco, N., and G. Tierney 2020. "Impacts on Urban Design", in *Multilevel Impacts of Emerging Technologies on City Form and Development*, ed. by Howell, A., H. Tan, A. Brown, M. Schlossberg, J. Karlin-Resnick, R. Lewis, M. Anderson, N. Larco, G. Tierney, J. Carlton, J. Kim and B. Steckler. Eugene, OR: Urbanism Next Center, University of Oregon, 115–141. https://cpb-us-e1.wpmucdn.com/blogs.uo-regon.edu/dist/f/13615/files/2020/01/NSF-Report_All-Chapters_FINAL_013020.pdf (15/10/2020).

Lehmann, T. 2011. "Der Bahnhof der Zukunft – Alternativen zum traditionellen Bahnhofsempfangsgebäude | Entwicklung eines modularen Entréesystems für kleine und mittlere Bahnhöfe", PhD thesis, TU Berlin, Faculty VI – Planning Building Environment. DOI: 10.14279/depositonce-2920.

Lenz, B., and E. Fraedrich 2015. "Neue Mobilitätskonzepte und autonomes Fahren: Potenziale der Veränderung", in *Autonomous Driving: Technical, Legal and Societal Aspects*, ed. by M. Maurer, J. C. Gerdes, B. Lenz and H. Winner. Berlin/Heidelberg: Springer Vieweg, 175–195.

Leszczynski, A., and R. Kitchin 2019. "UBER. The Seduction of UberCity", in *How to Run a City like Amazon, and Other Fables*, ed. by M. Graham, R. Kitchin, S. Mattern, and J. Shaw. London: Meatspace Press, 1179–1195.

Lewis, R., and M. Anderson 2020. "Impacts on Land Use", in *Multilevel Impacts of Emerging Technologies on City Form and Development*, ed. by Howell, A., H. Tan, A. Brown, M. Schlossberg, J. Karlin-Resnick, R. Lewis, M. Anderson, N. Larco, G. Tierney, J. Carlton, J. Kim and B. Steckler. Eugene, OR: Urbanism Next Center, University of Oregon, 97–113. https://cpb-us-e1.wpmucdn.com/blogs.uoregon.edu/dist/f/13615/files/2020/01/NSF-Report_All-Chapters_FINAL_013020.pdf (14/5/2020).

Marsden, G., I. Docherty and R. Dowling 2020. "Parking futures: curbside management in the era of 'new mobility' services in British and Australian cities", in *Land Use Policy* 91, 104012. DOI: 10.1016/j.landusepol.2019.05.031.

Mitteregger, M., A. Soteropoulos, J. Bröthaler and F. Dorner 2019. "Shared, Automated, Electric: the Fiscal Effects of the 'Holy Trinity'", *Proceedings of the 24th REAL CORP, International Conference on Urban Planning, Regional Development and Information Society*, 2–4/4/2019, Karlsruhe.

Mitteregger, M., E. M. Bruck, A. Soteropoulos, A. Stickler, M. Berger, J. S. Dangschat, R. Scheuvens and I. Banerjee 2022. *Connected and Automated Driving: Prospects for Urban Europe*, trans. M. Slater and N. Raafat. Berlin: Springer Vieweg. DOI: 10.1007/978-3-662-64140-8.

NACTO (National Association of City Transportation Officials) 2017. "Curb Appeal. Curbside Management Strategies for improving transit reliability". https://nacto.org/wp-content/uploads/2017/11/NACTO-Curb-Appeal-Curbside-Management.pdf (14/5/2020).

NACTO 2019. "Blueprint for Autonomous Urbanism: Second Edition". https://nacto.org/publication/bau2/ (14/5/2020).

Nehrke, G., and W. Loose 2018. "Nutzer und Mobilitätsverhalten in verschiedenen CarSharing-Varianten", project report. Berlin: Bundesverband CarSharing e.V.

orange edge 2016. "Klimaschutzteilkonzept Mobilität. Stadt Königs Wusterhausen: Endbericht". https://www.koenigs-wusterhausen.de/817028/KW-Endbericht.pdf (22/10/2020).

Papa, E., and A. Ferreira 2018. "Sustainable Accessibility and the Implementation of Automated Vehicles: Identifying Critical Decisions", in *Urban Science* (2) 1, 5.

Ram, M., Jaffri, S., Gerretsen, P., Rigter, D., Chorus, P., and M. Wiers-Faver Linhares 2013. *Maak Plaats! Werken aan knooppunt ontwikkeling in Noord-Holland*. Rotterdam: Vereniging Deltametropool, provincie Noord-Holland.

Ritter, E.-H. 2017. "Strategieentwicklung heute. Zum integrativen Management konzeptioneller Politik", in *PNDonline* 1, 12. http://archiv.planung-neu-denken.de/images/stories/pnd/dokumente/pnd-online_2007-1.pdf (14/5/2020).

Ruchinskaya, T., K. Ioannidis and K. Kimic 2019. "Revealing the Potential of Public Places: Adding a New Digital Layer to the Existing Thematic Gardens in Thessaloniki Waterfront", in *CyberParks – The Interface Between People, Places and Technology: New Approaches and Perspectives*, ed. by C.

Smaniotto Costa, I. Šuklje Erjavec, T. Kenna, M. de Lange, K. Ioannidis, G. Maksymiuk and M. de Waal. Cham: Springer International Publishing, 181–195.

SAE International 2018. "Taxonomy and Definitions for Terms Related to Driving Automation Systems for On-Road Motor Vehicles – J3016", 15/6/2018. www.sae.org/standards/content/j3016_201806/ (20/4/2020).

Schaller, B. 2017. "Empty Seats, Full Streets. Fixing Manhattan's Traffic Problem." http://schallerconsult.com/rideservices/emptyseats.pdf (14/5/2020).

Schlump, C., T. Wehmeier, B. Helff, G. Reesas, H. Wohltmann, T. Schäfer, A. Kindl and I. Luchmann 2014. "Neue Mobilitätsformen, Mobilitätsstationen und Stadtgestalt", *ExWoSt-Information* 45/1. Bonn: Bundesinstitut für Bau-, Stadt und Raumforschung im Bundesamt für Bauwesen und Raumordnung. https://www.bbsr.bund.de/BBSR/DE/veroeffentlichungen/exwost/45/exwost45_1.pdf;jsessionid=-FAC8F85DD18A6113963D34B4C80F4CAC.live11293?__blob=publicationFile&v=1 (20/10/2020).

Schmid, C. 2016. "Urbanität und urbane Qualitäten", in *Urban Qualities. Ein Handbuch am Beispiel der Metropolitanregion Zürich*, ed. by S. Kretz and L. Kueng. Zurich: Edition Hochparterre.

Schnieder, L. 2018. "Netzplanung", in *Betriebsplanung im öffentlichen Personennahverkehr: Ziele, Methoden, Konzepte,* ed. by L. Schnieder. Berlin/Heidelberg: Springer Vieweg, 21–43.

Schwartz, S. 2017. "New Mobility Playbook. Appendix B: Shared Mobility Study Technical Report". Seattle Department of Transportation. https://www.seattle.gov/Documents/Departments/SDOT/NewMobilityProgram/AppendixB.pdf (14/5/2020).

Sinner, M., P. Khaligh and U. Weidmann 2018. "Consequences of automated transport systems as feeder services to rail: SBB fund for research into management in the field of transport. Report", ETH Zurich – Research Collection. DOI: 10.3929/ethz-b-000266025.

Sinner, M., and U. Weidmann 2019. "How does rail perform against autonomous buses? Two case studies in Switzerland", 19th Swiss Transport Research Conference 2019, Ascona. DOI: 10.3929/ethz-b-000342826 (15/10/2020).

Soike, R., J. Libbe, M. Konieczek-Woger and E. Plate 2019. "Räumliche Dimensionen der Digitalisierung. Handlungsbedarfe für die Stadtentwicklungsplanung. Ein Thesenpapier", Difu special publication. https://repository.difu.de/jspui/bitstream/difu/256328/1/DM19101469.pdf (14/5/2020).

Soteropoulos, A., A. Stickler, V. Sodl, M. Berger, J. Dangschat, P. Pfaffenbichler, G. Emberger, E. Frankus, R. Braun, F. Schneider, S. Kaiser, H. Wakolbinger and A. Mayerthaler 2019. "SAFiP – Systemszenarien Automatisiertes Fahren in der Personenmobilität". Vienna: bmvit.

Sousa, N., A. Almeida, J. Coutinho-Rodrigues and E. Natividade-Jesus 2017. "Dawn of autonomous vehicles: Review and challenges ahead", in *Proceedings of the Institution of Civil Engineers – Municipal Engineer* (171) 1, 3–14.

Stadt Wien 2018. "Leitfaden Mobilitätsstationen. Die Umsetzung von Mobilitätsstationen in Stadtentwicklungsgebieten am Beispiel Zielgebiet Donaufeld, Wien", Werkstattberichte der Stadtentwicklung Wien 179. Vienna: MA 18 – Stadtentwicklung und Stadtplanung und MA 21 – Stadtteilplanung und Flächennutzung. https://www.digital.wienbibliothek.at/urn/urn:nbn:at:AT-WBR-575386 (06/12/2022).

Stiglic, M., N. Agatz, M. Savelsbergh and M. Gradisar 2015. "The benefits of meeting points in ride-sharing systems", in *Transportation Research Part B: Methodological* 82, 36–53.

Taut, B. 1977. *Architekturlehre. Grundlagen, Theorie und Kritik, Beziehung zu den anderen Künsten und zur Gesellschaft.* VSA, Hamburg

verkehrsplus 2015. "Ve3 – Planung von Verknüpfung an Verkehrsstationen", ÖBB infra. Download at (14/5/2020).

Wefering, F., Rupprecht, S., Bührmann, S., and S. Böhler-Baedeker 2013. *Guidelines. Developing and Implementing a Sustainable Urban Mobility Plan*. Rupprecht Consult – Forschung und Beratung GmbH. http://www.rupprecht-consult.eu/uploads/tx_rupprecht/Revised_SUMP_Guidelines_final_web_Jan_14.pdf (14/5/2020).

Weidmann, U., R. Dorbritz, H. Orth, M. Scherer and P. Spacek 2011. "Einsatzbereiche verschiedener Verkehrsmittel in Agglomerations". Swiss Confederation: Federal Department of the Environment, Transport, Energy and Communications DETEC. (14/5/2020).

Zhang, W. 2017. "The interaction between land use and transportation in the era of shared autonomous vehicles: A simulation model", dissertation, Georgia Institute of Technology, Atlanta, GA. https://smartech.gatech.edu/bitstream/handle/1853/58665/ZHANG-DISSERTATION-2017.pdf?sequence=1&isAllowed=y (14/5/2020).

Zhang, W., S. Guhathakurta, J. Fang and G. Zhang 2015. "Exploring the impact of shared autonomous vehicles on urban parking demand: An agent-based simulation approach", in *Sustainable Cities and Society* 19, 34–45. DOI: 10.1016/j.scs.2015.07.006.

Zhang, W., and K. Wang 2020. "Parking futures: shared automated vehicles and parking demand reduction trajectories in Atlanta", in *Land Use Policy* 91, 103963. DOI: 10.1016/j.landusepol.2019.04.024.

9 Transformations of European public spaces with AVs

Robert Martin, Emilia M. Bruck, Aggelos Soteropoulos

Robert Martin
JAJA Architects ApS, Copenhagen & Aalborg University CPH, Department of Planning

Emilia M. Bruck
TU Wien, future.lab Research Center and Research Unit of Local Planning (IFOER)

Aggelos Soteropoulos
TU Wien, future.lab Research Center and Research Unit Transportation System Planning (MOVE)

© The Author(s) 2023
M. Mitteregger et al. (eds.), *AVENUE21. Planning and Policy Considerations for an Age of Automated Mobility*, https://doi.org/10.1007/978-3-662-67004-0_9

1. INTRODUCTION

Connected and automated driving is one of several emerging mobility trends that will fundamentally impact the use and design of public spaces in the coming decades. The uptake of transportation network companies (TNCs), such as Uber, has shown that a greater use of shared modes adds more vehicles to the road and shifts pick-up and drop-off locations onto the street, i.e. increasing activity at the kerb (Larco 2018: 50; Erhardt et al. 2019). Similar effects were caused by recent waves of dockless micromobility options, such as free-floating bikes or e-scooters, which temporarily led to congested pavements and increased spatial demands in public space (Polis 2019). In effect, cities are challenged to rethink the exclusive rights given to cars within their mobility network. Ongoing mobility innovations and expected developments in automated mobility require a reallocation of public space and render existing categories of traffic division and regulatory frameworks outdated (Polis 2019: 12–13).

This article highlights possible trajectories for redesigning public spaces in a European context in order to illustrate urban futures in light of new mobility developments, such as automated mobility and a greater mix of traffic modes. To this end, this article views public space holistically, encompassing traffic infrastructure, public open spaces as well as adjacent buildings. Considered as such, public spaces may comprise a variety of qualities, functions and interests that differ, even diverge at times, depending on urban structure and street typology (Bendiks/ Degros 2019, Marsden et al. 2020, Karndacharuk et al. 2014). With automated mobility on the horizon, urban planners need to rethink whose interests they place at the centre of their designs and what transport modes are given priority. While industry and policy representatives emphasize traffic advantages, such as safety and efficiency gains, spatial and social implications of automated use cases remain highly uncertain (see Chap. 2 by Bruck and Soteropoulos in this volume).

While a number of design studies have been made that envision how public spaces could be transformed with automated vehicles, the majority of them refer to North American cities or no specific urban context at all (e.g. NACTO 2019, Schlossberg et al. 2018, Luo 2019, Sasaki 2018, Meyboom 2019). As a result, there is a lack of contextual design studies that highlight the specificity of urban form, mobility culture and planning rationale. Just as "total designs" of the modern and postmodern era denied the incremental growth of cities and pluralist decision-making (Venturi et al. 1977: 149), design visions for the ongoing mobility revolution need to take contextual factors into account in order to elucidate local implications – opportunities and risks – of new mobility technologies.

In contrast to most North American cities, many cities in Europe have high-density urban structures and compact historic cores. Many of those cities have urban transportation networks that are well integrated into their urban fabric, providing the backbone of urban mobility. Beyond that, cities such as Amsterdam or Copenhagen are known for having high percentages of cyclists and pedestrians. While this applies to inner-city districts, it is less the case in urban extension areas developed since the 1950s and '60s or low-density suburban developments where public transport is often difficult to reach and basic services are less accessible by bicycle or foot (van Essen et al. 2009: 13; Alessandrini et al. 2015: 146; Gavanas 2019: 4). Finally, while North American cities are known for expansive off-street car parks that enclose suburban shopping malls or carve voids into inner-city urban fabrics, European cities are faced with spatial constraints within their inner-city historical districts, where the existing intensity and diversity of uses put pressure on already limited public space (Marsden et al. 2020).

As the early euphoria around automated vehicles' (AVs') near-term market introduction wore off due to technological setbacks, it became more apparent that a longer-term period of mixed traffic conditions lies ahead in which automated vehicles share roads with conventional vehicles and rely significantly on connected services (Mitteregger et al. 2022, Backhaus et al. 2019). During this transition period, AVs will not be operating on the entire road network, but rather on designated streets or confined (geofenced) areas at limited speeds, i.e. special operational design domains that define the functional boundary of Level 4 AVs (SAE International 2018). As of yet few urban design studies have been made for European cities (e.g. Dijkstra/Ionescu 2019, ARUP 2018); they largely show visions of Level 5 AVs that assume AVs would operate within the entire traffic network and do not consider mixed traffic scenarios. It is, however, critical that urban planners and designers take into account a possibly long-term transitional period where there will likely be a need for strategies to manage the reallocation of kerb space, a reclassification of street typologies and mode distribution, and the creation of transition zones where vehicles shift from automated to manual modes (Backhaus et al. 2019: 17).

To that end, design visions are a vital tool to support coordinated planning, decision-making and development and ensure that public spaces remain a common spatial infrastructure contributing to quality of life in cities. This article introduces design experiments on possible public spaces with AVs, conducted by the Danish architectural firm JAJA Architects. Set in three varying urban areas within Copenhagen, Denmark, the designs build upon the specificities of local neighbourhood structures and mobility requirements. Through plans and three-dimensional images, possibilities of integrating AVs into a sustainable transportation system are explored. By doing so, varying urban futures unfold.

2. COPENHAGEN DESIGN EXPERIMENTS ON THE SUSTAINABLE DEPLOYMENT OF AVS

The following design experiments take place within the northern European capital city of Copenhagen, Denmark. The city is an exemplary context in which to investigate how AVs may impact urban form as part of a sustainable transportation system because Copenhagen is already a model of green mobility. Within the Municipality of Copenhagen, 29% of all trips that either begin or end within its boundary occur by bicycle, 70% of households are car-free, and it has one of the most accessible public transport systems in Europe (City of Copenhagen 2017a, Scheurer 2013). While the city's comparatively sustainable transportation system is enviable, it did not happen overnight. Copenhagen has benefited from a rich planning tradition starting with the Finger Plan from 1947, where urban development proceeded parallel to five "fingers" centred on commuter rail lines, which extend from a "palm" of dense urban fabric within the Copenhagen municipal boundary (Fig. 1). Subsequent investments in an underground metro system, as well as an extensive bicycle path network in the city centre, have led to the Municipality of Copenhagen having one of the lowest per capita car emissions in the world (City of Copenhagen 2016). However, despite this, its current transportation system is far from secure. Political tensions in Copenhagen over the space allocated for cycling, cars and public transport create continual backlashes and conflicts over street space, and the introduction of new mobility modes means that modal distribution is in constant flux (Henderson/Gulsrud 2019). How the introduction of AVs into this debate will affect modal share will be a result of social acceptance, policy, and spatial intervention.

Figure 1: Copenhagen metropolitan plan with project locations. Municipality of Copenhagen highlighted in dark purple with commuter rails (unbroken lines) and metro (lines).

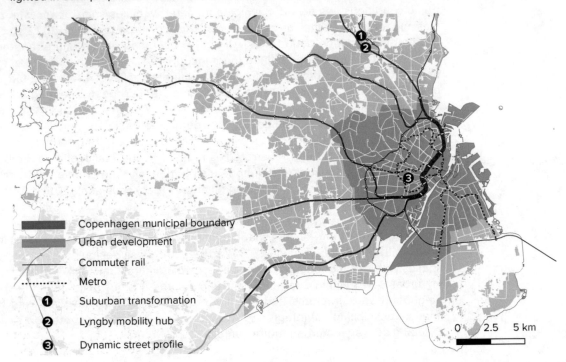

Copenhagen municipal boundary
Urban development
Commuter rail
Metro
❶ Suburban transformation
❷ Lyngby mobility hub
❸ Dynamic street profile

0   2.5   5 km

Source: JAJA Architects

As one moves along the fingers outside the municipal boundary, one finds a significantly different urban environment. Whereas only 7% of the residential building stock in the Copenhagen Municipality are single-family dwellings, this figure rises to 44% in the surrounding metropolitan region (Statistics Denmark 2019a). This dramatic change in spatial typology reflects a higher rate of car ownership (Statistics Denmark 2019a), sparser population density (Statistics Denmark 2020), and double the amount of space dedicated to road infrastructure per capita (Statistics Denmark 2016). While the primary consideration of AV introduction in the inner city will regard preserving and promoting active forms of transport, the real spatial transformative potential of AVs lies in the surrounding suburbs.

To understand how urban form may be affected throughout the Copenhagen metropolitan region by the introduction of AVs, the authors have chosen a future scenario that is radically different from how transport is today. In this scenario, privately owned automobile use has been virtually non-existent in the dense inner city since the Copenhagen Municipality banned private car use. Instead, residents and commuters move through a combination of public transport, fixed-route AV shuttles that run along arterial roads, and micromobility devices that range in size from kick scooters to electric cargo bicycles. Residents living in less dense suburbs outside of the inner city still have the option to own a car. However, most have chosen to adopt a tailor-made Mobility-as-a-Service (MaaS) package that includes, among other offerings, an on-demand, free-floating AV shuttle that provides a last-mile connection to nearby public transport nodes. The technological development of AVs has reached a bottleneck and, therefore, they have only been deployed with Level 4 capabilities (SAE International 2018). This technical barrier means that AVs may only operate within geofenced areas where the density allows for the commercial viability of creating and maintaining the high-definition 3D maps required for AVs to function safely. Therefore, motorized/conventional cars remain necessary for edge-case situations where AVs cannot operate, and traffic may be a mix of AVs and traditional automobiles.

To visualize what effect this scenario may have on existing public spaces and streetscapes in Copenhagen, the authors offer three design studies in different urban contexts within the city. The first takes place in the suburb of Lyngby, approximately 10 km north of the city centre, and investigates how a shift to a shared AV system may offer spatial opportunities to dissolve spatially segregated boundaries and provide communal amenities in an otherwise highly privatized monofunctional area. The second design study explores how the existing commuter rail station in Lyngby could be adapted to integrate an AV shuttle system with adequate space for pick-up and drop-off that supports an efficient multimodal transport system. The final design study investigates a modal space reallocation in an inner-city street where an increase in micromobility traffic places pressure on the spatial demands of a traffic artery used for a fixed-route AV shuttle.

2.1   RETHINKING THE SUBURB

While inner Copenhagen enjoys low car use, this dramatically changes as one moves into the surrounding suburbs where population density falls as single-family dwellings replace apartment buildings. The site of this exploration, the northern suburb of Lyngby, is a typical example. Despite enjoying excellent commuter rail connections and a decent bus service, this suburb still has over double the inner city's car ownership rate at 549 cars per 1,000 residents (Statistics Denmark 2019b). Compared to the inner city, which hosts an array of public and semi-public amenities on its streets, the suburbanization of Lyngby has created an urban condition wherein all functions occur within the boundary of the block, hidden behind high hedges or fences (Fig. 7). This clear separation between public and private arenas has left the public realm somewhat vacant. Whereas in historical contexts, suburban streets would be full of playing children, now due to safety concerns the road lays empty, with only the occasional passing car, idling service van or visitor's parked car (Fig. 2). The division between private property and the public realm has become so stark that the only interface between the two is the driveway. A resident may, therefore, never actually physically touch the public domain, entering their vehicle within the boundary of their property before driving away to their destination.

Figure 2: Existing residential street. High hedges and narrow sidewalks represent a public space that is designed only for automobile use.

Figure 3: Proposed residential street. Life returns to the street as redundant road areas are transformed into communal amenities. Fenced boundaries are dissolved as properties reconnect with the street's activities rather than blocking them out.

Source: JAJA Architects

JAJA's proposed adaptation to the street attempts to dissolve the suburban rationality of separation by spatially repurposing the abundant space given to automobiles in the road for new public amenities (Fig. 3). The primary motivation behind this redistribution of space comes from both a radical decrease in traffic demand as residents shift from privately owned vehicles to shared AV shuttles and the technological ability of AVs to safely navigate intricate driving lines, always obey speed limits and give way to pedestrians and children. Instead of providing a lane in each direction with enough room to overtake a parked car adjacent to the kerb (Fig. 4.), the road width is limited to that of a conventional single-vehicle lane for both traditional and autonomous vehicles.

The road then undergoes a series of manipulations to ensure that a right of access to all existing driveways remains so that residents still have the option to own a private car, and that there is space for vehicles travelling in opposite directions to give way or pass each other (Fig. 5). The residual space provides opportunities to install fixed amenities that both foster community, such as vegetable gardens, outdoor dining areas, community houses or sports facilities, and support the new multimodal transport system, such as a covered waiting area for AV shuttle services and parking space for shared micromobility devices. The boundary of these new facilities is not limited to a demarcated area. Instead, through safely negotiated and temporal use, the facilities can spill out into the road area, better utilizing the space for active functions that can stop when a vehicle passes.

Through an increase in public amenities, an opportunity arises to renegotiate the threshold between public and private. With more functions becoming shared, the abundance of open space behind individual boundaries, especially adjacent to the street, is rezoned to create new ancillary dwellings (Figs. 6 and 8). These new dwellings vary in ownership models and typology, with many of the functions outsourced to the communal facilities to attract a diverse range of new residents not suited to the homogenous rows of single-family dwellings otherwise found in the area. The increase in population would drive demand for AV shuttles, reducing the operating costs of the system while increasing the efficiency and desirability of the system.

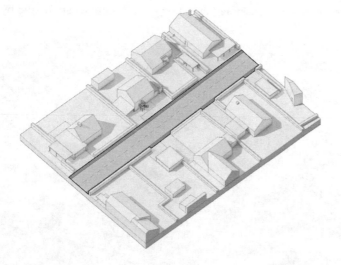

Figure 4: Existing street design

Existing street axonometric. The majority of the streetscape is dedicated to car use.

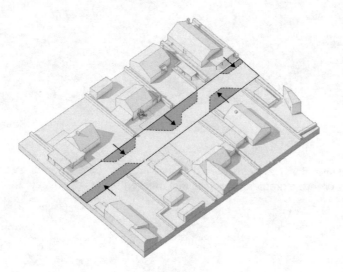

Figure 5: Expansion of public space

By reducing the street profile to one-way, but still providing spaces to overtake and connect to driveways, new pockets of space can be designated for communal amenities.

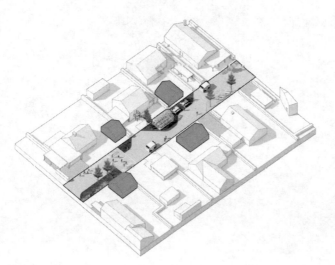

Figure 6: Programming

New ancillary dwellings are placed adjacent to communal activity areas to dissolve the boundary between public and private along the street.

Source: JAJA Architects

Figure 7: Existing site plan. A grid of single-family homes and garages separated by a field of fences and streets.

Figure 8: Proposed site plan. Newly inserted buildings and functions operating at different scales disrupt the grid and create a gradient of zones with different levels of privacy.

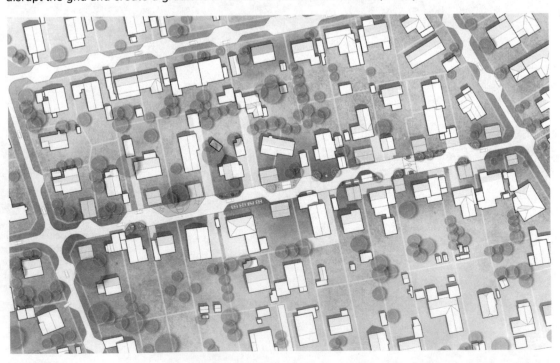

2.2 FROM TRAIN STATION TO MOBILITY HUB

Multimodal transport routes are often proclaimed to be the sustainable alternative to car trips, where commuters shift between higher and lower-capacity modes to reach their destination. However, this system is reliant on the proximity to transport nodes and available connecting routes only found in higher-density urban fabrics. The challenge of transporting commuters to network nodes in lower-density suburbs is referred to as the first/last-mile gap. Shared AV shuttle systems, as used in the previous design example, are often discussed as one solution to this common problem. Conceptually, this system operates similarly to already established car-pooling services such as Uber, Lyft and Via, where users' ride requests are bundled and assigned into trips with similar pick-up and drop-off points. However, the success of these services is highly dependent on population density, the concentration of users and the similarity of users' departure and arrival points and times. By focusing the departure or arrival point around public transport nodes, the shared AV shuttle system's efficiency is improved by accumulating similar trips. Nevertheless, points of friction are likely to occur at the interchange between modes as existing transport infrastructure has not been designed to enable AVs. The following design explores how adaptations to the existing train station at Lyngby can spatially support this new technology as users seamlessly transfer between AV shuttle and high-capacity train.

The existing Lyngby station is a train station on the Hillerød radial of the Finger Plan. It is centrally located within the suburb but is spatially segregated from the suburb's high street and mass of urban functions by a large bus terminal, two lanes of traffic, car parks and an elevated highway to the east (Figs. 9, 11 and 14). The station's entrance is located underneath the highway, where it is also connected to a shopping centre with 15 retail stores, including two supermarkets.

Figure 14: Existing site plan. The station is separated from the suburb's high street (running from top left to right of image) by multiple roads, parking lots, a bus terminal and an elevated highway.

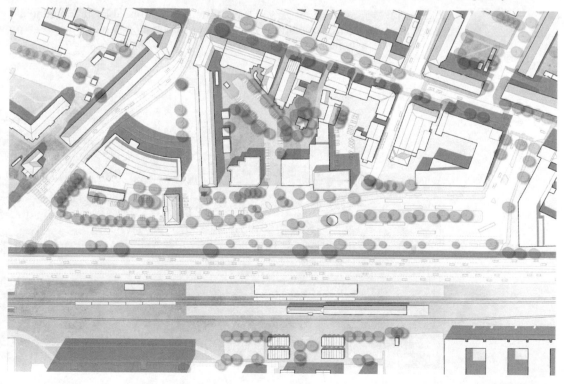

Figure 9: Existing street view at Lyngby station. Commuters are separated from the station entrance by a series of roads that must be crossed in sections.

Figure 10: Proposed street view at Lyngby station. A permeable station edge allows commuters to enter the stations from multiple points while an information-rich digital screen provides wayfinding connections to standing-by AV shuttles.

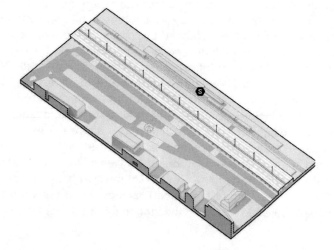

Figure 11: Existing station design

The existing station axonometric highlights the many obstacles to entering the station.

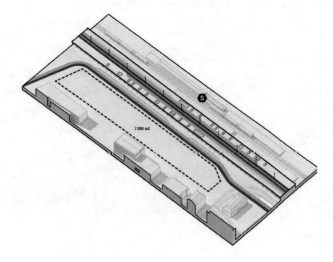

Figure 12: Redistribution of infrastructure

Pick-up/drop-off areas are condensed into two areas: the first area lies adjacent to the station entrance while the second is located on the elevated highway.

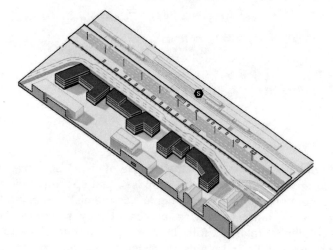

Figure 13: Urban infill and densification

New mixed-use development is situated in the publicly owned former bus terminal. The new development not only adds spatial qualities and increases density, but the revenue from the development can be utilized by the municipality to fund public programmes.

Source: JAJA Architects

The primary design challenge for this proposal was to create adequate space for the pick-up and drop-off areas for commuters arriving by AV shuttles. While many advocates for AVs suggest that excess parking space will be released by sharing these vehicles, studies have shown that the spatial requirements for pick-up and drop-off areas will be high as they should be designed to accommodate maximum inflow at peak times (Sinner et al. 2018). Therefore, the main decision made in the design is to consolidate the seven lanes of traffic that run in both directions adjacent to the station into one 150 m long designated area for transfers (Fig. 12). This area follows design principles found at airport kiss-and-ride locations, where one lane is used for parking (coloured light orange), one is used to wait for a free space (coloured orange) and the final one is used to pass by when finished (coloured red). Due to a dramatic decrease in traffic demand from sharing and AV platooning, as well as increased safety from connected vehicles, this principle is replicated on the elevated highway, which runs directly above the train station entrance. Cuts in the structure create vertical movement between the highway and the station, allowing more accessible routes to the station for residents who have to access it from the west.

Figure 15: Proposed site plan. A new mixed-use development completes the urban block, utilizing the former bus terminal. AV shuttle pick-up and drop-off areas have been consolidated to be directly adjacent to the station and on the reduced-capacity highway.

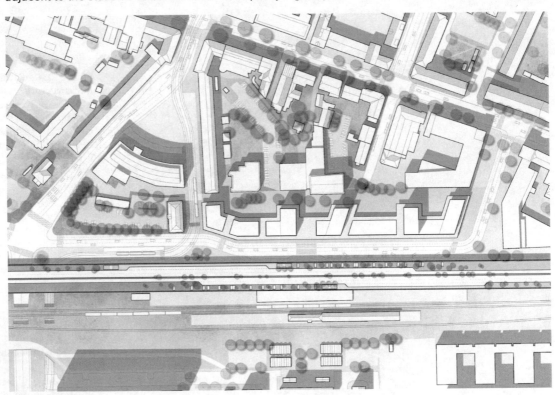

Source: JAJA Architects

The spatial benefit of this consolidation is the release of over 7,000 m² of publicly owned land directly adjacent to the train station. In this proposal, that space is utilized by constructing a mixed-use development of residential apartments, commercial space, public amenities, as well as parking facilities for micromobility devices adjacent to new separated bicycle paths (Fig. 13). The proposed development takes its form by closing the urban block to the east, creating a series of public and semi-public courtyards of varying scales that respect the existing pathways between the station and the high street (Fig. 15). The final move is to relocate the shopping centre from underneath the highway to the new mixed-use development. The now vacant space is transformed into a permeable covered thoroughfare that gives access to the station platforms directly from the pick-up/drop-off area. There are also seated waiting areas and digital wayfinding screens that help commuters find their designated shuttle (Fig. 10).

2.3 A NEW DYNAMIC STREETSCAPE

Unlike the suburbs of Copenhagen, where road space is abundant due to car-centric planning principles since World War II, the inner city has to negotiate modal allowance within a narrow spatial context designed centuries before the invention of the car. Subsequent additions of transport modes have constrained pavements and cycle paths to minimal widths. At the same time, two-way roads, car parks and bus stops occupy the majority of the space between buildings. Within inner Copenhagen, only 7% of citywide road space is taken up by cycle paths. In contrast, road space for cars amounts to 66% (City of Copenhagen 2017b), even though modal trips are split almost evenly between bicycles and cars. Overcrowding on cycle paths is already a severe problem in Copenhagen and a significant impediment to increasing the city's incredibly high levels of cycling (Danish Parliament 2016). Unfortunately, it is simply not an option to widen cycle paths on arterial roads as the constrained context is filled by the spatial provision of on-street car parking. AVs promise to release this space through the logic of never having to park (Duarte/Ratti 2018). However, this logic ignores the new spatial demands of AVs. We expect that AVs will increase door-to-door mobility and will, therefore, require equal space to embark or alight from the vehicle.

The conflict between AVs and cycle paths has given rise to significant design considerations in JAJA's urban scenario below, where a projected substantial increase in modal share by micromobility devices has resulted from the banning of privately owned vehicles in the city centre. The street under investigation is Gammel Kongevej (Fig. 16), which is one of the principal shopping streets in Copenhagen and dates back to the beginning of the 17th century. The street extends for 1.8 km from the western edge of the city centre and provides a direct connection to the western suburbs. The street is only 18 m wide from one building façade to the other, so it currently utilizes a three-lane system to accommodate all the spatial demands from different modes. One lane each is dedicated to vehicle traffic in either direction; a third lane is located in an alternating manner on either side to allow for kerbside parking and for buses to stop (Fig. 18). While this system provides space for vehicle modes, it is an underutilization of space (Fig. 19), and the spatial implication of these fixed infrastructures means that cycle lanes along the street are below the legal minimum width at only 1.5 m (Fig. 21; City of Copenhagen 2013). How then could pick-up/drop-off areas be integrated into this already crowded street while allowing an extension in the width of cycle lanes to meet increased travel demand by micromobility services?

Figure 16: Existing street view along Gammel Kongevej. A man quickly enters a stopped bus moments before cyclists are due to pass

Figure 17: Proposed street view along Gammel Kongevej. A man safely departs his AV shuttle onto the dynamic street surface, knowing that the coming cyclist will pass outside the boundary of his designated area.

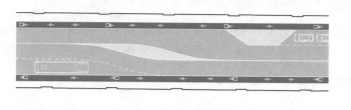

Figure 18: Existing street design

Existing zone plan of the street highlighting the spatial preference for automobiles over bicycles regardless of their equal modal share.

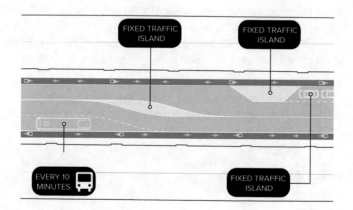

Figure 19: Functional requirements

Analysis of function demand. Fixed spatial infrastructures underutilize the space.

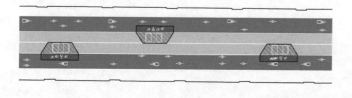

Figure 20: Dynamic street design

Rather than have a fixed shuttle stop, dynamic hop-on/hop-off areas can pop up along the street as user, vehicle and road surfaces are connected through the IoT.

Source: JAJA Architects

This design proposal utilizes advancements in the Internet of Things (IoT), where embedded sensors, lights and transmitters allow vehicles to communicate with the road infrastructure. Rather than having fixed street infrastructure that designates where certain functions should occur, the streetscape is enhanced with a grid of LEDs that can reallocate space in accordance with changing traffic volumes. Fixed on-street parking and bus stops are removed, allowing the third lane of the street to no longer be needed, and that space is redistributed to widen the cycle lanes to 3.5 m in both directions (Fig. 22). AV shuttles do not have fixed stopping points but are free to stop anywhere along the road (Fig. 20).

When a user makes a request to be picked up or dropped off, GPS coordinates of the location are communicated between the mobile device, AV shuttle and the road in preparation for the stop. As the AV shuttle approaches the destination, the road surface changes at the threshold between the road and cycle path to indicate a buffered area where passengers will alight and gives safe notice to incoming micromobility devices to avoid the buffered area. Modes using the cycle path will continue to have right of way, although half of their expanded lane will now be demarcated as a buffered passenger zone (Fig. 17). Enforcement of this buffered zone is enabled through sensors in the road that track infringements through in-vehicle unique identifiers (UID). These road sensors monitor the user's duration in the buffered zone, and the road surface only returns to normal once the user has left the area. It is important to note that in this design, priority is given to modes using the cycle path, so this form of traffic is not halted due to AV service. Modes using the cycle path will have right of way, while AV shuttles will stand on the road rather than adjacent to the kerb, knowing that other connected AV shuttles will anticipate the stop and wait or reroute if necessary.

Figure 21: Existing street plan. Commuters entering and exiting the bus are placed in conflict with cyclists. Although cyclists are required to give way to commuters, this law is often ignored because of lost inertia.

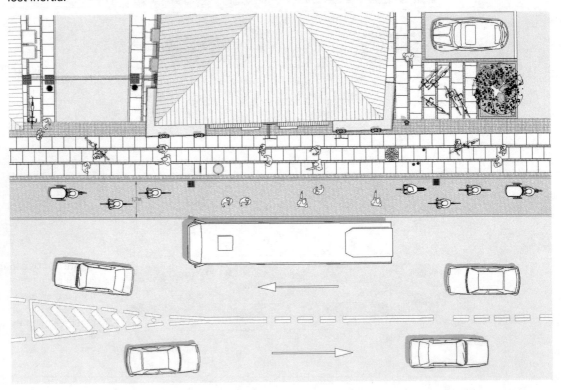

Source: JAJA Architects

Figure 22: Proposed street plan. The removal of one lane of traffic has allowed the bicycle lane to be doubled in width. The new road surface is embedded with IoT-connected LEDs that can create temporary buffered zones to allow users to safely enter and exit AV shuttles that still allow bicycles to pass by.

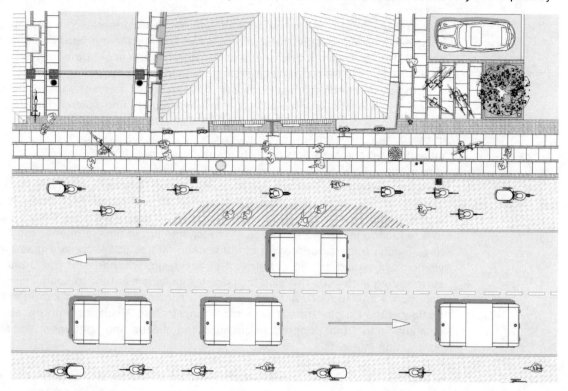

Source: JAJA Architects

3. CONCLUSION

This article presents design experiments on possible public spaces with AVs, i.e. how AVs may contribute to changes in urban form if integrated as part of a sustainable transportation system. The design experiments were set in three different areas within Copenhagen, Denmark, and focused on:

1. How a shift to a shared AV system could present an opportunity to dissolve spatially segregated boundaries and provide communal amenities in an otherwise highly privatized monofunctional area in the suburb of Lyngby

2. How an existing commuter railway station in Lyngby could be adapted to integrate an AV shuttle system with adequate pick-up and drop-off areas that support an efficient multimodal transport system

3. How the reallocation of space towards active travel modes could take shape in an inner-city street of Copenhagen, where increasing micromobility traffic aggravates the pressure of spatial requirements on a traffic artery used for a fixed-route AV shuttle.

These design experiments highlight that changes in urban design and infrastructure development related to the introduction of automated mobility services may vary significantly according to urban form and street typology. The functional requirements of a street design vary within a city and are determined by factors such as adjacent land use types, position within the urban street network, diversity of travel modes and users, as well as designated speed limits.

Due to the expectation that automated vehicles could generate greater demand, the pressure on street designs to facilitate higher numbers of vehicles per hour could increase (Larco/Tierney 2020). Thus, competing demands for street space might be aggravated in the future. The urban design challenge will rest even more than today on finding a suitable balance between catering to demands for more efficient movement and demands for attractive spaces. This is especially the case in inner-city streets where competing spatial demands are already high and heterogenous. Dynamic solutions, e.g. demand-based hop-on/hop-off areas, as presented in the design experiment for an inner-city street in Copenhagen, could pose a design-based measure that complements mobility management policies.

However, determining factors for adequate design interventions are highly contextual, both in material and political terms, and therefore require well-attuned solutions. To this end, local design experiments are critical in envisaging how to reallocate potentially freed-up space – due to a reduction in on-street parking – and tap into urban development potentials. As cities need to re-evaluate the prioritization of modes and find solutions to safe mode interaction, design visions can elucidate the benefits of street design changes for the urban environment and surrounding land use. This also includes the question of how to spatially integrate modes and enhance a multimodal transport system, as shown in the design experiment on the transformation of the Lyngby train station into a mobility hub. Visualizing potential changes can serve as a critical tool that supports negotiation and collaboration between affected stakeholders or contrasting interests.

In addition, considering the long-term transition period leading towards automation, it is critical to reflect upon which changes could be implemented irrespective of vehicle automation and which changes need further investigation. Short-term issues that cities should address include strategies for kerb management, prioritizing pick-up and drop-off zones over on-street parking, increased cycling and micromobility lanes, as well as enhancing the integration between shared mobility and transport networks. While the influx of new mobility options is at a peak and the prospect of automated mobility does not appear to be fading, uncertainties regarding any trend's durability prevail. As a means of acting in uncertainty, cities are increasingly adopting pilot projects. Not merely to test AVs (see Chap. 6 by Soteropoulos et al. in this volume), but also in order to test night-time pick-up and drop-off zones (Washington DC), clearing kerbs of commercial loading during designated times of day (New York City) or geofencing streets with high levels of active mobility interaction so as to avoid conflict with ride-sourcing services (San Francisco; Schaller 2019).

What is lacking are more comprehensive programmes that would usher in the transition from public spaces characterized by parked cars and travel lanes to those that can be used flexibly and cater to shared modes. However, in order to develop guidelines on the spatial requirements of new mobility options such as shared automated vehicles, further research and more comprehensive studies are necessary. Questions regarding the spatial demand for pick-up and drop-off activities or short-term parking need more thorough investigation through simulation and modelling. However, these should be developed in collaboration with design methodologies and visualizations that are better able to integrate the context of local development goals and neighbourhood characteristics.

REFERENCES

Alessandrini, A., A. Campagna, P. Delle Sitte, F. Filippi and L. Persia 2015. "Automated Vehicles and the Rethinking of Mobility and Cities", in *Transport Research Procedia 5*, 145–160. https://doi.org/10.1016/j.trpro.2015.01.002.

ARUP 2018. *FlexKerbs. Evolving Streets for a Driverless Future*, report, London. https://www.arup.com/-/media/arup/files/publications/f/flexkerbs_roads-for-the-future_arup.pdf (23/3/2020).

Backhaus, W., S. Rupprecht and D. Franco 2019. *Road vehicle automation in sustainable urban mobility planning*, practitioner briefing, Cologne. https://www.eltis.org/sites/default/files/road_vehicle_automation_in_sustainable_urban_mobility_planning_0.pdf (19/8/2020).

Bendiks, S. and A. Degros 2019. *Traffic Space is Public Space: A Handbook for Transformation*. Zurich: Park Books.

City of Copenhagen 2013. *Focus on cycling: Copenhagen guidelines for the design of road projects*, trans. J. Høberg-Petersen, Copenhagen. https://kk.sites.itera.dk/apps/kk_pub2/index.asp?mode=detalje&id=1133 (4/8/2020).

City of Copenhagen 2016. *CPH 2025: Climate Plan. Roadmap 2017–2020*, Copenhagen. https://urbandevelopmentcph.kk.dk/artikel/cph-2025-climate-plan (4/8/2020).

City of Copenhagen 2017a. *Copenhagen: City of Cyclists. The Bicycle Account 2016*, Copenhagen. https://kk.sites.itera.dk/apps/kk_pub2/index.asp?mode=detalje&id=1698 (4/8/2020).

City of Copenhagen 2017b. *Cycle Track Priority Plan (2017–2025)*, Copenhagen. https://idekatalog-forcykeltrafik.dk/wp-content/uploads/2019/05/cykelstiprioriteringsplan-2017-2025pdf-_1620.pdf (4/8/2020).

Dijkstra, R. and A. I. Ionescu 2019. "Streets and Robocars", in *Robocar and Urban Space Evolution: City Changes in the Age of Autonomous Cars*, ed. by A. I. Ionescu, V. M. Sanz and R. Dijkstra. Delft: TU Delft, 31–39.

Erhardt, G. D., S. Roy, D. Cooper, B. Sana, M. Chen and J. Castiglione 2019. "Do transportation network companies decrease or increase congestion?", in *Science Advances* (5) 5, eaau2670. https://doi.org/10.1126/sciadv.aau2670.

Gavanas, N. 2019. "Autonomous Road Vehicles: Challenges for Urban Planning in European Cities," in *Urban Science* (3) 61, 1–13. https://doi.org/10.3390/urbansci3020061.

Henderson, J. and N. M. Gulsrud 2019. *Street Fights in Copenhagen: Bicycle and Car Politics in a Green Mobility City.* London/New York: Routledge.

Karndacharuk, A., D. J. Wilson and R. Dunn 2014. "A Review of the Evolution of Shared (Street) Space Concepts in Urban Environments", in *Transport Reviews* 34 (2), 190–220. https://doi.org/10.1080/01441647.2014.893038.

Larco, N. 2019. "Urbanism Next: Autonomous Vehicles and the City", in *Robocar and Urban Space Evolution: City Changes in the Age of Autonomous Cars*, ed. by A. I. Ionescu, V. M. Sanz and R. Dijkstra. Delft: TU Delft, 47–53.

Larco, N. and G. Tierney 2020. "Impacts on Urban Design", in *Multilevel Impacts of Emerging Technologies on City Form and Development*, ed. by A. Howell and K. Lewis Chamberlain. Portland, OR: University of Oregon, 115–141.

Luo, Y. 2019. "From Transportation Infrastructure to Green Infrastructure – Adaptable Future Roads in Autonomous Urbanism", in *Landscape Architecture Frontiers* (7) 2, 92–99. https://doi.org/10.15302/J-LAF-20190209.

Marsden, G., I. Docherty and R. Dowling 2020. "Parking futures: Curbside management in the era of 'new mobility' services in British and Australian cities", in *Land Use Policy* (91), 104012. https://doi.org/10.1016/j.landusepol.2019.05.031.

Meyboom, A. L. and L. Vass 2019. *Driverless urban futures: A speculative atlas for autonomous vehicles.* New York, NY/London: Routledge.

Mitteregger, M., E. M. Bruck, A. Soteropoulos, A. Stickler, M. Berger, J. S. Dangschat, R. Scheuvens and I. Banerjee 2022. *AVENUE21. Connected and Automated Driving: Prospects for Urban Europe* trans. M. Slater and N. Raafat. Berlin: Springer Vieweg. https://doi.org/10.1007/978-3-662-64140-8.

NACTO 2019. *Blueprint for Autonomous Urbanism: Second Edition*, New York. https://nacto.org/publication/bau2/ (19/8/2020).

POLIS 2019. "Macro Managing Micro Mobility: Taking the long view on short trips", discussion paper. https://www.polisnetwork.eu/document/macromanaging-micromobility/ (19/8/2020).

SAE International 2018. "Surface vehicles recommended practice. J3016. Taxonomy and Definitions for Terms Related to Driving Automation Systems for On-Road Motor Vehicles." https://www.sae.org/standards/content/j3016_201806/ (19/8/2020).

Sasaki 2018. *Shifting Gears: An Urbanist's Take on Autonomous Vehicles.* https://issuu.com/sasakiassociates/docs/shifting_gears_20180531_-_issue (19/8/2020).

Schaller, B. 2019. *Making the Most of the Curb: Managing Passenger and Parcel Pick-up and Drop-off on Congested City Streets*, report, Brooklyn, NY. http://www.schallerconsult.com/rideservices/makingmostofcurb.pdf (19/8/2020).

Scheurer, J. 2013. *Measuring Copenhagen's public transport accessibility and network performance in a European context.* https://www.researchgate.net/publication/286814648_Measuring_Copenhagen%27s_public_transport_accessibility_and_network_performance_in_a_European_context (4/8/2020).

Schlossberg, M., W. Riggs, A. Millard-Ball and E. Shay 2018. *Rethinking the Street in an Era of Driverless Cars*, Urbanism Next, University of Oregon. https://doi.org/10.13140/RG.2.2.29462.04162.

Sinner, M., P. Khaligh and U. Weidmann 2018. "Consequences of Automated Transport Systems as Feeder Services to Rail. SBB Fund for Research into Management in the Field of Transport", in *IVT Schriftenreihe* (184). https://doi.org/10.3929/ethz-b-000266025.

Statistics Denmark 2016. *AREALDK1: Land by land cover, region and unit* (DISCONTINUED). https://www.statbank.dk/statbank5a/default.asp?w=1920 (4/8/2020).

Statistics Denmark 2019a. *BIL800: Families disposal of vehicles by region and pattern of disposal.* https://www.statbank.dk/statbank5a/default.asp?w=1440 (4/8/2020).

Statistics Denmark 2019b. *BIL800: Families disposal of vehicles by region and pattern of disposal.* https://www.statbank.dk/statbank5a/default.asp?w=1920 (4/8/2020).

Statistics Denmark 2020. *FOLK1A: Population at the first day of the quarter by region, sex, age and marital status.* https://www.statbank.dk/KM1 (4/8/2020).

Van Essen, H., X. Rijkee, G. Verbraak, H. Quak and I. Wilmink 2009. "EU Transport GHG: Routes to 2050? Modal split and decoupling options", Paper 5, draft. https://www.eutransportghg2050.eu/cms/assets/4823DraftPaper-5.pdf (19/8/2020).

Venturi, R., D. S. Brown and S. Izenour 1982. *Learning from Las Vegas*, 5th edition. Cambridge, MA: MIT Press.

10 At the end of the road: Total safety

How the safety concept of connected and automated driving systems is changing the streetscape

Mathias Mitteregger

Mathias Mitteregger
TU Wien, future.lab Research Center

© The Author(s) 2023
M. Mitteregger et al. (eds.), *AVENUE21. Planning and Policy Considerations for an Age of Automated Mobility*, https://doi.org/10.1007/978-3-662-67004-0_10

" A major aspect of media effects and development appears in the case of the road as a means of transportation. Like writing or radio the 'content' of the road is always another medium or other media, whether pedestrians, equestrians, wagons or cars. Depending on the type of vehicle-medium, the nature of the road-medium alters greatly" (McLuhan 1960, Part III, 15).

1. ROAD SAFETY AS A DRIVING FORCE

Presentations that promote the development of self-driving cars often begin with what in Michel Foucault's words could be called a "theatre of pain" (Foucault 2012: 42).[1] While photos show demolished school buses and cars torn in half, the presenter intersperses these drastic images with figures: 1.2 million people are killed on the roads every year, making road accidents the leading cause of death for 15- to 29-year-olds worldwide (WHO 2015: 2). Automated driving systems are to put an end to end this tragedy: humans must hand over control of the vehicle to learning algorithms that are superior to human skills and are never tired, distracted or drunk.

Road safety is the main argument put forward – above and beyond economic interests – in asserting the added value bestowed on society as a whole by connectivity and automation of the transport system. Numerous acceptance studies throughout the world attest the fundamental importance of safety for implementation of the overall technology, and it would be difficult to find a policy paper in which this aspect is not repeatedly emphasized. Not even accidents with test vehicles or overrated assistance systems in production vehicles can detract from this view.

This chapter looks at road safety as driven by connectivity and automation, from the perspective of the streetscape – with all its participants. It is argued that this development, which has already set in, is bringing about a turnaround that could in fact turn the concept of road safety "upside down". Such a reorientation would not only affect road safety as such. In this chapter, it is argued that this could undermine the principle of the public sphere, which is based on visibility, and replace it with a new form of curated coexistence.

1.1 MOBILE ROBOTS AS THE KEY TO SAFETY

Connected and automated driving systems must be safe – or at least safer than today's cars. This is seen as a basic prerequisite for broad-based social acceptance of this technology (Lazarus et al. 2018). This aspect transcends cultural boundaries and has been demonstrated on a global scale. "Safety" was the most commonly used term in all studies surveyed in a review of the literature (Jing et al. 2020). On the one hand, this perpetuates the current situation, since vehicle safety is already today a major factor in decisions to purchase a new car (Vrkljan/Anaby

---

1   The most dramatic presentation of this kind for me was "Advancing the AV opportunity" by Mark R. Rosekind, Chief Safety Officer of ZOOX, at the Automated Vehicle Symposium in San Francisco on 12th July 2018. The presentation is not available online; the content is described in part in Shladover et al. 2019.

2011). On the other hand, the bar is not set very high when one considers how much more dangerous passenger cars are, for example, than buses used in public transport in the European Union (ERSO 2019: 26).

Even scientific studies are at times unreservedly optimistic; this is especially true of older studies. Fully automated vehicles are idealized and popularized as the "crashless car" (KPMG 2012, Allesandrini et al. 2015). They became the technological embodiment of "vision zero", the goal of eliminating road fatalities entirely. This idea was soon rejected in view of the unchanging physical limits (Winkle 2015). Even simulation studies that documented increased traffic volumes as a result of fully automated car-sharing vehicles nevertheless insisted on the claim that "improvements in road safety are almost certain" (ITF 2015: 6). Here too, a fundamental contradiction was ignored. The relationship between the frequency with which road users are on the move or encounter each other and the risk of accidents, referred to as "exposure", has been well documented for several decades. More activity leads to more accidents (Elvik et al. 2009: 35). Greater restraint has now been called for, since the idea that "autonomous" vehicles would bring about absolute road safety has already fuelled expectations among future users that are seen as untenable and highly problematic (Georgieva/Kolodege 2018).

These days, the wording is normally more restrained: in policy and strategy papers and in technical development publications, a connection between automation and increased (not absolute) road safety is no longer taken for granted, but on the contrary is seen as a prerequisite for the former's approval. Many policy papers emphasize that the expected benefits of automation can only be realized with additional connectivity (in the context of C-ITS – Cooperative Intelligent Transport Services; cf. "Declaration of Amsterdam" 2016, European Commission 2017a, STRIA 2019, Meyer 2019). The authors point out that one should first of all speak of potentials that can also bring about new risks and misgivings (Feigenbaum et al. 2018). In terms of misgivings, the field of "cybersecurity" is usually emphasized. Hopes remain high despite these more recent relativizations, and each and every new technological add on increases the need for investment and development. The view thus continues to be expressed that market entry should not be postponed for too long. Since connectivity and automation could potentially already save lives (if vehicles equipped in this way were already somewhat safer than conventional cars), compromises must also be made: "We can't wait for the perfect" (Foxx in Shladover et al. 2019: 4). While this standpoint applies to assistance systems that support drivers, as will be shown below it cannot be transferred to levels of automation in which people are mere passengers.

1.2   PERSPECTIVE ON ROAD SAFETY: WHO BENEFITS?

Scientific discourse has now begun to focus not only on the potentials, but also specifically on the technological limitations (Mitteregger et al. 2022, Soteropoulos et al. 2020). An automated driving system that can reliably perform all driving tasks mastered by humans is now seen also by the industry as becoming feasible only many years into the future, if at all (Krafcik in Marx 2018). It follows that possible contributions to road safety are likewise unevenly distributed.

The advance of new technologies and the accompanying sociotechnological transformation is a complex sociological process (Schumpeter 1939, Geels/Schot 2007) and in particular a communicative process (Rogers 2003), which to date has repeatedly been accompanied by new spatial and social inequalities (Grübler 1992, and Chap. 19 by Dangschat in this volume). In the case of automated driving systems, small-scale disparities are also evident that arise due to the differing technological requirements of streets and situations with varying degrees of complexity. The more homogeneous and monitored a road section is and the more is invested in its maintenance, the better suited it is for automated driving systems. In other words, motorways – preferably newly built, in highly developed industrial societies and with a good data network – are their ideal field of application. Slow-moving shuttles used as an extension of public transport services are an exception here, but they also call for accompanying infrastructural measures (cf. Chap. 14 by Allmeier et al. in this volume). These limitations can be compared against road accident statistics to differentiate the effectiveness of the "crashless car". It has already been pointed out that for operation as an extension of public transport services, the bar of road safety is set incomparably higher than for a mere continuation of automobility as such. The European rail system would even fall into the category of "ultra-safe systems", for which fundamentally different, at times paradoxical conditions for the use of new technologies would apply (Amalberti 2001).

Of the more than 1.2 million fatal road accidents that are repeatedly cited to highlight the social benefits of automated and also connected driving systems, young, poor people in low- to middle-income countries are disproportionately represented. This group accounts for 90% of road fatalities worldwide (WHO 2015: 4). The situation is the worst in Africa, above all as a result of the relatively low level of motorization. Pedestrians and cyclists are the most vulnerable group there, with a combined share of more than 43% of road traffic fatalities (WHO 2015: 8).

Figure 1: Symbolic image from the WHO of vulnerable persons in road traffic, and Google's test operations in Chandler, Arizona.

The images shown here are explicitly excluded from the Creative Commons licence of the text. The rights remain with the authors. Sources: left: WHO (2015: VIII); right: Google (2020).

Automation

The development of connected and automated vehicles is diametrically opposed to this initial situation. In the countries and regions most affected, no significant testing is carried out, nor are virtual test environments in place for the training of learning algorithms. For the development of sensor technology, only one publicly available dataset exists, which however is exclusively oriented towards optical sensors (Mitteregger et al. 2020, Kang et al. 2019).

Differentiation is also possible with regard to potential areas of application in existing road networks. In Austria – one of the leaders in road safety – motorways and expressways account for 8.8% of road fatalities. The majority of fatal accidents occur on former federal roads (204 fatalities), state roads (104) and other roads (66; BMI 2020). The total length of the Austrian road network in 2010 was 114,590 kilometres, of which about 2,185 kilometres were motorways or expressways, corresponding to a share of around 2% (BMVIT 2012). In Austria, and especially in other countries with a high level of road safety, it is also evident that the safety level of vehicle occupants has increased, while that of road users outside motorized vehicles has decreased or remained constant. The number of fatal accidents involving cyclists, for example, has also been increasing in Austria in recent years, while the corresponding figure for passenger cars has fallen (Statistik Austria 2017: 11). The accident rate – the number of accidents per kilometres travelled – is 6.7 times higher for pedestrians and 9.4 times higher for cyclists than for drivers of cars (Elvik 2009: 56).

In summary, it can be said that a great deal of technological and economic effort is only likely to bring about an increase in road safety in the medium term, only in countries that are already privileged, and here again only on a fraction of the road network. From a current viewpoint, the genuine global problems of road safety lie entirely outside the projected development of automated driving systems. Furthermore, no serious efforts are discernible to take the actual initial situation into account in technological developments.

Connection
The connection of vehicles in networks is intended to increase road safety in two areas. Firstly, connected and automated driving systems will be supported in detecting their environment. On certain stretches of road (e.g. at intersections, on motorways or at roadwork sites) or in conditions under which a driving system is overburdened (e.g. snow, rain or an accident scene), the sensors installed in the infrastructure, or those of other vehicles, provide additional information on the surroundings so that the vehicle can continue to be driven (Carreras et al. 2018, STRIA 2019). The use of sensor technology, especially on motorways, is currently under discussion. The aim is to specifically enhance the suitability of the road network or to close gaps on individual routes by investing in the infrastructure (Fig. 2). If, for example, the performance of automated driving systems deteriorates on a section of road that is in principle suitable for their use, the infrastructure steps in to compensate (Fig. 2, right). This necessitates networking of the vehicle and installation of appropriate sensors as part of the digital infrastructure.

Although a uniform standard is still lacking, some stretches of road are already equipped with sensors in order to provide so-called "day-1 services" (e.g. information about roadworks or vehicle breakdowns). The relevant information is passed on to the driver via the networked vehicle (ASFINAG 2019, European Commission 2017a).

Day-1 services are also paving the way for the second area in which networking is to enhance the road safety of automated vehicles. With their help, accidents or incidents (and the driving system's reactions to them)[2] – and, in the more distant future, the current state of the vehicle surroundings and the driving system – can be comprehensively documented in real time. With communication between the driving systems or via a control centre, current information could be exchanged and taken into account for traffic control or route planning. In addition, connected and automated driving systems learn from each other (Casademont et al. 2019). It would thus be possible for the connected and automated vehicles from a specific manufacturer or those used by a specific transport operator – or even the entire transport system – to be optimized step by step. Such networked systems would become increasingly superior to humans with every kilometre covered, since according to the vision "all the unborn cars get born with the full wisdom of their forefathers" (Thrun in Shakland 2016).

---

2   It appears questionable as to whether the reaction mechanisms and decision-making principles of a system as complex as a highly automated vehicle or a networked fleet, which are based on learning algorithms, can ever be fully understood (Castelvecchi 2016).

In any case, the necessary digital infrastructure would generate considerable costs that could be passed on to the general public via the authorities (Polis 2018, Mitteregger et al. 2019). A comparable dynamism of externalizing the costs of an elitist system also existed at the beginning of the automotive era (McShane 1994: 203–228). Also on the part of the digital infrastructure, above all problems in the area of cybersecurity have been highlighted so far (Landini 2020).

Figure 2: Making a section of road suitable for automated driving systems by means of digital infrastructure

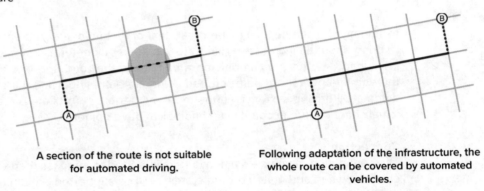

A section of the route is not suitable for automated driving.

Following adaptation of the infrastructure, the whole route can be covered by automated vehicles.

Source: the authors, based on Alkim in STRIA (2019: 21).

1.3   AN ENTIRELY NEW APPROACH

"The future of this new technology is so full of promise. It's a future where vehicles increasingly help drivers avoid crashes. It's a future where the time spent commuting is dramatically reduced, and where millions more – including the elderly and people with disabilities – gain access to the freedom of the open road. And, especially important, it's a future where highway fatalities and injuries are significantly reduced" (Elaine L. Chao in NHTSA 2017: i).

The path that has now been taken with connectivity and automation of road vehicles is being described as a fundamentally new concept of road safety. The new goal is to use technical systems to prevent accidents occurring at all (cf. Rosekind in Shladover et al. 2019: 4). In essence, this amounts to a transition from passive to active safety systems: passive safety systems such as seat belts, bumpers or airbags reduce the severity of an accident (for the occupants), while active safety systems such as emergency brake assist or adaptive cruise control prevent an accident from occurring in the first place.

This logic corresponds to that of aviation, which makes comprehensive use of this principle. For the safety of passengers in the event of potentially catastrophic events, the seat belt or the characteristics of the fuselage play a merely subordinate role. Regular civil aviation owes its high level of safety to a system that has been created to detect the principal external risk factors and to prevent known causes of accidents resulting from human error. The main components of this system are seamless air traffic control and comprehensive weather data. Specifically, this means that passengers are not first and foremost adequately secured for flying through a thunderstorm, but that the thunderstorm is detected or anticipated and then avoided. Comprehensive assistance systems also support the pilots, who in many cases only assume a monitoring role. What would this logic mean for road traffic, and for the streetscape in particular?

Figure 3: Safety as a shell (passive safety) and safety as attentiveness (active safety)

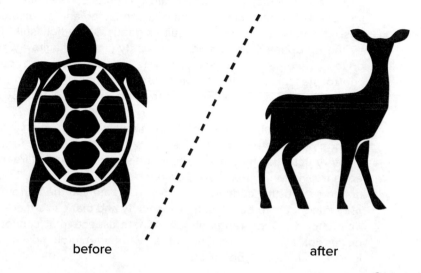

before   after

Source: the authors

Connected and automated driving systems are still only in the early, formative phase of technological development (Anderson/Tushman 1990; Bergek et al. 2008; Mitteregger et al. 2022: 67f.). When we speak today of the effects of connected and automated vehicles, a discussion arises about the potential of conceptual designs: their actual function corresponds to that which would be expected for this early phase of development. Even under ideal conditions, the performance of sensors is either equivalent or inferior to that of humans in most aspects. In addition, their functioning is greatly impaired when the surrounding conditions are not ideal (poor visibility, worn road markings, etc.; Schoettle 2017). But even under favourable conditions, their performance is still limited: a study that evaluated tests in California demonstrated that rear-end collisions are more likely under test conditions when the automated driving system is activated and no human safety driver is behind the wheel (Boggs et al. 2020). This discussion carried out in the media and among the public at large is relevant, because fundamental aspects of operation are negotiated even at this early stage in the propagation of new technologies (Rogers 2003, Foucault 1981; cf. Chap. 4 by Manderscheid and Chap. 19 by Dangschat in this volume).

Current discourse on road safety is placing fundamental demands on connected and automated driving systems; these can be summarized as follows:

1. High theoretical capabilities (potentials) are ascribed to the concepts of connected and automated driving systems in terms of road safety.

2. This is supported by the assumptions that automated driving systems:

   a. could function more reliably than human drivers and

   b. will be cognitively superior to humans.

3. Connectivity could enhance surroundings detection and compensate for possible weaknesses of an automated driving system.

4. Finally, connectivity enables comprehensive reporting (of accidents, near-accidents, incidents and near-incidents), which would allow all parts of a system to learn from each other for further improved performance.

Finally, connectivity enables comprehensive reporting (of accidents, near-accidents, incidents and near-incidents), which would allow all parts of a system to learn from each other for further improved performance. The idea of a connected and automated transport system as a perfect passive safety system would require a genuine paradigm shift. This has been clearly formulated by the European Commission: road safety (and traffic flow) were for a long time organized by drivers and other road users. The system relied on observance of traffic regulations and traffic control measures. Connected and automated transport is turning this logic upside-down: a bottom-up system is now becoming a top down system (STRIA 2019: 8).[3]

It is inconceivable that such a fundamental reorientation would not lead to corresponding changes in the streetscape. This safety system, designed for perfection, would also influence all other activities that take place in the streetscape apart from transport. Herein lies the difference to aviation, whose safety systems are being adopted. This aspect of connected and automated road traffic concerns not only road safety in the narrower sense. What is at issue is the principle that is to be chosen by a society and that can secure safety in the public streetscape. This affects a wide range of subordinate aspects that cannot be reduced to a traffic-related discussion alone, nor can they be solved by technological means – as is the frequently problematized field of cybersecurity.

2. ROADS: TRANSPORT ROUTES AND LIVING SPACE

Reducing the streetscape to a function of mere physical circulation is a reductionism with well-known far-reaching consequences. Streetscapes are also valuable living space, and this dual significance inextricably links them to the very concept of "city" (Marshall 2009). The significance attributed to roads in an evaluation of the history of the city is outlined below.

The movement of things through and in streets enables the metabolism of dense human settlements: it has a primarily biological component when it comes to the steady flow of consumer goods. But it also has a decidedly cultural component when it comes to an exchange of works that are designed to outlive their makers. And finally, cities do not only live from the movement of objects. The flow of ideas in and through streets allows things to be communicatively called into question and one's own circumstances to be created – the social component, which arises in an exchange with one's vis-à-vis, the traveller (Arendt 1958; Reki 2004; Simmel 1908: 509–512). The flow (and stagnation) of people, things and ideas organized via streets sustains cities and urban societies and makes it necessary to constantly redefine one's own position.

Without the public space of the streetscape, dense conurbations would be uninhabitable or – as alternative concepts revealed by archaeologists show – would have to be fundamentally rethought (Hodder/Pels 2010). There is a functional aspect here too, because the inhabitants of dense urban spaces need attractive public spaces (the importance of which

---

3   "In road transport, e.g. where safety and efficiency have been organized for long time [sic] with the driver and other road users in charge of complying with traffic rules and traffic management, connected and automated road transport turns this concept from bottom-up to top-down: If the electronic control systems embedded in the vehicle take decisions instead of the human driver, the cognitive capabilities of an automated vehicle are determined by the performance of its perception systems, algorithms and knowledge base" (STRIA 2019: 8).

is further increasing in view of the global climate crisis and was recognized again during the Covid-19 pandemic) which can be used as an extension of residential and living space for sitting, talking or playing (Gehl 2009, EEA 2009). These two often competing demands on the streetscape – transport on the one hand and lingering on the other – have invariably shaped the development of streets and cities. However, this conflict of usage ultimately enables streets to be seen as "institutionalized human movement" (Rykwert 1986), with their design and usage revealing dominant power structures, identities and ways of life (Sheller/Urry 2006, Cresswell 2011: 551).

2.1 THE SIGNIFICANCE OF THE STREETSCAPE FOR URBANITY

> *"[C]ities are their streets. Streets are not a city's veins but its neurology, its accumulated intelligence" (Gopnik 2016).*

In terms of quality of life, the significance of streetscapes as living space is not adequately described by a merely functional attribution. Ever since ancient times, streetscapes have been seen as part of the public space, which is what transforms the city ("polis") from a collection of stones ("urbs") into a community of people ("civitas") that acts according to certain principles (Fustel de Coulanges 1979). Entering the public space of the street means venturing out from the controlled security of private space: what is in the public domain can be looked at, criticized and modified – provided it is perceived by the public eye (Arendt 1958: 95). The social space of the streetscape is thus constantly created anew and modified (Massey 2005). It changes with its protagonists over the course of the day, through the seasons, on the basis of legislation and also with technologies that enable new ways of living (Gerhardt 2012: 32f.). This is the foundation of the "open, readily mutable nature of streets" (Appleyard 1987: 1).

As part of the public space, the streetscape is the scene of the formalized and spontaneous events and happenings of changing urban societies – where executions, music, protest, a football match or love can take place. Every modification to the streetscape thus has consequences for the city as a whole and for its society. The street thus becomes the stage of cultural struggles: wherever the right to protest is restricted, or access to the streetscape is denied to sections of society or the space is redistributed between pedestrians, cyclists and cars, the entire concept of the city is affected. Accordingly, a critical discourse concerning new technologies in public space is more than warranted, since – taking into account specific local characteristics – these can globally transform the streetscape through space demands, emissions and necessary new regulations. The passenger car is the best-known example.

2.2 SAFETY AND PUBLIC LIFE

In the search for "anthropological commonalities of mobility behaviour", Cesare Marchetti makes use of a biological determinism: humans live with an inherent tension that arises between a "cave instinct" on the one hand and the "fundamental instinct to expand their territory" on the other (Marchetti 1994: 75). Leaving the cave is thus invariably "arduous", since striving to move outside not only entails physical effort, but also carries the "danger of being attacked by predators or enemies" (ibid.). This bleak view of human existence raises the question of how animals endowed with such instincts were able to create settlements and develop villages into metropolises, which Marchetti then uses to substantiate his theory in the course of his text.

Some of the most frightening scenes in literature are based on an upheaval in the "mutable nature of the street", whereby the mass becomes a mob and turns against the individual or a minority. And even the freedom of public space always remains a privilege that can never be granted to all (Arendt 1958: 51). The cruel reality that results is that while homeless people, members of minorities or discriminated groups of people are in public space, their presence is ignored and their actions – and also their safety – are accordingly restricted (cf. Simmel 1903).

Marchetti's notion of a Hobbesian natural state of mobility remains questionable. However, the theory of constant time budgets thereby supported has resurfaced in discussions of the possible impact of automated vehicles (Almeida Correia et al. 2016, Maia/Meyboom 2018, Newman et al. 2016). What it shows, however – regardless of its inherent agoraphobia – is the importance of safety for the use of public space, because as significant as the streetscape may be as a transport and living space for cities, the protection it offers remains fragile.

"The street has always been the scene of [...] conflict, between living and access, between resident and traveler, between street life and the threat of death" (Appleyard 1987: 9). Contrary to Marchetti's thesis, the exposure or visibility necessarily associated with the public sphere has been linked to a certain form of security, based namely on density and diversity. Are humans not social animals that cannot survive on their own (Aristotle, Politics 1253a1–11)? And would the street bustling with cafés and bars not be preferable to the dark alley on one's way home at night in the vast majority of cases?

Immanuel Kant went so far as to declare the public sphere to be a constituent principle of his philosophy, according to which it functions as a critical "audience" and exposes all behaviour in which the individual acts to their own advantage and restricts or endangers others in their actions. All that must be done only in private, according to Kant, has a "fear of light": if such actions were to become public, there would be a risk that "the resistance of all would be provoked against my intention" (EwF 391, EwF 386; Gerhardt 2012: 163f.). It is thus necessary to encounter each other on an equal footing. Venturing into the public arena means taking a certain risk, since I myself will be dependent on the attention of others, and my actions will be critically examined. In return, with my attention I determine what is scrutinized and who is protected. Equal conditions only prevail where the watchful gaze can be returned.

2.3 EYES ON THE STREET

The best-known proponent from the sphere of urban planning of the position that visibility, safety and the public realm are intertwined is Jane Jacobs. She reminds her readers that safe streets cannot be the product of a centralized system of power, but are created by the individuals who use them. "Sidewalks and those who use them are not passive beneficiaries of safety or helpless victims of danger. Sidewalks, their bordering uses, and their users, are active participants in the drama of civilization versus barbarism in cities" (Jacobs 1961: 30). For Jacobs, there must be "eyes on the street, eyes belonging to those we might call the natural proprietors of the street" (ibid.: 35), i.e. people in the streetscape and the surrounding buildings. Together they provide the safety that makes life on the street – and thus in the city – at all possible.

Jacobs, too, bases her argumentation on a parity principle: it is the public itself that creates civilized coexistence. Where such a coexistence of equals has been breached, "no number of policemen, however large, can restore civilized coexistence" (ibid.: 31). Jacobs' fundamentally democratic stance is evident in her insistence on the principle of equality – and in her view that this also applies to the streets of New York's upper class, which are populated by all manner of servants, porters and dog-sitters (and now surveillance cameras to an increasing extent), who however are only there because they are paid for their activity. In truth, according to Jacobs, these places lack all incentives that would draw anyone into the streetscape of their own free will (ibid.: 40). The crucial point is that safety and security can only lead to "civilization" if they are generated by the public rather than being enforced by an institutionalized apparatus of power. All technical mechanisms and institutional bodies violate this principle.

In *Being and Nothingness*, Jean-Paul Sartre presents a detailed argumentation that is in keeping with Jacobs' subsequent observations on the streets of Greenwich Village in New York. Sartre insists that equality and liberty only exist where a gaze is returned (1962: 356). This dynamic of power loses its equilibrium when, for example, a person peers through a keyhole and sees without being seen. The architectural expression of this principle is the panopticon, which Foucault referred to as an icon of modern surveillance mechanisms (Foucault 2012).

2.4   THE END OF EQUAL CONDITIONS

Jacobs' aversion to the passenger car is closely linked to this line of argument. Her participation in the protests against urban motorways such as the Lower Manhattan Expressway, which the city planner Robert Moses wanted to cut through Manhattan, is legendary (Gratz 2010).

Jacobs' insistence on a connection between visibility on an equal footing, publicity and safety makes it clear that a car-centred urban structure not only affects road safety in the narrower sense, but also reveals a more profound effect that was responsible for "perhaps the greatest transformation of the city in the last thousand years" (Marshall 2005: 3): motorized individual transport has undermined the principle of parity of public space in practically all cities. While the occupants of a car of course perceive people in the streetscape, all visual encounter is drastically shortened. The form of a person in the vehicle is partly obscured from others by reflections in the windows. Voices heard from the outside are muffled by the vehicle's body; all sound that enters the interior has to compete with the noise of the engine. The person inside the vehicle is protected and lays claim to a space of at least ten square metres, while all oth-

Figure 4: Lower Manhattan Expressway, New York City (model)

Photo: Paul Rudolph, Lower Manhattan Expressway, New York City, c. 1970. Model, perspective; Library of Congress, www.loc.gov/item/2010647138/.

ers are confronted by tonnes of steel. Under these unequal conditions, the situation that had existed up until the onset of motorized individual transport, in which the busiest streets were invariably also the most important places of social encounter in a city, came to an end (ibid.: 3f.).

A serious attempt to restore the safety of the road that was lost due to the car thus would not only bring about a reduction in road accidents, it would help restore equal conditions throughout large parts of the road network and thus a spatial situation in which safety encourages public activity. A look at the past, however, tells a different story. Since changing concepts of freedom have been accompanied by changing opportunities for mobility, new mobility has gone hand in hand with new forms of surveillance throughout the course of history.

3. FROM SURVEILLANCE TO SOCIAL ENGINEERING

"[The] movement of persons and of 'things' (goods) will become the focal points of the transport system. All people will be connected to the transport system, as will be all goods (via the 'Internet of Things'), and they will collect and share information" (European Commission 2017b: 10).

In the feudal system, all those who could not be assigned to a particular lord were branded. Statistical census methods and prisons became widespread at a time when nation states were emerging and mobility was increasing, even for the poor. The passenger car, as a private space on public ground, is said to have helped the judiciary make increasing inroads into previously private spheres of life (Cresswell 2006, Foucault 2007, Seo 2019).

With increasing mobility, the principle based on reciprocity of social controls in societies was gradually taken over by institutions and technological innovations. In a future scenario published in part in *Mobilities* (Urry 2007), John Urry reflects on connected and automated vehicles and mobility in a "digital panopticon". Urry later also regarded this scenario as "increasingly necessary" in view of the global climate crisis (Adey/Bissel 2010: 6) and speculated that Singapore could become the first place to attain this condition. Contrary to Urry's thesis, it is argued here that not a sudden appreciation of the fragile ecosystem, but a reformation in road safety could be the driver of such a scenario.

3.1 THE NEVER-PERFECT SYSTEM

What would such a system look like? As Urry also emphasizes, this connected and automated transport system would be exclusive (cf. Urry 2008: 273f.). As pointed out above, firstly, the considerable investments necessary on the part of the public sector would be a limiting factor, and secondly, medium-term technological feasibility would restrict its use to a mere fraction of the existing road network. For John Urry, automobility is losing significance in favour of the climate. The line of development reconstructed here does not give high priority to such a transformation.

According to the current discourse, automation and networking are contributing to the disappearance of established boundaries – such as those between public and private transport or between freight and passenger transport – in the course of a "hybridization" (Lenz/Fraedrich 2015, Mitteregger et al. 2022: 44). Ultimately, the question "Does a certain route require the presence of a human, or can it be delegated to machines?" will therefore take precedence over today's fundamental question "Which means of transport do people choose for a certain route?" (Mitteregger et al. 2022: VIII). Already today (and especially as a result of the Covid-19 pandemic), a wide variety of mobile robots are in use, with which people delegate not only routes but also tasks – and here especially in the area of safety – to machines (Mitteregger 2020). In this connection, the technological limitations described above arise from the complexity of the streetscape, but apply to the small, slow-moving robots used on pavements only to a much lesser extent if at all.

However, this is not a "trend book", but the excess of existing principles. One goal that is already emerging is that of uniting the production and quality standards of the automotive sector with the capabilities of IT companies, thereby creating a new standard for all manner of automated mobile applications. Safety is seen as an essential factor here and is mentioned as a possible USP – "automotive safety" – at conferences of the technology developers (cf. Kopetz 2020).

A traffic system designed for active safety, in which vehicles can comprehensively access historical and current data, could be designed to avoid intersections, streets or neighbourhoods for example that are not considered safe. These would be bypassed, just like thunderstorms in air traffic. Ideal routes would pass along streets with easily predictable conditions. In the streetscape, the probability of predicting the behaviour of other people and objects plays a similar role to that of physics for autopilots in aviation. The entrance to a school, where masses of people not normally guided by reason alone are encountered twice a day, would be given a wide berth by this system. What approach should be taken towards individuals or groups of people that the safety system classifies as displaying problematic behaviour?

Paradoxically, in such a system more traffic could actually lead to more safety. To make better predictions of people's behaviour in the streetscape, the density of measuring points within the space would have to be increased so that behaviour could be better predicted or manipulated.

Figure 5: A selection of current applications of connected and automated driving systems

The images shown here are explicitly excluded from the Creative Commons licence of the text. The rights remain with the authors. Sources: 1: Australian Centre for Field Robotics (2017), 2: the authors, 3: knightscope (2021), 4: Casei (2018), 5: peloton-tech (no date)

If such measurements are not performed by sensors installed in the infrastructure, they could be carried out by a large number of mobile robots.

In this connection, the purely traffic-related aspect of safety in the streetscape has long played a subordinate role. New fissures in society would be inevitable as a result of targeted avoidance or as a result of a multitude of sensors at neuralgic points (or times). These would no longer be comprehensible to humans, however, since they have their origins in opaque datasets that serve for the training of artificial behaviour (cf. Castelvecchi 2016).

For security systems that rely on comprehensive reporting, it has been shown that in the course of the complete recording of data, every incident and even every near-incident can be presented "apparently convincingly" as a problem (Amalberti 2001: 113). The underlying understanding here has long been critically viewed: a technological concept promises highly theoretical potential for performance and safety, while faulty human behaviour has a negative effect on performance and must be controlled or eliminated. To date, in the transport sector this logic has been restricted to delimited fields such as industrial, mining or logistics locations, and air and rail traffic. With connected and automated driving systems, it would affect the public space of roads for the first time.

3.2   A NEW ROAD MEDIUM

The streets of the modern age, which still constitute most of our built environment today, are in many ways the perfect stage for the innate striving of this era for "absolute movement" (Jormakka 2002). Our epoch has been shaped by the tension inherent in this paradigm – the veneration of speed, rationality, grand narratives and plans on the one hand, and a constant drifting of the desired order into chaos through ever-increasing movement on the other. The street of

Figure 6: Modified choice of route for avoiding unsafe traffic situations

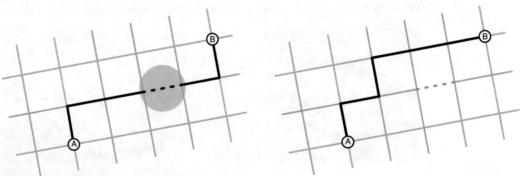

A section of the route causes unsafe traffic situations.

Following modification, the entire route can be covered with automated driving.

Source: the authors, based on Alkim in STRIA (2019: 21).

the 21st century may emerge under the paradigm of "total safety" (Zuboff 2019: 398–415). As already stated, this is not something fundamentally new, a departure from existing paths driven by external parameters, but is the exaggeration of a familiar principle. Connected and automated mobility services, which enable situation-dependent, spontaneous and flexible forms of usage, are exacerbating the rift between chaos and control. The development path shown here does not lead in a direction that sets out to solve existential problems of our time such as climate change or social and economic inequality. The streetscape of this new safety system would not be built in concrete as in our modern era. Control and order are achieved through data and endless "nudging" (Thaler/Sunstein 2009; for a critique see Stickler/Sodl 2019) – a gentle influencing of people's behaviour in the streetscape. The public space of the street in this curated world would be unrecognizable.

Hannah Arendt was at pains to point out that a coexistence of people that is designed for communality needs a "common world" which can be seen as a basis and point of departure for critical reflection (Arendt 1958: 52–55; Madanipour 2003: 114–151). This common basis can comprise things, conventions, or a shared history and laws. There is a need for a "we" that structures coexistence and that can never be dogmatic, but serves as a basis for critical reflection. Where order is established through the targeted, personalized influencing of behaviour, it may be assumed that this basis will dissolve. Whoever defines the goals of such a system must accept that their attainment will remain turbid. People in the public space of the streetscape would become mere objects, and the asymmetry of the knowledge generated would be immense (Zuboff 2019).

The public sphere and resilience
Venturing out into the public sphere calls for an "experimental attitude", since "life is problem-solving" (Gerhardt 2012: 221). The goal of total safety, implemented in a top-down safety system in the streetscape, undermines this principle in a hitherto unknown quality. Automobility has already shaken the foundation of this principle. Streets that are seen as vibrant and diverse nevertheless continue to offer equality, freedom and safety, since it can be expected that all behaviour will be subject to critical public scrutiny.

Similar to Urry's scenario, the conditions that favour implementation of such a system can be created in the "competition of cities" based on a reward system of "city rankings". A curated juxtaposition, familiar to date mainly in the form of gated communities, could even fare better in these rankings than a public sphere characterized by hustle and bustle and disorder. Ever since the emergence of sedentary cultures, human development has invariably been linked to the development of large cities. The constantly increasing complexity of settlements has made the search for solutions to urban problems a major driving force (Hall 1998: 7). Cities with advanced economies have departed from manufacturing and the turnover of goods in favour of service production, and have further developed into information societies (Castells 1989, Hall 1995, Hall/Pain 2009). We speak today of complexity and creativity as being the central resources of these cities. They are distinguished from cities and regions that are dependent on raw materials, are specialized and susceptible to market dynamics, and usually prosper for only a short time. Where public space is deprived of its mutable character and an external entity holds sway – extensively shaping, evaluating and exploiting – these places likewise become the plaything of external forces.

REFERENCES

Adey, P., and D. Bissell 2010. "Mobilities, meetings, and futures: an interview with John Urry", in *Environment and Planning D: Society and Space* (28) 1, 1–16.

Alessandrini, A., A. Campagna, P. Delle Site, F. Filippi and L. Persia 2015. "Automated vehicles and the rethinking of mobility and cities", in *Transportation Research Procedia* 5, 145–160.

Almeida Correia, G. H. de, D. Milakis, B. van Arem and R. Hoogendoorn 2016. "Vehicle automation and transport system performance", in *Handbook on Transport and Urban Planning in the Developed World*, ed. by M. C. J. Bliemer, C. Mulley and C. J. Moutou. Cheltenham: Edward Elgar Publishing, 489–516.

Amalberti, R. 2001. "The paradoxes of almost totally safe transportation systems", in *Safety Science* (37) 2–3, 109–126.

Anderson, P., and M. Tushman 1990. "Technological discontinuities and dominant designs: A cyclical model of technological change", in *Administrative Science Quarterly* 35, 604–634.

Appleyard, D. 1987. "Foreword", in *Public Streets for Public Use*, ed. by A. V. Moudon. New York: Van Nostrand Reinhold, 1–14.

Arendt, H. 1958. *The Human Condition*. Chicago: The University of Chicago Press.

ASFINAG 2019. "Motorways will start to 'talk' to modern vehicles using WLAN". https://tinyurl.com/yclc2whx (3/6/2020).

Australian Centre for Field Robotics (ACFR) 2017. "Our robots". Mantis and Shrimp. https://confluence.acfr.usyd.edu.au/display/AGPub/Our+Robots (1/2/2021).

Bergek, A., S. Jacobsson, B. Carlsson, S. Lindmark and A. Rickne 2008. "Analyzing the functional dynamics of technological innovation systems: A scheme of analysis", in *Research Policy* (37) 3, 407–429.

BMI (Federal Ministry of the Interior) 2020. "Transport statistics 2019". Vienna. https://tinyurl.com/y9vt-dxc2 (3/6/2020).

BMVIT (Federal Ministry of Transport, Innovation and Technology) 2012. "Verkehr in Zahlen. 2011 edition". Vienna. https://tinyurl.com/ycggn5db (4/6/2020).

Boggs, A. M., B. Wali and A. J. Khattak 2020. "Exploratory analysis of automated vehicle crashes in California: a text analytics & hierarchical Bayesian heterogeneity-based approach", in *Accident Analysis & Prevention* 135, 1–21. DOI: 10.1016/j.aap.2019.105354.

Carreras, A., X. Daura, J. Erhart and S. Ruehrup 2018. "Road infrastructure support levels for automated driving," Proceedings of the 25th ITS World Congress, 17–21/9/2018, Copenhagen, Denmark. https://tinyurl.com/yb8ffjry (3/6/2020).

Casademont, J., A. Calveras, D. Quiñones, M. Navarro, J. Arribas and M. Catalan-Cid 2019. "Cooperative-Intelligent Transport Systems for Vulnerable Road Users Safety", in *7th International Conference on Future Internet of Things and Cloud*, Istanbul, 141–146. DOI: 10.1109/FiCloud.2019.00027.

Casei 2018. "Athenaeum in Chicago zeichnet autonomen Konzepttraktor als revolutionären Schritt in die Zukunft des Traktordesigns aus". Press release. tinyurl.com/4pew59zx (1/2/2021).

Castells, M. 1989. *The Informational City: Information Technology, Economic Restructuring and the Urban-Regional Process*. Oxford: Basil Blackwell.

Castelvecchi, D. 2016. "Can we open the black box of AI?", in *Nature News* (538) 7623, 20.

Cresswell, T. 2006. *On the Move: Mobility in the Modern Western World*. New York/London: Routledge.

EEA (European Environment Agency) 2009. *Ensuring quality of life in Europe's cities and towns*, EEA Report 5 2009. Copenhagen: European Environment Agency.

Elvik, R., A. Høye, T. Vaa and M. Sørensen 2009. *The Handbook of Road Safety Measures*. Bingley, UK: Emerald Group Publishing Limited.

ERSO (European Road Safety Observatory) 2019. "Annual Accident Report 2018". https://tinyurl.com/y94vaeo4 (29/5/2020).

European Commission 2017a. "C-ITS platform PHASE II – Final Report", Brussels. https://tinyurl.com/y7hzgxt9 (4/6/2020).

European Commission 2017b. "Connected and Automated Transport. Studies and reports". Brussels: European Commission, Directorate-General for Research and Innovation. Download at: https://tinyurl.com/y89radja (4/6/2020).

Feigenbaum, B., G. Goodin, A. Kim, S. Kimmel, R. Mudge and D. Perlman 2018. "Policymaking for Automated Vehicles: A Proactive Approach for Government", in *Road Vehicle Automation 4*, ed. by G. Meyer and S. Beiker. Cham: Springer, 33–41.

Foucault, M. 1981. *Archäologie des Wissens*. Frankfurt am Main: Suhrkamp.

Foucault, M. 2007. *Security, Territory, Population: Lectures at the Collège de France 1977–1978*. London: Palgrave Macmillan.

Foucault, M. 2012. *Überwachen und Strafen: Die Geburt des Gefängnisses*. Frankfurt am Main: Suhrkamp.

Fustel de Coulanges, N. D. 1979. The ancient city: a study on the religion, laws, and institutions of Greece and Rome. Gloucester, MA: Peter Smith.

Geels, F. W., and J. Schot 2007. "Typology of sociotechnical transition pathways", in *Research Policy* (36) 3, 399–417.

Gehl, I. 2009. *Public Spaces & Public Life*. Seattle: City of Seattle.

Georgieva, T., and K. Kolodege 2018. "Bridging the Automated Vehicle Gap: Consumer Trust, Technology and Liability", paper presented at the 2018 Automated Vehicle Symposium, 10/7/2018. https://tinyurl. com/yba9tx9s (29/5/2020).

Gerhardt, V. 2012. *Öffentlichkeit*. Munich: C. H. Beck.

Gopnik, A. 2016. "Street Cred", *The New Yorker* (92) 30.

Gratz, R. B. 2010. *The Battle for Gotham: New York in the Shadow of Robert Moses and Jane Jacobs*. New York: Bold Type Books.

Grübler, A. 1992. "Technology and Global Change: Land Use, Past and Present", IIASA Working Paper WP-92-002, Laxenburg.

Hall, P. G. 1995. "Towards a General Urban Theory", in *Cities in Competition: Productive and Sustainable Cities for the 21st Century*, ed. by J. Brotchie, M. Batty, E. Blakely, P. Hall and P. Newton. Melbourne: Longman Australia, 3–31.

Hall, P. G. 1998. *Cities in Civilization*. London: Weidenfeld & Nicolson.

Hall, P., and K. Pain 2009. *The Polycentric Metropolis: Learning from Mega-City Regions in Europe*. London: Routledge.

Hodder, I., and P. Pels 2010. "History houses: a new interpretation of architectural elaboration at Çatalhöyük", in *Religion in the Emergence of Civilization: Çatalhöyük as a Case Study*, ed. by I. Hodder. Cambridge, UK: Cambridge University Press, 163–186.

ITF (International Transport Forum) 2015. "Urban Mobility System Upgrade. How shared self-driving cars could change city traffic", Corporate Partnership Board Report. Paris: ITC. https://tinyurl.com/yaeylous (29/5/2020).

Jacobs, J. 1961. *The Death and Life of Great American Cities*. New York: Random House.

Jing, P., G. Xu, Y. Chen, Y. Shi and F. Zhan 2020. "The Determinants behind the Acceptance of Autonomous Vehicles: A Systematic Review", in *Sustainability* (12) 5, 1719, 1–16.

Jormakka, K. (ed.) 2002. "Absolute Motion", Datutop 22. Tampere, Finland: Tampere University of Technology.

Kang, Y., H. Yin and C. Berger 2019. "Test your self-driving algorithm: An overview of publicly available driving datasets and virtual testing environments", in *IEEE Transactions on Intelligent Vehicles* (4) 2, 171–185.

Kopetz, Hermann 2020. "An Architecture for Driving Automation". https://www.the-autonomous.com/news/an-architecture-for-driving-automation/ (24/8/2020).

knightscope 2021. "Knightscope Credited for Reducing Crime". Blog post. https://www.knightscope.com/blog-1/2021/01/07/crime/ (1/2/2021).

KPMG 2012. "Self-driving cars: the next revolution". https://tinyurl.com/y6woqoks (29/5/2020).

Landini, S. 2020. "Ethical Issues, Cybersecurity and Automated Vehicles", in *InsurTech: A Legal and Regulatory View*, ed. by P. Marano and K. Noussia. Cham: Springer, 291–312.

Lazarus, J., S. Shaheen, S. E. Young, D. Fagnant, T. Voege, W. Baumgardner and J. S. Lott 2018. "Shared automated mobility and public transport", in *Road Vehicle Automation 4*, ed. by G. Meyer and S. Beiker. Cham: Springer, 141–161.

Lenz, B., and E. Fraedrich 2015. "Neue Mobilitätskonzepte und autonomes Fahren: Potenziale der Veränderung", in *Autonomes Fahren. Technische, rechtliche und gesellschaftliche Aspekte*, ed. by M. Maurer, J. C. Gerdes, B. Lenz and H. Winner. Berlin/Heidelberg: Springer Vieweg, 175–196.

Madanipour, A. 2003. *Public and Private Spaces of the City*. London/New York: Routledge.

Maia, S. C., and A. Meyboom 2018. "Understanding the effects of autonomous vehicles on urban form", in *Road Vehicle Automation 4*, ed. by G. Meyer and S. Beiker. Cham: Springer, 201–221.

Marchetti, C. 1994. "Anthropological Invariants in Travel Behavior", in *Technological Forecasting and Social Change* 47, 75–88.

Marshall, S. 2009. *Cities, design & evolution*. New York: Routledge.

Marx, P. 2018. "Self-Driving Cars Are Out. Micromobility Is In", Medium, 15/11/2018. https://tinyurl.com/y7g88778 (3/6/2020).

Massey, D., 2005. *For Space*. London: Sage Publications.

McLuhan, M. 1960. *Report on Project in Understanding New Media*. Washington: National Association of Educational Broadcasters.

McShane, C. 1994. *Down the Asphalt Path: The Automobile and the American City*. New York: Columbia University Press.

Meyer, G. 2019. "European roadmaps, programs, and projects for innovation in connected and automated road transport", in *Road Vehicle Automation 5*, ed. by G. Meyer and S. Beiker. Cham: Springer, 27–39.

Mitteregger, M. 2020. "Pandemien beschleunigen den Ersatz von Menschen durch Maschinen", *Wiener Zeitung*, 21/4/2020. https://tinyurl.com/y7k42vma (3/6/2020).

Mitteregger, M., A. Soteropoulos, J. Bröthaler and F. Dorner 2019. "Shared, Automated, Electric: the Fiscal Effects of the 'Holy Trinity'", in *Proceedings of the 24th REAL CORP*, International Conference on Urban Planning, Regional Development and Information Society, 627–636. https://tinyurl.com/yamvccho (3/6/2020).

Mitteregger, M., M. Berger and A. Soteropoulos 2020. "Algorithmen von morgen, die in einer Welt von gestern lernen: Das Inselwissen automatisierter Fahrsysteme", in *Mobilität – Erreichbarkeit – Ländliche Räume ... und die Frage nach der Gleichwertigkeit der Lebensverhältnisse,* ed. by M. Herget, S. Neumeier and T. Osigus. Braunschweig: Thünen-Institut für Ländliche Räume, 75–77.

Mitteregger, M., E. M. Bruck, A. Soteropoulos, A. Stickler, M. Berger, J. S. Dangschat, R. Scheuvens and I. Banerjee 2022. *AVENUE21. Connected and Automated Driving: Prospects for Urban Europe, trans. M. Slater and N. Raafat*. Berlin: Springer Vieweg. https://doi.org/10.1007/978-3-662-64140-8.

Newman, P., L. Kosonen and J. Kenworthy 2016. "Theory of urban fabrics: planning the walking, transit/public transport and automobile/motor car cities for reduced car dependency", in *Town Planning Review* (87) 4, 429–458.

NHTSA (National Highway Traffic Safety Administration) 2017. "Automated Driving Systems 2.0: A Vision for Safety". https://tinyurl.com/y87nzq3q (3/6/2020).

peloton-tech   (n.d.)   https://peloton-tech.com/wp-content/uploads/2017/04/Peloton1-1024x576.jpg (1/2/2021).

POLIS (European Cities and Regions Networking for Innovative Transport Solutions) 2018. "Road Vehicle Automation and Cities and Regions". Brussels: Polis. https://tinyurl.com/yb8ppy9h (3/6/2020).

Reki, B. 2004. *Kultur als Praxis. Eine Einführung in Ernst Cassierers Philosophie der symbolischen Formen*. Berlin: Akademie Verlag.

Rogers, E. M. 2003. *Diffusion of innovations*. New York: Free Press.

Rykwert, J. 1986. "The street: the use of its history", in *On Streets*, ed. by S. Anderson. Cambridge, MA: MIT Press, 14–27.

Sartre, J. 1962. *Being and Nothingness*. Reinbek: Rowohlt.

Schoettle, B. 2017. "Sensor Fusion: A Comparison of Sensing Capabilities of Human Drivers and Highly Automated Vehicles", *Report No. SWT-2017-12*. Ann Arbor, MA: University of Michigan, Transportation Research Institute.

Schumpeter, J. A. 1939. *Business Cycles. A Theoretical, Historical and Statistical Analysis of the Capitalist Process*. New York/Toronto/London: McGraw-Hill.

Seo, S. A. (2019). *Policing the Open Road: How Cars Transformed American Freedom*. Cambridge, MA: Harvard University Press.

Shakland, S. 2016. "AI expert: Super-smart cars are just a glorious beginning", CNET, 21/10/2016. https://tinyurl.com/jv7yo8y (3/6/2020).

Sheller, M., and J. Urry 2006. "The new mobilities paradigm", in *Environment and Planning A: Economy and Space* (38) 2, 207–226.

Shladover, S. E., J. Lappin and R. P. Denaro 2019. "Introduction: the Automated Vehicles Symposium 2017", in *Road Vehicle Automation 5*, ed. by G. Meyer and S. Beiker. Cham: Springer, 1–14.

Simmel, G. 1903. "Die Großstädte und das Geistesleben", in *Die Grossstadt. Vorträge und Aufsätze zur Städteausstellung*, Jahrbuch der Gehe-Stiftung Dresden, ed. by T. Petermann, vol. 9, 185–206.

Simmel, G. 1908. *Soziologie. Untersuchungen über die Formen der Vergesellschaftung*, 1st ed. Berlin: Duncker & Humblot.

Soteropoulos, A., M. Mitteregger, M. Berger and J. Zwirchmayr 2020. "Automated drivability: Toward an assessment of the spatial deployment of level 4 automated vehicles", in *Transportation Research Part A: Policy and Practice* 136, 64–84.

Statistik Austria 2017. "Straßenverkehrsunfälle mit Personenschaden. Jahresergebnisse 2016". https://tinyurl.com/y8qakyxv (3/6/2020).

Stickler, A., and V. Sodl 2019. "Nudging als Ansatz zur Förderung von nachhaltiger Mobilität? Potentiale und Risiken von verkehrspsychologischen Ansätzen zur ökologischen Nachhaltigkeitstransformation", in *Jahrbuch Raumplanung 2019*, ed. by M. Berger, J. Forster, M. Getzner and P. Hirschler. Vienna: NWV Verlag, 75–96.

STRIA 2019. "Roadmap on Connected and Automated Transport: Road, Rail and Waterborne". Brussels: European Commission.

Thaler, R. H., and C. R. Sunstein 2009. *Nudge: Improving decisions about health, wealth, and happiness.* Munich: Penguin.

Urry, J. 2007. *Mobilities.* London: Polity.

Urry, J. 2008. "Climate change, travel and complex futures", in *The British Journal of Sociology* (59) 2, 261–279.

Vrkljan, B. H., and D. Anaby 2011. "What vehicle features are considered important when buying an automobile? An examination of driver preferences by age and gender", in *Journal of safety research* (42) 1, 61–65.

Winkle, T. (2015). "Sicherheitspotenzial automatisierter Fahrzeuge: Erkenntnisse aus der Unfallforschung", in *Autonomes Fahren. Technische, rechtliche und gesellschaftliche Aspekte*, ed. by M. Maurer, J. C. Gerdes, B. Lenz and H. Winner. Berlin/Heidelberg: Springer Vieweg, 351–376.

WHO (World Health Organization) 2015. "Global Status Report on Road Safety 2015". Geneva: World Health Organization.

Zuboff, S. 2019. *The age of surveillance capitalism: The fight for a humane future at the new frontier of power.* London: Profile Books.

11 Integration of cycling into future urban transport structures with connected and automated vehicles

Lutz Eichholz, Detlef Kurth

Lutz Eichholz
TU Kaiserslautern, Department of Spatial and Environmental Planning

Detlef Kurth
TU Kaiserslautern, Department of Spatial and Environmental Planning

© The Author(s) 2023
M. Mitteregger et al. (eds.), *AVENUE21. Planning and Policy Considerations for an Age of Automated Mobility*, https://doi.org/10.1007/978-3-662-67004-0_11

1. INTRODUCTION

One of the greatest challenges facing the development of connected and automated vehicles (CAVs) today is that of anticipating the movements of other road users (Kahn 2018, Mingels 2019). Carlos Ghosn, former CEO of Renault-Nissan-Mitsubishi, commented as follows on the matter of cycling and the ongoing automation of motor vehicles: "One of the biggest problems is people with bicycles [...], [they] don't respect any rules usually [...] from time-to-time they behave like pedestrians and from time-to-time they behave like cars" (Reid 2018). This statement describes the conflicts that can arise from the close encounter between two modes of transport that differ greatly in terms of both propulsion and practical operation.

How the upcoming challenges for bicycle traffic can be solved if automated driving becomes much more prevalent in cities is being investigated by TU Kaiserslautern in the interdisciplinary research project "Concepts for the integration of cycling into future urban traffic structures with autonomous vehicles – RAD-AUTO-NOM". The Institute for Mobility & Transport (imove), the Division of Electromobility and the Chair of Spatial and Environmental Planning are jointly analysing the effects of this development in terms of traffic, technology and urban space. The research project is being funded by the Federal Ministry of Transport and Digital Infrastructure (BMVI) in Germany from June 2019 to May 2022.

This chapter discusses the first partial results of the work packages carried out at the Chair of Spatial and Environmental Planning. It specifically deals with the opportunities and risks that arise for urban bicycle traffic in interaction with CAVs. Based on analyses of the current traffic situation and of the design of traffic areas and public space, we derive the conditions under which automated vehicles can be integrated without conflict. A crucial question is how urban planning can ensure that liveable cities with a high quality of stay and of urban design can be maintained for all in a future characterized by CAVs and a high share of bicycle traffic.

2. THE FRAMEWORK CONDITIONS FOR CYCLING

For CAVs to operate accident-free and without unnecessary stopping, traffic must be predictable. The "edge case", a situation that in fact rarely occurs, describes the intuitive riding style of cyclists in the open streetscape; CAVs can only react to this with difficulty.

In German metropolises, the share of bicycles in all journeys made rose from 9% to 15% between 2002 and 2017; for other large cities the increase was from 10% to 14% (Nobis 2019: 21). It is still difficult to determine what consequences the Corona pandemic will have on transport. The current trends towards active transport are expected to intensify, with more people using motorized private transport (cf. Amelang 2020, Klein et al. 2020). These trends lend emphasis to the fact that CAVs must learn to interact with cyclists without conflict.

In the following section, we shall outline how cycling is legally regulated in Germany, how riding styles are to be classified, what types of cyclists there are and what conflicts and accidents occur in the current traffic scenario. We partly differentiate here between bicycles and pedelecs[1]. Other types of cycles such as cargo bikes, bicycles with trailers or tandems have requirements similar to those of classic bicycles.

2.1 THE LEGAL SITUATION

Where a dedicated infrastructure for bicycles is lacking, cyclists in Germany must travel on the far right of the road. In this mixed traffic with motor vehicles, a distance of 0.8 to 1 metre from the edge of the road is appropriate. This distance can be modified in hazardous situations (e.g. car doors being opened). Cyclists may overtake other road users on the left as long as a sufficient clearance can be maintained. In safe situations, they may also overtake motor vehicles on the right (ADFC 2018a). On some one-way streets, cyclists may travel in the opposite direction. It is permissible to ride faster than walking pace in traffic-calmed areas. Up to the age of eight, children must ride their bicycles on the pavement; to the age of ten, they may then choose whether to use the cycling infrastructure or ride on the pavement (section 2 par. 5 StVO).

Where infrastructure for cycling is provided, a distinction is made as to whether it is located on or next to the carriageway. Protected lanes and cycle lanes are marked on the carriageway. Protected lanes are indicated by a broken line, and cycle lanes by a solid line. Protected lanes constitute part of the carriageway and may also be used by other vehicles if necessary, but only as long as cyclists are not endangered. Cycle lanes may not be used by other vehicles (ADFC 2018a). Some protected cycle lanes and dedicated cycle paths are physically separated from the lanes for motor vehicles (ADFC 2019). Cycle lanes with compulsory use must always be indicated by an appropriate sign and have a minimum width of 1.6 metres (ADFC 2018a: 8; FGSV 2006: 84).

Cyclists have two legal options for turning left: they can leave their lane in good time ahead of the intersection (even in the case of compulsory cycle lanes) and turn from the centre of the carriageway; in doing so, they may adhere to cycle guidance markings where provided, although this is not mandatory. Alternatively, they can turn indirectly by first crossing the intersection straight ahead and then turning from the right-hand edge of the road (ADFC 2018a; section 9 StVO).

2.2 DRIVING BEHAVIOUR AND ACCEPTANCE OF RULES

The statement "As a device the bike is so closely attuned to the body that it is almost as agile" (Fleming 2017: 57; cf. Chap. 12 by Fleming in this volume) aptly describes cyclists' approach to riding: they are fast, agile and manoeuvrable, normally have no turn indicators or brake lights and can navigate through motorized and pedestrian traffic. In addition, they sometimes do not behave in accordance with the rules. In a survey, 62% of cyclists stated that they did not always obey all traffic rules (BMVI 2019a: 29). The fact that cycling infrastructure is often in poor condition or even lacking entirely is one reason for the fluctuating acceptance of rules on the part of cyclists (Huemer/Eckhardt-Lieberam 2016, Schreiber/Beyer 2019).

Cycle lanes, and protected lanes that comply with the standard widths, are used by 86% and 88% of cyclists respectively. Where the lanes are narrower than standard, the frequency of use

---

1   Pedelecs ("pedal electric cycles", e-bikes) are bicycles with an auxiliary motor that only cuts in during active pedalling. The electric motor provides support up to a speed of 25 km/h. The same regulations apply to pedelecs as to bicycles without motor assistance (ILS 2013: 9f.).

falls by around 15% (Richter et al. 2019: 77). Long red phases lead to more traffic-light violations (Schwab 2019). Riding on the left of the road in violation of the rules occurs mainly in the vicinity of significant destinations, e.g. shopping centres (Alrutz et al. 2009: 30).

The highly agile and manoeuvrable nature of bicycles is also evident in a comparison of braking distances for motor vehicles, e-scooters and bicycles. Assuming that motor vehicles travel at a speed of approx. 50 km/h in normal traffic and bicycles at a maximum of approx. 20 km/h, the stopping distance of a car is around five times that of a bicycle due to their speed and mass. Even an e-scooter travelling at the same speed as a bicycle requires a longer stopping distance (ADAC 2019):

- motor vehicle (50 km/h): reaction distance 14 m + braking distance 14 m = 28 m

- e-scooter (20 km/h): reaction distance 5.5 m + braking distance 8 m = 13.5 m

- bicycle (20 km/h): reaction distance 5.5 m + braking distance 4 m = 9.5 m

It can be assumed that the reaction distance of CAVs is shorter than that of vehicles with drivers. Nevertheless, the stopping distance from a speed of 50 km/h is still significantly greater than for bicycles. The bicycle is thus the transport medium with the shortest braking distance. This is one factor that leads to unpredictable driving manoeuvres.

In summary, the following factors have an influence on the riding behaviour and acceptance of rules on the part of cyclists: cycling infrastructure, complexity of the traffic space, subjective safety, type of cyclist and category of area (cf. Fig. 1).

Figure 1: Factors influencing cyclists' acceptance of rules and riding style

Cycling infrastructure

Cycling traffic facilities must be not only safe, but also attractive and purposeful in order to be accepted and used.

Complexity of the traffic space

Intuitive comprehensibility vs.
many choices and excessive complexity.

Subjective safety

Infrastructure that is perceived as unsafe is used less often and encourages cyclists to ride where they perceive the safety level as high, regardless of regulations (pavement instead of narrow protected lane, pedestrian zone instead of road).

Type of cyclist

Different types of cyclist have divergent riding styles: from slow and compliant to fast, taking the direct route regardless of legal requirements.

Area category

Unlawful riding on the left or crossing of roads occurs more frequently in the vicinity of buildings that are important destinations for cyclists.

Source: the authors

2.3   TYPES OF CYCLIST

The various types of cyclist represent an important indicator in assessing riding style and the demands placed on infrastructure and public space. In the Netherlands, cyclists are classified into six groups: the defining criteria are riding skills and motivation to cycle, along with physical and occupational factors. The groups can be described as follows (Woolsgrove/Armstrong 2020):

- "Everyday Cyclist": rides to work or school, takes direct routes, wants to ride undisturbed with as little stopping as possible;

- "Sporty Cyclist": sees the bicycle as an item of sporting equipment (also mountain bike, racing bike), rides long distances and fast;

- "Recreational Cyclist": rides for pleasure, rides together with others, often stops at cafés or attractions;

- "Attentive Cyclist": wants to cycle safely, knows and observes the traffic rules, wants clear traffic signs and safe intersections;

- "Vulnerable Cyclist": (mostly children, the elderly and people with disabilities) wants to cycle safely and not be overtaken by other road users (including other cyclists); infrastructure must be forgiving of mistakes;

- "Courier Cyclist": wants to reach their destination very quickly (time pressure), needs more space (cargo bike, trailer).

Roger Geller, Bicycle Coordinator at Portland Bureau of Transportation, also analysed the stress tolerance of cyclists in traffic. This results in four types of cyclists that take into account different attitudes of the general public towards cycling:

- Strong and Fearless,

- Enthused and Confident,

- Interested but Concerned,

- No Way, No How (NRVP 2019).

These "four types of cyclist" are now recognized categories. Each category is associated with specific conditions that must be fulfilled for the respective group of users in order for them to be mobile on a bicycle (ADFC 2019, Dill 2015). Geller's typology has been scientifically confirmed in several studies at Portland State University and further developed. According to this scheme, 60% of the population are "Interested but Concerned", 5% are "Enthused and Confident" and 2% are "Strong and Fearless". The remaining 33% are non-cyclists in the "No Way, No How" category (Geller 2009).

In many classifications, acceptance of rules on the part of the individual types of cyclist is not explicitly addressed. In the Dutch categorization, acceptance of rules can be considered the lowest for the "Everyday Cyclist" and the "Courier Cyclist". In Geller's typification, the designation "Strong and Fearless" already suggests a flexible attitude towards adherence to rules. Traffic ecologists at TU Dresden attest "Enthused and Confident" cyclists a low level of rule acceptance (Francke et al. 2018: 10). This can be applied to many further cyclist type models.

It is difficult to predict to what extent the algorithms of CAVs will be able to identify the various types of cyclist in future. However, as the different types have different riding styles, CAVs could possibly predict which riding style (e.g. offensive, defensive, compliant) to expect on this basis.

2.4 ACCIDENTS INVOLVING CYCLISTS

Accident statistics
According to the Federal Statistical Office, a total of 88,472 accidents involving bicycles or pedelecs occurred in Germany in 2018. From 2017 to 2018, the number of accidents involving cyclists in Germany increased by 11% (Destatis 2019). The percentage rise in fatalities among cyclists and pedelec riders is even greater, with an increase from 382 in 2017 to 455 in 2018 (+16%; ibid.). According to accident researcher Siegfried Brockmann, the overall risk of death while riding a bicycle or pedelec is three times higher than in the case of cars or motorbikes. The likelihood of serious injury is seven times higher (Burger 2019).

Accident patterns
90% of accidents involving cyclists occur in built-up areas (Destatis 2019). Accidents occurring outside built-up areas usually have much more severe consequences: 40% of these accidents have fatal consequences (BMVI 2012: 28). A large proportion of cycling accidents recorded by the police occur at junctions, or when turning into or crossing the carriageway (DLR 2015, Schreck 2016, GDV 2016). Cycling accidents involving a single direction of travel are caused by tailgating, overtaking by other road users – often with insufficient clearance distance – or obstruction by parked vehicles along cycle infrastructures (LAB 2014, Richter et al. 2019, Tagesspiegel 2019, Schreiber/Beyer 2019, DVR 2019). Further frequent causes of accidents are failure to give way and misjudgement of the speed of other road users (GDV 2015). When a collision occurs, the passenger car is the main cause of the accident in 75% of cases (Destatis 2019).

Accidents involving parked vehicles
A study by the German Insurance Association (GDV) from 2019 found that along with marked cycle infrastructure (cycle lanes and protected lanes), parked vehicles have a significant influence on the frequency of accidents involving bicycles. According to this study, 65% of accidents in cycle lanes or protected lanes occur as a result of car doors being opened (so-called "dooring"; Schreiber/Beyer 2019): the fact that 16% of car drivers fail to look back over their shoulders when opening their doors greatly increases the accident risk for passing cyclists (DVR 2019). On these sections of road with parking spaces, the accident frequency is about four times as high as in areas without parked vehicles.

Even without accidents, however, parked vehicles are a major impediment for cyclists, especially when they are parked on cycle paths – and thus incorrectly. Such obstruction hinders efficient and safe cycling: according to a study, 40% of 40,000 evaluated cyclists in protected lanes are affected by such hindrances. In a survey of 35 cycle lanes and 47 protected lanes, numerous parking violations by cars were detected: 6.1 parked and 10.1 temporarily stationary cars were registered per kilometre and hour in protected lanes, and 1.1 parked and 4.5 temporarily stationary cars in cycle lanes (Richter et al. 2019). Overall, the GDV found that almost one in five accidents involving cyclists were related to parking (GDV 2020).

3. FRAMEWORK CONDITIONS FOR CONNECTED AND AUTOMATED VEHICLES

It is difficult to predict when CAVs can be introduced in Germany. Statements from IT and automotive companies regarding technical feasibility are extremely varied in nature and are influenced by the competitive situation.

In a study published in early 2019 by the Fraunhofer Institute for Systems and Innovation Research, it is expected that new automation technologies will initially find more widespread use in local public transport (LPT) vehicles and lorries, and that higher-priced private cars will be the first to use them. Regardless of transport mode, the study assumes that even in 2050, Level 4[2] automated vehicles will still be in the minority (ISI 2019: 109–119). CAVs are expected to find widespread use throughout Germany, but their distribution will vary greatly between cities and other areas: in complex environments, CAVs will only be able to travel slowly or not at all (cf. Soteropoulos et al. 2020; Mitteregger et al. 2022: 80–83).

3.1 DRIVING BEHAVIOUR OF CONNECTED AND AUTOMATED VEHICLES

Connected and automated vehicles provide an opportunity for full enforcement of traffic regulations (Beckmann/Sammer 2016: 4; Rothfuchs/Engler 2018: 569). How such vehicles should react when there is a risk of collision is an ethically controversial matter that has not been conclusively determined (cf. Bonnefon et al. 2016). With regard to the safety of cyclists, consideration must be given as to how CAVs should be programmed for situations in which they have to decide between protecting themselves and protecting others. This can give rise to new points of conflict and accident risks for cyclists, if for example a CAV has to swerve to avoid an oncoming lorry and enters a protected or cycle lane (Awad et al. 2018).

In the research project, it is assumed that CAVs in Germany will have the following driving characteristics:

- Speed limits are not exceeded; this also applies during overtaking.

- The legally prescribed safe clearance distance to other road users is maintained at all times. Where this is not possible, the vehicle stops or slows down until the distance can be adhered to.

- Exceptions to traffic regulations can only be made when their observance would cause an accident. In critical situations, stopping is given the highest priority.

- CAVs cannot communicate digitally with all other road users. They must visually perceive other road users and obstacles.

- The algorithms used in CAVs aim to accord the safety of their own occupants the same value as that of all other road users.

---

2   In this chapter, "level" refers to the respective automation level according to SAE standard J3016 (SAE International 2018).

In view of the legally prescribed clearance distances for overtaking, CAVs will adopt a defensive driving style. Since the 2020 amendment to the Road Traffic Act, a minimum clearance of 1.5 metres applies in Germany for overtaking cyclists in built-up areas (section. 5 par. 4 Road Traffic Act). Furthermore, the introduction of a new traffic sign has been resolved that prohibits cars from overtaking bicycles in certain places (BMVI 2020).

3.2 EFFECTS OF CONNECTED AND AUTOMATED TRANSPORT ON TRAFFIC VOLUMES AND ON NUMBERS OF PARKED PRIVATE VEHICLES

To what extent public space and streetscapes can be redesigned in the "century of the seeing vehicle" (Dickmanns 2020) largely depends on whether motorized private transport (MPT) volumes are expected to increase or decrease in future. Many studies dealing with this topic currently assume that private transport volumes will increase with increasing automation (Heinrichs 2015, 2017; Botello et al. 2019; Millard-Ball 2019; Lee et al. 2019; Weert/Ruhrort 2019; Mitteregger et al. 2022: 37). The reasons for this include higher demand on the part of newly acquired user groups (e.g. people with driving impairments), increased attractiveness of private transport and, in the future, empty trips (Heinrichs 2015, 2017; Botello et al. 2019). The increasing scepticism towards shared transport media or public transport as a result of the Covid-19 pandemic in 2020 could lead to a rise in the use of MPT in the medium term — depending on the future development of the virus and the possibilities of containing it.

With traffic consisting only of connected and automated vehicles, urban traffic areas would theoretically have higher capacities than with driver-operated vehicles. This is a result of faster reaction times when moving off from standstill or travelling in a convoy, and in dense configurations with only a small distance between individual vehicles (Heinrichs 2015; Beckmann/Sammer 2016: 4; Rothfuchs/Engler 2018). A drawback is that these advantages only come to bear for CAVs when they do not have to interact with driver-operated vehicles and other road users. This would require widespread installation of new infrastructure. An alternative would be to exclude other road users in certain areas, which would mean a return to the principle of the car-friendly city (cf. Dangschat 2018).

The demand for parking space for CAVs could be greatly reduced, although it is difficult to predict to what extent. Studies that assume a theoretical value of 100% CAVs forecast a reduction in the number of parked vehicles by up to 90% if CAVs are shared by several users at the same time (Heinrichs 2015; Rothfuchs/Engler 2018; Ritz 2018; Lemmer 2019: 26–29; Millard-Ball 2019: 5; Mitteregger et al. 2022: 73). Other researchers expect a significantly lower level of savings (Skinner/Bidwell 2016, Alessandrini et al. 2015). It should be noted for the various studies that significant reductions in parking volumes are only expected to come about once high proportions of Level 4 and 5 vehicles are in operation with high occupancy rates.

Even though these assumptions regarding cost and space savings with CAVs relate to a relatively distant future, cities must already plan for this future today. Measures should be developed that promote active traffic, electric mini-vehicles and public transport, taking into account the increasing proportion of CAVs (ISI 2019: 157). Appropriate measures would be the provision of more and better-quality space for active mobility, toll systems for core and inner city areas, designation of roads or lanes that favour cars with more occupants, and extension of public transport systems — especially of PT on demand. A major opportunity for CAVs lies in the fact that their systems will first become established in high-priced cars, in buses and in lorries (ibid.: 109–119). In this transitional phase, public transport can develop much greater user-friendliness and cost efficiency, since much less personnel is needed and on-demand systems can be introduced more easily with an automated public transport system. In addition, gaps in the public transport network can be closed by means of on-demand services, with reduced dependence on both automated and driver-operated motorized individual transport.

3.3 SAVINGS IN TRAFFIC SPACE THANKS TO CONNECTED AND AUTOMATED VEHICLES

As described in the previous section, especially spaces currently used for parking will no longer be required when CAVs are in operation. This can only come about if more trips are shared in future. To illustrate these possible changes to traffic space, the current distribution of spaces must be determined.

Only a few surveys relating to the distribution of road space in cities exist. In 2014, an analysis in Berlin set out to determine what portions of traffic space are allocated to the various transport media. The outcome was that 58% is reserved for cars, including 19% for parked vehicles, and 33% of the space is taken up by footpaths, while only 3% of the traffic space is allotted to bicycles (Agentur für clevere Städte 2014). In Copenhagen, the bloggers from Copenhagenize.com examined how the traffic space is divided up on Hans Christian Andersen Boulevard. The outcome was that most of the space is allocated to motorized traffic (54%, including 12% for roadside parking); pedestrians use 26% and cyclists 7% of the space (Copenhagenize.com 2017). Random sampling carried out in our research project using online city maps yielded similar results on seven streets and at four intersections in Kaiserslautern, Mannheim and Karlsruhe.

In the Guidelines for the Design of Urban Roads (RASt 06)[3], parking spaces are included in 59 of the 97 recommended road cross sections. Areas for parked vehicles are only lacking in the sample cross sections for residential streets and roads with no building development. On average, 14% of the traffic area of all sample cross sections is reserved for stationary vehicles (parking and clearance areas; own evaluation of RASt 06, based on FGSV 2006: 36–62, cf. Table 1).

Table 1: Recommended road cross sections according to RASt 06 (FGSV 2006)

	Carriageway	Parking
Residential street	4–6.5 m	Partly without (4 of 9) 14–30% of overall width
Collector road	5.5–7.5 m	Partly without (3 of 8) 23–31% of overall width
Neighbourhood street	5.5–7.5 m	Partly without (1 of 5) 19–26% of overall width
Local entrance street	6.5–7.5m	Partly without (2 of 7) 13–23% of overall width
Main shopping street	6–13 m	Partly without (4 of 13) 11–21% of overall width

Source: FGSV (2006), the authors and assessment based on Louen (2020)

---

3   "Richtlinien für die Anlage von Stadtstraßen" (RASt 06) is a set of technical regulations valid for Germany, published by Forschungsgesellschaft für Straßen- und Verkehrswesen (Road and Transportation Research Association) in Cologne; they are recommended for use in all federal states.

On the basis of these few surveys and the evaluation of RASt 06, it can be assumed that in German cities around 15% of traffic surface is available for parked vehicles (cf. Fig. 2). The proportion varies greatly depending on the type of street. There are hardly any streets without parking areas.

Figure 2: Distribution of selected metropolitan road spaces according to road users

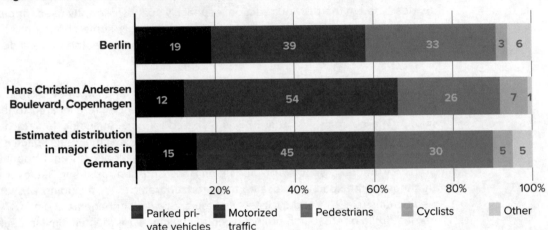

Source: Berlin – Agentur für clevere Städte (2014); Copenhagen – Copenhagenize.com (2017); the authors (estimated distribution for major cities in Germany)

Parking spaces could be dispensed with where CAVs are used for ride-sharing and are thus almost constantly on the move. However, this scenario is not likely to become reality in the near future. It should therefore be assumed that the gradual introduction of CAVs will initially only mitigate the space requirements of parked and moving vehicles to a limited extent. An inner-city traffic concept geared towards avoidance of parked vehicles can already begin to promote street profiles with a higher quality of stay and an equitable distribution of space between public transport, private motorized transport and active mobility. This can free up space for decentralized, high-quality open areas along with spaces for cycling infrastructure. Rather than only making space available to private motor vehicles, it would also be possible to provide sport and recreation spaces, playgrounds and greenery, along with extensive inner-city outdoor catering areas.

4. KEY OBJECTIVES OF THE GERMAN FEDERAL GOVERNMENT FOR CYCLING AND FOR CONNECTED AND AUTOMATED VEHICLES

4.1 KEY OBJECTIVES FOR CYCLING

The promotion of cycling makes an important contribution to attaining environmental goals and solving traffic problems in inner cities. Many European cities are recording increases in the share of cycling (Schreiber/Beyer 2019); such an increase is also to be achieved in Germany. The National Cycling Plan (NRVP) 2020 lists safe infrastructures and a pronounced sense of safety among cyclists as important aspects. The NRVP lists eight key objectives. Many of these are aiming for a significant increase in the share of cycling among all age groups and forms of usage (leisure, work, goods transport). A further goal calls for a comprehensive cycling network with low-conflict infrastructures in order to increase the subjective and objective sense of safety and to improve the quality of stay in cities. These goals are to be reached by means of digitalization and effective planning (BMVI 2019b).

4.2 KEY OBJECTIVES RELATING TO CONNECTED AND AUTOMATED VEHICLES

The Federal Government's strategy for connected and automated driving sets out to establish Germany as the lead provider and lead market. Testing of CAVs is to be made possible at an early stage in trials and regular operation. The Federal Government expects that connected and automated mobility (CAM) will increase efficiency and safety, with lower emission levels, and will strengthen Germany as a location for innovation and industry. To make this possible, the following key objectives were defined in 2015: "adaptation of the national legal framework, in particular amendments to the Road Traffic Act; adoption of a plan of measures to create ethical rules for driving computers; establishment and coordination of test fields for connected and automated driving in real traffic; support for research and development of CAV solutions from basic to applied research; and active shaping of regulations and standards in administrative bodies at European and international level" (BMVI 2015).

To avoid negative effects of CAM, the Federal Government's advisory body, the National Platform Future of Mobility, demands that private motorized transport should not increase with the spread of automated mobility. The reasons mentioned for a possible increase in the use of private vehicles include empty trips and transfer effects from other transport modes such as cycling and public transport (NPM 2019: 5).

The Federal Government has formulated the goal that Germany is to be the world's first country to allow the use of autonomous motor vehicles in regular operation, throughout the national scope of validity (Seibert 2020).

5. CRITERIA FOR THE DRIVABILITY OF CONNECTED AND AUTOMATED VEHICLES IN INTERACTION WITH CYCLISTS

Whether automated driving functions can be used depends to a great extent on the environment in which a particular vehicle is travelling (Soteropoulos et al. 2020, Mitteregger et al. 2022, SAE International 2018). The criteria that allow CAVs to operate in a setting shared with cyclists are derived from the previous sections.

The following criteria result from a consideration of the riding behaviour of cyclists and the associated demands on the infrastructure: road width, type of cycling infrastructure, speed of road users, possibility of assessing the behaviour of cyclists (clear lines of sight), type of cyclist, and the number and type of road users (cf. Fig. 3 on the next page).

Road width, the number of cyclists, type of cycling infrastructure and speed of traffic have the greatest influence on a successful coexistence of CAVs and cyclists. With a uniform speed of

20 to 30 km/h, safety clearance distances can be greatly reduced; and unobstructed lines of sight must be ensured over short stretches of road. According to RASt 06, lateral clearance distances within the space can be halved if the speed of travel is reduced from 50 to 30 km/h (FGSV 2006: 88).

CAVs cannot overtake cyclists on narrow roads and must adapt their speed accordingly. The way cyclists ride is influenced not only by the infrastructure, but also by the surroundings. If there are houses or other buildings adjacent to the road that are important destinations for cyclists, the likelihood of unpredictable riding manoeuvres increases. This makes it difficult for CAVs to operate without conflict and in accordance with regulations.

The heterogeneous nature of streetscapes presents CAVs with challenges of varying complexity. When many cyclists with different riding styles are on the move at the same time and cannot always be detected due to a lack of clear lines of sight in narrow streetscapes, drivability is severely limited. Streetscapes of this type are to be found above all in historically evolved, densely populated urban structures with a varied usage profile. More spacious roads used almost exclusively by motor vehicles are much better suited for CAVs. This type of road, on which only a small number of cyclists are currently encountered, is mainly found in outer suburbs and rural areas; examples are commercial, industrial, trunk and local access roads. This differentiation is similar to the drivability categorization complied by Soteropoulos based on the example of Vienna, according to which older, above all centrally located neighbourhoods pose a greater challenge for CAVs than peripheral, car-friendly urban areas (Mitteregger et al. 2022: 80–83). In summary, it can be stated that in dense inner-city neighbourhoods with a high proportion of cyclists, CAVs must adapt their speed to that of the bicycles moving freely in the streetscape.

Figure 3: Overview of criteria for the drivability of connected and automated vehicles in interaction with cyclists

Road width

 Safety clearance distances and compliance with the 1.5 m overtaking distance possible

 Number of lanes

Cycling infrastructure

 None, mixed traffic (protected lane, cycle lane), surmountable and insurmountable separation

 Quality of infrastructure (standard width provided, pavement quality)

Speed of road users

 Braking distances

Assessment of cyclists' driving behaviour possible

 Clear view (no obstruction from parked cars, vegetation infrastructure)

 Adherence to rules (presence and quality of cycling infrastructure, type of area, type of cyclist)

Type of cyclist

 Riding style (defensive, offensive), type of bicycle

 Vehicle (pedelec, cargo bike, bicycle with trailer)

Number and type of road users

 Cyclists, electric mini-vehicles, goods deliverers, pedestrians

Source: the authors

6. SUITABILITY OF CYCLE LANES FOR CAVS: MIXED OR SEPARATED TRAFFIC?

In Germany, cycling in streetscapes is regulated above all by four forms of traffic management: without a specific infrastructure (A), on marked protected lanes (B) or cycle lanes (C), or on separated cycle lanes (D; cf. Fig. 4).

Figure 4: Forms of traffic management for cyclists

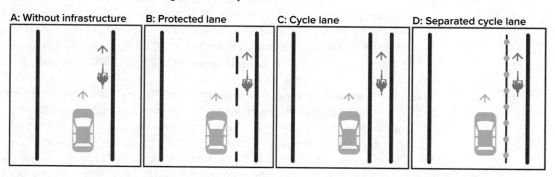

Source: the authors

Major differences in interaction between CAVs and cyclists only arise when cycling lanes are separate from the remaining carriageway. In addition to the various forms of traffic management, different streetscapes and neighbourhoods are also differently suited for the coexistence of cyclists and CAVs (cf. previous section).

The advantages and disadvantages of separated lanes for cyclists are the subject of controversial discussion among planners and accident researchers. The possible effects of the two traffic systems on the coexistence of CAVs and cyclists are outlined in the following section (cf. Table 2). First, the consequences of separating cycling and motorized traffic are described. Cities with a distinct separation between these transport modes are to be found for example in Denmark and the Netherlands (Gehl 2010: 213; Bernold et al. 2017; Milde 2017). In most other countries in Western Europe, bicycles are largely integrated into mixed traffic. Cities such as Paris, Brussels or Berlin are now starting to increasingly provide separate cycle lanes, especially on main roads.

For cyclists, a positive effect of separating bicycles from motorized traffic is that it increases their subjective safety (NRVP 2019, FixMyCity 2020); this can motivate a large portion of the population to cycle (Sinus 2017: 67, 70–72, 140), which in turn can lead to further safety gains for cyclists. The numbers of accidents involving cyclists are significantly lower in countries with a high proportion of cycling. This is "safety in numbers": motorists who expect to encounter cyclists are less likely to cause accidents involving them (Jacobsen 2003; Dutch Cycling 2019; Gehl 2010: 216; Woolsgrove/Armstrong 2020: 6; Greenpeace 2018: 10). To what extent this also applies to CAVs remains to be clarified.

For types of road frequented by a large number of different road users, structurally separated cycle lanes are the only way for CAVs to autonomously reach their maximum permissible speed (Taub 2019, Heinrichs 2015). This is due to the legally prescribed minimum clearance distance for overtaking and the difficulty for CAVs to anticipate cyclists' movements (Randelhoff 2017, Mingels 2019).

Separated cycle lanes are largely lacking in inner cities in Germany, so new infrastructure would have to be installed on a large scale. Depending on how a structural separation is effected, this can restrict the permeability of streetscapes and quality of stay and accessibility in cities, and thus detract from the guiding principle of a mixed and lively city. The public space and traffic space would also have to be significantly modified. New areas for separate cycle lanes can be created by reallocating the areas currently allotted to parked vehicles and moving motorized traffic.

Mixed traffic has the advantage that the infrastructure and public space require only little modification or reconstruction. Open streetscapes, in which all road users are active on an equitable basis and show mutual consideration, reflect the diversity of the bustling European city. They enable the liveliness and openness that characterize liveable cities, and thus contribute to a high quality of urban life. In mixed traffic, all road users that adhere to the rules must adapt their speed to each other, which would result in significantly lower speeds overall. The risk of serious or fatal accidents drastically decreases with slower traffic. The City of Oslo, which specifically slows down its motorized traffic and provides separated lanes for cyclists on main roads, did not record any road fatalities among cyclists or pedestrians in 2019 (Bliss 2018, Walker 2020).

Mixed traffic situations are very challenging for CAVs, as they have problems reacting to the intuitive movement patterns of driven vehicles and bicycles in the open streetscape. The defensive riding style of cyclists often slows them down, and overtaking is rarely possible while maintaining the legally required clearance. For cyclists to be correctly overtaken, either a second lane or a single lane with a width of 5.5 metres is needed (cf. FGSV 2006: 27, section on legal classification of bicycle traffic). Analysis of existing roads and the recommendations from RASt 06 show that existing road networks often make overtaking in compliance with the rules difficult. Correct overtaking is normally impossible in two-way traffic (cf. Table 2).

According to the current state of research on the advantages and disadvantages of separated and mixed infrastructure, different road types can be specifically developed to enable efficient transport for liveable cities. A separate cycling infrastructure is the best option on relatively wide main roads, especially where motor vehicles travel faster than 30 km/h. As far as possible, the space for this infrastructure should be reallocated from surfaces currently assigned to parked vehicles and moving motorized traffic. Mixed traffic is suitable on secondary roads where the speed of motorized vehicles is limited to 30 km/h. With an increasing share of bicycles, this would cause CAVs to adapt their speed to that of cyclists due to a lack of opportunities for overtaking, making the entire traffic situation safer.

Table 2: Cycling and connected and automated vehicles in mixed or separated traffic: strengths and weaknesses

Separated traffic	
Strengths	**Weaknesses**
• CAVs have fewer problems interacting with other road users	• Large-scale installation of new infrastructure required
• High subjective safety for cyclists	• Much space required; not possible on streets with small cross sections
• More people are motivated to cycle; this can increase safety for all cyclists	• Depending on the type of separation between transport modes, accessibility in cities is restricted

Mixed traffic	
Strengths	**Weaknesses**
• No new infrastructure needed	• Major problems for CAVs in urban traffic
• "Open" streetscapes with a high quality of stay	• In limited streetscapes with mixed traffic, highly and fully automated vehicles will probably have to switch to manual mode or travel at walking pace
• All road users adhering to the rules must travel more slowly: fewer serious accidents and increased potential for active mobility	• Low subjective safety among cyclists; however, this can increase with a high share of CAVs, as these travel and overtake according to the rules
• Mutual consideration is essential and makes for traffic on an equal footing	• Correct overtaking of cyclists is not possible at all, or only with difficulty, on most inner-city roads in Germany

Source: the authors

7. THE SIGNIFICANCE OF INTERSECTIONS

Figure 5: Intersection with provision for indirect left turns

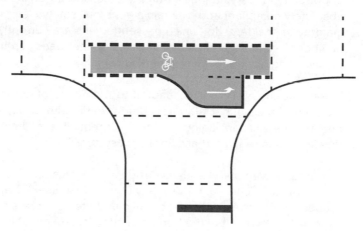

Source: the authors, based on RASt 06 (FGSV 2006)

Intersections are often places of encounter, with great significance for urban life. At the same time, they are black spots for accidents, often involving bicycles and motor vehicles. Accidents can be prevented with design variants such as protected intersections[4], or intersections where cyclists are more readily visible due to stop lines positioned in front of those for motorized

---

4   An intersection where cyclists and pedestrians are separated from motor traffic and cross using separate lanes at the side of the carriageway (cf. Darmstadt fährt Rad 2020).

traffic in combination with an earlier green traffic light phase. As the sensors of CAVs are positioned in such a way as to eliminate blind spots, this could prevent many accidents. Just like driver-operated vehicles, CAVs need a clear field of vision in order to detect cyclists under all conditions. It is therefore advisable to prohibit parking at intersections and instead find configurations that allow a clear view of the entire intersection for all road users. In addition, CAVs could better anticipate the riding behaviour of cyclists if these are restricted in their possibilities. Intersections where indirect left turns are supported (cf. Fig. 5 on the next page) or where cycling infrastructure comes to an abrupt end make it difficult to anticipate cyclists' movements.

8. CONCLUSION

Cyclists are agile, nimble and are subject to special traffic regulations. Accordingly, the sharply growing share of cyclists on the roads, not only since the Corona pandemic, presents CAVs with great challenges. Provisional results of the work packages of the Chair of Urban Planning in the research project "Concepts for the integration of cycling in future urban transport structures with autonomous vehicles – RAD-AUTO-NOM" show that without modifications to the infrastructure or to speed of travel, it will be difficult for CAVs to move through dense neighbourhoods frequented by a large number of cyclists.

The rising frequency of accidents involving cyclists, with far too often fatal consequences, shows that there is urgent need for action in order to make urban traffic space safer. This is a great opportunity for CAVs to make traffic less hazardous: unlike driven vehicles, their systems will be absolutely compliant and maintain a generous safety clearance to cyclists. This can prevent the majority of accidents for which passenger cars are currently largely accountable. Reducing the number of stopped and parked vehicles on the roads can have the same effect, since one fifth of all cycling accidents involve parked cars.

Where the infrastructure is adapted to the needs of cyclists, they are more willing to accept the rules. This is especially important for traffic with CAVs, as a lack of acceptance of rules can become a significant point of conflict for regulated and controlled automated driving (intuitive driving versus regulated and intolerant systems).

The urban environment and the design of streetscapes and public space have a major impact on quality of life in cities and the safety of their road users. Dense, compact cities can avoid a high proportion of private motorized traffic and create more shared-use zones in the public space. Especially as a result of the Corona pandemic, urban mobility is being rethought worldwide. The streetscape is seen as an "urban terrace" – a large "shared space" where catering, queues and large crowds have their place on at least an equal footing with car traffic. This transformation can only succeed where there is more equity regarding space in cities and the lion's share of the traffic area is no longer set aside for motorized private transport. In many streetscapes, the space allocated for parking alone would be sufficient to provide adequate areas for more active mobility.

Bicycle-friendly cities in Denmark and the Netherlands, which are seen as European pioneers in terms of cycling, are better prepared for a future characterized by bicycle and CAV traffic. Due to their partly separated infrastructure specifically installed for bicycles, the acceptance of rules among cyclists is higher; this could make their behaviour more predictable for CAVs.

Another way to make cycling more predictable is to modify roads in such a way that the legally prescribed overtaking clearances to cyclists can be maintained. Barriers for cyclists and pedestrians can also discourage crossing on selected roads, although such measures would have a decidedly negative impact on streetscapes and make cities significantly less liveable and diverse. This would realize the "[f]undamental contradiction: high-quality public space and CAM at higher speeds" (Mitteregger et al. 2022: 134) and be tantamount to a new car-friendly city.

Although key objectives have been defined for cyclists and for CAM in Germany's transport policy, the ways in which cities must adapt to CAVs and to the generally changing framework conditions of transport depend to a great extent on future legal regulations. An important goal for urban planners is not to allow "islands of autonomy" – areas or streets where automated driving becomes possible at an earlier stage – to spontaneously arise. The aim should be to take the planned use of CAVs as an opportunity to design streetscapes in an equitable way for all road users, especially in settlements on the outskirts of a city. Automated public transport should also be taken into account here. Lower maximum speeds for private cars and more room for road users with smaller space requirements (public transport, cyclists, pedestrians and electric mini-vehicles) are important measures for ensuring liveable urban neighbourhoods with public spaces open to all parts of the population.

The increase in cycling traffic and the gradual introduction of connected and automated mobility have little influence on the fundamental priorities for the shaping of lively cities with safe transport spaces. Significant guidelines are for example:

- maintaining clear lines of sight for all road users,

- sufficient space for the various transport media that travel at low speeds, and

- a compact, mixed-function urban structure that makes for reduced traffic volumes.

To enable automated driving in urban spaces with a high share of cycling, maximum speeds should normally be adapted accordingly. This applies above all to narrow streetscapes frequented by many different road users. In the course of the mobility transformation, especially as a result of the Corona pandemic, speed reductions have already been introduced in some inner cities (e.g. Brussels, Mainz or Bilbao). The entire public space could thus serve as an "urban terrace" and a shared space, with a significant reduction in accident figures.

9. OUTLOOK AND REQUIREMENT FOR RESEARCH

In the further course of the research project, specific recommendations and design approaches are being drawn up to prevent conflict between cyclists and CAVs. To this end, the Institute for Mobility & Transport (imove) at the Technical University of Kaiserslautern is conducting surveys and driving tests to determine functional safety requirements using methods from the transport sciences. The Division of Electromobility is setting up a simulation environment to evaluate the interaction between traffic planning concepts and technical systems. For this purpose, movement profiles of cyclists are analysed using machine learning methods to enable recognition and prediction. The Department of Spatial and Environmental Planning will develop urban planning concepts for the design of cycling infrastructure in public spaces under CAV conditions.

The matter of how cities take account of the increasing share of CAVs on the roads in their planning is giving rise to a number of additional research requirements: in this article, little mention has been made of the increasing differentiation of micromobility (cyclists, electric mini-vehicles, pedelecs, cargo vehicles, etc.) and the impact of this trend on CAVs. Intersections are also of enormous significance for conflict-free traffic. Particularly in view of their importance for urban space and traffic, they should be the subject of a differentiated examination. Differences between the various types of intersection in terms of automated drivability are an important factor here. It should be discussed how the levels of automated driving can be more strongly adapted to different streetscapes and the coexistence of different traffic modes (cf. Stayton/Stilgoe 2020). This would be helpful in planning where and how CAVs can operate.

REFERENCES

ADAC 2019. "Digitale Hilfe für den Schulterblick Ausstiegswarner gegen Dooring-Unfälle im Vergleich", 20/8/2019. Munich. https://tinyurl.com/yuanftyc (9/10/2019).

ADFC (Allgemeiner Deutscher Fahrrad-Club) 2018a. "Verkehrsrechte für Radfahrende", 28/4/2020. Berlin. https://tinyurl.com/5xzzbs7r (3/9/2019).

ADFC 2018b. "Sichere Kreuzungen durch mehr Fahrradstellplätze", 24/4/2018. Berlin. https://tinyurl.com/jfet72f8 (18/9/2019).

ADFC 2018c. Positionspapier geschützte Radfahrstreifen, 13/4/2018. Berlin. https://tinyurl.com/6hzva-7du (20/6/2020).

ADFC 2019. "So geht Verkehrswende – Infrastrukturelemente für den Radverkehr", May 2019. Berlin. https://tinyurl.com/jyfbr9tt (9/10/2019).

Agentur für clevere Städte (Hrsg.) 2014. "Wem gehört die Stadt? Der Flächen-Gerechtigkeits-Report", 5/8/2014. Berlin. https://tinyurl.com/rmttt8 (28/12/2019).

Alessandrini, A., A. Campagna, P. D. Site, F. Filippi and L. Persia 2015. "Automated vehicles and the rethinking of mobility and cities", in *Transportation Research Procedia* 5, 145–160.

Alrutz, D., W. Bohle, H. Müller and H. Prahlow 2009. *Unfallrisiko und Regelakzeptanz von Fahrradfahrern*. Bergisch Gladbach: Wirtschaftsverlag NW.

Amelang, S. 2020. "Corona crisis shakes up shift to sustainable urban mobility", Clean Energy Wire, 29/5/2020. www.cleanenergywire.org/news/corona-crisis-shakes-shift-sustainable-urban-mobility (4/6/2020).

Awad, E., S. Dsouza, R. Kim. J. Schulz, J. Heinrich, A. Shariff, J. Bonnefon and I. Rahwan 2018. "The Moral Machine experiment", in *Nature* 563, 59–64.

Beckmann, K. J., and G. Sammer 2016. "Autonomes Fahren im Stadt- und Regionalverkehr, Memorandum für eine nachhaltige Mobilitätsentwicklung aus der integrierten Sicht der Verkehrswissenschaft", 25/10/2016. DOI: 10.13140/RG.2.2.21205.42721.

Bernold, M., M. Pintner and A. Felczak 2017. "Niederlande: Erfolgsgeheimnisse aus dem Fahrrad-Paradies", in *Drahtesel. Das österreichische Fahrradmagazin* 3, 9–11. https://tinyurl.com/6k3c5ntt (6/10/2020).

Bliss, L. 2018. "The War on Cars, Norwegian Edition", Bloomberg CityLab, 3/5/2018. www.citylab.com/transportation/2018/05/oslos-race-to-become-a-major-bike-haven/559358/ (4/1/2020).

BMVI (Bundesministerium für Verkehr und digitale Infrastruktur) 2012. "Nationaler Radverkehrsplan 2020: Den Radverkehr gemeinsam weiterentwickeln". Berlin. https://tinyurl.com/2uj8nkw2 (30/8/2019).

BMVI 2015. "Automatisiertes und vernetztes Fahren". Berlin. https://tinyurl.com/7xw8atmj (18/11/2019).

BMVI 2019a. "Bundesministerium für Verkehr und digitale Infrastruktur: Fahrradverkehr – Radfahren schützt das Klima und die Umwelt". Berlin. https://tinyurl.com/sjpxzzzt (19/9/2019).

BMVI 2019b. "Die acht Leitziele des nationalen Radverkehrsplans". Berlin. www.zukunft-radverkehr. bmvi. de/bmvi/de/home/info/id/15 (19/9/2019).

BMVI 2020. "Wir machen den Straßenverkehr noch sicherer, klimafreundlicher und gerechter". Berlin. www.bmvi.de/SharedDocs/DE/Artikel/K/stvo-novelle-sachinformationen.html (19/6/2020).

Bonnefon, J., A Shariff and I. Rahwan 2016. "The social dilemma of autonomous vehicles", in *Science* 6293, 1573–1576.

Botello, B., R. Buehler, S. Hankey, A. Mondschein and Z. Jiang 2019. "Planning for walking and cycling in an autonomous-vehicle future", in *Transportation Research Interdisciplinary Perspectives* 1, 1–4. DOI: 10.1016/j.trip.2019.100012.

Burger, J. 2019. "Radfahren: Neben der Spur", in *Zeitmagazin* 26, 18/9/2019, 14–20.

Copenhagenize.com 2017. "Arrogance of Space – Copenhagen – Hans Christian Andersen Boulevard", 9/5/2017. https://tinyurl.com/2az63v42 (2/1/2020).

Dangschat, R. 2018. "Automatisierung des (urbanen) Verkehrs – Neu-Erfindung oder Widerspruch zur 'Europäischen Stadt'", in *Urbanität im 21. Jahrhundert*, ed. by N. Gestring and J. Wehrheim, Frankfurt am Main/New York: Campus, 313–335.

Darmstadt fährt Rad 2020. "Wunderlösung Schutzkreuzung? – Teil 2", 19/1/2020. www.darmstadtfaehrtrad.org/?p=2594 (22/1/2020).

Destatis (Statistisches Bundesamt) 2019. "Verkehrsunfälle: Kraftrad- und Fahrradunfälle im Straßenverkehr 2018", 19/8/2019. Wiesbaden. https://tinyurl.com/y2kn3a7c (20/9/2019).

Dickmanns, E. 2020. "Dickmanns Pionierarbeit zum autonomen Fahren", press release, 28/2/2020, Karlsruhe: Fraunhofer-Institut für Techno- und Wirtschaftsmathematik ITWM. https://tinyurl.com/pm2284pv (4/6/2020).

Dill, J. 2015. "Webinar, Part II: Four Types of Cyclists: A National Look", TREC Webinar Series, Book 5, 11/8/2015. www.core.ac.uk/download/pdf/81253510.pdf (19/9/2019).

DLR 2015. "XCYCLE Projekt – Advanced Measures to Reduce Cyclists' Fatalities and Increase Comfort in the Interaction with Motorised Vehicles", funded by The Horizon 2020 Framework Programme of the European Union, Università di Bologna. www.xcycle-h2020.eu/, www.cordis.europa.eu/project/id/635975 (12/7/2020).

Dutch Cycling 2019. "Die Niederlande – Eine Entwicklungsgeschichte des Radfahrens", *The Dutch Cycling Embassy*. www.dutchcycling.nl/images/downloads/Brochure_German.pdf (10/10/2019).

DVR (Deutscher Verkehrssicherheitsrat) 2019. "Dooring-Unfälle: Mehr als jeder zehnte Auto Fahrende beim Aussteigen unaufmerksam", 9/9/2019. Bonn. https://tinyurl.com/6xbz7x46 (9/10/2019).

FGSV (Forschungsgesellschaft für Straßen- und Verkehrswesen) 2006. "Richtlinien für die Anlage von Stadtstraßen – RASt 06", Issue 2006. Download at https://tinyurl.com/3j2jdvsx (19/8/2020).

FixMyCity 2020. "Studie zur subjektiven Sicherheit im Radverkehr – Ergebnisse und Datensatz einer Umfrage mit über 21.000 Teilnehmenden", 6/7/2020. Berlin. www.fixmyberlin.de/research/subjektive-sicherheit (10/7/2020).

Fleming S. 2017. *Velotopia: The Production of Cyclespace in Our Minds and Our Cities*. Rotterdam: nai010.

Francke, A., J. Anke and S. Lißner, 2018. "Sag mir, wie du radelst und ich sage dir, welche Infrastruktur du dir wünschst – Darstellung erster Ergebnisse einer Radfahrtypologie", in *26. Verkehrswissenschaftliche Tage: Grenzenlos(er) Verkehr!?, Tagungsband*, ed. by TU Dresden, 139–152.

GDV (Gesamtverband der Deutschen Versicherungswirtschaft) 2015. "Geschwindigkeitswahrnehmung von einspurigen Fahrzeugen", 12/6/2015. Berlin. Download at www.udv.de/de/publikationen/forschungsberichte/geschwindigkeitswahrnehmung-einspurigen-fahrzeugen (1/7/2020).

GDV 2016. "Typische Unfälle zwischen Pkw und Radfahrern". Berlin. Download at www.udv.de/download/file/fid/10069 (1/4/2020).

GDV 2020. "Unfallrisiko Parken für Fußgänger und Radfahrer", *Unfallforschung kompakt* No. 98. Berlin. Download at https://udv.de/download/file/fid/12576 (12/7/2020).

Gehl, J. 2010. *Städte für Menschen*, 3rd edition. Berlin: Jovis.

Geller, R. 2009. "Four Types of Cyclists". Portland, OR: Portland Bureau of Transportation. www.portlandoregon.gov/transportation/article/158497 (9/5/2019).

Greenpeace 2018. "Radfahrende schützen – Klimaschutz stärken: Sichere und attraktive Wege für mehr Radverkehr in Städten", 1/8/2018. Hamburg. https://tinyurl.com/fwfp7jy8 (10/2/2020).

Heinrichs, D. 2015. *Autonomes Fahren und Stadtstruktur*. Berlin/Heidelberg: Springer Vieweg.

Heinrichs, D. 2017. "Autonomes Fahren fordert die Planung heraus", *Treffpunkt Kommune*, 5/10/2017. www. treffpunkt-kommune.de/autonomes-fahren-fordert-die-planung-heraus/ (20/9/2019).

Huemer, A., and K. Eckhardt-Lieberam 2016. "Regelkenntnisse bei deutschen RadfahrerInnen: Onlinebefragung unter Erwachsenen und SchülerInnen", in *Zeitschrift für Verkehrssicherheit* (62) 5, 21–27.

ILS (Institut für Landes- und Stadtentwicklungsforschung) 2013. "Einstellungsorientierte Akzeptanzanalyse zur Elektromobilität im Fahrradverkehr", *ILS-Forschung* 1/2013. Dortmund. www.repository.difu.de/jspui/bitstream/difu/220522/1/DM13110843.pdf (10/7/2020).

ISI (Fraunhofer Institute for Systems and Innovation Research) 2019. "Energie- und Treibhausgaswirkungen des automatisierten und vernetzten Fahrens im Straßenverkehr", Wissenschaftliche Beratung des BMVI zur Mobilitäts- und Kraftstoffstrategie. Karlsruhe. https://tinyurl.com/u587csvs (15/12/2019).

Jacobsen, P. 2003. "Safety in numbers: more walkers and bicyclists, safer walking and bicycling", in *Injury Prevention* (9) 3, 205–209.

Kahn, J. 2018. "To Get Ready for Robot Driving, Some Want to Reprogram Pedestrians", Bloomberg News, 16/8/2018. https://tinyurl.com/ymajj9vv (15/9/2020).

Klein, T., D. Köhler, T. Stein and E. Süselbeck 2020. "Radverkehr im Ausnahmezustand. Mit Rückenwind aus der Krise?", difu Deutsches Institut für Urbanistik, 2/6/2020. www.difu.de/nachricht/radverkehr-im-ausnahmezustand-mit-rueckenwind-aus-der-krise (4/6/2020).

LAB (The League of American Bicyclists) 2014. "Bicyclist safety must be a priority: Findings from a year of fatality tracking – and the urgent need for better data". Washington, DC. www.bikeleague.org/sites/default/files/EBC_report_final.pdf (10/6/2020).

Lee, P., J. Loucks, D. Stewart, G. Jarvis and C. Arkenberg 2019. "Technology, Media, and Telecommunications Predictions 2020", Deloitte.Insights. https://tinyurl.com/24f5dh9e (2/1/2020).

Lemmer, K. 2019. *Neue autoMobilität II: Kooperativer Straßenverkehr und intelligente Verkehrssteuerung für die Mobilität der Zukunft*. Munich: Utz.

Louen, C. 2020. "Veränderungspotential im Straßenraum durch autonomes Fahren", Lecture at the Conference on Autonomous Driving and Urban Development, Kassel.

Milde, M. 2017. "Radverkehr in den Niederlanden und Deutschland. Ein kritisch-optimistischer Vergleich", Lecture, AGFS Congress 2017: 200 Jahre Fahrrad. https://tinyurl.com/2tw8fauy (6/10/2019).

Millard-Ball, A. 2019. "The autonomous vehicle parking problem", in *Transport Policy* 75, 99–108. https://tinyurl.com/wm4a8ypy (7/11/2019).

Mingels, G. 2019. "Das Auto darf selbst fahren – so, wie ich es will", spiegel.de, 1/11/2019, www.spiegel.de/plus/alex-roy-der-schrillste-kritiker-der-selbstfahrenden-autos-a-00000000-0002-0001-0000-000166735193 (7/11/2019).

Mitteregger, M., E. M. Bruck, A. Soteropoulos, A. Stickler, M. Berger, J. S. Dangschat, R. Scheuvens and I. Banerjee 2022. *AVENUE21. Connected and Automated Driving: Prospects for Urban Europe, trans. M. Slater and N. Raafat*. Berlin: Springer Vieweg. DOI: 10.1007/978-3-662-64140-8.

Neufert, E. 2009. *Bauentwurfslehre*. Wiesbaden: Vieweg+Teubner.

Nobis, C. 2019. "Mobilität in Deutschland – MID. Analysen zum Radverkehr und Fußverkehr", study by infas, DLR, IVT and infas 360 on commission from the Federal Ministry for Digital and Transport (FE-Nr. 70.904/15), Bonn/Berlin. https://tinyurl.com/3psam2vk (5/1/2020).

NPM (National Platform Future of Mobility) 2019. "Zweiter Zwischenbericht: Handlungsempfehlungen zum Autonomen Fahren. Arbeitsgruppe 3: Digitalisierung für den Mobilitätssektor". Berlin. https://tinyurl.com/k9kuxdh5 (7/12/2019).

NRVP (Nationaler Radverkehrsplan) 2019. "Keine Angst beim Radfahren: Subjektive Sicherheit im Radverkehr", Fahrradportal, 5/6/2019. Berlin: Federal Ministry for Digital and Transport. www.nrvp.de/21242 (5/6/2019).

Randelhoff, M. 2017. "Zukunftsszenarien für das automatisierte Fahren in der Stadt – Chancen und Risiken für den Radverkehr", lecture, 27/12/2017. www.zukunft-mobilitaet.net/166701/verkehrssicherheit/automatisiertes-fahren-radverkehr-zukunftsszenarien-stadtverkehr-vortrag/ (18/9/2019).

Reid, C. 2018. "Cyclists Don't Respect Rules, Said Nissan CEO Carlos Ghosn Arrested On Suspicion Of Fraud", *Forbes*, 19/11/2018. https://tinyurl.com/32fu2k5d (15/9/2019).

Richter, T., O. Beyer, J. Ortlepp and M. Schreiber 2019. "Sicherheit und Nutzbarkeit markierter Radverkehrsführungen", *Forschungsbericht Nr. 59*. Berlin: Gesamtverband der Deutschen Versicherungswirtschaft. Download at www.udv.de/download/file/fid/11903 (10/10/2019).

Ritz, J. 2018. *Mobilitätswende – autonome Autos erobern unsere Straßen: Ressourcenverbrauch, Ökonomie und Sicherheit*. Wiesbaden: Springer.

Rothfuchs, K., and P. Engler 2018. "Auswirkungen des autonomen Fahrens aus Sicht der Verkehrsplanung", in *Straßenverkehrstechnik* 8, 564–571.

SAE International 2018. "Taxonomy and Definitions for Terms Related to Driving Automation Systems for On-Road Motor Vehicles – J3016", 15/6/2018. www.sae.org/standards/content/j3016_201806/ (20/4/2020).

Schreck, B. 2016. "Radverkehr – Unfallgeschehen und Stand der Forschung", in *Zeitschrift für Verkehrssicherheit* (62) 2, 63–77.

Schreiber, M., and O. Beyer 2019. "Sicherheit und Nutzbarkeit markierter Radverkehrsführungen", in *Straßenverkehrstechnik* (12), 585–865.

Schwab, A. 2019. "Radwege und niederländische Kreuzungen: Keine Wunderlösung", in *mobilogisch!* 2, 18–23.

Seibert, S. 2020. "Gestärkt aus der Krise, gemeinsam die Mobilität der Zukunft gestalten", *Pressemitteilung 316*, 8/9/2020. Berlin: Federal Press Office (BPA). https://tinyurl.com/hv4vctxd (10/9/2020).

Sinus 2017. "Fahrrad-Monitor Deutschland 2017: Ergebnisse einer repräsentativen Online-Umfrage", 25/10/2017. Heidelberg/Berlin: Sinus Markt- und Sozialforschung. https://tinyurl.com/way95epp (18/2/2020).

Skinner, R., and N. Bidwell 2016. "Making Better Places: Autonomous vehicles and future opportunities", WSP | Parsons Brinckerhoff in association with Farrells. www.wsp.com//media/Sector/Global/Document/Making-better-places.pdf (6/10/2020).

Soteropoulos, A., M. Mitteregger, M. Berger and J. Zwirchmayr 2020. "Automated drivability: Toward an assessment of the spatial deployment of level 4 automated vehicles", in *Transportation Research Part A: Policy and Practice* 136, 64–84.

Stayton E., and J. Stilgoe 2020. "It's Time to Rethink Levels of Automation for Self-Driving Vehicles", SSRN, 13/5/2020. Download at https://tinyurl.com/4sfj6efa (7/10/2020).

Tagesspiegel 2019. "Radmesser". https://interaktiv.tagesspiegel.de/radmesser/ (3/9/2019).

Taub, E. 2019. "How Jaywalking Could Jam Up the Era of Self-Driving Cars", *New York Times*, 1/8/2019, www.nytimes.com/2019/08/01/business/self-driving-cars-jaywalking.html (6/10/2019).

Walker, A. 2020. "Oslo saw zero pedestrian and cyclist deaths in 2019. Here's how the city did it", *Curbed*, 3/1/2020. https://tinyurl.com/56sy6yjk (7/1/2020).

Weert, C., and L. Ruhrort 2019. "Mobilitätsatlas: Daten und Fakten für die Verkehrswende", ed. by Heinrich-Böll-Stiftung and VCD Verkehrsclub Deutschland e.V. www.boell.de/sites/default/files/2019-11/mobilitaetsatlas.pdf?dimension1=ds_mobilitaetsatlas2019 (6/11/2019).

Woolsgrove, C., and J. Armstrong 2020. "Best Practice Guide: Safer Cycling Advocate Program". Brussels: European Cyclists' Federation. https://tinyurl.com/53r5zn9p (15/1/2020).

12 Against the driverless city

Steven Fleming

It's the first rule of marketing: don't start with the product, start with the customer's need. The customer seeking a drill bit, we are reminded, is really just seeking a hole.

We could likewise say the person who we think wants a driverless car, doesn't want that, or even a helicopter or teleporter for that matter. They want to be somewhere else.

If that somewhere else is a place of retreat — a house in the country or isolated beach — then maybe, just maybe, an autonomous car is the answer. But how many trips are to get away from other people? Most trips are not to places per se. Most trips are to meet people or join groups. We move for the sake of new unions.

None of us wants a driverless car, any more than we want a new drill bit. What we really want is to regroup. From the group we're a part of at breakfast — our family, let's say — we might want to join a work team, crew, gang, committee, cohort, jury, class, board, panel, etcetera, in time for the start of the workday. Before returning to the group we began with, we might have joined audiences, choirs, class reunions, sports teams, or any of hundreds of possible groupings.

Many groupings, of course, involve only two people: a server and a diner, a doctor and a patient, a customer and a shop assistant, or simply two friends. Whatever their size, the possibility of many such groupings in a day is the raison d'être for cities. The form of any city, along with its mobility system, is a result of the inhabitants' wish to group and regroup. It's why central land in a city costs more: times taken to get there are shorter.

Artists' impressions of future cities, if cities were reshaped to suit driverless cars, appeal to our emotions — they are sublime. They hypnotize us with technological possibilities, the way a new drill can. If we fall under the spell of those visions, we risk treating our cities like Masonite sheets in the hands of weekend handymen with new drill bits, turning them into pegboard for no reason except that their drill bits allowed.

We've seen how our ancestors got carried away with the car. They went crazy remodelling human settlements to suit the new tool, forgetting what settlements are built for, and that is to help people group and regroup. We repeat their mistake if we get too excited about the new kind of cars, the ones we don't steer. We start thinking these shiny new tools will let us build places of meeting even further apart, never mind what intuition would tell us, that the regrouping function is served by compactness. Since we won't have to drive them, we start thinking of cars as buildings of sorts, in which families might sleep or staff might sit in face-to-face meetings. Since we're not concerning ourselves with life-cycle analyses, public health or

---

Steven Fleming
Cycle Space

M. Mitteregger et al. (eds.), *AVENUE21. Planning and Policy Considerations for an Age of Automated Mobility*, https://doi.org/10.1007/978-3-662-67004-0_12

the poor, we might as well go all the way and imagine our bodies as playing cards, and every grouping as a dealt hand, and every building as a card dealer, in a world operating as a giant machine, built to shuffle people around.

There are cities like Dubai that function as symbols of wealth. In cities like those, the re-sorting of people into teams, boards, crews and the like, could cease altogether and the economy would barely be touched. The meeting rooms, classrooms and shops aren't the engines of Dubai's wealth. The oil fields and gulf are, and those don't depend on people coming together. If any city could sink its surplus into a gigantic experiment that involves driverless cars, and all the infrastructure they require, it would be Dubai.

Most cities aren't like that. They depend on mobility, even more than the basics, like food and water. There's the countryside for those who are hungry or thirsty. What brings people to cities is the opportunity of connecting with people. You only have to look at what social distancing did to most economies during the Covid-19 pandemic to know cities' fortunes depend on face-to-face groupings, and regroupings, of people. Most cities don't have the luxury of experimenting with city forms that proceed, teleologically, from a new gadget. The gadget's success could be the city's demise. Thinking must proceed from the problem of regrouping itself.

Accepting that regrouping is the city's paramount function, leads us to take a step back for a moment and look at the city in such terms alone. As best as we can, we need to look past the pegboard mistake of our immediate ancestors who got carried away building settlements dependent on cars. We should also look beyond whatever form our cities took as a result of the invention of trams and trains. At a time when some are rushing to imagine how driverless cars might reshape the city, we should pause to think in the opposite manner and not even assume that historic town centres,, shaped over centuries by the most basic mode, walking, are necessarily more than historical blunders.

It would be nice to suggest that we think about building a city from scratch, to be the most fluid city ever, when it comes to sorting people into new groups. However, we know from the arguments caused by Thomas More's *Utopia* or Plato's projection of an ideal polis in the *Laws* that discussions that aren't grounded in tangible models can throw up more stumbling blocks than they provide answers.

What we need are examples of actual places, specifically designed to group and regroup large numbers of people, that are sufficiently like cities for us to see how their lessons might be transferred. High schools and Olympic parks would both qualify. However, for their sophistication and comparability to cities, airports do the most for our thinking.

The biggest airports (Hartsfield-Jackson Atlanta, Beijing, Dubai, O'Hare, etc.) regroup more people per day than cities of a million inhabitants. At holiday time they manage more than 300,000 connections per day. If we disregard security checkpoints, which cities don't have, and luggage, which city commuters don't carry and which in airports is transported underground, then we can see the airport as a place that lets Bob and Mary leave the group of passengers they arrived with on a plane from Bogota, and a few minutes later—fifteen at the most — become members of a group who will be departing on the next plane to New York. That's the time it will take Bob and Mary to walk to the Skytrain, take a short ride, then walk from the train to their gate for departure.

On the way, Bob and Mary will have crossed paths with more people than cross their path when they commute in their city. However, those strangers will not have turned traffic lights red on their way. Those strangers will not have caused crowds to thicken so much that Bob and Mary were unable to pass. If Bob and Mary noticed any of those strangers at all, it would have been because they thought they were attractive or strange. None would have been a hindrance, like traffic.

It's just as well elected governments don't operate airports. Imagine if the mayor of your city was given the reigns. Come hell or high water, she would find some way of opening the concourse to all the car addicts who gave her their vote, and wouldn't they love her for that! Instead of a concourse where people must walk, they will have one that allows them to drive. They could drive right the way through and collect their loved ones at the gate. They could drive around looking for parking spots right outside any restaurant or shop. They will have a concourse in which there are taxis hustling to drive Bob and Mary from the gate they arrived at to the gate they must get to before the plane they must catch goes to New York without them on board.

It would be madness, but don't laugh: this very thing happened to the city you live in.

I'm from Australia where streets use to have porticos extending halfway out to the middle, but traffic engineers, using words borrowed from the field of hydraulics, wanted streets that had frictionless edges. There's a sad story in Charles Montgomery's book *Happy City* about the pioneer of traffic engineering, Miller McClintock. He completed a PhD on the topic in 1924, filled with arguments for the removal of cars from the city, but with a family to feed, he took a position financed by Studebaker right after university to champion an antithetical line. The new line involved the abolition of porticos. Regardless of how they protected pedestrians, the way pedestrians in airports have roofs to walk under, the columns of porticos were vulnerable in the event of a car crash. Studebaker wanted their customers to have the freedom to crash.

To different degrees, it's a similar story all over the world. Right when walking was getting hard enough anyway because cities were growing, traffic engineers exposed walking to the weather and stopped it with red lights in the shape of stationary men. Now, every few minutes a green walking man flashes for a few seconds, then it's back to vroom vroom.

Movement got put into cars, which in the city don't move. It was as though the city was viewed as a parade ground for the worship of rich peoples' automobiles, crawling at a snail's pace for something to do.

In cities where prosperity derives from the length of a river, the depth of a harbour, a natural fortification or access to natural resources, you might argue that mobility doesn't matter much, anyway. What does it matter if cars cause congestion if they are useful as symbols of status!

But cities, these days, don't function that way. Their fortunes ride on their capacity to assemble and reassemble groups sourced from potentially millions of people, all in the course of a single workday. Families relocate to cities for the opportunity of sending one member to a specialized school, another to a specialized job and of putting another's specialized service or product in reach of millions of potential customers – and all of this during a single workday. Some come from rural areas, where there are no jobs, markets or schools. Others come from cities where commute times from most neighbourhoods (the poor outer districts, which the bourgeoisie tend to ignore) are so slow the city's millions lack access to opportunities all of them would naturally seek.

The main objection to this definition of cities, as places where millions struggle daily for the chance to connect, comes from the modern-day patrician class. Occupying central neighbourhoods, a short stroll from the best schools, shops and jobs, they bemoan the plebeians and their movement. They define cities as places in which to be still. They say the masses don't need all their movement. They need better cafés, or, as Marie Antoinette might say, instead of mobility, they can eat cake.

Enough cities in this world are burgeoning despite their terrible coffee, and in some cases, barely a park or museum, for us to be certain that niceties like these are not the reason that cities exist. Cities exist for the possibility of getting from district to district for education or trade. Cities without the finer things – fine public space, for example – may see the disappearance of a handful from the upper echelons of society, seeking cities with culture or sun, but cities with gridlock lose the struggling masses, who by their sheer weight of numbers matter far more. The cities that win them, and the companies keen to employ them, are the cities with decent commute times. The airport analogy helps us understand why.

The world's biggest airport would be in New York, if hot destinations were what made certain airports turn into international hubs. However, scanning the field of international airports, we see the really large ones – Dubai, Atlanta, O'Hare – had something else going for them. There was efficiency in the way they helped people regroup that wasn't lost as new terminals kept being added. Whichever airport in a region of the world can keep on providing quick transfers as more and more airlines choose it as a base, will become a honeypot for even more airlines, which will keep coming as long as the concourse and Skytrain can shuffle the volumes of people effectively.

The honeypot lures all sorts of airlines, from those flying small planes to obscure destinations to those filling Airbuses headed to capital cities. The bigger the airport, the easier it is to fully book flights. The bigger the airport, the bigger the market to which an airline will have access.

But the airport must be efficient. If one airport can't guarantee swift passenger connections, there are others that airlines can choose.

The busiest airport in the world is in Atlanta. That's not because Atlanta is unrivalled in terms of its central location to receive international and domestic air traffic, and it's certainly not because millions are clamoring to visit Atlanta itself. It's because the terminal design was efficient and remained efficient as its efficiency led to more growth.

The same mobility system underpins airports as underpins transit-oriented cities. A mass rapid transit system – a Skytrain – moves between terminals, leaving people to walk the last quarter mile. To save them cluttering the concourse while schlepping their luggage, suitcases get handled by others downstairs. Meanwhile, to speed up their walking, moving walkways are often provided.

But stop for a moment and think about the ways many staff move around in these places. Because they are familiar with the environment, they can be trusted on bikes, or in some countries Segways or scooters. Whatever the choice of personal mobility device, it will be faster than walking, yet equally agile. If passengers were familiar with the environment, they might all be trusted to cycle as well.

For staff with duties covering a whole airport, having fast wheels is essential. In some airports it can be as far to walk from one extremity to the other as it is to walk from midtown to downtown in Manhattan. Walking would consume their whole day. But agility is important as well. If staff made these trips in the golf carts used to help the disabled move around airports, they would be slowed down the way carts are slowed down, by pedestrians in their paths. What staff need are nimble wheels that allow them to slip between gaps.

We've imagined what would happen to airports if shrinking, populist, grovelling mayors were in charge, but what would happen if the CEOs of airports became the mayors of our cities? Could they ban cars, remove traffic signals and kerbs, reinstate weather protection and allow as many people as could be trusted to speed up their walking with bikes? The first of many questions this raises, is whether or not this would be a recipe for crashes.

There is a live experiment in the central borough of Amsterdam – a district where 68% of all trips are by bike – that absolutely dispels fears that lots of bikes in an airport-like city would cause injuries due to bike crashes. In the middle of the craziest crossroads, where north/south and east/west bike and pedestrian traffic gets funnelled, the city of Amsterdam has created a "shared space", the gedeelde ruimte between Amsterdam Centraal and the city's free ferries to Amsterdam Noord. The nearest thing you could liken it to are those figure-8 sessions they used to run at skate rinks, the ones that always seemed to end with some poor kid breaking their arm. The shared space behind Amsterdam Centraal is a pure cluster of madness that no planner would want for their city (especially if they cared about the vision-impaired), but it does prove one thing without doubt: pedestrians and cyclists mix without hitting.

That's not to say no cyclist anywhere in the world has ever crashed into somebody walking; those incidents get so much attention from people who think bikes aren't legitimate transport but toys, that no one could be unaware that such accidents happen. What is being argued is simply that, where both pedestrians and cyclists are prevalent and known to each other, like in the shared space behind Amsterdam Centraal, they miss each other with surprising aplomb. They are doing, naturally, what engineers are striving to achieve with autonomous cars. They're automatically engaging in lane formation by following the rider ahead, and they are making subtle adjustments to their speeds in order to anticipate, and just miss, all the others crossing their paths.

To anyone with an interest in two and two, and in putting them together with a mind to reaching new and amazing conclusions, a few possibilities arise from this observation. We start to imagine cities that don't have any cars, but that do have weather protection, like supersized airports. If we like, we can imagine what might have happened to the world's prototypical city, Rome, if the ancient world had had bikes.

The previously walled centre of Rome was 13.7 square kilometres and was traversed at walking speed only. Linearly, cycling is five times faster than walking (15 to 20 km/h, as opposed to 3 to 4 km/h) and therefore capable of encompassing areas 25 times greater in the same time. Had the Romans all cycled, their city could have been 68.5 square kilometres in area and still have had the same average trip times. Instead of housing just 450,000 people, the ancient city of Rome could have housed more than ten million.

Let's not quibble about densities, streets widths or how ancient Roman cyclists would have made it up the Palatine Hill with no gears. Let's just acknowledge that if we were starting from scratch, with a new city, we could achieve a hell of a lot if we planned it around cycling and banished the cars.

What would a city made from scratch, around bicycle transport, be like? Our imaginations wouldn't be too far off target if we pictured 18th and 19th-century cities built around the speed and dynamics of horses. What is Amsterdam, if not a city designed around horses, where the horses got swapped for bikes?

We have licence, though, to imagine something tailor-made for bicycle transport. We are well within our rights, for example, to imagine a city where the streets are protected from rain. Why not! We no longer need rain to wash away horse poo. These days, we have sanitary bicycles that we can use – not to mention a plethora of tiny, electric-powered, personal mobility devices. It's about time we thought about streets that provide pedestrians and cyclists with weather protection.

It's time too that we thought about spiralling buildings, buildings that would allow us to cycle all the way up and through them. The time is right to think about undulating ground surfaces, where places of stopping and slowing are raised, and travelling paths are all lowered. That way, we would not have to brake to slow down or pedal any harder when getting our bikes back to speed.

What gives us licence to think in such fanciful terms, about some imaginary, future, bike city? We are granted this licence by all those who have been imagining the city of the future if it had driverless cars. If it is OK for some to speculate about cities being even more sprawling, or to picture the street as a virtual train track, filled with cars electronically coupled so they speed along bumper-to-bumper, it is fair that we be allowed to imagine a city planned around bicycle transport.

It may just be the case, too, that a city designed around cycling will have even faster connections, and provide even more comfort, than a city shaped by driverless cars.

In a moment (on the next pages) we'll look at three images imagining the city of the future. Though they haven't been published before, the thinking behind them is from my 2017 book *Velotopia*. It was promoted as the book Ebenezer Howard might have written had he cared more about cycling than trains, or the book Le Corbusier may have written had he not been besotted by cars.

In *Velotopia* I express my bemusement that neither of those authors ever mentioned bicycles as part of their mobility plans. There was actually a bike boom in full swing when Howard was writing about garden cities. As for Le Corbusier, he was a cycling enthusiast himself; you would think the bicycle would figure in some measure in his vision for future cities, but like Howard, his mind went to enclosed and powered machines.

In *Velotopia* I said the problem they must have seen with bikes was all of the horse shit. Streets used to be caked in the stuff, meaning people whose jobs were indoors preferred to travel indoors, in trains, omnibuses and cars. However, I have come to see an even larger, if less visible, factor at play: the factor of money and power.

It was industrialists who lured millions to cities in the late 19th century, so it was industrialists, not politicians or planners, who decided how they would commute. Industry in Victorian London took off thanks to steam engines and railways, so as planning proceeded for the population to grow to six million by the year 1900, it was done hand in hand with the train. Producers of oil and cars were behind the post-war prosperity of American cities like Dallas and Detroit, so those cities grew according to a car-dependent model of urban growth.

Bicycling doesn't have a small cartel of rail or oil magnates who stand to get rich from more cycling. The only ones standing to gain are … everyone else!

The benefactors of a bicycling city are those of us who are not oil or car companies. We are all the small and large businesses, and all the consumers, carers, students and workers. Our interests don't include trains, cars, or even bikes for that matter. All we want is a city where everyone gets around quickly, preferably cheaply. Can we squeeze into the cigar room of a businessmen's club and conspire about highways or railways? Hardly! We number in the millions. We are the masses at large. All we have are representative bodies: the governments that we elect.

The time has come for governments to open their eyes to something no industrialist is likely to tell them: that fast machines are obstructions to a fast city. To think fast machines could make cities faster, is to misunderstand what cities are. Cities are sorting devices, built to sort, and re-sort, people into all manner of temporary groups.

The city can be likened to a giant bowl full of cornflour and water into which each of us can lower a spoon. All of us, simultaneously, can slowly move our spoons to any point in the bowl we may wish, and will get to that point in good time. All it takes though, is for one fool to try and stir quickly, and the entire bowl of cornflour will freeze.

That fool is the person who brings a car into a city. A hundred years ago, their presence created the need for traffic signals, turning movement in the city into a stop-and-start matter. In the future that fool will be in a driverless car, thinking each car should be coupled to cars fore and aft, all speeding around bumper-to-bumper. It will not occur to this fool that he has turned the street into a chainsaw. Lest they be minced, pedestrians will need to be stopped from ever crossing the road, with fencing like you see around train tracks. Will the fool care? Of course not! He is an heir to the Filippo Tommaso Marinetti spirit, not seeing the city as a delicate re-sorting device to make ever-changing groupings of people, but as a stage for rocketing bodies between removed points on the map. He will never understand that rocketing bodies is a business for NASA and SpaceX. He would rather get frozen in oobleck.

The three large plates in this chapter provide an alternative vision of the city of the future to those offered by believers in driverless cars. The first is of an apartment block, designed around a bike-friendly access gallery.

Plate 1: An apartment building designed to make use of things that have wheels.

Image: Steven Fleming

The idea here is to eliminate a portion of time in each trip that the automotive industry has chosen to discount: the time it takes to get on the road after walking out your front door. Average car trip times for cities are collated from data recorded in logbooks that study subjects keep in the glove compartments of their cars. They don't capture the time that can be spent schlepping children and groceries between cars homes, and destinations. In the building we're about to imagine, that schlepping all disappears. If you wanted to, you could take your children in your cargo bike and use the cargo bike as a trolley when doing your shopping.

Figure 1: An apartment designed to make use of the bike.

Designer: Matt Sansom

It would not be essential to include an elevator in a building like this. If there was one, it would only be for taking you up. Coming down would be faster by bike – the bike you would keep inside your apartment if you lived in a building like this. You wouldn't leave home without it, at least not if you were concerned about time. That is because there would be no faster way out of this building than on a bike. With the whole building designed around a spiralling escape ramp of sorts, leaving would be as simple as riding out your front door and letting go of your brakes. The only faster way out of a building would be via a fireman's pole!

Figure 2: Most shops could make use of the bike as a trolley.

Photo: Matt Sansom

Similarities can be drawn between this proposed new kind of building and: a building called the 8 House in Copenhagen that has maisonette apartments accessed via a ramped aerial street; Park Hill in Sheffield that has horizontal aerial streets extending to meet the hillside where the building is sited; and the West Village Basis Yard in Chengdu, a perimeter block with scissored bike ramps connecting the access galleries back to the ground.

Bike access eventually had to be banned in the 8 House due to inadequacies in the detailed design, but Park Hill and the West Village Basis Yard remain popular with people wanting to ride from high-level apartments directly down onto the ground. The building we're imagining would be even more popular among cyclists, given it provides them the most direct path to the street.

Note how it spirals clockwise from bottom to top. That is because the example we're looking at has been designed for countries like England and Australia, where we drive on the left. The idea is to put cyclists who are on their way down, and thus going faster, on the side furthest away from the apartments' front doors.

However, finessing of that sort doesn't completely negate the need to make those ramps wider. If they're to handle people walking, people in wheelchairs and people on bikes, the regular width of an access corridor would at least need to be doubled.

To save a doubling of the width, causing a halving of the plan's efficiency (in other words, an inordinate ratio of public circulation space to the private floor area of the apartments), it would make sense if bike-access apartment blocks had twin-level, maisonette, units. That way, access would only need to occur on alternate levels. To really boost efficiency, access could be stretched out to every third level. The architect would just need to configure the apartments in a manner pioneered by architects Peter and Alison Smithson in a building called Robin Hood Gardens. The following drawing shows how two-storey apartments can sit over and under floor levels largely given over to ramp space. The only private space on the access level would be entry vestibules and single offices or bedrooms connected by private stairs to apartments, one stretching over, its neighbour stretching under, the level of entry and access.

Figure 3: Access galleries can be widened without reducing floor plan efficiency.

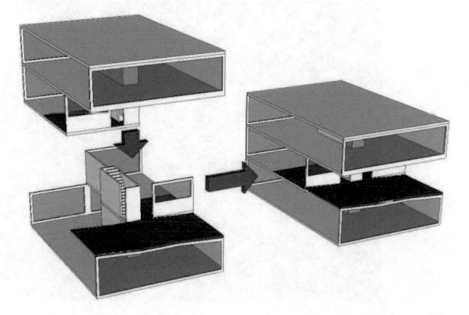

Velotopia works through the thinking and planning of various types of ramped bike-access apartment blocks, with apartments as small as one bedroom ranging to large ones that interlock like bayonets. There are also slab blocks organized around aerial streets that switch back on themselves, the way roads switch back when they're leading up mountains.

Lest any scenario be forgotten, there are sections on all kinds of buildings, from office blocks to retail and even country retreats. Has there been uptake? As an author, you don't always know. I can say that as a consultant, I have slipped these ideas into masterplans I've worked on in Norway and China. Also, the City of Bogota had me work there to deliver a competition brief for a school and bicycle industry incubator. Initial plans are to give the poor district of Bosa a building derived from "Velotopian" thinking.

The real-world dimension to the work gives it a practical edge, missing from city-of-the-future visions with driverless cars. Ramp gradients are not randomly chosen but conform to guidelines developed for wheelchairs. Similarly, hard lessons learned in the 1960s, when there was no passive surveillance of access galleries, have put limits on what can be imagined as real.

If there is one contemporary orthodoxy that can't be acquiesced to, it is New Urbanism's moratorium on big buildings. An atrium needs at least a 60-metre circumference for a ramp to fit in it, and double that if the ramp is to skip alternate stories. There is no way, in old cities, for buildings of that size to be squeezed onto small lots, at least not without the amalgamation of sites or some inelegant planning.

Really though, this represents a fairly insignificant point. Old city centres aren't where humanity is struggling to provide housing. Urbanization is playing out on former industrial land, or, in poor countries where urbanization is happening at the greatest pace, new buildings are being constructed on cleared forests and farmland. Either way, new buildings are big enough to be designed around ramps, instead of high-maintenance lifts. The second of our large plates shows what could await a person on a bike as they ride out of their building.

Plate 2: A street that is easy to cycle on, yet hard to drive on, meaning it is easy to walk there and use wheelchairs too.

Image: Steven Fleming

The first thing to note is that the street isn't flat but something more analogous to a skate park. Only this skate park is for adults and on the scale of a city.

The concept derives from a desire to prevent fast-moving, heavy machines from ever dominating the city. A car or a motorbike would be shaken to bits if it went faster than bike pace on a ground surface that goes up and down.

With fast machines out of the picture, we can imagine the perfect environment for cycling, while not forgetting the needs of pedestrians or people in wheelchairs. The inspiration comes partly from skate parks, where children stop on elevated plateaus. The elevated plateaus of *Velotopia* are where cyclists slow down before dismounting or riding into the buildings.

One metre below those plateaus are elevated mounds, positioned at all of the crossroads. The idea is to slow cyclists a little where their paths intersect.

The lowest plane of this undulating terrain is where cyclists would be moving the fastest. Test-riding has taught us that a two-metre elevation, from the lowest/fastest point of the road to the top of the mounds, is just the right height to bring a loaded cargo bike to a standstill without braking, and to get it back to full speed when it heads down.

Figure 4: A stop-motion photo showing ideal elevations for slowing and starting without having to pedal or brake.

Photo: Charlotte Morton

The book *Velotopia* has ideas about grade separation between cyclists, pedestrians and disabled for the busiest parts of our cities — around central train stations, for instance. But in a city without transit nodes (subway stations, for example) and where cyclists aren't funnelled onto the few streets remaining after driving has taken the others, movement would be diffused. Few streets would have so many cyclists that they couldn't operate as shared spaces, with pedestrians, disabled people and bicycle riders all crossing paths without hitting.

Based on that assumption, the vision of the future city shown in Plate 2 has paths on diagonals, between the plateaus, at a gradient to suit people in wheelchairs or using mobility scooters.

Figure 5: An undulating ground surface would thwart driving, be better for wheelchairs and unleash the convenience of bicycle transport.

The second thing to note, is that bike routes are covered. Some of the artists' impressions in *Velotopia* show giant saw-tooth roofs covering all open space: the whole of the street and all the courtyards within buildings. The idea would be to let people make beelines, oblivious to buildings held aloft on thin columns, and oblivious to rain or snow falling.

The vision shared here is more modest. Only the travel routes are protected, leaving cyclists exposed to a few seconds of rain at the beginning and the ends of their journeys. In this case we're imagining awnings that are parabolic, to capture and harvest rainwater on wet days, and, in sunny weather, to tilt to track the sun and focus all the energy on one point for collection.

Where one might expect the most vociferous critics of covered streets to be drivers, poopooing public expenditure on anything except for more parking, the real hatred for this idea has come from Dutch bicycle advocates. The same people who in one breath say cycling is transport, not recreation, insist that cyclists want the rain on their face to remind them that they are alive. What they are saying is that cyclists are different from patrons of other modes, such as car or rail travel, all of which provide weather protection. Given that country's Calvinist roots, it is tempting to guess that what these tough guy (and gal) cyclists are saying, is that they are God's chosen ones, predestined for heaven, because they ride bikes in the rain. It's an us-and-them attitude with no place in a city that is lacking any drivers to mock.

We end this series of three image plates with a view of a plateau with no buildings. It's a raised space, which means our wheel-borne population naturally slow, or come to a stop, when they get here. While they are on this stopping plateau, they are storing potential energy, by virtue of their slight elevation. When they leave and head down, that potential energy will be converted into the kinetic energy they will use to join their next group.

Plate 3: Their elevation would make cyclists naturally slow down before arriving at building entries and public spaces like this.

Image: Steven Fleming

Like all three of the plates, it is rendered in high enough definition that details in the background ask to be seen. Key among them is the permeability of a city, where buildings can all be passed under.

Figure 6: Cyclists' and pedestrians' inherent ability to cross paths means buildings could be raised and everyone could make near-perfect beelines to their destinations in a bike-centric city.

Image: Steven Fleming

Perhaps the most important chapter in *Velotopia* is the chapter estimating trip times. They come out better for a city of six million cycling than for the six million driving in the Dallas–Fort Worth metroplex – an example often held up in support of car transport. Granted, my trip time calculations are for an "ideal" bicycling city, having no hills, a high population density and nothing to make cyclists slow down. What I would say about my ideal, though, is that it is far easier to achieve in real life than the ideal of a car city. It doesn't need a lot of things added, like billion-dollar freeways or cross-city tunnels. The main task is taking things away, namely the cars, which since they have wheels, should not cost very much to remove!

All this leads us to the first of two basic questions: which should we build for, the bike or the driverless car? Where cities are rapidly expanding, or being built out of nothing, the opportunity exists to make cities cheaply, with no elevators or heavily engineered roads, and where the costs to individuals are equally low – bikes retail for US$50 in poor countries, about as much as motorists spend every time they refuel their cars. Once you factor in the cost to life from having populations sitting, not exercising, when they commute, and the cost to the planet of heavy modes of transport compared to light-weight bicycling, the car city looks like a hideously expensive proposal. When it comes to new urban districts, therefore, or whole new cities constructed from scratch, the case for purpose-building for cycling begins to look strong.

The main objections to cycling – that it is slow and puts people out in the weather – are moot when we construct our cities for cycling from scratch. When we think about bike trips being all under cover, then using the bike to schlep children and loads the whole way, and not obstructing cycling with traffic lights or the detours we subject it to by routing it clear of car traffic, we begin to imagine a city that is just as comfortable, convenient and time-saving as any we can build around driverless cars, but without the price tag or externalized costs to public health and the planet.

Next comes the more vexed question of adapting existing human settlements so they group and regroup people more quickly. If any role can be imagined for the driverless car, it might be in completing the task of making the countryside a part of that story, in other words, letting people live in the country but still have access to the opportunities that come from fast connections to millions of people.

What an absurd proposition! Houses built for tilling the land will not be drawn together – at least not in their millions – until we actually do have personal flying contraptions, or teleporters. The cost of the roads would be greater than the cost of the buildings – that is, if we're serious about giving people in the countryside equal access to the specialized education and jobs that populations of millions can offer. For the sake of a relatively small percentage of the world's current dwellings, connecting the countryside using driverless cars would be as wasteful of cities' resources as it would be wasteful of car manufacturers' R&D funding.

The fact that R&D into driverless car technology is persisting, is evidence, on its own, that the automotive industry has its eye on more than the country. It expects its products will be welcome in actual cities, on streets that need to accommodate all people's movement between all the densely packed buildings. The cost of that R&D work is itself proof that the automotive industry isn't angling to provide driverless car technology on a supplementary scale. It must work at an economically viable scale to the automotive industry. In other words, they are hoping existing urban areas will be retrofitted so that driverless cars become the dominant mode.

Anyone who might be concerned about the slight dangers inherent in a rise in bicycle transport, would have to view the car industry's plan as an absolute nightmare. The polite visions they're promoting right now, of cars stopping for kids chasing puppies out onto the street, are to establish a foothold for driverless cars in the city. Once they have that, their profit motive will compel them to lobby, year after year, for measures that will make driverless cars more prolific. Since capitalism knows no limits to increased profits, the goal must be for driverless cars to take over. By the time car companies have recouped the R&D outlay that they're spending right now, and turned a speculative investment into real profit, they will have convinced us to fence off our streets the way we've fenced off our railways, so cars can be coupled like rail-cars. The only opportunities for crossing the street, since intersections will be optimized for cars to sift through without touching, will be via over- and underpasses, paid for by the public and herding us via circuitous routes. The most practical means of just crossing the street, will be to pay for a ride in a driverless car. We'll be slaves to an industry that doesn't sell holes, it sells drill bits.

I would urge anyone being swept down this path by arguments pointing to the short-term benefits of driverless cars, to pause and engage with the *Velotopia* vision. While bicycle-centric, it is pro-walking and pro-public transport in ways the car city cannot be. Also, while it starts with a discussion of a city for six million, built out of nothing on a flat plain, it is ultimately focused on actual cities, with their existing car dependence, rail infrastructure and hills.

The lofty idea of a perfect bike city for six million people is offered merely to balance the scales. When a highway is planned, we all know the vision we are building towards. It was pushed on the world right through the interwar period in Le Corbusier's Plan Voisin, Frank Lloyd Wright's Broadacre and, most influentially, Norman Bel Geddes's Futurama exhibition of 1939. When we build a railway with transit-oriented settlements clustered around the new stations, we have Ebenezer Howard's garden city model to tell us what we are setting out to achieve. We similarly know, when we plan a pedestrian-centric new township, that a plethora of ideal models, ranging from those systematized by Kevin Lynch and Jane Jacobs, to actual medieval town centres, are the long-term goal of our efforts. Despite being fast, safe, cheap, healthy, sustainable, nimble and able to carry things for us, the bicycle never had a city model defined as a goal. *Velotopia* has corrected that error and shown in the process that bicycle-centric built environment planning could lead to faster average commute times than any of the other models just mentioned.

Velotopia's downfall is that no powerful group is ever likely to champion it for their own profit. Selling a bike to each person will not make the bike industry as rich as the car industry became by selling each family one, two or three cars. Getting everyone cycling won't make a few influential landowners rich the way landowners get rich when train stations are built near their holdings. In a bicycle city, all but the most peripheral land is equally accessible, in terms of average trip times. We also know from the way land prices evened out in cities in China after the arrival of dockless bike sharing systems, that anyone owning land in a central location, who might be expecting it to exponentially appreciate as the city expands, could turn against cycling when they see its levelling effect on land values. None of these traditional profiteers from mobility systems, or the lack of them, will use their power and influence to champion the *Velotopia* model.

The challenge in promoting a bicycle city, is not finding the ears of the rich and influential but of the people at large, because it is they who will reap the fast trip times and comfort. We just have to show that cycling is worth building for, to the exclusion of shiny distractions, like driverless cars.

PART III
Spatial development

Mathias Mitteregger, Rudolf Scheuvens

Mobility, transport and settlement structure are inextricably linked. That it is not the technologies alone that shape space and spatial structure but that planning and mobility culture, for example, play a decisive role is revealed by a comparison: despite their many similarities, the car has not had anywhere near as profound an influence on the European continent as it has on North America. Generalizing terms like "suburbanization" serve only to obscure this fact. Consequently, the chapters in Part III, which focuses on spatial development, always take both factors into account: technology and planning. On different levels, they show what targeted spatial development approaches might look like and which influential variables need to be considered.

In their article *Strategic spatial planning, "smart shrinking" and the deployment of CAVs in rural Japan*, Ian Banerjee and Tomoyuki Furutani show how connected and automated driving systems are embedded in what is probably the world's largest transformation strategy in this context. With its national spatial strategy, Japan is tackling the island nation's major problems: the radical depopulation of entire swathes of the country, population ageing and the associated impacts on both society and the economy. Automated mobility services have a fundamental part to play here and are thus explored in conjunction with transport infrastructure. The authors reveal the affinity between this transformation strategy, which is organized on three hierarchical levels, and Walter Christaller's central place theory: the mobility hub becomes the nucleus of settlement development, whether as a small station, as an integrated regional hub or in collaborative core urban areas. In rural regions, all the services offered by a central town are concentrated in such hubs, thousands of which are due to emerge throughout the country over the coming years. The catchment areas around these nodes are supplied by automated vehicles running on local transport routes.

In *Integrated strategic planning approaches to automated transport in the context of the mobility revolution: A case study in suburban and rural areas of Vienna/Lower Austria*, Mathias Mitteregger, Daniela Allmeier, Lucia Paulhart and Stefan Bindreiter present strategic planning approaches created during a transdisciplinary process in which representatives of four example communities and regions worked together with planners and automated mobility experts. In every example the aim was to make attractive mobility services possible in suburban and rural areas by connecting and automating transport. The result was multilevel strategic planning approaches that highlight the opportunities – as well as the limitations – of this technology. In each case, connected and automated driving was promoted hand in hand with the development of active mobility. On the basis of this transdisciplinary work, the authors make the case for a strategy that is problem-centred and tailored to the target group; in all the transport and spatial constellations considered, this leads to connected and automated vehicles first being used for specific mobility needs.

In their chapter *Opportunities from past mistakes: Land potential en route to an automated mobility system*, Mathias Mitteregger and Aggelos Soteropoulos contemplate another spatial aspect of transforming the mobility system. The authors argue that the much-discussed reduced need for parking spaces due to the advent of automated mobility services has thus far only taken into account a small proportion of the land that could potentially be converted. A move away from owned mobility and towards spontaneously used mobility services would trigger a systemic change that would affect the entire product life cycle of motorized private transport and all its associated functions – from the car dealership to the car wash. Here too, the area under consideration is the metropolitan region of Vienna/Lower Austria; the authors investigated every site that might come into question with the aid of a GIS-based analysis. They show that significant opportunities arise in the area of brownfield development, but that for these to be tapped a major overhaul of planning processes is required.

13 Strategic spatial planning, "smart shrinking" and the deployment of CAVs in rural Japan

Ian Banerjee, Tomoyuki Furutani

Ian Banerjee
TU Wien, Research Unit Sociology (ISRA)

Tomoyuki Furutani
Keio University, Faculty of Policy Management

M. Mitteregger et al. (eds.), *AVENUE21. Planning and Policy Considerations for an Age of Automated Mobility*, https://doi.org/10.1007/978-3-662-67004-0_13

1. INTRODUCTION

The experiments taking place around connected and automated vehicles (CAVs) in global innovation networks today are largely technological in nature. This research takes a relational view of CAVs by investigating how they can be conceptualized within the larger context of strategic spatial planning. To do so, it takes Japan as a case study and explores how the current government is applying the tools of its new National Spatial Strategy (NSS) to strategically steer the development of its main economic and social sectors, including transport and the deployment of CAVs. All sectors of national innovation programmes, including those for CAVs, are guided by the ideas and principles of the grand narrative of change envisioned by the new NSS (see Fig. 1).

In the first AVENUE21 publication, Tokyo was presented, along with San Francisco, London, Gothenburg and Singapore, as one of the pioneering regions for CAVs (see Mitteregger et al. 2020: 85–90). This research returns to Japan to take an in-depth view of the dynamic relations co-evolving between the country's new instruments of the NSS and the societal context of deploying CAVs. The main aim of this analysis is to find out if and how the device of a national spatial strategy can be helpful to conceptualize the potential deployment of CAVs. This research offers less a critique than an analysis of the rationale behind the newly adopted approach towards spatial planning in Japan, and the "new thinking" the government is struggling to embrace in order to cope with the country's daunting socio-economic crisis. It spans an arc from the macro level of national spatial planning to the micro level of CAV tests conducted in rural Japan.

Before looking at an exemplary set of experiments conducted with automated mobility in rural Japan, we take an expansive view of the thematic issues and policy positions in the country. We argue that in order to understand the logic underlying the conceptualization of the potential deployment of CAVs in Japan, it is essential to understand the country's overall demographic, spatial and economic predicament, and how the government wants to address the country's challenges with new ways of thinking about its future, new planning concepts and new ideographic narratives.

Strategic spatial planning has often been criticized for being inherently neoliberal and autocratic, promoting gentrification and large-scale, profit-oriented urban development projects. However, planning theorists have also pointed to its potential to be inclusive, emancipatory and innovative in socio-territorial terms (Swyngedouw et al. 2002, Moulaert et al. 2003, Bornstein 2007). This research seeks to explore its emancipatory potential, albeit in the very specific societal and cultural context of Japan. Despite the fact that the concept of strategic spatial planning has existed for many decades and has been proficiently practised in countries of both the Global North and the Global South, it is difficult to find a shared definition. For this research, we adopt the view of Oosterlinck et al. (2011) who define strategic spatial planning as "[...] a method for collectively re-imagining the possible futures of particular places and translating these into concrete priorities and action programmes" (Oosterlinck et al. 2011: 1; see also Albrechts 2004, 2006; Healey 2004). We look at strategic spatial planning not as conventional master planning pursued through passive control and zoning (Albrechts 2006) but as a "[...] transformative and integrative, (preferably) public sector-led socio-spatial process through which visions, coherent actions and means for implementation and co-production are developed, which shape and frame both what place is and what it might become" (Oosterlinck et al. 2011: 3, adapted from Albrechts 2006; see also Healey 1997, 2007). We also pursue an "institutionalist" understanding of planning, which is based more on a social-science-oriented planning theory than one that is planner-client oriented. This approach draws more attention to planning practices and the complex institutional dynamics connected to a place, changes in social relations and the ways of collective decision-making (for more, see van den Broeck 2011: 54). We assert that the conceptualization of the deployment of CAVs is shaped by the dynamic interlinkages between global, national and particularly local factors. Its analysis on the local level is a challenging exercise in disentangling its cultural and institutional threads, and not to forget the threads of power and vested interests (see Chap. 18 by Stickler in this volume).

The reason for choosing Japan for this exploration is due to the opportunity it offers to study a nation's ongoing response to the severe demographic crisis it is facing today, with both the instruments of a comprehensive national spatial strategy (NSS) and the resolve to deploy CAVs. The way the conceptualizations of new forms of mobility, including those of CAVs, are embedded in the larger framework of the country's new NSS makes Japan stand out among the pioneering regions for CAVs.

The demographic crisis that Japan, the third-largest economy in the world, is facing today is expected to reduce the country's current population by 30% by 2060. Given the current trend and pace of depopulation in most parts of the country, the provision of utilities such as water, gas, public transport, etc. may become impossible to maintain in the coming decades. Ageing and radical depopulation of most of the countries' prefectures are already creating a substantial decrease in agglomeration potential and the loss of workforce. The crisis is unprecedented in history in terms of its magnitude and complexity. This seeming impasse has fuelled the Japanese authorities' fear of economic downfall and partial collapse of the country's regional infrastructure. As in many other countries, response to depopulation has been slow; however, the current government has finally placed the topic as a key item on its national policy agenda. Japan's main political objective today is to secure its national prosperity by revitalizing its economy and to create sustainable settlement patterns that react to depopulation in creative ways. The political resolve to respond to the challenge manifested itself in the watershed year of 2014, when a number of ambitious economic and spatial restructuring strategies were simultaneously presented by various governmental institutions. This new thrust forward by the Japanese government has sparked considerable global interest in its policies. Japan is believed to be at the forefront of innovations addressing the challenges of ageing and shrinking in concerted ways – with mobility strategies and CAVs playing key roles therein (Roland Berger 2018, OECD 2016a).

This paper examines the government's response to the country's challenges along three analytical dimensions: spatial planning, governance and mobility. It puts them in relation to each other after taking a historic view of the country's evolving approach towards planning, particularly taking into account the country's 60-year-long experience with the instruments of national spatial planning. The first section of this paper outlines the demographic challenge faced by Japan and the process of depopulation on the regional level. The second section reviews the evolution of national spatial strategies with a focus on the new grand narrative of a radically "compacted and networked" country inscribed in its two most comprehensive strategic responses: the National Grand Design 2050 (2014) and the 2nd National Spatial Strategy (2015). The third section examines the changing notion of governance by looking at how a consensus has been found among policymakers to transform Japan's traditionally centralized form of government into a multilevel form of governance by encouraging nationwide processes of participation from the bottom up. Finally, the fourth section looks at the new institutional set-up created for the development of CAVs with a focus on their deployment especially in rural areas. It looks at how locally specific tests are conducted in rural and mountainous regions of the country, and at how a variety of factors are influencing the experiments with CAVs in these areas.

The research for this paper is based on a literature review and field experiences. We have drawn particularly on the comprehensive review of the new policies in Japan made by the OECD (2016a), the resources of the Ministry of Land, Infrastructure, Transport and Tourism (MLIT 2014, 2015, 2017, 2019) and the learnings from experiments conducted with CAVs at the Keio

Figure 1: Flowchart showing how the current national crisis has pressured the government into responding with new visions and policies and how these are successively being translated into concrete innovation and action programmes (e.g. for CAVs)

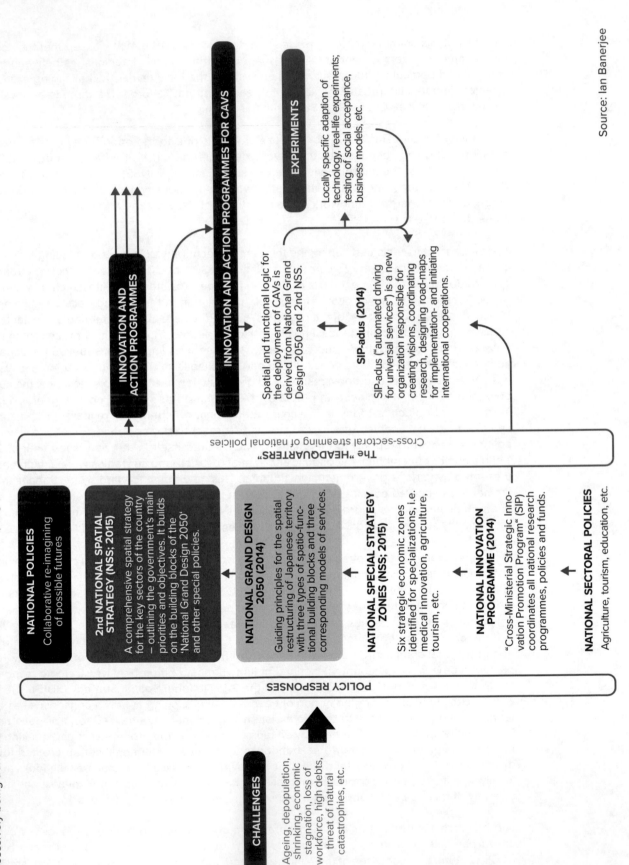

Source: Ian Banerjee

University in Tokyo. The presentations and discussions at the international conference SIP-adus Workshop in Tokyo in November 2019 (SIP-adus 2019a) have also helped to sharpen the authors' picture of Japan's approach to connected and automated vehicles in comparison with other proactive countries in the field.

On the discursive level, this paper wants to contribute to the exchange of ideas between Japan and Europe. The necessity of learning from each other is timely for two pertinent reasons: both regions are facing the challenge of ageing (Reuters Graphics 2020) and the difficulty of serving their elderly population in rural areas; and both regions are interested in revitalizing their rural areas by making them more attractive places to live and work by, for example, deploying new mobility services such as those with CAVs.

2. JAPAN'S DEMOGRAPHIC CHALLENGE

2.1 POPULATION IMPLOSION

After the devastations of World War II, Japan surprised the world by swiftly rising to become the second-largest economy in the world, only to be recently overtaken by China. Here are some indicators of its success (for more, see Diamond 2019: 294–298): today, Japan accounts for 8% of global economic output; in the last decades, it has continuously managed to be among the top ten in the Global Competitiveness Index of the World Economic Forum; it is the world's leading creditor nation with the second-highest foreign exchange reserves (even though its high domestic debt draws more global attention); it has the world's best roads and one of the densest infrastructure systems; it has the second-highest number of patents per million inhabitants; it makes the world's third-largest absolute investment in R&D (after China and the USA); it is one of the cleanest and safest countries in the world, and the third-most equal country in terms of wealth distribution (after Denmark and Sweden).

After five decades of its spectacular display of techno-economic success, Japan now faces a composite crisis of a magnitude it has not seen since the beginning of the Meiji Restoration in 1868 (for more, see Diamond 2019). The 11th-most populous country in the world (as per 2019) has today the highest proportion of elderly citizens of any country in the world, followed by Italy and Germany (Population Reference Bureau 2019). Ageing, the highest life expectancy in the world (80 years for men, 86 for women), the dramatic fall in fertility rates and the subsequent shrinking of its regions have trapped the country in a series of chain reactions that are threatening to disrupt its economy, social contract and geopolitical influence. Japan's population grew from 33 million in 1868 to 128 million in 2008 (Fig. 2). After 60 years of growth, it is expected to shrink to 107 million (16%) by 2040, to 97 million (24%) by 2050 and to 86 million in 2060 (IPSS 2014, Funabashi 2018). After peaking around 2010, Japan's population is expected to fall back to around 50 million by the end of the century (Funabashi 2018); 40% of that population would be over 65. The increase in the "oldest-old", namely those aged 75 and above, will more than double from 11% in 2010 to 27% in 2060 (IPSS 2014, 2017). The demographic transition taking place in the country is unprecedented in human history (OECD 2016).

After at least two decades of neglect, the so-called Masuda Report appeared in 2014 to "galvanize a large part of the political elite" (OECD 2016: 194), particularly those living in the regions suffering population decline. The Masuda Report, with the original title of "Stop Declining Birth

Figure 2: Population trend by age group in Japan (1920–2060)

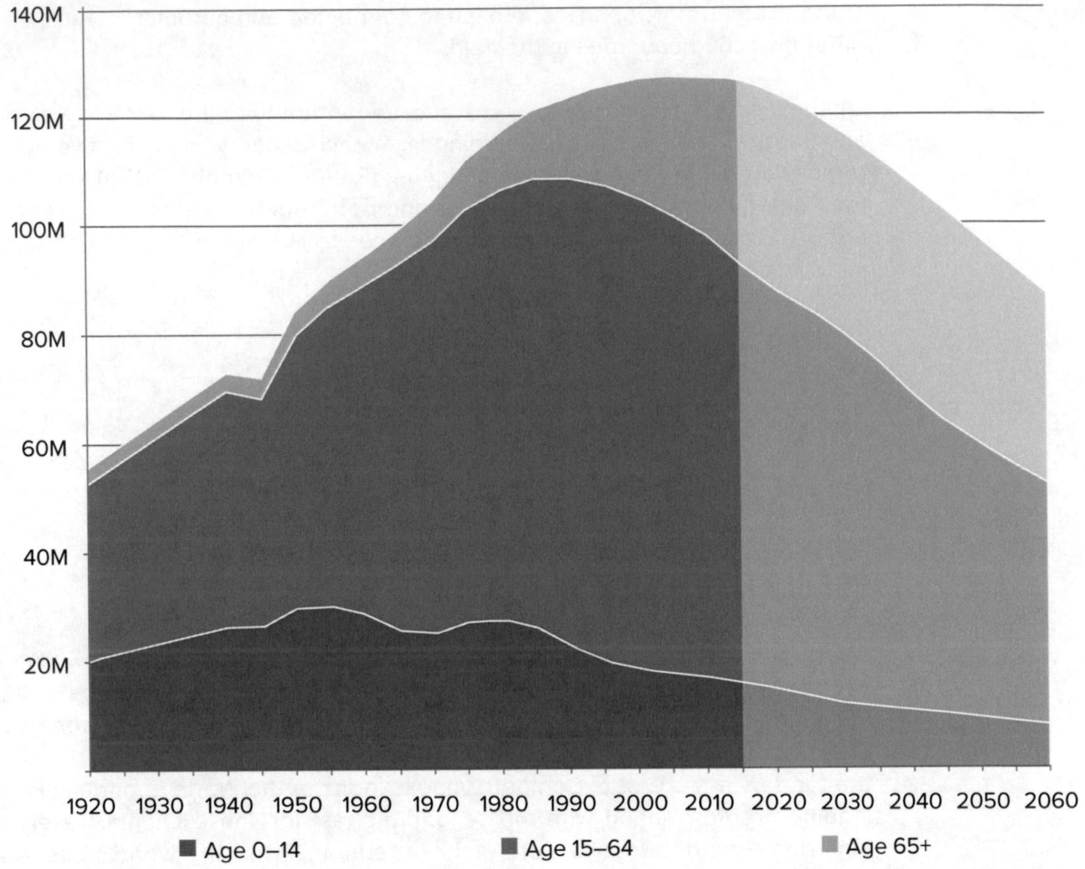

Source: IPSS (2014, 2017)

Rates: The Local Revitalization Strategy", was named after the chairman of the Japan Policy Council (JPC) Hiroya Masuda, the former head of the Ministry of Internal Affairs and Communications (MIC). It triggered an impassioned nationwide debate and a call for radical action on governmental and institutional levels. The report sent out the warning that 826 local governments, which is roughly half of the total, risked "extinction" by 2040 (OECD 2016: 195). The Masuda Report succeeded in transmitting its key messages with the help of succinct narratives. Its title "Stop Declining Birth Rates" pointed to the problem while at the same time making a plea for change. The second part of the title, "The Local Revitalization Strategy", offered a solution to the problem and a recommendation for action, which, in principal, was about making "regional cities attractive to young people" by building "new structures of agglomeration" (OECD 2016: 195). It advocated the creation of "core regional cities" with 40,000 inhabitants connected to others with new information and transport technologies. These cities were to be attractive enough to function as "dams" against the outflow of young people into larger cities. The idea closely resembles the building blocks of the new National Grand Design 2050 that would follow that same year (see 3.4.2).

Four years after the Masuda Report, a scientific publication followed up on the same topic, carrying the title: "Japan's Population Implosion: The 50 Million Shock" (Funibashi et al. 2018). The editor, Yoichi Funabashi, backed by over a dozen authors, identified the policy failures of the last three decades and the lost opportunities to avert the impasse, making a passionate appeal for radical structural measures. It is one of the most comprehensive accounts of the demographic and social dilemma currently threatening Japan. The authors painted a dystopian picture of a coming catastrophe, showing how it had already begun in many regions. They pointed out that even though the population decline is tied to a multitude of other problems in Japan, it is not merely one problem among many but the key to understanding the principal structural deficiency afflicting the country (Webb 2018).

While there are many reasons for population decline (see Diamond 2019), some, like Funibashi et al. (2018), link the central cause of the chronically low fertility rate in Japan to the lack of work-life balance and the intense stress factors citizens are exposed to in their everyday lives, leading to "[...] a sense of fatigue [...] in Japanese workers, especially young workers with long commutes, extensive overtime, and ambiguous personnel evaluation systems" (Funibashi et al. 2018: 104). They argue that Japan's birth rate will not recover without a fundamental reform of the social structures that are threatening interpersonal relationships and deterring citizens from marriage and childbirth (Funibashi et al. 2018: 103).

2.2   "DISAPPEARING MUNICIPALITIES"

Socio-spatial data shows that the population decline in Japan has been relatively slow and spatially uneven in the last decades. Between 1985 and 2010 it was less than 10% in five prefectures with the highest rate of decline. Some prefectures even saw an increase in their pop-

Figure 3: Population projections for municipalities from 2010 to 2060. Figures after 2040 were extrapolated by the IPSS authors from the same hypothetical figure.

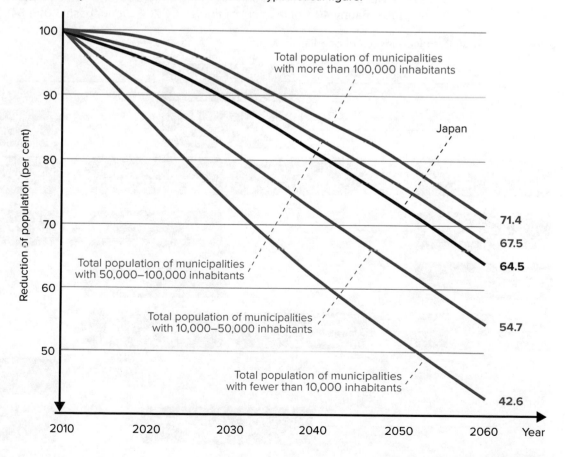

Source: IPSS (2017)

ulation during this period. It was only after 2008, when the population started to plummet as a whole, that it became prevalent for almost all of Japan. The term "disappearing municipalities" is commonly used in the discourse to concisely depict the fate of a large number of municipalities affected by depopulation and ageing. Most prefectures in Japan will only have 50% to 70% of their 2010 population count by 2060 (Funibashi 2018: 52). According to Japan's Regional Population Projections (IPSS 2014), the rate of decline in 42 prefectures (out of 47) will be more than 10% between 2020 and 2040.

The reason for the drop in regional population, in addition to the general drop in fertility rates, is largely due to rural-to-urban migration. After World War II, a large number of young people started to migrate to three large metropolitan regions: (1) Tokyo, Chiba, Saitama and Kanagawa; (2) Aichi, Gifa and Mie; and (3) Osaka, Kyoto, Hyogo and Nara. Today, roughly half of Japan's population lives in these three regions. With their rapid industrial growth and global economic success, these three regions started to attract large numbers of young people from rural areas – a trend that continued until around 2005. During the high growth era of the booming 1960s when Japan grew into a global economic powerhouse, more than 600,000 graduate students would flock to these regions every year (Funibashi 2018: 54). Today, in 2020, the number is around 100,000 per year. The aged population ratio will start to affect these mega metropolitan areas slightly later than rest of Japan, that is, from around 2030 (Kaneko/Kiuchi 2018: 11). This will be the time when Japan will have "a real sense of population decline" (Funabashi 2018: 52). By 2060, the population is expected to be less than or equal to half of the current level in about half of the municipalities nationwide. 70% of all municipalities are likely to see a reduction of around 20%. Forecasts show that the smaller the size of the municipality, the faster the speed of decline will be (Fig. 3). Towns and villages with 10,000 or fewer inhabitants are expected to have approximately 40% of the current figures by 2060 (Funabashi 2018: 54).

Figure 4: The possible consequences of population decline

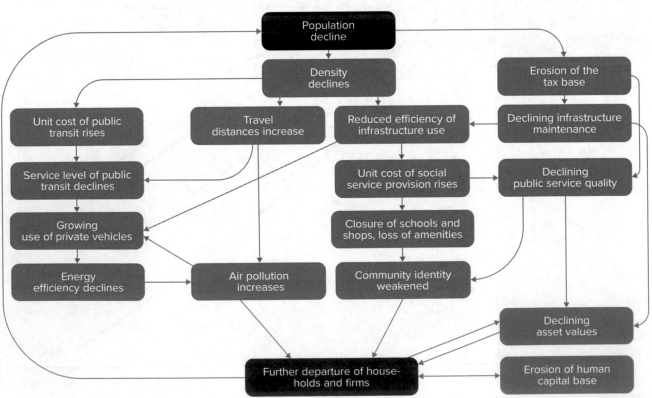

Source: OECD (2016)

Urban population decline is known to create a dangerous chain of events that usually triggers a sustained downward spiral (OECD 2016: 233). It can lead to migration, erosion of human capital, the downsizing of services, reduction of opportunities available locally, reduction of real estate values, etc. The pressure on financing public services will mount as local tax revenues decline and the per person cost of providing services rises. This will then lead to the closure of schools, hospitals and public facilities (see Fig. 4).

3. EVOLUTION OF NATIONAL SPATIAL PLANNING

3.1 ACHIEVING BALANCED GROWTH: COMPREHENSIVE NATIONAL DEVELOPMENT PLANS (1962-1998)

Over the past 60 years, Japan has experienced profound changes in its socio-economic condition including rapid acceleration of industrialization and a fast increase in its population. During this period, while the country soared to become the second-largest economy in the world, the Japanese government practised national-level spatial planning with the aim of alleviating spatial disparities and steering the country's economy in a balanced way. The first Comprehensive National Development Plan (CNDP) was published in 1962, with three subsequent CNDPs following in 1969, 1977 and 1987. The strategic plans generated idealized visions of the Japanese territory, addressing issues like land use, industrial locations, social infrastructure, culture, tourism, natural resources, human resources, etc. For 50 years, the CNDPs remained the main steering instrument for Japanese national and regional land policies (Ono 2008), leading the country into an era of unprecedented wealth.

3.2 FIRST RESPONSES TO DEPOPULATION: GRAND DESIGN FOR THE 21ST CENTURY (1998-2008)

Fifty years of high growth rates and relative regional stability of the population and settlement structures in Japan ended in the 1990s. After the collapse of the dot-com bubble in 1993, the Japanese economy started to slide into stagnation and by the end of the century, the spectre of depopulation had started to emerge in the public discourse. At this turning point, the concept of CNDP was sidelined and a new framework for the country's spatial strategy was presented in 1998: the Grand Design for the 21st Century. The Grand Design was not only a spatial strategy but also addressed a broader range of issues such as the consequences of globalization, the revolution in information technologies and very importantly, the country's demographic situation (OECD 2016: 86). Also, it advocated a multiaxial development plan as a long-term vision, marking for the first time a departure from the excessive dependence on the economic engine of the Tokyo megaregion (Ono 2008). Also for the first time, it began to promote wider citizen participation (Ono 2008).

3.2.1 Compact city and transport-oriented design (TOD)

The 1998 "Act on Vitalization in City Centers" was the first attempt to strategically incentivize the concentration of urban functions and invigorate commercial activities in city centres while striving to minimize the role of the central government.

The first tangible and design-oriented spatial policy that responded to depopulation started with the increasingly popular metaphor and concept of the "compact city". While national and local governments started debating various ways to enact new laws to reinvigorate shrinking city centres, Aomori City (population: 287,000) and Toyama City (population: 415,000) were the first cities to respond proactively to the situation (Kikodoru et al. 2008: 16). Based on the idea of the compact city, they drew up master plans in 1998 that would incentivize the concentration of urban functions in city centres in order to save the costs of maintaining public facilities, conserve energy and prevent further urban sprawl (OECD 2012, Kaneko/Kiuchi 2018). These attempts demonstrate the early adoption of transport-oriented design (TOD) in Japan, which focuses on creating public transport and new urban centres along the transportation axes of railways and buses and encourages the building of housing and medical services near stations and bus stops. Toyama City in particular promoted "active mobility" by encouraging urban development that facilitates access to functions within walking distance.

In 2005, the Sectional Committee on Urban Planning and Historic Landscape of the Panel on Infrastructure Development wrote an initial report on the need for "urban restructuring" by replacing existing urban structures with "concentrated" urban structures. They proposed the concentration of commercial, administrative, medical, cultural and other functions to ensure accessibility without reliance on cars (Kaneko/Kiuchi 2018). Three revised acts strengthened the involvement of local governments to set and evaluate measurable targets.

3.3 A TURNING POINT: FIRST NATIONAL SPATIAL STRATEGY (2008)

Despite the initial efforts to concentrate urban functions, the outcomes of the new measures remained limited and urban sprawl continued. After the first Grand Design, which had sketched out a broader view of the country's future, in 2005 the 1962 legislation of CNDPs was fundamentally revised and renamed the National Spatial Planning Act. The new act constituted an important shift: from a growth-driven planning regime to one concerned with demographic decline and the sustainable use of national territory (OECD 2016). Most significantly, the new strategy also started to look at how to promote more autonomy for local governments. In 2006, eight local regions were defined and asked to prepare their own regional spatial plans (RSPs; ONO 2008), which were to be subsequently integrated into the National Spatial Strategy (NSS). The 1st National Spatial Strategy was finally adopted in 2008. With the depopulation debate in full swing and the economic stagnation turning into a serious problem, this marked a major turning point for Japan.

3.3.1 Low-carbon city (2012)

In November 2012, the Ministry of Land, Infrastructure, Transport and Tourism (MLIT) presented directions for a new policy with basic guidelines towards sustainable development by promoting "regional intensification" by encouraging the construction of green buildings and the use of low-carbon and recyclable systems. This led to the enactment of the Act on Promotion of Low-Carbon Cites in December 2012 (Kaneko/Kiuchi 2018). Thus, two policy narratives, the "compact city" and the "low-carbon city", began to shape the urban discourse in the country. The new plan allowed local governments to create their own ways to promote the low-carbon city in so-called "Urbanization Promotion Areas". Reduction in income tax, relaxation of certain regulations and financial support through subsidies were offered as incentives. By October 2014, 16 cities had been designated to be low carbon.

The two cities in Japan that are often mentioned as a reference in this context are Maebashi City and Kochi City. Maebashi City produces around 40% more annual CO_2 emissions per person in the transportation sector than Kochi City. The reasons for this difference lie in the different forms of the cities and their different degrees of spatial compactness (Kaneko/Kiuchi 2018: 22). This comparison is often brought forward as an argument for promoting a combination of the low-carbon city and the compact city concepts.

3.4 THE SEARCH FOR A GRAND NARRATIVE OF CHANGE

3.4.1 Scenarios of hope: Reframing the problem

The reluctance to accept the grim prospects of shrinkage as something permanent, as observed in other parts of the world, may be due to electoral reasons or simply because of a natural human unwillingness to believe oneself to be part of a declining community (Schlappa/ Neill 2013). A large number of medium-sized cities in OECD countries are facing similar challenges, yet there is a remarkable lack of good practices to be found in the field. The dominant narrative of growth has created an all-pervasive paradigm and institutional mindset that drives the vested interests of the infrastructure industry of most countries through the "iron triangle" linking bureaucracy, politics and business (Adams 1981). This may also be a contributing factor to why governments are slow to respond to the situation.

As the image of the forbidding reality of depopulation started to trickle down into the consciousness of Japanese society, policymakers found themselves compelled to create scenarios that conveyed some elements of hope to bring back confidence into a society growing progressively pessimistic. The year 2014 marked a watershed moment for the generation of such scenarios of hope. That year, while the scathing Masuda Report was circulating, the government proposed a number of ambitious spatial and functional restructuring strategies. It is interesting to see how these strategies could reframe the bleak forecasts of recent years with narratives of opportunities and potentials. For example, "disappearing municipalities" and "shrinking cities" became "smart shrinking" or "rightsizing cities". Population decline, undoubtedly a daunting challenge, was framed in a way that it could offer new opportunities. This narrative shift was an important step towards sparking community action and institutional change within a wide range of social networks.

3.4.2 A culmination point: National Grand Design 2050 (2014)

Two decades of debates on depopulation and economic stagnation culminated in a framework for a coherent and long-term spatial development strategy that, finally, squarely addressed Japan's socio-demographic crisis: the "National Grand Design 2050: Creation of a country generating diverse synergies among regions" was unveiled by the Ministry of Land, Infrastructure, Transport and Tourism (MLIT) in 2014 (OECD 2016, Funabashi 2018). The new Grand Design identified the main trends and challenges as the ageing population, depopulation, ageing infrastructure, increasing competition between cities, natural disasters, technological change and threats to food, water and energy supplies (MLIT 2014). It formulated its key strategies based on a new population count founded on a national grid, of which more than 60% was assumed would lose half its population. 20% of these grid squares were forecasted to become uninhabited by 2050 (MLIT 2015). Based on the identified trends and challenges, the Grand Design 2050 created the principles for a long-term spatial strategy up to 2050. These were subsequently integrated into all other government plans, including the most important – the 2nd National Spatial Strategy – to be adopted the following year.

Smart shrinking: The principles of the National Grand Design 2050

The two main ideographic narratives of the strategy are "compact" and "networks". The new Grand Design suggests the "rightsizing" of cities and the creation of new types of functionally connected regions. It proposes a spatial restructuring of the entire Japanese territory with three types of spatio-functional building blocks and three corresponding models of services (Amano/Uchimura 2018, OECD 2016). These are: (1) villages with "small stations", (2) mid-sized cities with "integrated regional hubs", and (3) a "super megaregion" around Tokyo with "collaborative core urban areas". Before elaborating on the spatial building blocks, we shall have a look at the key components and principles underlying the strategy.

Compacted and networked
The two principal ideas underlying the strategy are: (a) to make the country spatially more "compact" in order to make public service delivery efficient and supportable; and (b) to strategically interlink the compacted areas into larger functional "networks" with the hope of creat-

Figure 5: "Smart shrinking": diagram of the key ideas and principles of the National Grand Design 2050 and the 2nd NSS

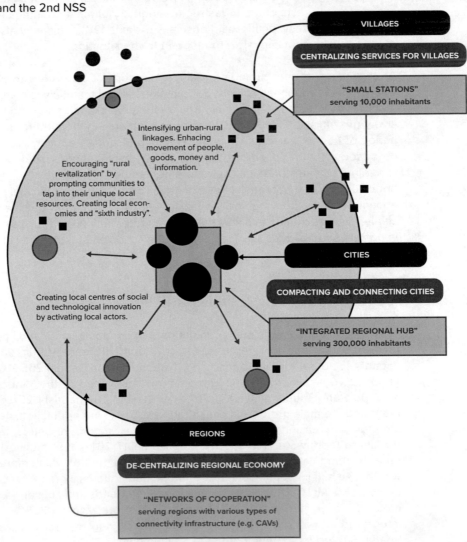

Source: Ian Banerjee, adapted from MLIT (2014)

ing more innovation by intensifying processes of exchange between neighbouring regions. It is assumed that this will help to maintain or regain the agglomeration effects lost by depopulation and to nurture the engines of growth by radically improving the connectivity infrastructure between the regions.

The principal ideas of the Grand Design mainly take their cue from studies in economic geography and location theories. They are based on research showing that the doubling of the population living within a radius of 300 km can increase productivity by 1.0% to 1.5% through various forms of spillover (OECD 2015). While there is overwhelming evidence that economic benefits can increase with city size, there is also evidence that smaller cities can benefit by "borrowing" agglomeration from neighbouring cities (OECD 2016: 84). Agglomeration effects are believed to result not only from population density but from "the ease with which agents can interact and transact with a large number of other agents" (OECD 2016: 84). Also, studies show that these interactions do not necessarily have to take place only in close vicinity – many highly innovative regions are successful by being part of international linkages despite having moderate populations.

Figure 6: The hierarchic ordering of space according to the central place theory of Walter Christaller

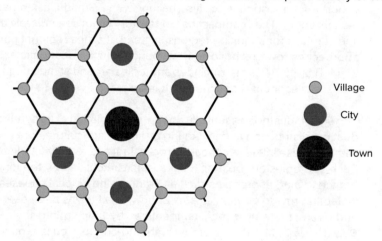

Source: Ian Banerjee, adapted from Christaller (1933)

The principles of the Grand Design also remind us of central place theory (CPT), a highly influential yet disputed theory put forward by the German geographer Walter Christaller in the 1930s (Christaller 1933). Based on the analysis of existing human settlements in Southern Germany at the time, Christaller proposed an idealized model for planning settlements based on the concept of centralization as the ordering principle. He asserted that settlements existed as "central places" to host services for surrounding areas. According to Christaller, people purchased goods and services from the closest places, and in the transportation model this ordering could minimize the network length and maximize the connectivity of the centres to be served (Agarwal 2009). Building on this, he argued it would be possible to calculate the size and number of settlements, evenly distributed across the entire territory, using the hierarchical order of space built on hexagons. While there are striking similarities between the principles recommended by the new spatial strategy in Japan and CPT, it is notable that the precondition of Japan is radical shrinking, while CPT was conceptualized in times of growth.

Diversity and collaboration
As the subtitle of the Grand Design suggests, it is about the "Creation of a country generating diverse synergies among regions" (MLIT 2014). Diversity and collaboration are seen here as pre-

conditions for regional development in times of depopulation (MLIT 2014; OECD 2016b: 10). Regions and towns are encouraged to identify their specific natural, cultural, economic and social assets and understand their potential to attract people and investment: "This very diversity of endowments and strategies creates the possibility for collaboration, because it gives rise to the possibility of identifying potential complementarities among places and building strategies to exploit them" (OECD 2016b: 10).

Economic diversification in regions along with intensified interactions are expected to lead to more "constructive dialogues" between people and places (MLIT 2015: 31): "As the population declines, competition among regions and cities for people and resources will intensify, largely because they have similar endowments, needs and aspirations. However, it is their diversity that may offer the best hope for the future" (OECD 2016b: 10). The new strategy recommends experiments with models of governance that foster new types of political cooperation between regions (see Section 4).

Rural revitalization and the "sixth industry"
The main reason for migration to big cities in Japan (and most other countries) is the lure of jobs and lifestyle. The only way to prevent this is believed to be the revitalization of rural areas with the creation of non-agricultural jobs and the provision of lifestyle options attractive to the younger cohorts. As a response to this challenge, the Grand Design has taken up the narrative of "rural revitalization". The "compacting" of territory is not seen merely as a defensive measure to streamline services for shrinking regions; instead, it urges regional governments to embrace innovative approaches towards building new foundations for economic activities in rural areas (OECD 2016: 81; MLIT 2015). This also includes strategies around stimulating senior entrepreneurship known as the "silver economy" and the "sixth industry".

The sixth industry is essentially about "the formation of integrated value chains encompassing production, distribution and marketing by linking agriculture, forestry and fisheries producers to those with expertise in the secondary (processing) and tertiary (marketing) sectors [...]" (OECD 2016: 213). Such strategies are built on local assets exemplified by the logic of strategies like those used in France to market products such as French wine or local cheeses. A number of lighthouse projects in Japan provide good examples of how small towns have been able to reinvent their economies and even create new regional identities by tapping into their local resources. For example, the Seiwa area in Mie Prefecture was able to create a buzz around environmental innovation; Ama-Cho in Shimane Prefecture has been able to shine with food technology and education; and very prominently, Kamiyama in Tokushima Prefecture has found success with IT companies and art. All have established a basis for future prosperity and created conditions that are attractive to young people, without the unrealistic desire to "bounce back" to their previous population counts.

Attractive rural landscapes and amenities, combined with good external connectivity, are leading to the rise of a growing number of start-ups in the field known as "knowledge-intensive service activities" (KISA). This trend can be also observed across many OECD countries (OECD 2016b: 17). The sixth industry and KISA are not expected to have a substantial macroeconomic impact in Japan, but it is anticipated that "small niche activities could still make a big difference in thinly populated areas" (OECD 2016a: 211). A considerable part of the process of reimagining the future of Japan will be about how to create prosperity with fewer people.

Intensifying urban-rural linkages
Making rural life attractive to younger cohorts is seen as an essential task by the government. The Grand Design recommends strengthening relationships between rural areas and urban residents by creating stronger "urban-rural linkages". The experience of new types of rural life is actively promoted by initiatives such as "Attractive Rural Areas – Make Coming Back to Rural Areas Real", put forward by the Ministry of Agriculture, Forestry and Fisheries (MAFF) in 2015. It is broken down into very simple projects, such as enhancing children's understanding of rural life by offering experi-

ence-rich tours or encouraging direct exchanges of goods and services between various sectors and industries.

3.4.4 The three spatio-functional building blocks

The Grand Design 2050 reimagines the reconfiguration of Japanese territory with three types of spatio-functional building blocks based on three spatial scales and three models of social services. They are conceptualized as follows.

Villages served by "small stations" (serving 10,000 inhabitants)
It is recommended that the villages, where most of the population decline is expected to take place, are interlinked into networks of villages of a defined size and served by centralized community service delivery hubs called "small stations". These stations have two functions: to offer services and to help gather people, goods, money and information to create new value. They offer life services such as childcare, health/elderly care, government information, etc. Each of these small stations could serve up to 10,000 inhabitants in the village network. Estimates show around 5,000 to 7,000 of them would need to be constructed across the country (OECD 2016: 82). For example, Kochi Prefecture on the island of Shikoku plans to build 130 such stations,

Figure 7: "Small stations" are places offering centralized public services for surrounding villages with around 10,000 inhabitants. They are also places for gathering, bundling resources and creating new values. New models of transport and mobility (including CAVs) are conceptualized in an ongoing process.

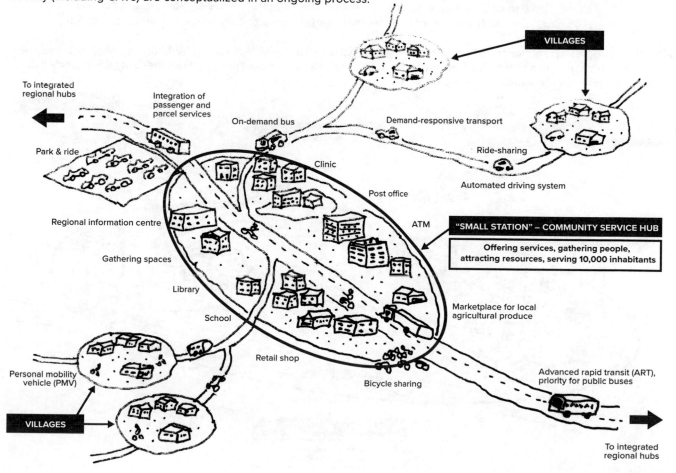

Source: Ian Banerjee; adapted from MLIT (2014)

positioned 4 to 5 km from each other, within a catchment area of 54 km². Studies by the National Institute for Land and Infrastructure Management (NILIM) have mapped out the measures necessary to push forward the "systematic shrinking of urban areas", which might be necessary when restructuring the urban areas through these small stations (Kaneko/Kiuchi 2018: 22). The politically and socially difficult question of relocation has often been discussed but mostly rejected by citizens (SIP-adus 2019a).

The deployment of CAVs with different capacities and ranges is expected to help create mobility services for these small stations. It is important to note that the aim of the strategy is to reduce car dependency and urban sprawl; the strategy promotes the concentration of key urban functions along with the extension of public transport networks (OECD 2016).

Mid-sized cities served with "integrated regional hubs" (serving 300,000 inhabitants)

The strategy recommends the integration of mid-sized cities into functional urban areas (FUAs) with a combined population of at least 300,000. This rationale is based on the assumption that a population of 300,000 is necessary to offer and maintain high standards in education, medical care and employment opportunities. They will be served through "integrated regional hubs", also called "urban compact hubs". These hubs will be connected with each other within one hour of travel ("high-grade city links"). The strategy estimates Japan would need 60 to 70 such regional hubs (Amano/Uchimura 2018: 23).

A "super megaregion" served by "collaborative core urban areas"

The aim of creating a "super megaregion" is to enhance the agglomeration potential of Tokyo. Against the backdrop of the country's stagnating economy, this is of particular national interest.

Figure 8: "Integrated regional hubs" are integrated functional urban areas (FUAs) serving 300,000 inhabitants. These are places offering high-end services such as education, health, etc. Better physical and digital connectivity are expected to make these places more innovative and productive.

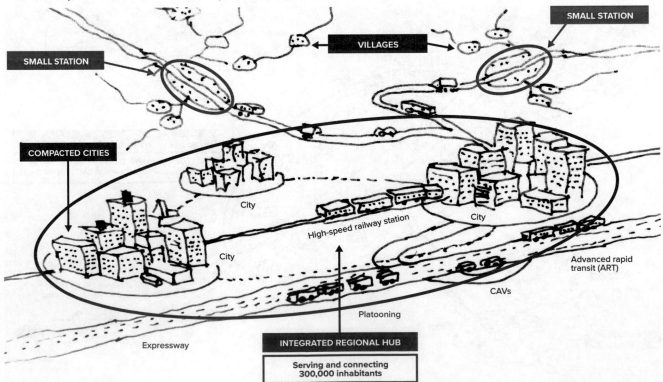

Source: Ian Banerjee; adapted from MLIT (2014)

As often mentioned in the urban discourse, along with New York and London, Tokyo is one of the world's top three "global cities" (Sassen 2001). It represents a crucial aspect of the country's economic competitiveness and its cultural landscape. The future of Tokyo is an important and sensitive topic for national spatial planning in Japan. The idea put forward by the strategy is to create more intense physical and digital connectivity between the three metropolitan regions of Tokyo, Nagoya and Osaka by connecting them with high-speed internet and an ultra-high-speed train (maglev) called Chuo-Shinkansen. This will make them physically reachable within an hour. With 64 million inhabitants, this will be the largest functional urban region (FUR) in the world. To maintain their competitive edge in logistics, innovation and knowledge creation, these regions are expected to need the services of a wider range of specialized forms of financial, business and producer services. The new strategy aims to achieve a higher concentration of such high-level urban services by building "collaborative core urban areas" situated near railway stations and other central locations (OECD 2016: 158).

3.4.5 2nd National Spatial Strategy: A comprehensive planning device (2015–)

The 2nd NSS, adopted by the national government in 2015, currently serves as the most important socio-economic planning document for Japan (for an English summary, see MLIT 2015). It integrates the building blocks of the National Grand Vision 2050 and outlines a comprehensive spatial strategy addressing all sectors of development. Also, with reference to Japan's geopolitical interest, it recommends the intensification of transnational exchanges with other regions in Asia, particularly with Southeast Asia (for more, see MLIT 2015).

Its overarching objectives are: (a) revitalizing regional and rural economies, (b) creating new settlement patterns by "smart shrinking", and (c) strengthening the global competitiveness of the major metropolitan areas. The architecture of the 2nd NSS reflects a consistent, long-term vision of the government's response to population decline. It formulates the nation's priorities and objectives while ensuring the coherence of the main sectoral policies, including the above-mentioned eight regional spatial strategies. In particular, the NSS includes the country's economic strategies, such as the National Special Strategic Zones (NSSZ). In 2014, the same year as the Grand Design was presented, Japan approved the creation of six NSSZs with different specializations, such as medical innovation, agriculture, tourism, etc. (see Fig. 1). The six zones are flanked with "regional innovation strategy promotion areas" such as the Industrial Cluster Programme, the Knowledge Cluster Initiative, and the Regional Innovation Strategy Programme. Notably, with regard to depopulation an explicit aim of the NSS is to stop the net inflow of currently around 100,000 persons per year from the rural areas to the Tokyo region (OECD 2016, MLIT 2015) and to encourage flows back to villages and smaller towns.

Despite the accomplishments attained by Japan's top-down national strategic spatial planning that has been applied over the last 60 years, there are growing doubts about the efficacy of its practice in the new millennium. The recurrent phenomena of industrial centralization and urban sprawl have shown how the location choices of both industry and households have consistently challenged the government's objectives of balanced growth and sustainable settlement patterns. This outcome has indeed created some degree of disillusionment with national spatial plans (OECD 2016).

However, withstanding all critique, the importance of coordinated, national-level strategic planning remains an imperative in Japan. In fact, at the beginning of the 21st century, in times of rapid change and insecurity, and "[...] at a time when government is devoting enormous energy to overcome traditional sectoral approaches to policy in favour of an integrated, government-wide approach to the challenges of demographic change, such a coordinating device is indispensable" (OECD 2016: 87). Also, it is believed that when national strategies are the outcome of extensive participatory processes, they can strengthen communication between stakeholders (OECD 2016). Drawing from this belief, the process of developing a national strategic plan is now considered in Japan to be as important as the plan itself – if not more so. This is also one of the reasons for the government's search for governance models that will enable more participatory decision-making processes in planning.

4. GOVERNANCE

The need to engage with local municipalities and the requirement to work with multiple levels of government have contributed to two specific shifts in viewing governance in Japan. This section briefly outlines these two shifts. It is by no means a critique or a comprehensive review of citizens' (dis)contents; however, it points to a major shift taking place in this field (for more on new approaches towards governance, see Chap. 16 by Hamedinger in this volume).

4.1 STRUCTURE OF TERRITORIAL GOVERNMENT

The structure of the territorial government in Japan is built on the complex relationship between the central government and the subnational governments (for details, see OECD 2016). The Japanese archipelago consists of the main island (Honshu) and three major islands (Kyushu, Shikoku, Hokkaido) with 6,848 lesser islands, totalling 377,960 km² in area. Administratively it is divided into 47 prefectures and 1,742 municipalities as of 1st January 2013. The three major metropolitan areas of Kanto (around Tokyo), Kinki (around Kyoto and Osaka) and Chukyo (around Nagoya) contain 51% of Japan's total population (IPSS 2014). Cities in Japan have different statuses. With 9.6 million inhabitants (metropolitan region: 37.6 million), Tokyo has the status of a prefecture divided into 23 wards, which are seen as independent cities. The so-called "designated cities" have a population of over 500,000, the "core cities" more than 300,000 and the "special cities" over 200,000 people.

Figure 9: Structure of territorial government in Japan

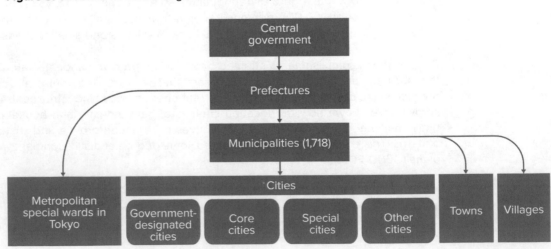

The enmeshment of fiscal structure and the commitment to interregional equity has created a very strong influence of the central government on subnational governments (OECD 2016). It is beyond the scope of this study to review the complexity of Japanese government and its evolution in recent decades; however, for this review, it is relevant to note that the current government is in search of a new approach to government (and governance), which will essentially instigate the devolution of power and the empowerment of local municipalities.

4.2 TWO SHIFTS IN THE APPROACH TO GOVERNANCE

The search for new models of governance intensified in Japan after the new Grand Design and the 2nd NSS were presented in 2014 and 2015 respectively. The main cause for this search was the institutional realization that the new national vision of a "compacted and networked" country could only be realized with the intense involvement of local actors and the production of local knowledge (OECD 2016). The central government is not expected to have the capacity be able to fully understand the specificities, assets or constraints of local regions, let alone have the resources to innovate on that level. Two essential moves implied in the strategy are: (a) breaking away from the notion of transforming every region into a growth machine, and (b) starting from the places and aspirations of the local community (and not solely investing in profit-oriented projects implemented in strategic locations).

Such discursive reflections have led to two shifts in looking at governance: (a) the shift away from a traditionally top-down approach to planning to one that encourages more bottom-up initiatives, and (b) the move away from sectoral to integrated policies.

4.2.1 Top-down meets bottom-up

While the old Grand Design for the 21st Century from 1998 had begun to promote wider citizen participation and partnerships, the first National Spatial Planning Act went on to institutionalize it in 2005. One of the intentions of this new act (see Section 3.3) was to incorporate mechanisms of decentralization into Japan's national spatial planning system. It adopted a two-tier structure: the overarching National Spatial Plan (NSP) and eight Regional Spatial Plans (RSPs; Kenji 2008: 511). This act consequently marked the beginning of the integration of local municipalities in the planning process. Based on this, many local authorities began to experiment with home-grown strategies, marking the growth of an experimental culture that would later prove supportive in developing local strategies for the deployment of CAVs.

After the "wake-up call" of the Masuda Report in 2014, a new consensus started to emerge across Japan: in order to respond to the diversity of the country's geographic, economic and social challenges, it was assumed that while maintaining strong leadership from the centre, a systematic acceleration of bottom-up processes in local communities was a matter of necessity. An important idea put forward by the new strategic approach is to encourage "healthy" competition and simultaneously foster strategic collaboration between subnational governments. To this end, the government has embarked on creating instruments to ensure that prefectural and local authorities have enough resources, know-how and authority to design and implement their own ideas. It has also started a discussion on how new governance structures could improve the traditional tax allocation mechanisms enacted through the country's intergovernmental transfer system (OECD 2016).

4.2.2 Breaking down the sectoral silos

The second recommendation made by the new NSS is to break down the sectoral walls between governmental institutions.

These efforts are predicated in part on building on a sense of crisis (*kikikan*) in order to overcome institutional inertia and the tendency of bureaucratic structures to operate within narrowly defined sectoral policy "silos". There are important potential complementarities among different strands of public policy that can only be realized with a whole-of-government approach, and that is precisely what the government has been working to realize (OECD 2016: 79).

"Breaking down the silos" seems to be an idea that had found its moment in Japan. It corresponds with another major institutional innovation initiated in the same year by the Council for Science, Technology and Innovation (CSTI) – something that would directly affect the programmes for CAVs (see Section 5.1).

4.2.3 The "Headquarters": Coordinating sectoral policies

To enable the walls of sectoral silos to be broken down, the Japanese government created a new organization, also in the year 2014, called the "Headquarters for Overcoming Population Decline and Vitalizing of the Local Economy", in short "Headquarters". The mandate of this organization is to coordinate the national and regional policy interventions across all relevant policy domains and all sectors of the country. It was an organization put in place to build bridges across the deep rifts that exist or tend to grow in due time between sectors, organizations and institutions.

The statement made by Prime Minister Shinzō Abe at the first meeting of the Headquarters in September 2014 denotes the importance placed in the role of this organization: "The most important aspect of measures to be implemented by this Council is that they eliminate vertical segmentation and employ one-stop responses [...]" (OECD 2016: 88). To further emphasize the importance of the organization, a new minister, directly under the aegis of the Cabinet Secretariat, was installed to head the organization. The Headquarters functions on two levels: (a) at the political level, it operates through a Council on Overcoming Population Decline and Revitalization of the Local Economy chaired directly by the prime minister; and (b) at the working level, it operates through a staff of civil servants who support the minister with day-to-day operations (OECD 20016: 88). The minister is also in charge of the National Special Strategic Zones (NSSZ; see Section 3.4.4); thus, he has the responsibility of aligning the country's population strategy, spatial strategy and economic strategy. Also, he has to push and coordinate the local governments' mandate to create their own long-term visions and revitalization plans within the framework of the 2nd NSS in a participatory way – involving all key actors, including experts and civil society.

What makes the Headquarters interesting is that it is instructed to offer both information (including big data) and financial support to local and prefectural authorities. If requested by the municipalities, the Headquarters is also obliged to appoint a civil servant with ties to the local region, who can for instance act as a consultant and advisor to the respective municipality (OECD 2016).

5.   CAVS IN JAPAN

Japan is one of the key actors in the global innovation landscape for CAVs. Together with the USA and the European Union, it plays a strong political role in shaping the international discourse unfolding around CAVs. This is reflected in the yearly trilateral meetings conducted jointly by the three political regions. In this section we shall outline the country's institutional setting that enables the development of CAVs, its relation to NSS and the experiments taking place in rural and mountainous regions.

5.1   THE SIP: AN INSTITUTIONAL CATALYST FOR CROSS-SECTORAL RESEARCH

In 2014, an organization called SIP (Cross-Ministerial Strategic Innovation Promotion Program) was created by the Council for Science, Technology and Innovation (CSTI), under the auspices of the prime minister and the minister of state for science and technology policy, to facilitate transdisciplinary research by cutting through the ministerial silos (SIP 2017, Amano/Uchimura 2016). The key idea was to create a portfolio of programmes and projects derived from an over-all set of strategic goals and ambitions. The portfolio is divided into three general policy areas and 11 specific fields such as energy, disaster prevention, cyber security, etc. Each of these 11 fields is equipped with its own budget and ascribed a "programme" led by a programme di-rector. The programme directors' main function is to facilitate the cross-sectional coordination of the triple-helix movement between government (cross-ministerial), industry and academia. They are responsible for the entire chain of activities from initiating basic research to a clear exit strategy (i.e. application and commercialization). One of these 11 policy areas, Next-Gen-eration Infrastructure, is dedicated to automated driving systems (ADS). With the growing im-portance of ADS, SIP created a new section of its body called Automated Driving for Universal Services, in short SIP-adus, in 2014. The main objective of SIP-adus is to coordinate and cat-alyse research between a wide spectrum of actors, with the broader aim of integrating social innovation with technological innovation (Amano/Uchimura 2016). SIP-adus builds on the al-ready existing national portfolios of Intelligent Transport Systems (ITS), Dynamic Map Planning, Human Machine Interface, Advanced Rapid Transit technologies (ART) and the projects of Next Generation Transport (for more, see Amano/Uchimura 2018).

5.2   NSS AND CAVS

As argued earlier, the reason for the detailed elaboration of the national spatial strategy is to explore how the spatial and functional logic for the deployment of CAVs has been derived from its comprehensive ideas of socio-territorial innovation.

Like most countries in the world, Japan's mobility sector is experiencing deep technological changes along with rapid changes in its citizens' mobility behaviour. However, in addition to the globally shared challenges of decarbonization, lack of finance and accessibility, Japan's trans-port system is confronted with the challenge of serving a dramatic number of "disappearing municipalities" – a condition that is causing more and more transport services to close down, particularly in rural areas. In this dire situation, the promises of CAVs have greatly fuelled the imagination of mobility experts in the country. CAVs in Japan are seen as a promising solution to many of its ailments. This circumstance makes Japan stand out sharply in the global race to deploy CAVs. The deployment of CAVs in Japan is not seen as a luxury but a necessity; its value is seen less as a private vehicle than as a public service.

As indicated in the sections above, the new Grand Design and the new NSS have also set the stage for conceptualizing the future of mobility in Japan. The new grand notion of a "compacted and networked" country requires an overall conception of new mobility systems – mainly on two levels: on one level, it needs to address intraregional mobility within "compacted" settlements, involving measures such as the reduction of cars and the promotion of public transport. On another level, it needs concepts for interregional mobility – those that will connect the compacted regions into strategic "networks". This will necessitate site-specific transit systems of very different types and capacities ranging from "platooning" to advanced rapid transit (ART), from personal mobility vehicles (PMV) to automated demand-responsive transport (ADRT).

5.3 CAVS IN RURAL JAPAN

5.3.1 Demand-responsive transport (DRT)

Long before the advent of automated vehicles, Japan had been testing alternative concepts of mobility in rural areas (OECD 2016a). The proactive attitude of policymakers towards CAVs in Japan can partly be explained by the fertile ground that these experiments have cultivated. The most significant of these experiments have been conducted around the concept known as demand-responsive transit (DRT).

The mobility challenge for rural Japan (as also for similar low-density regions in many other parts of the world) is how to deploy quick, reliable, affordable transport options without requiring many public subsidies. To meet this challenge, municipalities started experimenting with the DRT model almost 40 years ago. In Japan, its system has evolved into one with three types of buses: (a) those with fixed routes, operating only if there is demand; (b) those that offer some flexibility in the fixed route; and (c) those that offer relatively free routes. Some places integrate taxis with flat rates for door-to-door services, and others offer bus services combined with shared taxis. Some 200 municipalities offer DRT services in Japan today. The rise of digital platforms led to a doubling of the provision of DRT between 2006 and 2013 (OECD 2016a: 238). The 2007 Act on Revitalization and Rehabilitation of Local Public Transportation Systems, and its amendment in 2014, gave municipalities more freedom to self-organize. This further "unleashed a great deal of local experimentation" in public transport, putting Japan at the "forefront of efforts to adapt public transport for rural areas" (OECD 2016: 236). Looking at the experiments conducted with CAVs in rural Japan, in fact, reminds us strongly of the DRT experience collected in the last decades.

5.3.2 The main drivers for CAVs in rural regions

Japan's proactive attitude towards CAVs has mainly been sparked by (a) the hope of serving the mobility needs of an ageing society ("active ageing"), and (b) the hope of counteracting the problem of a growing shortage of bus and lorry drivers.

About 330,000 people aged over 65 years returned their driving licences in 2016, after which it was found in a survey that about 70% of them felt a general reluctance to leave their homes (MLIT 2017; Amano/Uchimura 2016, 2018). In the context of this situation, the promotion of active ageing is, firstly, about how to deploy CAVs in ways that could enable the elderly to play an active and productive role in society for as long as they wish. Secondly, it is about reducing age-related fatalities, as 54% of victims of fatalities are 65 years of age or older (Amano/

Uchimura 2016, 2018). In fact, they are not only the victims of accidents but also the cause of accidents. This makes road safety involving the elderly one of the highest priorities in Japan.

The average share of the aged (65+) in rural areas is currently about 31%, while the national average is about 23% (MLIT 2017). This poses the question of how to maintain the provision of food, care services, etc. to the elderly in these increasingly depopulating areas of the country. As the public transportation market is liberalized in Japan, it is natural that companies should withdraw from the places where they cannot maintain profits. More than 13,000 km of bus routes have been cancelled since 2007 (MLIT 2017). This situation is aggravated by the fact that both bus and taxi companies are finding it more and more difficult to recruit drivers in depopulating areas. The shortage of lorry drivers (more than 40% of whom are aged over 50) has also started to make the delivery of daily provisions and other goods increasingly difficult. This has made both transit and delivery of goods highly interesting fields for site-specific and tailored applications for CAVs.

5.3.3 Experiments with CAVs

Research and field operational tests (FOTs) with CAVs are initiated by SIP-adus along with national governmental organizations like the Cabinet Office, the Ministry of Land, Infrastructure, Transport and Tourism (MLIT), the Ministry of Economy, Trade and Industry (METI) or by local municipalities together with private companies and universities (for a full report on all field operational tests from 2014 to 2018, see SIP-adus 2019b). In 2018, 14 research locations were identified by the national government and 30 locations by local governments, universities and private companies (MLIT 2019). A series of extensive tests built on fluid alliances forged between diverse sectors are currently underway in rural areas (MLIT 2019).

Figures 10–13: CAV tests at Keio University, Tokyo

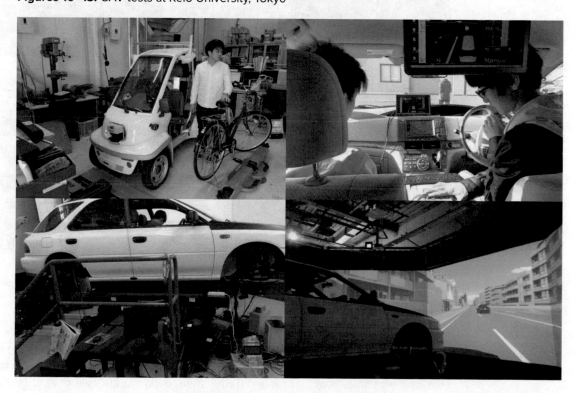

Photos: Ian Banerjee

Experimental settings in rural areas

Most of the experiments conducted by MLIT take place with four types of vehicles: two types of buses with a capacity of six passengers (10 km/h) and 20 passengers (35 km/h); a passenger vehicle for six passengers (12 km/h) and a passenger car for four people (40 km/h; see Table 1, Figs. 14–17). They are guided by four different combinations of sensors: electromagnetic induction wire (EIW), 3D maps, global positioning systems (GPS) and inertia measurement units (IMU; MLIT 2019).

Table 1: Details of vehicles and sensors for tests in rural Japan by MLIT

		Level 4	Level 2	GPS	IMU	3D map	EIW*	Passengers	Speed
Type I	Bus	●		●	●	●		6	10 km/h
Type II	Bus	●	●	●			●	20	35 km/h
Type III	Passenger car			●			●	4–6	12 km/h
Type IV	Passenger car	●	●	●			●	4	40 km/h

Type I   Type II

Type III   Type IV

Source: MLIT (2019)

The main purpose of the so-called "social experiments" conducted by MLIT (2017) are to understand the operational design domains (ODDs) and social needs involving (a) the necessary existing road environments for CAVs; (b) the existing transportation systems; (c) local specificities such as climate or telecommunication infrastructure; (d) the potential costs for scaling up the services; (e) social acceptance; and (f) the impacts on the lifestyle of especially elderly citizens. From a technological viewpoint, the main focus is on how to verify, develop and maintain the dedicated travelling spaces, and also how to improve communication between vehicles and road infrastructure (V2I). From a business viewpoint, the feasibility of combining public transport and logistics is of special interest, as is the potential profitability of the respective cases.

"Roadside stations" in mountainous regions

A series of CAV tests conducted in the mountainous regions of Japan exemplify the experimental setting of the tests in rural Japan. A number of them are located around "roadside stations" called "michi-no-Eki" (compare with the concept of "small stations" proposed in the Grand Design 2050 and 2nd NSS, Section 3.4.4). Located along main trunk roads, these roadside stations are more than just highway rest stops: they offer medical amenities, shopping facilities and other essential services for everyday life. In Japan, the total number of roadside stations amounts to 1,167, of which more than 80% are located in rural areas (MLIT 2017). The experimental setting (Fig. 18) simulates the real-life needs of the elderly in rural areas. The procedure is simple: the automated vehicles wait for local residents at the roadside station – also the place where the control centre is located. Once an order has been placed via an app, a CAV picks up the client(s) from their home (or any given location) and takes them to do their errands, such as going to the clinic, shopping or leisure activities. The customers are then transported back to their desired location.

Figure 18: Schematic diagram of the setting of social experiments with CAVs in roadside stations in rural regions

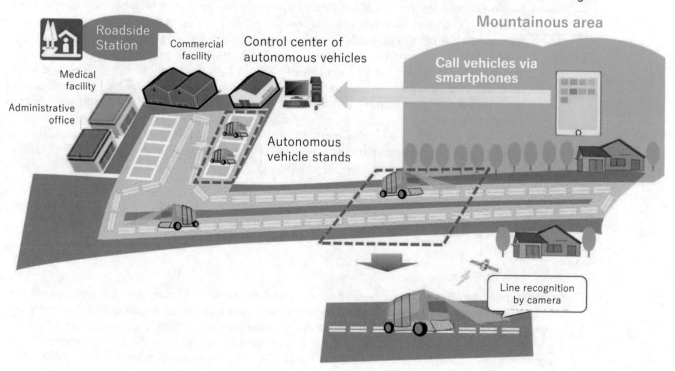

Source: MLIT (2017)

5.3.4   Findings

An interim report published by MLIT (2019) on the social experiments conducted in rural areas with hybrid forms of public transport and delivery systems comprises an extensive evaluation of the tests undertaken from 2014 to 2018. While no one can say how long it will take to scale up the tests, the findings give a clear picture of the thematic issues surrounding the social, technological, and financial feasibilities of the deployment of CAVs in rural areas. They can be summarized as follows:

1. **Social needs:** regarding the possible operational route for automated vehicles as public transport, three major types of needs for services are identified: (1) demand-type services that are personalized and completely flexible; (2) fixed-type services that are semi-flexible, for example, by allowing people to hop on and hop off the vehicles at flexible stops; and (3) the need for transfers to existing high-speed transit (first mile/last mile).

2. **Social acceptance:** (a) the perceived reliability of self-driving technology is significantly higher after boarding the vehicle than before; (b) people have feelings of uneasiness if the amounts of automated delivery are relatively small.

3. **Road space:** (a) operating on steep slopes in mountainous regions is technically possible; (b) narrow streets without pavements tend to hold up CAVs; (c) in mixed traffic, two-lane streets are better for CAVs to handle than single lanes.

4. **Technology:** under snowy or foggy conditions, and in mountainous areas where GPS accuracy is reduced, CAVs utilizing vehicle-to-infrastructure technologies (V2I) such as electromagnetic induction wires (EIW) or magnetic markers showed better results than those using LiDAR.

5. **Business models:** (a) to make CAV more socially acceptable, it is necessary to make the operating entity, service models and the definition of roles of providers and users more transparent; (b) the cooperation and coexistence with existing public transport services is absolutely essential (as complementary components); (c) annual budget calculations indicate that it is necessary to consider financing CAVs not only through fares and shipping charges, but also through various collaborations with private companies and subsidies from local governments; (d) collaboration with existing delivery and food distribution companies can improve the efficiency and cost of delivery of services in a given region; (e) it is necessary to educate users about the novel possibilities of using CAVs; (f) it is important to illustrate to communities how CAVs can improve their local quality of life.

6. CONCLUSION

This paper has offered an analysis of how the current Japanese government is addressing the country's challenge of an ageing and radically shrinking population with a unified set of policy responses. The authors elucidated how the government is applying the instrument of a new national spatial strategy (NSS) as a navigational tool to orchestrate the development of the country's key economic and social sectors, including the deployment of technological novelties such as CAVs. The country's problems have led the government to align its entire technological funding and innovation programme with the overall orientation of its new NSS. The research has shown that a national spatial strategy can be a helpful device to conceptualize a public-value-oriented deployment of CAVs – if the strategy has a cross-sectoral and multi-scalar approach, and if the strategy is the product of a broad and collaborative reimagining process.

Having peaked in 2010, Japan's population is expected to plunge 30% by 2060. After two decades of intensifying debates on shrinking and mounting accusations of neglect on the side of the government, the nation's policymakers have finally found the political resolve to engineer a long-term vision for the country's future and a strategic plan for multidimensional change. Contrasting with the pessimistic mood prevailing in the country, the latest grand vision effuses new hope and an optimistic view of the future based on a plausible future scenario of "rightsized" cities and the "smart shrinking" of regions. Two intertwined policy frameworks put forward by the Japanese government between 2014 and 2015 formulated an integral system of ideas and principles that were designed to fundamentally transform the country's approach to planning for the coming 40 years. The new strategy is framed by (a) a set of guiding principles that underlie the country's spatial development plan till 2050 (Grand Design 2050), and (b) an integrated national spatial strategy that stipulates the direction for the development of all the major sectors in the country till around 2025 (2nd NSS). All of Japan's future plans adhere to the system of ideas presented in these two key policy frameworks.

The authors investigated two defining features in the Japanese approach that make it an interesting case for studying the deployment of CAVs. The first feature pertains to how the country stands out among other proactive countries by the way it defines its experiments around CAVs as essential responses to its severe socio-economic challenges — underlining the argument that CAVs are not a luxury for the few but potentially a common good for all. This makes public investment in the risk-prone technologies of automated driving systems (ADSs) more arguable than in countries where relevant challenges are either not politically recognized or CAVs are not adequately defined as something that could create public value. The Japanese government's proactive attitude towards CAVs is mainly driven by the hope of meeting the needs of a rapidly ageing society, particularly in rural areas, and of solving the country's acute problem of its growing shortage of bus and lorry drivers. In this context, it is interesting to note that the experiments taking place around CAVs are usually described as "social experiments" and not as technological experiments (MLIT 2017, 2019).

The second distinctive feature highlighted in the study pertains to how the new spatial strategy significantly shapes the rationale for the deployment of CAVs in Japan on both local and interregional levels. The case of Japan demonstrates that the deployment of CAV-based operations can be easier to conceptualize if a unified policy framework follows a principles-based and socio-spatially differentiated logic that responds to the challenges on all spatial scales of the country — from villages to megaregions. Such a unified and spatially differentiated logic has also made it easier for the Japanese government to deduce the logic for the countrywide deployment of CAVs and to offer guidance to municipalities and local communities that aspire to co-create their own CAV-based services. They are encouraged to act locally but think nationally. This study concludes by asserting that if an NSS is crafted strictly with the aim of addressing the country's challenges and of serving its public interest, then the deployment of CAVs can potentially be realized without the fear of commercial appropriation of the technology by a monopoly.

Summary of the main elements of the strategy
The authors have identified the following seven points that summarize (a) the aims, principles and recommendations put forward by the National Grand Design 2050 and the 2nd NSS and (b) the key interim results from the experiments conducted with CAVs:

1. The principle aim of the unified national strategy is to ensure and maintain the country's high level of prosperity with fewer people. The National Grand Design 2050 makes recommendations about to how to reach this aim by (a) centralizing public services to make them more sustainable during the coming period of radical shrinking, and (b) decentralizing the economy by creating incentives to build a broad range of diversified and localized economic activities. The key ideas underlying the strategies' principles are presented as ideographic narratives that can be easily comprehended by all citizens (Fig. 5).

2. The narratives of the two main policy-related principles employ the concepts of "compacting" and "networks". The long-term aim is to densify cities by "compacting" them ("rightsizing") and subsequently interlinking them into "networks" of cooperation ("smart

shrinking"). The compacted centres are encouraged to become centres of social and technological innovation, networked by connectivity infrastructure of various kinds — this includes the infrastructure for CAVs.

3. Additional policy catchwords used in the strategy are: "revitalization", "diversity", "collaboration" and the "sixth industry". These terms refer to incentives created by the central government to motivate local municipalities to generate non-agricultural jobs in rural areas by encouraging them to tap into their own unique resources. The government has created a competitive situation where municipalities have to generate innovative ideas to vie for the limited resources provided to support their projects.

4. The unified spatial strategy envisions the restructuring of Japan with three types of spatio-functional building blocks based on three spatial scales, and three models of public services (Figs. 6–7). These are: (a) villages served with "small stations" (for 10,000 inhabitants), (b) mid-sized cities served with "integrated regional hubs" (for 300,000 inhabitants), and (c) "super megaregions" served with "collaborative core urban areas" (for over 60 million people). These three spatial models can be seen as the "spatial DNA" of the strategic plan: their permutations and combinations are designed to produce and reproduce the socio-economic spaces of human settlements and subsequently shape the entire territory of new Japan. It is important to note that this strategy looks at the urban-rural linkages of the country as a seamless spatial continuum.

5. New concepts for mobility are currently being reimagined both within and between the compacted settlements. The logic of deploying CAVs is closely intertwined with the systemic notion of spatial reconfiguration undertaken with the three aforementioned spatial building blocks. This can involve operational settings ranging from "platooning" to advanced rapid transit (ART), from personal mobility vehicles (PMV) to automated demand-responsive transport (ADRT).

6. Experiments with CAVs are currently being conceptualized with the following aims in rural Japan: to serve the ageing population; to attract young people by offering high quality of life; to make up for its growing scarcity of professional drivers; and to offer sustainable public transportation and freight services. Co-creative experiments with various use cases are being conducted in rural areas by the central government, municipalities, universities and private companies. An interim report indicates that while there are serious "objective" needs for CAVs, there are many challenges to be met regarding road infrastructure, technology, operations, business models and social acceptance.

7. Finally, Japan's approach to governance is changing in fundamental ways. The two ongoing shifts are: (a) the move away from a traditionally top-down approach to one that encourages more bottom-up initiatives, and (b) the move away from sectoral to cross-sectoral policies in order to create more synergies between the sectors. A bridging organization called the "Headquarters" was formed on the ministerial level to coordinate the task of orchestrating and streamlining the entirety of national and regional policy interventions and action programmes across all relevant domains and sectors of the country.

While the 2nd NSS may arguably be the most comprehensive and ambitious spatial strategy plan proposed by a democratic government today, the results are yet to be seen. Bringing together all the country's key actors to implement the strategy remains a challenging exercise for the government. However, for the international discourse — particularly for Europe — it will be an interesting exercise to reflect and critically discuss the implementation of the Japanese strategy at this early stage as it responds to demographic and socio-spatial challenges that have started to afflict many parts of the continent in similar ways and with comparable severity.

REFERENCES

Adams, G. 1981. *The Politics of Defense Contracting: The Iron Triangle*. New York: Routledge. https://doi.org/10.4324/9780429338304.

Agarwal, P. 2009. *Walter Christaller: Hierarchical Patterns of Urbanization, Centre for Spatial Integrated Social Science*. Santa Barbara: University of California. https://web.archive.org/web/20091105054923/ http://www.csiss.org/classics/content/67 (13/5/2020).

Albrechts, L. 2004. "Strategic (spatial) planning re-examined", in *Environment and Planning* (31) 5, 743–758. https://doi.org/10.1068/b3065.

Albrechts, L. 2006. "Bridge the gap: from spatial planning to strategic projects", in *European Planning Studies* (14) 10, 1487–1500. https://doi.org/10.1080/09654310600852464.

Amano, H., and T. Uchimura 2016. "A National Project in Japan: Innovation of Automated Driving for Universal Services", in *Road Vehicle Automation 3*, ed. by G. Meyer and S. Beiker. Cham: Springer, 15–26. https://doi.org/10.1007/978-3-319-40503-2_2.

Amano, H., and T. Uchimura 2018. "Latest Development in SIP-Adus and Related Activities in Japan", in *Road Vehicle Automation 4*, ed. by G. Meyer and S. Beiker. Cham: Springer, 15–24. https://doi.org/10.1007/978-3-319-60934-8_2.

Bornstein, L. 2007. "Confrontation, collaboration and community benefits. Lessons from Canadian and US cities on working together around strategic projects", paper presented at the ISoCaRP conference, Antwerp, Sep 2007.

Broeck, P. van den 2011. "Analysing social innovation through planning instruments", in *Strategic Spatial Projects – Catalysts for Change*, ed. by S. Oosterlynck, J. van den Broeck, L. Albrechts, F. Moulaert and A. Verhetsel. New York: Routledge, 52–78.

Christaller, W. 1933 (2009). *Die zentralen Orte in Süddeutschland*. Vienna: wbg Academic.

Diamond, J. 2019. *Upheaval: How Nations Cope with Crisis and Change*. UK: Random House.

Healey, P. 1997. "The revival of strategic spatial planning in Europe", in *Making Strategic Spatial Plans: Innovation in Europe*, ed. by P. Healey, A. Khakee, A. Motte and B. Needham. London: UCL Press.

Healey, P. 2004. "The treatment of space and place in the new strategic spatial planning in Europe", in *International Journal of Urban and Regional Research* (28) 1, 45–67. https://doi.org/10.1111/j.0309-1317.2004.00502.x.

IPSS (National Institute of Population and Social Security Research) 2014. "Chapter 1 Overview of Population Trends in Japan", in *Social Security in Japan 2014*. http://www.ipss.go.jp/s-info/e/ssj2014/001.html (22/5/2020).

IPSS 2017. *Population Projections for Japan (2017): 2016 to 2065*. https://bit.ly/3buXAW2 (22/5/2020).

Kaneko, H., and N. Kiuchi 2018. "National Grand Design and Spatial Policy in Depopulating Period", in *KRIHS 36th Anniversary International Seminar*, 6–38. http://www.nilim.go.jp/lab/jbg/depopulation/others/natgradesspapol.pdf (12/5/2020).

Kidokoro, T., H. Noboru, L.P. Subanu, J. Jessen, A. Motte and E.P. Seltzer (eds.) 2008. *Sustainable City Regions: Space, Place and Governance*. Japan: Springer.

Kenji, O. 2008. "Challenges for a balanced and sustainable development in Japan", in *Informationen zur Raumentwicklung* (Heft 8). Berlin: Bundesinstitut für Bau-, Stadt- und Raumforschung BBSR, 507–514.

Mitteregger, M., E. M. Bruck, A. Soteropoulos, A. Stickler, M. Berger, J. S. Dangschat, R. Scheuvens and I. Banerjee 2022. *AVENUE21. Connected and Automated Driving: Prospects for Urban Europe*, trans. M. Slater and N. Raafat.. Berlin: Springer Vieweg. https://doi.org/10.1007/978-3-662-64140-8.

MLIT (Ministry of Land, Infrastructure, Transport and Tourism) 2014. *Grand Design of National Spatial Development Plan towards 2050 Japan*. Tokyo. https://www.mlit.go.jp/common/001088248.pdf (22/5/2020).

MLIT 2015. *National Spatial Strategy (National Plan)*. Tokyo: Ministry of Land, Infrastructure, Transport and Tourism. https://www.mlit.go.jp/common/001127196.pdf (13/4/2020).

MLIT 2017. *Autonomous Vehicle Services in the Mountainous Regions around Roadside Stations (michino-eki)*. Tokyo. https://www.mlit.go.jp/common/001178887.pdf (19/4/2020; in Japanese).

MLIT 2019. *Interim Report on Autonomous Vehicle Service in the Mountainous Regions around Road-side Stations*. Tokyo. https://bit.ly/30tPrei (19/4/2020; in Japanese).

Moulaert, F., A. Rodriguez and E. Swyngedouw 2003. *The Globalized City: Economic Restructuring and Social Polarization in European Cities*. Oxford: Oxford University Press.

OECD 2012. *Compact City Policies: A Comparative Assessment, OECD Green Growth Studies*. Paris: OECD Publishing. https://doi.org/10.1787/9789264167865-en.

OECD 2015. *Compendium of Productivity Indicators*. Paris: OECD Publishing. https://doi.org/10.1787/b2774f97-en.

OECD 2016a. *OECD Territorial Reviews: Japan*. Paris: OECD Publishing. http://dx.doi.org/10.1787/9789264250543-en.

OECD 2016b. *OECD Territorial Reviews: Japan Policy Highlights*. Paris: OECD Publishing. https://www.oecd.org/regional/regional-policy/Japan-Policy-Highlights.pdf (13/4/2020).

Oosterlynck, S., J. van den Broeck, L. Albrechts, F. Moulaert and A. Verhetsel (eds.) 2011. *Strategic Spatial Projects: Catalysts for Change*. New York: Routledge

Population Reference Bureau 2018. *Which country has the oldest population? Based on Aging Demographic Data Sheet 2018 of International Institute for Applied System Analysis IIASA*. Washington, DC: Population Reference Bureau. https://bit.ly/38oAIFM (25/5/2020).

Roland Berger 2018. *Reconnecting the Rural: Autonomous Driving as a Solution for Non-Urban Mobility*. Munich: Roland Berger. https://www.rolandberger.com/de/Publications/Reconnecting-the-rural-Autonomous-driving.html (14/12/2019).

Reuters Graphics 2019. *Going Grey*. https://graphics.reuters.com/JAPAN-AGING/010091PB2LH/index.html (18/5/2020).

Sassen, S. 1991. *The Global City: New York, London, Tokyo*. New York: Princeton University Press.

Schlappa, H., and W. J. V. Neill 2013. *From Crisis to Choice: Re-Imagining the Future in Shrinking Cities*. Saint-Denis: URBACT. https://bit.ly/3qyLv6o (23/7/2020).

SIP 2017. *What is the Cross-ministerial Strategic Innovation Promotion Program?* https://www8.cao.go.jp/cstp/panhu/sip_english/5-8.pdf (2/5/2020).

SIP-adus 2019a. "SIP-adus Workshop 2019." https://en.sip-adus.go.jp/evt/workshop2019/ (2/5/2020).

SIP-adus 2019b. *2nd Phase of SIP-adus Project Reports 2014–2018*. http://en.sip-adus.go.jp/rd/ (2/5/2020).

Swyngedouw, E., F. Moulaert and A. Rodriguez 2002. "Neoliberal Urbanization in Europe: Large-Scale Urban Development Projects and the New Urban Policy", in *Antipode* (34) 3, 542–577. https://doi.org/10.1111/1467-8330.00254.

Toshihiko, H. 2015. *A Shrinking Society: A Post-Demographic Transition in Japan*. Japan: Springer.

Warren, T. 1930. *Population Problems*. New York: McGraw Hill. https://doi.org/10.2307/1229783.

Webb, J. 2017. "Book Review: Japan's Population Implosion", *Tokyo Review*, 2/5/2018. https://www.tokyoreview.net/2018/05/japans-population-implosion/ (2/5/2020).

14 Integrated strategic planning approaches to automated transport in the context of the mobility transformation

A case study in suburban and rural areas of Vienna/Lower Austria

Mathias Mitteregger, Daniela Allmeier, Lucia Paulhart, Stefan Bindreiter

Mathias Mitteregger
TU Wien, future.lab Research Center

Daniela Allmeier
TU Wien, future.lab Research Center and
Research Unit of Local Planning (IFOER)

Lucia Paulhart
TU Wien, future.lab Research Center

Stefan Bindreiter
TU Wien, simlab and Research Unit
of Local Planning (IFOER)

© The Author(s) 2023
M. Mitteregger et al. (eds.), *AVENUE21. Planning and Policy Considerations for an Age of Automated Mobility*, https://doi.org/10.1007/978-3-662-67004-0_14

1. TRANSFORMING MOBILITY WITH CONNECTED AND AUTOMATED VEHICLES

This article uses case studies to set out specific ways in which connected and automated vehicles can be applied to bring about the required revolution in mobility. More specifically, the aim is to present and discuss strategic planning concepts for a selection of local authorities and regions in the Vienna/Lower Austria metropolitan region. These strategic planning concepts have been developed to complement existing design concepts and/or action plans that largely remain abstract and generic. They are the outcome of a transdisciplinary process: urban planners and researchers formed focus groups with policymakers and local government administrators from the chosen local authorities and regions, as well as connected and automated mobility experts. This partnership produced multistage strategic planning approaches with the aim of establishing connected and automated vehicles (CAVs) as an attractive alternative to motorized private transport.

Mobility and settlement planners must urgently engage with this technology in a structured, proactive (and, at least initially, theoretical) way. A glance at the wide range of relevant initiatives being put into action demonstrates why. Around the world, CAVs are being intensively put through their paces (for more on this, see the articles in Part I "Mobility and transport" and Part IV "Governance"). Moreover, in Europe the legal foundations are already being put in place to enable this technology to be regularly used on our roads (Kugoth 2020).

However, there is still some way to go before we see these technologies in our daily lives, and the initial euphoria has lessened significantly, at least among some experts. In recent years, research on the possible impacts of connected and automated vehicles has become markedly more cautious (for more on the early euphoria surrounding "self-driving cars", see Dangschat/ Stickler 2020). Interviews carried out with experts across Europe confirm this lowering of expectations: many now think that this technology can only make a positive contribution to the mobility transformation if policy is designed to specifically regulate its use from the outset. Yet opinion remains divided on what form this regulation should take (Lenz/Fraedrich 2015; Mitteregger et al. 2022: 33–46).

These growing concerns have, however, been largely absent from the technology's ongoing development and promotion. Governments the world over have been supporting research and development in this field. In the EU, automated vehicles are seen as the key to a "digital single market" (Buchholz et al. 2020): funds are specifically being allocated to develop the technology, while the necessary infrastructural expansion, standardization initiatives and alignment of legislation are being pursued (STRIA 2019; for more on criticism and perceptions, cf. Dangschat 2020b for Germany and Manders/Klaasen 2020 for the Netherlands). Cybersecurity and general questions concerning liability are the only two areas where a regulatory framework has so far been actively put into place (Taeihagh/Lim 2019).

Only if these new technologies succeed in helping to replace privately used cars – the 20th century's dominant mode of transportation – will they have made a meaningful contribution to the transport and mobility transformation. The development of connected and automated transport (CAT) should therefore be used to specifically tackle the ecological, social and economic challenges we face. It is thus not a matter of encouraging the disruptive development of CAVs, but of shaping, nurturing and overseeing a mobility transformation that is partly made possible by the connection and automation of transport.

In Europe, steps have already been taken to make this a reality. In 2019, the European Commission's Sustainable Urban Mobility Plans (SUMP; Wefering et al. 2013) were revised to accommodate advances in the connection and automation of transport (Backhaus et al. 2019). There are also plans and speculative designs that focus – albeit still very optimistically – on the use of "self-driving" cars in cities (see, for instance, Harrouk 2020, Ratti et al. 2020).

Transport engineers have developed concepts that explore sophisticated applications. For instance, there are already examples of network plans for automated mobility services (Madadi et al. 2019) that only use parts of the road network and are reminiscent of personal rapid transit (PRT) networks from the 1960s (McDonald 2012). There is currently a lack of integrated planning approaches for integrated settlement and mobility designs that highlight the ways in which connected and automated vehicles can help achieve a mobility transformation and the conditions needed for this to happen in specific transport and spatial scenarios, especially in rural areas.

The challenges faced here are considerable. It is improbable that one European city or region can single-handedly design a mobility system that is increasingly shaped by automation and connectivity. The web of responsibilities is simply too tangled; for instance, regulations for road traffic are set at the national rather than the local level. The tasks and issues are so complex that it is almost impossible to tackle them adequately with existing resources. If this new mobility system is to be developed in harmony with the current objectives of the mobility transformation, a comprehensive dialogue must take place at various transport policy levels and involve both new and established actors in (urban) mobility markets, institutions and administrative bodies. This is one of the key requirements set out by transformation researchers (Wittmayer/ Hölscher 2017).

Following decades of voices repeatedly making the case for the integrated development of space and mobility, a change now needs to occur. Unregulated development that leads to additional traffic and thus takes up more land, produces more emissions and creates hazards – as cars have now done for the last hundred years – is a scenario that humanity can no longer afford at the start of the 21st century.

2. CURRENT DEVELOPMENTS IN CONNECTED AND AUTOMATED VEHICLES

At the beginning of the last decade, self-driving cars received a tremendous amount of attention in the media. This hype was driven by IT companies suddenly taking an interest in the mobility sector. As these actors were motivated by other business models, they were able to realize technological and organizational innovations, and thus further add to the excitement around this new technology. The mobility service providers that appeared during this period (Uber, Lyft, Bolt, etc.), numerous sharing companies offering bicycles and e-scooters as well as car-sharing services, which have been around slightly longer, are now commonplace in urban areas, especially larger cities. This wave of euphoria and uncertainty shaped expectations regarding connected and automated vehicles. But now even technology developers are taking a more sober approach: a revolutionary scenario whereby a single market actor is able to produce a fully automated driving system that can handle every situation as adeptly as a human is largely regarded as unlikely (cf. Beiker 2015, Shladover 2018, Mitteregger et al. 2022).

The many passages on sustainable mobility that featured in policy and strategy papers back then attest to the huge hopes that were pinned on the technology; now each line is considered to need careful review (for instance, STRIA 2019; Kirchengast et al. 2019: 58). The perceived positive effects, such as reclaiming parts of the streetscape by freeing up parking spaces, the reduction of vehicles by increasing the use of car or ride-sharing, improved traffic safety thanks to the superior cognitive abilities of machine learning, reduced energy consumption thanks to more efficient modes of driving as well as a broader scope for the use of Mobility as a Service (MaaS), are all currently the subject of debate (Dangschat 2019, 2020a; Dangschat/Stickler 2020; Soteropoulos et al. 2019; Mueller et al. 2020; Pangbourne et al. 2020).

Yet even today connected and automated vehicles largely exist only as concepts. That is why, given the current state of connected and automated vehicle development, a study such as this invariably includes several untested hypotheses. The theories that have been developed are outlined and examined in the following section.

2.1 EVOLUTION NOT REVOLUTION: AUTOMATED MOBILITY AND THE LONG LEVEL 4

Today's perception of the technological challenges inherent in the development of automated vehicles will likely mean a longer transitional phase. In other articles, we have coined the term "Long Level 4" to describe this phase. Here we mean "a gradual process [...] extending over several decades, during which CAVs will be deployed only in parts of the road network. During this transition period, conventional means of transport will continue to play an essential but increasingly specialized role" (Mitteregger et al. 2022: VIII).

As connected and automated vehicles are deployed as part of the Long Level 4, negative effects may become more visible and appear early in the process. This is primarily due to expectations that those areas of the road network that are already predominantly designed around cars will be more easily adaptable to CAV use (Shladover 2018; Mitteregger et al. 2022: 80–83; Soteropoulos et al. 2020). The roads that are expected to see CAV roll-out first are the result of a technological paradigm that has shaped urban development and transport policy since the 1950s. Such roads have been designed solely for use by motorized vehicles and their capacity is measured in vehicle throughput over a specific period (traffic density/volume) or speed limits (in km/h).

For this reason, it is expected that within the complex road networks of Europe's cities and regions, motorways and multi-lane trunk roads will see the first CAVs, followed by industrial roads and business routes. At low speeds, automated parking and services operating within a set boundary are conceivable. Until now, roads that serve as more than just transport routes, i.e. that are also used and revitalized as public spaces, have been too complex for CAV software, algorithm systems and sensors (Shladover 2016).

2.2 CONNECTED AND AUTOMATED VEHICLE APPLICATIONS

During this transition period, the fully automated vehicle is effectively fragmented into different highly automated applications that develop progressively and whose gradual deployment is contingent upon not only the developmental trajectory of the technology but the various complexities of the driving tasks and the infrastructural requirements (cf. ERTRAC 2019, Wachenfeld et al. 2015; Fig. 1). For this study, we examined five different highly automated vehicle applications, which were then used to guide discussions within the focus groups (see Table 1).

Table 1: Connected and automated transport system applications

PARKING ASSISTANT

Parking Assistant

Parking assistants (or automated valet parking) enable automated vehicles to drive to and from a nearby or far-away parking space. These systems are mainly being discussed, and are most likely to be imminently rolled out, within the context of park-and-ride facilities (at airports and train stations). A system of this kind makes it possible for passengers to call or drop off a vehicle in a designated pick-up/drop-off area even before changing their mode of transportation (e.g. before leaving the train). Another place where this application could be used are large car parks for shopping centres or retail parks. This technology could also enhance car-sharing vehicles. There is also the potential to charge battery-powered electric vehicles.

SITE ASSISTANT

Area assistant

Area assistants are designed for use within a limited, usually not publicly accessible area of land. There are some well-known applications that have been in place for just over a decade, including heavy goods vehicles in ports, in mines or automated reconnaissance vehicles that are used by military facilities to conduct security operations. Improved sensors enable new applications and access to new spaces: there are numerous companies that are either developing or already supplying mobile delivery and security robots for use in spaces within university campuses, theme parks, industrial estates, train stations and shopping centres. The most well known is the Waymo pilot scheme taking place in the suburbs of Phoenix, Arizona.

MOTORWAY ASSISTANT

Motorway assistant

Motorway assistants are systems that take over driving tasks on motorways or other trunk roads. The development of motorway assistants is primarily being encouraged by vehicle manufacturers and can also be considered an active safety system. This application is not just limited to individual transport: it is also a viable option for HGVs, utility vehicles (e.g. for haulage) or long-distance buses. Specially designed lanes are frequently discussed, especially for HGVs, to further reduce demands on the driving technology.

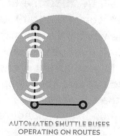

AUTOMATED SHUTTLE BUSES
OPERATING ON ROUTES

Automated shuttle buses operating on routes

Automated driving systems on specially constructed routes have been in use for several decades. One such example is PRT systems at airports (e.g. Heathrow) or as a last-mile solution to enable access to offices (e.g. Rivium close to Rotterdam). There is very little to distinguish automated shuttles on routes from shuttles deployed on public roads. It can be assumed that automated shuttles, at least at present, only drive on selected, clearly defined and approved routes (and lanes) that are protected and designated as such, and that vehicles only call at specified stops. On such routes, vehicles are able to operate without a driver; however, similar to air travel, they are monitored from a control room by pilots and can be deactivated. There is thus some technical crossover with area assistants.

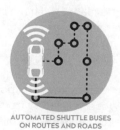

AUTOMATED SHUTTLE BUSES
ON ROUTES AND ROADS

Automated shuttle buses on public roads

Automated shuttles are currently being tested predominantly by public transport companies (see Chap. 6 by Soteropoulos et al. in this volume). However, the vehicles are yet to operate without a safety driver. These automated vehicles are mainly being tested as a possible shuttle service for underground and tramlines and thus as a complementary public transport service, for instance in suburban areas. The potential to save personnel costs, as well as the more flexible range of applications, is believed to make this a more economical option.

Source: Mitteregger et al. (2022), Kyriakidis et al. (2019), Shladover (2018), Perret et al. (2017), Wachenfeld (2015)

3. SUSTAINABLE MOBILITY AND SETTLEMENT DEVELOPMENT IN RURAL AREAS

The objective being pursued by transport policymakers is clear: the climate crisis demands a radical transformation of mobility – and past failings mean change must be achieved within a short space of time. In the EU, the transport sector has so far failed to reduce greenhouse gas emissions. In fact, since 1990, emissions have been rising rapidly (with a sharp surge between 2005 and 2015); outlooks also remain bleak (IEA 2020). Yet the changes required to successfully bring about the necessary transformation have long been known. As early as the 1970s (cf. Schwedes 2017), researchers and activists were formulating versions of a strategy that can be summarized as follows:

- traffic must be avoided,

- a shift needs to occur away from individual transport options and towards more environmentally friendly forms of mobility (walking, cycling and public transport) or zero-emission vehicles,

- and, lastly, steps must be taken to improve the quality of streetscapes. In towns, roads must become much more effective as public spaces (for more on the differences between transforming drive, traffic and mobility, see Chap. 4 by Manderscheid in this volume).

3.1 CO_2 EMISSION REDUCTION TARGETS IN AUSTRIA'S TRANSPORT SECTOR

The path to achieving greater ecological sustainability in the transport sector is still largely defined through targets. The possible approaches to achieving this aim across the EU's member states, however, remain vague, if they have been sketched out at all. By 2030 the European Commission aims to have reduced greenhouse gas emissions by 55% compared to 1990 levels (European Commission 2020). And from 2050, Europe should be the world's first "climate-neutral continent", a goal that will be pursued with a budget of €1 trillion (European Commission 2019). The Commission is also striving for the EU's transport sector to stop all net greenhouse gas emissions by 2050, all while stressing the economic viability and positive social impacts of the overall mobility shift. The current targets being pursued by Austria's policymakers even go one step further: the country aims to be climate neutral by 2040 and thus "play a pioneering role in climate protection in Europe" (Austrian Government Programme 2020).

Such long-term aims have repeatedly been called for by transformation researchers. However, given the world's limited carbon budget, it remains doubtful whether the planet will have by then long surpassed the Paris Agreement's aim to limit global average warming to between 1.5 and 2 °C. At the current rate of emissions, the global budget to limit warming to 1.5 °C will be reached by around 2027; for an increase of 2 °C, the limit will be reached by 2045 (IPCC 2018).

Roughly one quarter of Austria's greenhouse gas emissions are produced by the transport sector (Environment Agency Austria 2019). If we look at the greenhouse gases excluded from emissions trading schemes, transport is by far the largest contributor, making up 46% of emissions. Austria also exhibits a pattern seen at the wider EU level: while in recent years there have been visible reductions in all other sectors – from construction to agriculture – transport has in fact seen its emissions rise. Austria aims to reduce its annual transport sector CO_2 emissions by 7.2

million tonnes to 15.7 million tonnes by 2030 (at present, the figure stands at 22.9). Given that emissions have been rising in recent decades, this would mean that emissions must be reduced to 1991 levels (Kirchengast et al. 2019).

3.2 THE ROLE PLAYED BY CONNECTED AND AUTOMATED VEHICLES

Figure 1 compares the expected market availability of different CAV applications with the European Union's planned emission reductions. Here the theoretical availability of an application is being explicitly discussed and not its widespread use within society.

Figure 1: Necessary greenhouse gas reductions in the European Union's transport sector (Transport & Environment 2020) and the implementation of various CA services.

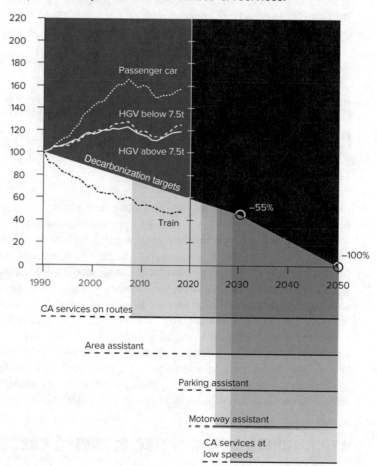

Source: the authors based on ERTRAC (2019), Kyriakidis et al. (2019) and Perret et al. (2017)

3.3 THE UNIQUE CHALLENGES OF RURAL SPACES

It is repeatedly stressed that the challenges of the transport and mobility transition will differ fundamentally in urban and rural areas (VCÖ 2016, Rudolph et al. 2017, Canzler et al. 2018). It is important to remember that this situation is, to a large extent, of governments' own making: in addition to the sectoral restructuring of economies in Europe and the associated social impacts, it is the transport and settlement policy decisions of recent decades that have led to the situation currently faced (Sieverts 2018). There is a lack of appealing alternatives to private cars in rural areas because public railways in these locations have been scaled back while the locality's further expansion is either tolerated by settlement policymakers or even intensified through incentives at local and national levels (e.g. by using the distribution of funds between the federal government, provinces and local authorities to encourage the rezoning of land designated for construction, encouraging owner-occupied homes and offering a commuter allowance). *Bahnstraßen* ("station roads") in extensive municipalities that no longer lead to working train stations bear witness to this development (see, for instance, Poysdorf in Lower Austria). Transforming transport in these locations, where an A-S-I strategy has only been pursued in recent years, has been made significantly more challenging.

The Verkehrsclub Österreich (VCÖ) has stated that 70% of the Austrian population live outside of large cities and that these areas are responsible for almost 80% of the greenhouse gas emissions produced by passenger mobility (cf. VCÖ 2019: 2). Thus the creation of alternative mobility options in suburban and rural areas is undoubtedly necessary for transport to be decarbonized.

4. APPROACHES TO AND THEORIES CONCERNING STRATEGIC PLANNING CONCEPTS IN FOUR AREA TYPES

In the following section, we outline strategic planning concepts developed in a transdisciplinary process and based on four selected area types. The aforementioned issues involved in bringing about a more sustainable mobility model in rural areas serve as a basis and have been consolidated into specific transport and spatial challenges in each area examined in our analysis. In focus groups, multistage development strategies were created to ensure the largest possible provision of public transport-type services across settlement zones in each area type in order to establish attractive alternatives to private cars. The planning concepts are integrated, i.e. transport issues are not only resolved with transport-based solutions. The innovative appeal of connected and automated mobility services is to be harnessed not to increase the acceptance of connected and automated vehicles but to boost the level of acceptance towards measures that will be necessary to bring about the transport and mobility transformation. This can be crucial, particularly in areas where people are heavily impacted by transport policy decisions made at the national level (e.g. an increase in fuel prices) due to a lack of real alternatives.

4.1 APPROACHING AND ADAPTING PLANNING CONCEPTS

Throughout the course of several workshops, specific areas around Vienna were selected and then further developed into prototypes identified by certain transport and spatial features. Once the areas had been chosen, the team examined current planning documents and strategy papers as well as obtained geographic and demographic information to paint an accurate picture of the current situation in each area type. Using the action plans from our previous study (Mitteregger et al. 2022: 141–158), we developed theories for the development potential of these areas. Both the assessment of the status quo in each space and its potential were discussed and reviewed in focus groups involving stakeholders from the municipalities, regions and the state of Lower Austria, as well as connected and automated mobility experts. The technical possibilities of connecting and automating transport during the time frame considered as part of the study were compiled and edited for the workshop with the help of existing literature (see Figs 2 and 3).

Figure 2: Work during focus groups with local experts in Vienna during the autumn of 2019.

Photos: Lena Hohenkamp

Building on the results of the focus groups (see list of participants in Fig. A1 in the Appendix), strategic planning concepts were defined, each including a multistage development process. These are presented as narratives in Section 5. Wherever possible and appropriate, the impact of the outlined planning stages was qualitatively and quantitatively evaluated. Due to restrictions imposed as a result of the Covid-19 pandemic in the spring and summer of 2020, the second stakeholder workshop had to be cancelled. The feedback provided by experts on a rough draft of this paper was incorporated into the final version.

Figure 3: Focus group discussions were aided by reference material compiled prior to the workshops

Photo: Lena Hohenkamp

4.2 SELECTING THE AREA TYPES AND CASE STUDY LOCATIONS

The typology put forward by Matthes and Gertz (2014) was used as a basis to categorize the chosen areas. This approach explicitly examines the extent to which residents in each location can complete everyday tasks without a car. The typology can also be easily applied to other studies and was developed using a broad empirical foundation (Table 2).

Table 2: Systematization, characteristics and hierarchy of case study area types

Hierarchy of transport reduction	Area type	Description (based on Matthes/Gertz 2014)	Case study
Less challenging ↑	**Outskirts, suburban centre**	Suburban centres and outskirts are comparatively heterogeneous spaces. Designing everyday mobility to enable the lowest possible number of car journeys is possible to some degree, especially in combination with options for cycling, but driving remains dominant in these area types, partly due to a lack of car restrictions (e.g. there is no shortage of parking spaces).	Vienna South (area type A) Mistelbach (area type B)
	Public transport axis	Public transport axes differ from the periphery, particularly with regard to the accessibility of service centres and workplaces by public transport and bicycle. Designing everyday mobility without or with just one car per multi-person household is challenging but possible.	Ebreichsdorf (area type C)
Challenging	**Periphery**	The periphery is characterized by a very low density of settlements and workplaces, a lack of local amenities and poor public transport access to centres and workplaces. Here it is either very challenging or impossible to manage without a car on a daily basis.	Bad Schönau (area type D)

Source: the authors based on Matthes/Gertz (2014)

The selection was chosen based on themes and issues arising in a Long Level 4. Known transport and settlement policy problems were addressed (see Table 4). The aim was to develop different automated driving applications as alternatives to "automobility". One key question formed the basis of analysis in each of the chosen area types (see Table 3).

Table 3: Area types, example areas and key questions addressed by the case study

Area type	Question	Case study
Outskirts (area type A)	How can the mobility transformation be realized in industrial zones and business centres located on the outskirts of towns and cities while simultaneously ensuring their commercial growth?	**Vienna South**
Suburban centre (area type B)	What development options can be made available for suburban centres that could become dormitory towns as a result of automated driving on motorways?	**Mistelbach**
Public transport axis (area type C)	What functions can last-mile automated mobility services fulfil along large-capacity regional public transport axes?	**Ebreichsdorf**
Periphery (area type D)	Can automated mobility services improve the mobility options open to the residents of peripheral spaces and enable the relocation of traffic?	**Bad Schönau**

Source: the authors

The following table lists the typical problems that characterize the various spaces. The challenges included in the table are the result of our analysis of the chosen municipalities/areas (see Table 4 on the next page). This information was then used to develop theories for potentially desirable connected and automated vehicle applications that are consistent with the aim of transforming mobility. This overview thus also shows that certain connected and automated transport use cases (see Table 1) should be given priority when faced with certain problems.

Table 4: Area types, challenges and theories on development potential involving the application of connected and automated driving systems

Case study (type of area)	Challenges characterizing the area	Theory on the development potential with Level 4 applications
Vienna South (outskirts)	• Dynamic industrial area with excessive land use • Transit area (traffic flows around large cities) • High level of traffic and pollution • Congestion and stress on infrastructure	Area and/or motorway assistants create relevant criteria to encourage businesses to relocate to the area and influence new forms of commercial mobility management. These spaces are important testing grounds. Developments here can be groundbreaking.
Mistelbach (suburban centre)	• Strain on existing housing stock • Lack of public transport links between the surrounding municipalities/urban districts and key functions located in the centre • Rising commuter traffic towards Vienna since the completion of a motorway • Urban sprawl in residential areas and business sites	The use of automated shuttles can improve public transport access to centrally located functions. Specific route planning and the careful repurposing of existing infrastructures would be necessary to make this happen.
Ebreichsdorf (public transport axis)	• Strain on existing housing stock • Creating transport links between the train station and the surrounding town and village centres (train station outside of the city centre) • Commuter traffic to Vienna • Disused railway line	Spatial features specific to the locality (in this case, a disused railway line) can become the pavements and cycle paths, or CAV access routes, of tomorrow and enable connected and automated services on specific routes. In this context, connected and automated services could also be used to help boost local development.
Bad Schönau (periphery)	• Insufficient and/or non-existent public transport access, especially for scattered settlements • High dependency on private cars	Connected and automated shuttles could be used in synergy with tourist sites to optimize accessibility.

Source: the authors

To respond to the question concerning area type A, a section within the ribbon development to the south of Vienna encompassing eleven municipalities was chosen. For area B, Mistelbach, a district capital to the north of Vienna, was examined. To answer our question concerning area type C, researchers looked at Ebreichsdorf, where a new train station offering an excellent transport link to Vienna is being built. Finally, researchers analysed Bad Schönau, a small municipality on the border between Lower Austria, Burgenland and Styria, to respond to the query posed for area type D (Fig. 4).

Figure 4: Location of the four area types within the Vienna/Lower Austria metropolitan region, including main transport links

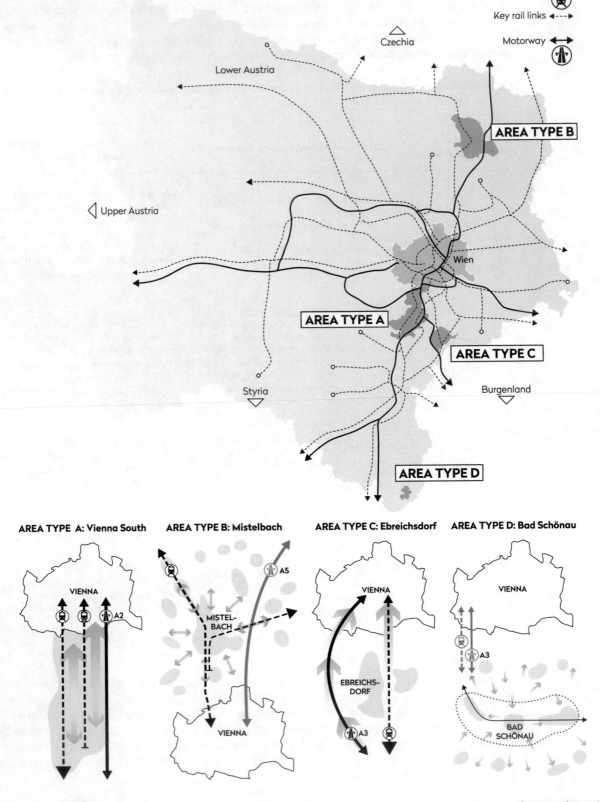

Key rail links ◄---►

Motorway ◄—►

Czechia

Lower Austria

AREA TYPE B

Upper Austria

Wien

AREA TYPE A

AREA TYPE C

Styria

Burgenland

AREA TYPE D

AREA TYPE A: Vienna South

VIENNA

A2

AREA TYPE B: Mistelbach

A5

MISTEL-
BACH

VIENNA

AREA TYPE C: Ebreichsdorf

VIENNA

EBREICHS-
DORF

A3

AREA TYPE D: Bad Schönau

VIENNA

A3

BAD
SCHÖNAU

Source: the authors

In the following section, the strategic development concepts created during the research project are presented as narratives. The concepts are considered in detail and the various stages of each transformation are discussed. Specific areas for action in various specialist fields of urban policy and planning are subsequently developed.

5. FUTURE CONCEPTS FOR INTEGRATED MOBILITY AND SETTLEMENT DEVELOPMENT

5.1 VIENNA SOUTH (AREA TYPE A)

The name "Vienna South" is used to encompass the ribbon development located to the south of the city, which is closely linked to the capital both in terms of its locality and functions. The area is directly to the south of Austria's capital and covers 11 municipalities in the districts of Baden and Mödling. A green space, the Vienna Woods (Wienerwald), can be found to the west of the area. To the east, the area borders a cultural landscape that is mostly undeveloped. The municipal boundary with Bad Vöslau was chosen as the area's southern border (see Fig. 5).

Given its proximity to Vienna, the availability of space and the area's good transport links, Vienna South is one of Austria's most dynamic economic zones. Housing in the historical town centres, and particularly on the slopes close to the Wienerwald, is in high demand. A controlled and coordinated settlement development of this ribbon zone is the key challenge for land use planners. Local regional planning and transport organizations are currently tackling this (these are Stadt-Umland-Management, Planungsgemeinschaft Ost and Verkehrsverbund Ost-Region); however, they have limited scope for action.

Vienna South is also a transnational transit space and part of the Trans-European Transport Network (TEN-T). Traffic flows from the surrounding area largely head towards the Austrian capital. The A2 Südautobahn (South Motorway) is a central transport axis in the east of Austria that runs towards Styria and onwards to Carinthia, Slovenia and Italy to the south and towards Bratislava and Brno to the north. The Südbahn (Southern Railway) links Vienna to Graz (where passengers can travel on to Ljubljana, Zagreb and the port city of Rijeka) or Klagenfurt and Villach (with connections to northern Italy and the port city of Trieste) in the south (and is currently being extended in two locations). The capital also has rail links to Brno and Prague in the north. Furthermore, the area is structured around a third transport link: a branch line ("Badner Bahn") connects central Vienna to the district capital of Baden.

Challenges
In this area type, the importance of historical centres has dwindled. Vienna South is home to Shopping City Süd (SCS), Austria's largest shopping centre (comprising 192,500 square metres of selling space and employing 5,000 staff), and Austria's top-performing commercial site in the shape of the Lower Austria—south industrial hub (IZ NÖ-Süd), which also has 11,000 employees. In addition to vehicles transiting through the area, these sites generate considerable volumes of traffic. Online shopping is noticeably putting pressure on brick-and-mortar retail (even big-name stores). Desperate attempts are being made to find opportunities for development. Efforts are also underway to secure the future of commercial sites such as the IZ NÖ-Süd, which

has two road links to the A2 as well as its own on-site depot. Competition between these locations has increased significantly and new logistics chains mean competitors are now abroad as well as at home. The change underway is demonstrated by the growing significance of site factors once considered "soft" (design quality, culinary offering) and, above all, public transport accessibility (cf. IHK 2020; GVA Mödling 2016; Görgl et al. 2017; Statistik Austria 2017, 2019; SUM 2020).

Figure 5: Structural plan of Vienna South with a particular focus on the area surrounding the IZ NÖ-Süd industrial site

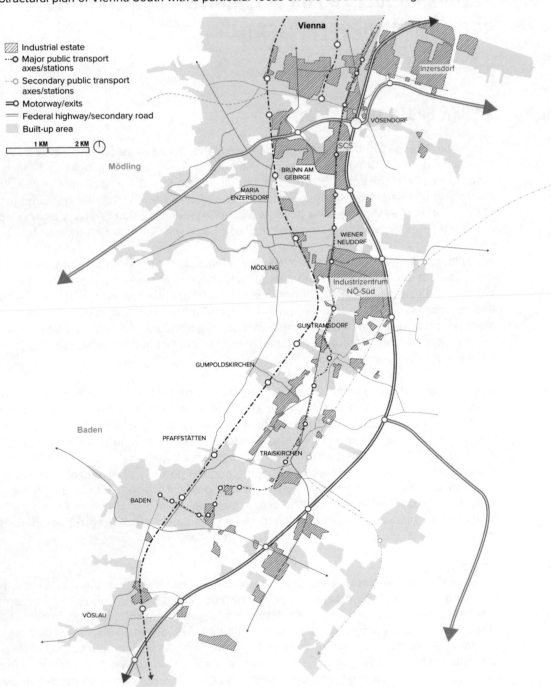

Source: the authors

TRANSFORMATION STAGE 1
ON-SITE MOBILITY PLATFORM

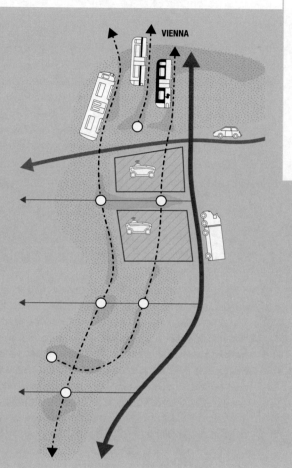

THE FIRST TRANSFORMATION STAGE WILL SEE IZ NÖ-SÜD BE-COME A TEST SITE FOR AUTOMATED MOBILITY. THE PROJECT WILL BE RUN BY BUSINESSES OPERATING ON THE SITE AS A PUBLIC-PRIVATE PARTNERSHIP WITH THE SUPPORT OF PUBLIC OPERATORS (FROM THE PROVINCE OF LOWER AUSTRIA) AND ACTORS FROM THE MOBILITY SECTOR. THE AIM OF THIS FIRST PHASE IS TO MEET THE DIFFERENT LOGISTIC'S NEEDS OF THE BUSINESSES ON SITE. SOME OF THE ROADS ON THE IZ NÖ-SÜD SITE WILL BE SPECIFICALLY ADAPTED TO FORM A NETWORK SUITABLE FOR AUTOMATED MOBILITY SERVICES (FIG. A3 IN THE APPENDIX).

THE ROADS DEVELOPED FOR THE AREA ASSISTANTS WILL ALSO BE ADAPTED FOR USE BY PEDESTRIANS AND CYCLISTS. PAVE-MENTS AND CYCLING LANES WILL BE SEPARATED FROM THE REST OF THE ROAD AND FEATURE AN ATTRACTIVE DESIGN. THE OPERATORS OF IZ NÖ-SÜD WILL WORK TOGETHER WITH LOCAL STAKEHOLDERS TO ENSURE THE LOCATION'S FUTURE. LAND AT THE SITE'S TRANSPORT LINKS (A2 MOTORWAY: WIENER NEU-DORF EXIT; IZ NÖ-SÜD EXIT TO THE BADNER BAHN: STOPS AT GRIESFELD, NEU GUNTRAMSDORF AND AT THE SITE'S DEPOT) WILL BE SECURED.

1 PRINCIPLES OF THE VIENNA SOUTH TRANSFORMATION

CA APPLICATION

SITE ASSISTANT

MOBILITY

CA SERVICES GO HAND IN HAND WITH SOFT MOBILITY

MOBILITY

LINK CA SERVICES TO OTHER SECTORS

TRANSFORMATION STAGE 2
DESIGNING INTERFACES

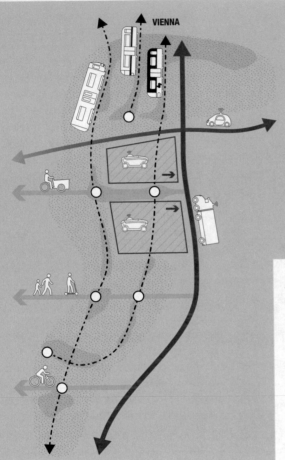

VIENNA

THE INTERFACES AT THE OUTSKIRTS OF THE SITE WILL NOW BE DEVELOPED INTO MOBILITY HUBS. IT WILL BECOME POSSIBLE TO OFFER ATTRACTIVE SERVICES, EVEN FOR WORKERS ON SITE. CA SHUTTLES NOW TRANSPORT PASSENGERS FROM THE BADNER BAHN'S STOPS. AN AREA ASSISTANT CAN NOW BE IMPLEMENTED ACROSS THE WHOLE OF THE IZ NÖ-SÜD SITE. PASSENGER TRANSPORT WILL PRIMARILY BE PROVIDED BY TWO AUTOMATED BUS LINES THAT CROSS THE IZ NÖ-SÜD SITE AND STOP AT THE MOBILITY HUBS (FIG. A3; THE EXISTING ROAD NETWORK'S SUITABILITY FOR AUTOMATED VEHICLES WAS CONSIDERED, FIG. A2).

A VISIBLE TRANSFORMATION WILL BE UNDERWAY: MORE GREEN SPACES APPEAR ALONGSIDE PAVEMENTS AND CYCLE PATHS AND ON OLD CAR PARKS. THE SITE THUS BECOMES MORE APPEALING FOR OTHER SECTORS. AFTER THIS PHASE IS COMPLETE, THE MOTORWAY CAN ALSO BE USED BY AUTOMATED VEHICLES. THE ON-SITE MOBILITY PLATFORM NOW ALSO ENABLES SYNERGIES ON THE MOTORWAY THAT LEAD TO GREATER EFFICIENCY AND LESS TRAFFIC CONGESTION. COMPANIES RETHINK THEIR APPROACH AND INSTEAD OF OFFERING EMPLOYEES COMPANY CARS, THEY PROVIDE FLEXIBLE MOBILITY. THE MOBILITY HUBS LOCATED AT MOTORWAY JUNCTIONS ARE USED, SIMILAR TO PORT AREAS, AS GOODS HANDLING SPACES AND ARE KEY INTERCHANGES FOR PASSENGER TRANSPORT.

2 PRINCIPLES OF THE VIENNA SOUTH TRANSFORMATION

CA APPLICATION	MOBILITY	SPACE

SITE ASSISTANT

ADAPTABLE MOBILITY HUBS DESIGNED FOR URBAN SPACES

CAR-FRIENDLY SITES CLOSE TO THE MOTORWAY BECOME PRIME SITES

TRANSFORMATION STAGE 3
A MOBILITY PLATFORM IS LAUNCHED AND INTEGRATED INTO THE VIENNA SOUTH RIBBON DEVELOPMENT

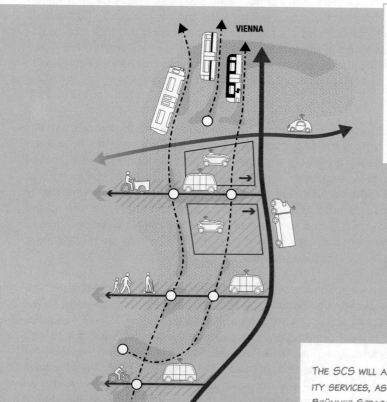

VIENNA

EXPERIENCES GAINED MAKE IT POSSIBLE TO INTEGRATE THE SITE INTO ITS SURROUNDING AREA. IT IS ALSO POSSIBLE TO REDUCE CAR DEPENDENCY IN OTHER LOCATIONS ALONG THE MOTORWAY. REGIONAL TRANSFORMATION OBJECTIVES ARE DEVELOPED JOINTLY WITH THE MUNICIPALITIES IN THE AREA. THE MOBILITY PLATFORM IS TRANSFORMED INTO A PUBLIC INFRASTRUCTURE SERVICE PROVIDER. THE AIM IS TO DEVELOP A REGION WITH CLEARLY ARTICULATED PLANNING GOALS AND IN WHICH A COMPLEX SPATIAL EXPERIENCE CAN BE CAREFULLY DESIGNED.

THE SCS WILL ALSO BE ACCESSIBLE VIA AUTOMATED MOBILITY SERVICES, AS WILL SMALLER BUSINESS SITES ALONG THE BRÜNNER STRASSE AND TRIESTER STRASSE. THE FOCUS NOW LIES ON LINKING PARALLEL LARGE-CAPACITY ROUTES VIA AN INTEGRATED LOCAL TRANSPORT NETWORK (ACTIVE MOBILITY AND AUTOMATED SHUTTLES ARE BEING JOINTLY DEVELOPED). SPACES RECLAIMED FROM PARKED CARS ARE THEN UTILIZED IN A RANGE OF WAYS BY REDESIGNING BROWNFIELD SITES.

3 PRINCIPLES OF THE VIENNA SOUTH TRANSFORMATION

CA APPLICATION

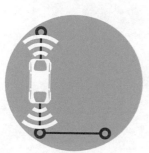

CA SHUTTLE BUS

MOBILITY

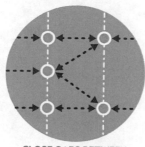

CLOSE GAPS BETWEEN TRANSPORT AXES WITH CA SERVICES

MOBILITY

UPGRADE FEEDER ROADS FOR CAV

Development strategy: a clear-cut course

Area types such as Vienna South should be among the first to witness the roll-out of connected and automated vehicles and to show visible signs of how effective the mobility transformation can be. Such sites can be found close to large cities everywhere and are chosen by businesses that are dependent on innovative logistics and are thus constantly moving towards more automation and connectivity. The following planning stages proceed on the assumption that these areas will serve as pioneers and thus play a key role in the transformation of the mobility system.

Discussion and reflection

Vienna South faces considerable volumes of traffic. This is largely a result of commercially successful sites such as IZ NÖ-Süd and the SCS that play a significant role both locally and nationally. These locations will, however, invariably and inevitably be transformed. A new host of characteristics are coming to the fore. Area types such as Vienna South could be the next niche zones from which connected and automated driving takes hold. These locations are thus highly relevant as role models and pilot schemes. The transformation of these areas, either brought about spontaneously or initiated in a targeted manner, will set the course for developments to come. If opportunities to build a transport system around greater sustainability are missed here, all further attempts to do so will face a much steeper uphill climb.

At present, there are hardly any pedestrians or cyclists at these sites. These streetscapes are also barely used as spaces for relaxation (for instance, by workers on their lunch breaks). These areas thus provide an opportunity to test the waters when it comes to new forms of mobility without putting undue strain on the surrounding area. Areas such as Vienna South could also give rise to a public freight transport network or an integrated public transport network for passengers and goods. The valuable experiences of users and operators could be collected. Shifting or pooling types of transport in this way (to avoid unnecessary journeys) can be an important step towards ensuring the competitiveness of these locations while also initiating a sustainable transformation.

5.2   MISTELBACH (AREA TYPE B)

Mistelbach is the capital of a district with the same name and is situated in the north-east of the Weinviertel, Austria's largest winegrowing region. The municipality comprises ten small towns, with five forming the functional centre. The Laaer Ostbahn (S-Bahn lines 2 and 7) provides a direct public transport link to Vienna. In 2015 an additional transport link was created in the east of the city: the A5 motorway is connected via a bypass.

In recent years Mistelbach has grown due to its good transport links and high reserves of building plots. With the targeted redevelopment of brownfield sites, the city aims to consolidate its existing compact and historical buildings. Overall, the settlement structure is developing towards the east in the direction of the motorway (e.g. business park, shopping centre; see Fig. 6).

Challenges

As a centre, Mistelbach fulfils key regional functions for a commuter belt with around 120,000 residents. The town provides shopping, service and administrative functions (for example, it is home to the district court, the regional hospital, and primary and secondary schools). Trains serving the S-Bahn line north of Mistelbach/Zaya station also operate at high frequency, making them attractive for many commuters (trains call at the town's second station, Mistelbach Stadt, less frequently). A branch line runs to the south of the Mistelbach/Zaya train station that once connected Mistelbach with a second district capital, Gänserndorf. Today this line is only used for goods transport and as a historical railway.

The regional bus network is unable to compete with individual transport. Although all key local municipalities are served, the frequency and timing of services are not convenient. Mistelbach's role as a regional centre will be undermined by a new motorway. Since the new junction was opened, commuter numbers (particularly to Vienna) have been rising (Stadt Mistelbach 2014; Statistik Austria 2017, 2019; Görgl et al. 2017).

Development strategy: two key criteria

There is a considerable risk that greater automation along the motorway will mean Mistelbach ceases to function as a regional centre and increasingly becomes a peripheral dormitory town with no specific function. In the case of Mistelbach, avoiding transport means strengthening its position as a centre to avoid as many long journeys to Vienna as possible. To make this a reality, this will involve, on the one hand, connecting the town's key functions and, on the other, improving the region's public transport network.

Figure 6: Structural plan of Mistelbach (Lower Austria)

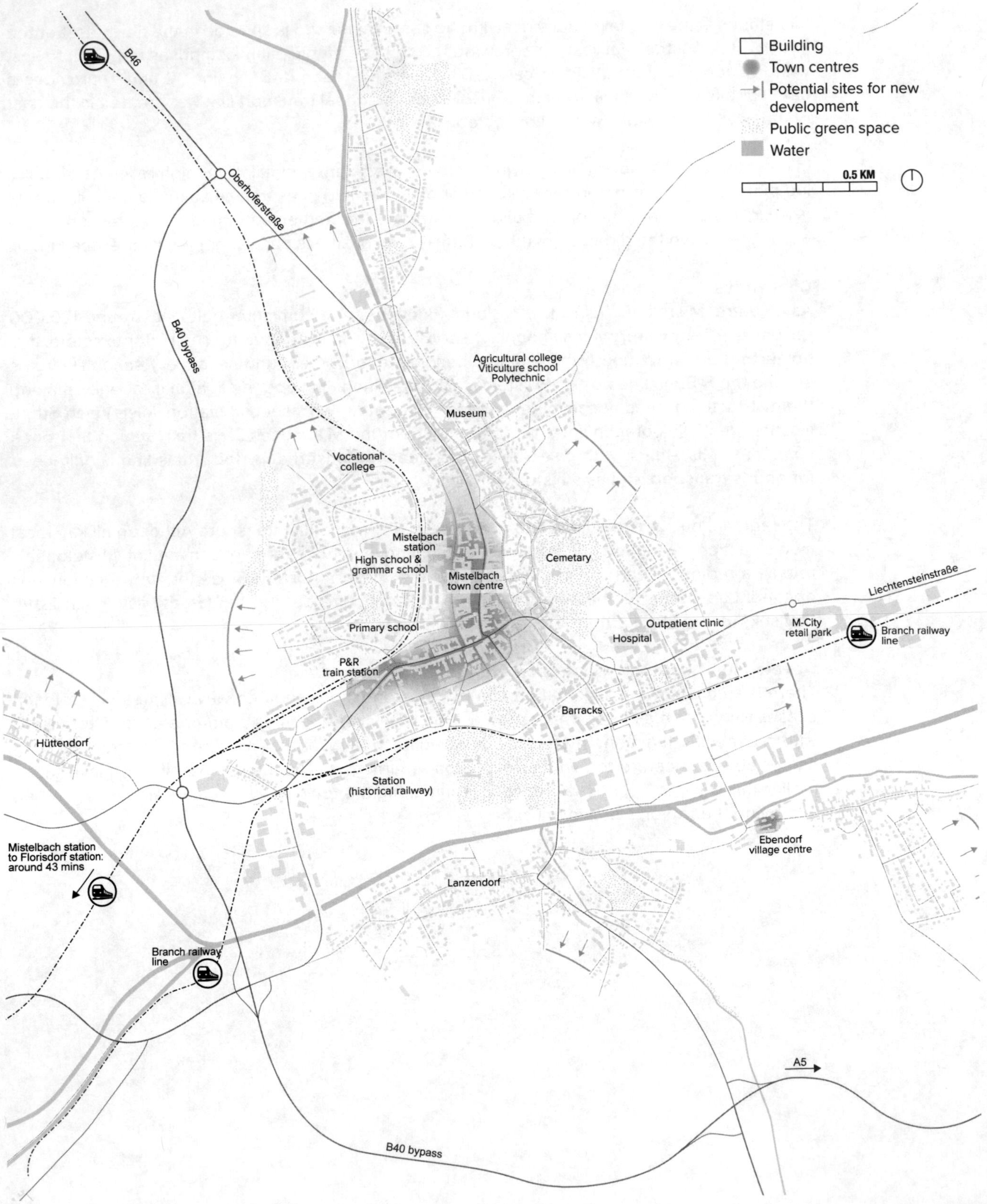

Building
Town centres
Potential sites for new development
Public green space
Water

0.5 KM

B46

Oberhoferstraße

B40 bypass

Agricultural college
Viticulture school
Polytechnic

Museum

Vocational college

Mistelbach station
High school & grammar school

Mistelbach town centre

Cemetary

Primary school

Outpatient clinic

Hospital

M-City retail park

Branch railway line

Liechtensteinstraße

P&R train station

Barracks

Hüttendorf

Station (historical railway)

Ebendorf village centre

Mistelbach station to Florisdorf station: around 43 mins

Lanzendorf

Branch railway line

A5

B40 bypass

Source: the authors

TRANSFORMATION STAGE 1
THE FIRST STEP TOWARDS AUTOMATED URBAN TRANSPORT IN MISTELBACH

THE FIRST AIM OF THE TRANSFORMATION IS TO IMPROVE THE INTERMODAL ACCESSIBILITY OF MISTELBACH'S KEY FUNCTIONS. AN INITIAL LINE FOR AN AUTOMATED SHUTTLE, WHICH WILL RUN FROM MISTELBACH/ZAYA STATION TOWARDS THE SOUTH ALONG THE SETTLEMENT BOUNDARY, IS TO BE DEVELOPED AS A PILOT PROJECT. THIS WILL LINK A SPORTS AND LEISURE CENTRE, AN INDUSTRIAL ESTATE AND, LASTLY, MISTELBACH-GÄNSERN-DORF REGIONAL HOSPITAL. PAVEMENTS AND CYCLE PATHS WILL ALSO BE ADDED AND/OR EXPANDED ALONG THE ROUTE. THE LINE'S TERMINUSES WILL OFFER CONNECTIONS TO THE REGIONAL BUS NETWORK.

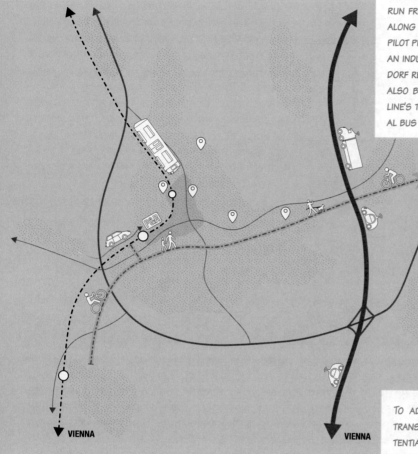

VIENNA

VIENNA

TO ADVANCE THE DEVELOPMENT OF THE REGIONAL PUBLIC TRANSPORT NETWORK, A SEARCH IS CONDUCTED TO FIND POTENTIAL SITES IN AND AROUND MISTELBACH FOR AN AUTOMATED SHUTTLE LINE. THIS WILL FOCUS ON THE TRAIN LINE RUNNING TOWARDS GÄNSERNDORF BUT PRIMARILY ALSO ROADS IN THE TOWN CENTRES THAT CAN BE MADE ACCESSIBLE BY LOWERING THE PERMITTED SPEED LIMIT (FIG. A4).

1 PRINCIPLES OF THE MISTELBACH TRANSFORMATION

CA APPLICATION	MOBILITY	MOBILITY	MOBILITY
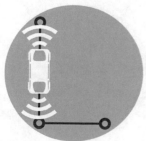			
AUTOMATED SHUTTLE BUSES	CA SERVICES GO HAND IN HAND WITH SOFT MOBILITY	BRANCH LINES ARE THE CAV ROUTES OF TOMORROW	LINES MUST LINK CENTRAL FUNCTIONS

TRANSFORMATION STAGE 2
FROM PILOT PROJECT TO TARGET GROUP-SPE-CIFIC EVERYDAY MOBILITY

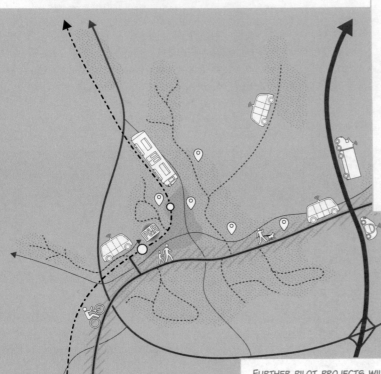

VIENNA

Once experience has been gained via the pilot project and a comprehensive analysis of the infrastructure for potential routes has been conducted, target group-specific services will be developed to access functions located centrally (in Mistelbach, see structural plan in Fig. 6). The aim is to create public services for everyday mobility to cover a range of different needs. The first step entails the expansion of automated services at off-peak times for commuters and increasingly for young people too. The infrastructural analysis has shown that high infrastructural investment would be needed to establish an extensive automated transport network. Work will then begin on an integrated mobility platform that can be used to access all available services.

Further pilot projects will begin so that many residents can experience the new services. Here the bypasses of recent decades come in useful. Lower speed limits are now desired in town centres, allowing a new type of mobility to be trialled. Shuttles are implemented in surrounding municipalities following a rotation plan. If investments are to be made into road infrastructure, then the focus should be on creating high-quality public spaces in town centres. A growing number of functions are linked within Mistelbach's urban zone. The focus is now on the mobility of the town's school pupils and the elderly. The first integrated mobility hubs are created at the train station and the regional hospital.

2 PRINCIPLES OF THE MISTELBACH TRANSFORMATION

CA APPLICATION
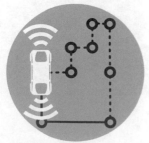
AUTOMATED SHUTTLE BUSES

MOBILITY

LINK CA SERVICES TO OTHER SECTORS

MOBILITY

LINE PLANNING SENSITIVE TO ADJACENT ACTIVITIES

SPACE

CA SERVICE CATCHMENT AREAS WITH HIGH DEVELOPMENT POTENTIAL

TRANSFORMATION STAGE 3
REGIONAL TRANSIT IN MISTELBACH

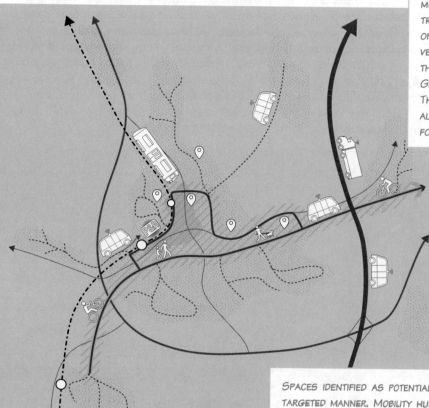

A REGIONAL DEVELOPMENT CONCEPT IS CREATED. THE MUNICIPALITIES NOT ONLY DEVELOP A JOINT PUBLIC TRANSPORT SERVICE BUT PLEDGE TO ALSO EMBARK ON A RADICAL LAND POLICY. THIS INVOLVES THE DEVELOPMENT OF AUTOMATED SERVICES RUNNING ON THE MOTORWAY TO PROVIDE EXPRESS ROUTES TO GÄNSERNDORF, LAA AN DER THAYA AND HOLLABRUNN. THE ENHANCED TECHNOLOGICAL POSSIBILITIES OF AUTOMATED DRIVING SYSTEMS NOW MAKE IT MORE AFFORDABLE TO DESIGN SUCH ROUTES.

SPACES IDENTIFIED AS POTENTIAL SITES ARE PRIORITIZED AND DEVELOPED IN A TARGETED MANNER. MOBILITY HUBS ARE INTEGRATED INTO EXISTING FUNCTIONS. INTERFACES WITHIN THE MOBILITY SYSTEM HELP BOOST CENTRAL LOCATIONS. THE FEATURES OFFERED WITHIN MOBILITY HUBS WILL DEPEND ON THE LOCATION AND DEMAND. THESE MAY BE CAR-SHARING SERVICES FOR LEISURE MOBILITY, ESPECIALLY AT WEEKENDS, AS WELL AS E-SCOOTER AND BICYCLE HIRE.

3 PRINCIPLES OF THE MISTELBACH TRANSFORMATION

CA APPLICATION	MOBILITY	MOBILITY	SPACE
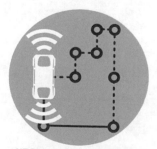			
AUTOMATED SHUTTLE BUSES ON ROUTES AND ROADS	CA SERVICES ONLY IN CERTAIN AREAS	PUBLIC TRANSPORT HUB AREAS SHOWCASE SERVICES	CA SERVICE CATCHMENT AREAS WITH HIGH DEVELOPMENT POTENTIAL

Discussion and reflection

Automated driving on a motorway could put immense strain on a regional centre such as Mistelbach. If more affordable, comfortable journeys are possible on large-capacity roads, smaller retail entities, leisure facilities and other establishments will have to contend with greater competition. Mistelbach's expansive commuter belt will likely hinder the large-scale roll-out of automated vehicles in the longer term. The pathway proposed in this paper is focused on allowing as many people as possible to experience mobility in a new way. If there is a sense that everyday journeys to important functions in Mistelbach can be easily completed without a car, this would be a crucial step. This case study once again demonstrates just how vital it is to generate intensive cooperation between a highly diverse range of actors.

5.3   EBREICHSDORF (AREA TYPE C)

The town of Ebreichsdorf is located in the district of Baden roughly 25 kilometres to the south of Vienna. The municipality comprises four small towns (Fig. 7). Mobility in this area is largely concentrated on routes towards Vienna: 80% of the working population commutes, with 50% travelling to the Austrian capital. Ebreichsdorf has two main regional transport links to Vienna: the A3 southeast motorway and the Pottendorfer S-Bahn line. The Pottendorfer line, which offers travellers a link to Austria's eastern railway line towards Hungary, passes through Unterwaltersdorf; however, passenger transport has been suspended on this line. Ebreichsdorf's city bus network connects the town with the other cadastral communities of Schranawand, Unterwaltersdorf and Wegelsdorf.

Challenges

As it is situated close to Vienna and has good transport links to other parts of the country, the region is an attractive location for companies from all sectors. Ebreichsdorf is home to four large and several small industrial estates. The area's economy is structured around small- and medium-sized craft and industrial companies, as well as service providers. Business traffic sometimes passes through the centre of towns and localities.

Ebreichsdorf is also a popular residential town. For years, its population has grown at a faster than average pace for Lower Austria. Settlements in the area largely comprise single-family homes. What little dense housing there is tends to be concentrated around the town centre and along the B16 road to Vienna. Key services are located here, including for residents living in the area around Ebreichsdorf. Along with the expansion of the Pottendorfer line, there are plans for the specific development of Ebreichsdorf. As a satellite town close to Vienna, some of the projected urban growth is to be realized close to public transport links (Görgl et al. 2017; Stadtgemeinde Ebreichsdorf 2014; Statistik Austria 2017, 2019).

Development strategy: Strengthening the development of existing structures with lively streets

In Ebreichsdorf, the first stage of the transformation has already begun. The Pottendorfer line is currently being expanded and should be finished by 2023. A large section will run on a new line roughly one kilometre further east. As part of this, Ebreichsdorf train station will be moved from its current central location to a greenfield site, which will also improve the link between the Pottendorfer line and Unterwaltersdorf. The former train station and rail line running through Ebreichsdorf will be closed down. Once it has been decommissioned, this railway line will have major potential for other uses.

The expansion and construction of the train line aims to accelerate growth and elevate Ebreichsdorf to a central hub in the area south of Vienna. The Ebreichsdorf Smart City project has also looked at different ways to integrate the new train station and future settlement development by drawing up a range of potential scenarios. This urban planning decision will initially mean a loss in public transport access for Ebreichsdorf town centre (the considerable drop in services is shown in Fig. A5 between transformation stages 0 and 1).

Figure 7: Structural plan of Ebreichsdorf (incl. new train station)

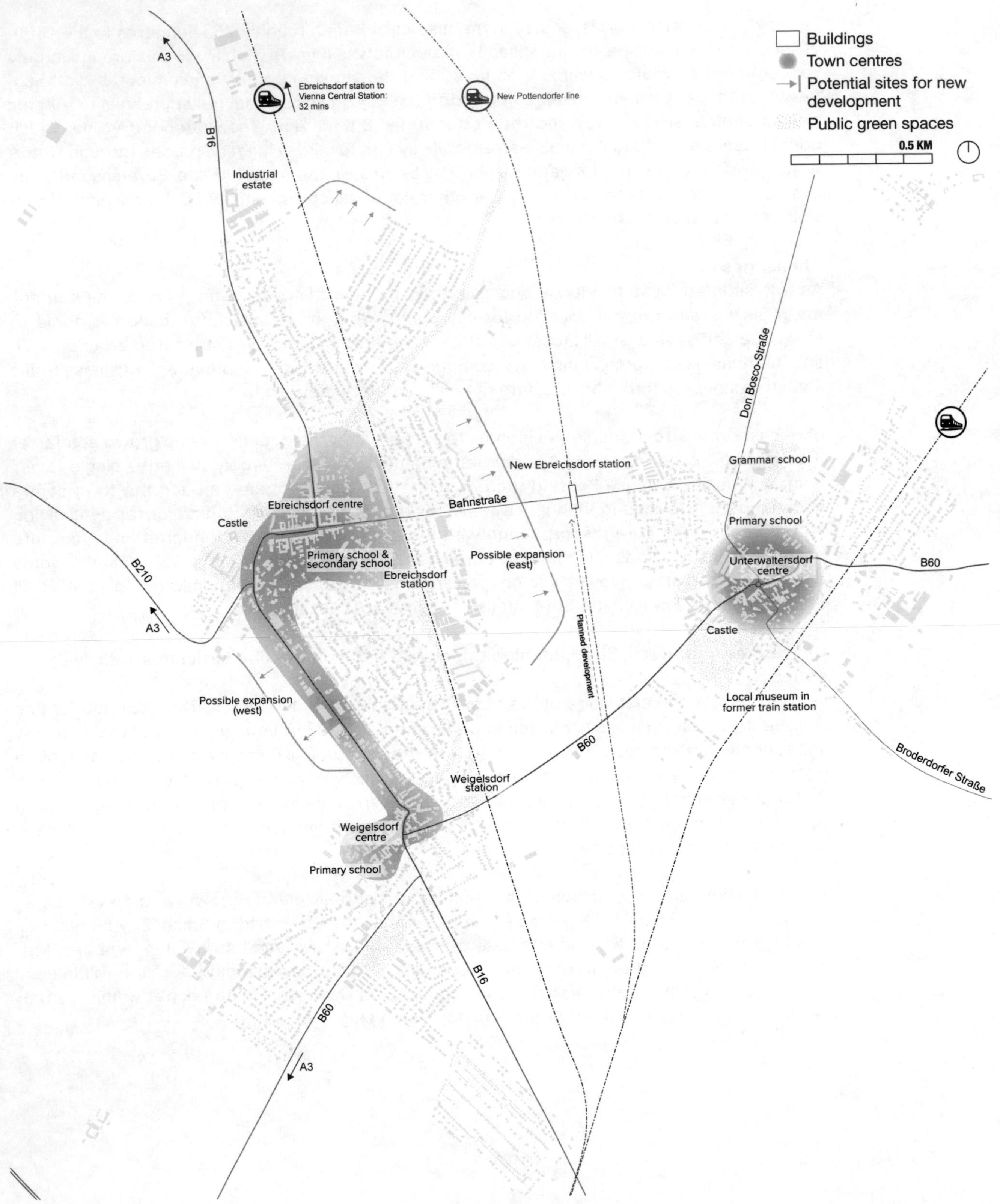

Buildings

Town centres

Potential sites for new development

Public green spaces

0.5 KM

A3

B16

Ebreichsdorf station to Vienna Central Station: 32 mins

New Pottendorfer line

Industrial estate

Don Bosco-Straße

New Ebreichsdorf station

Grammar school

Ebreichsdorf centre

Bahnstraße

Castle

Primary school

Primary school & secondary school

Ebreichsdorf station

Possible expansion (east)

Unterwaltersdorf centre

B210

B60

A3

Castle

Planned development

Local museum in former train station

Possible expansion (west)

B60

Broderdorfer Straße

Weigelsdorf station

Weigelsdorf centre

B60

Primary school

B16

A3

Source: the authors

TRANSFORMATION STAGE 0
PREPARATION WHILE THE NEW STATION IS BUILT

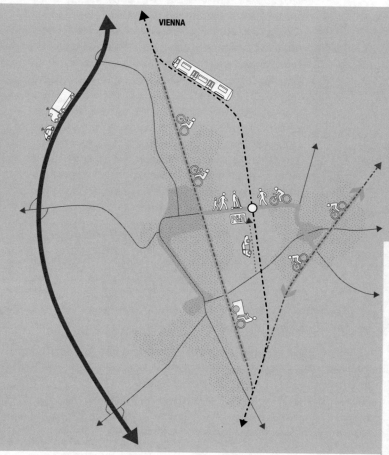

VIENNA

THE NEW STATION IS BEING BUILT TO SERVE AN AREA WITH A RADIUS OF ROUGHLY 25 KILOMETRES. A PARK-AND-RIDE SITE IS CURRENTLY STILL NECESSARY BUT WILL BE DESIGNED SO THAT IT CAN BE REMOVED AT A LATER DATE. IT WILL PLAY A ROLE DURING THE ENTIRE TRANSFORMATION PROCESS BUT ITS FUNCTION WILL CONSTANTLY BE CHANGING. THE FRONT OF THE TRAIN STATION WILL MAINLY BE DESIGNED AROUND THE NEEDS OF THOSE US-ING ACTIVE FORMS OF MOBILITY. THIS CONCERNS THE PROXIMITY OF PARKING SPACES (OR E-BIKE CHARGING STATIONS) TO THE PLATFORM, THE ROUTE BETWEEN THE TWO AND THE ATTRACTIVE DESIGN OF APPROACH ROADS FOR CYCLISTS AND PEDESTRIANS, WHICH WILL FEATURE AN ABUNDANCE OF LIGHTING AS WELL AS GREENERY TO OFFER PROTECTION FROM THE ELEMENTS. PAVE-MENTS AND CYCLE PATHS WILL BE BUILT ON A SEPARATE LANE.

0 PRINCIPLES OF THE EBREICHSDORF TRANSFORMATION

MOBILITY

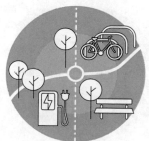

PUBLIC TRANSPORT HUB AREAS SHOWCASE SERVICES

MOBILITY

CLEAR PRIORITIZATION OF SOFT LOCAL MOBILITY AT PUBLIC TRANSPORT HUBS

SPACE

SECURE AREAS AROUND PUBLIC TRANSPORT HUBS FOR FURTHER DEVELOPMENT

TRANSFORMATION STAGE 1
THE FIRST INTEGRATED CA ROUTE IS BUILT

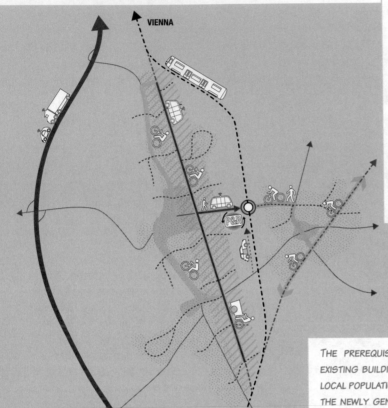

VIENNA

TRANSFORMATION STAGE 1 WILL SEE THE USE OF AUTOMATED SHUTTLES MADE POSSIBLE THROUGH TARGETED INFRASTRUCTURE DEVELOPMENT, AND INFRASTRUCTURAL REQUIREMENTS FOR ACTIVE FORMS OF MOBILITY WILL ALSO BE IMPROVED. CA SHUTTLES WILL START OPERATING ON THE APPROACH ROAD FROM THE EBREICHSDORF AREA TO THE NEW STATION. CLOSE TO THE OLD STATION, THE ACCESS ROAD WILL CROSS THE DISUSED RAILWAY LINE, WHICH TRAVERSES THE ENTIRE EBREICHSDORF RESIDENTIAL AREA FROM NORTH TO SOUTH. THERE WILL ALSO BE INTEGRATED DEVELOPMENT ON THE ROUTE TO ALLOW ACTIVE MOBILITY AND CA SHUTTLES. IN THE TOWN ITSELF, MICROHUBS WILL BE BUILT ALONG THE ACCESS ROAD AND THE OLD RAILWAY LINE THAT SERVE AS BRIDGES CONNECTING PREVIOUSLY SEPARATED RESIDENTIAL AREAS TO THE WEST AND EAST OF THE RAILWAY.

THE PREREQUISITES FOR THE TARGETED DEVELOPMENT OF EXISTING BUILDINGS ARE THUS IN PLACE: TOGETHER WITH THE LOCAL POPULATION, CONCEPTS WILL BE CREATED DETAILING HOW THE NEWLY GENERATED POTENTIAL CAN BE USED. THE ROADS RUNNING EITHER SIDE OF THE OLD RAILWAY EMBANKMENT WILL GRADUALLY BE TURNED INTO A LINEAR PARK. SELECTED VACANT AND FORMER PARKING SPACES ALONG THE RECHTE BAHNZEILE ROAD WILL BE SECURED FOR FUTURE DEVELOPMENT.

1 PRINCIPLES OF THE EBREICHSDORF TRANSFORMATION

CA APPLICATION	MOBILITY	MOBILITY	SPACE
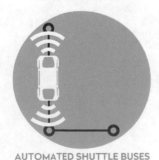		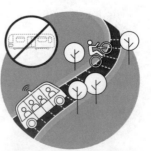	
AUTOMATED SHUTTLE BUSES OPERATING ON ROUTES	CA SERVICES GO HAND IN HAND WITH SOFT MOBILITY	BRANCH LINES ARE THE CAV ROUTES OF TOMORROW	CA SERVICE CATCHMENT AREAS WITH HIGH DEVELOPMENT POTENTIAL

TRANSFORMATION STAGE 2
FROM PASSENGER MOBILITY TO INTEGRATED LOCAL TRANSPORT

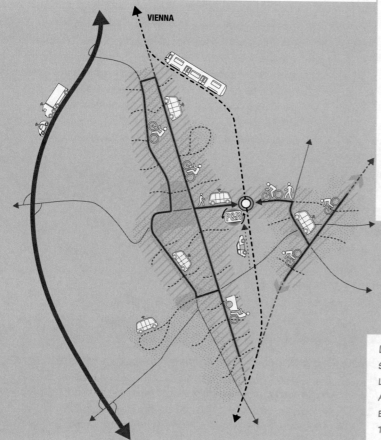

VIENNA

ONGOING TECHNOLOGICAL DEVELOPMENTS AND MUNICIPALITIES' EXPERIENCES OF RUNNING CA SHUTTLES WILL GIVE RISE TO NEW OPPORTUNITIES TO BOOST EXISTING CENTRES AND SITES VIA IMPROVED TRANSPORT LINKS. CA SHUTTLES WILL TAKE OVER A LARGE PART OF THE CITY BUS NETWORK, WHICH WILL THEN ONLY SERVE PERIPHERAL URBAN AREAS. AN EVALUATION OF THE EXISTING ROAD NETWORK'S SUITABILITY WILL BE USED TO PLAN ROUTES. THIS WILL ALLOW THE FREQUENCY AND TIMING OF SERVICES TO BE VASTLY IMPROVED. ONCE THE CA SHUTTLE IS OPERATING ON EBREICHSDORF'S MAIN SQUARE, SPACE HERE WILL ALSO BE RECLAIMED FROM CARS AND ALLOCATED TO ACTIVE FORMS OF MOBILITY. HIGH-QUALITY PUBLIC SPACE WILL ALSO BE CREATED.

DURING THIS STAGE OF THE EXPANSION, THE SYSTEM WILL SHIFT FROM SIMPLY TRANSPORTING PASSENGERS (I.E. ENABLING COMMUTER TRAFFIC) TO A LOCAL PUBLIC TRANSPORT AND ACTIVE MOBILITY NETWORK FOR PEOPLE AND GOODS. LOCAL OPERATORS WILL BENEFIT FROM BEING ABLE TO OFFER THEIR CUSTOMERS MOBILITY SERVICES. A PARTICIPATORY PROCESS WILL BE USED TO TURN EXISTING CONCEPTS FOR LOCAL DEVELOPMENT PLANNING INTO A POLYCENTRIC VISION THAT INTEGRATES ALL THE CADASTRAL MUNICIPALITIES.

2 PRINCIPLES OF THE EBREICHSDORF TRANSFORMATION

CA APPLICATION

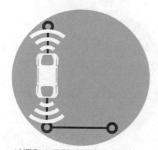

AUTOMATED SHUTTLE BUSES
OPERATING ON ROUTES

MOBILITY

LINES MUST LINK
CENTRAL FUNCTIONS

MOBILITY

LINE PLANNING SENSITIVE TO
ADJACENT ACTIVITIES

TRANSFORMATION STAGE 3
A REGIONAL CONCEPT

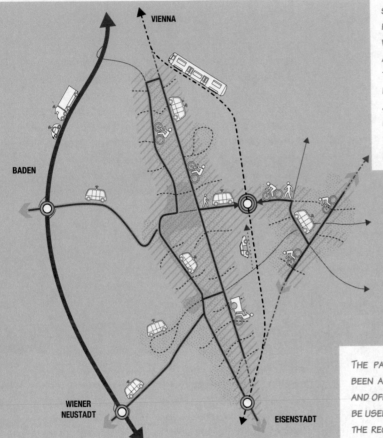

THE CA SHUTTLE NETWORK WILL BE FURTHER EXPANDED AND SERVE ALMOST ALL OF THE RESIDENTIAL AREA, SIMILAR TO THE FORMER CITY BUS NETWORK BUT WITH A MUCH IMPROVED SERVICE. PERIPHERAL SITES WILL HAVE ACCESS TO A CA RING-AND-RIDE TAXI SERVICE. AN INTRICATE, INTEGRATED LOCAL TRANSPORT NETWORK FOR ACTIVE MOBILITY AND CA SHUTTLES IS NOW A REALITY. IT WILL BE USED FOR SHOPPING AND LEISURE JOURNEYS MADE BY TOURISTS, SCHOOL-CHILDREN AND COMMUTERS AS WELL AS BY LOCAL BUSINESSES. UNLIKE CARS, AUTOMATED VEHICLES ONLY RUN IN CERTAIN PARTS OF THE ROAD NETWORK.

THE PARK-AND-RIDE SITE AT THE TRAIN STATION HAS NOW BEEN ALMOST FULLY REPURPOSED. IT NOW HOSTS BUSINESS AND OFFICE SPACES; A RANGE OF SHARING SERVICES CAN ALSO BE USED. THE NEXT STEP IS THE POLYCENTRIC DEVELOPMENT OF THE REGION, WHICH WILL HELP RELIEVE SOME OF THE CONSIDERABLE TRAFFIC FLOWS TOWARDS VIENNA. FOR THIS PURPOSE, A JOINT REGIONAL DEVELOPMENT CONCEPT WILL BE DRAWN UP WITH BADEN, WIENER NEUSTADT AND EISENSTADT.

3 PRINCIPLES OF THE EBREICHSDORF TRANSFORMATION

CA APPLICATION	MOBILITY	MOBILITY	MOBILITY
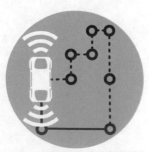			
AUTOMATED SHUTTLE BUSES ON ROUTES AND ROADS	LINK CA SERVICES TO OTHER SECTORS	ADAPTABLE MOBILITY HUBS DESIGNED FOR URBAN SPACES	CA SERVICES ONLY IN CERTAIN AREAS

Discussion and reflection

Ebreichsdorf is already a municipality comprehensively engaged in shaping its future. The strategic planning narratives presented in this paper demonstrate the multitude of options that exist to reduce car dependency. Figure A5 (see Appendix) shows the shift in areas accessible by public transport during the various phases of this transformation. Furthermore, at the end of the process we have set out here, large areas of the municipality (including the cadastral communities) will have access to a network of high-quality streetscapes that are inviting spaces for walking and cycling and served by automated shuttles.

We can once again see that comprehensive cooperation is essential for the success of a transformation process of this kind as well as the ability to transfer the model to other areas. The technology behind automated shuttles is demystified and becomes one of many building blocks. Above all, committed action needs to be taken by a range of actors, the local population must be involved early on and, most notably, the shared view must be held that the pressures caused by the climate crisis represent this century's biggest challenge in terms of mobility and settlement development.

5.4   BAD SCHÖNAU (AREA TYPE D)

Bad Schönau is a small spa town in the south-eastern part of the Industrieviertel, a historically industrial region of Lower Austria that borders Burgenland and Styria. The dispersed municipality comprises seven villages, only two of which currently have more than 50 inhabitants. With just under 500 inhabitants, Bad Schönau is by far the largest. The few central functions that exist are located in Bad Schönau. Residents often travel outside of the municipality to nearby localities such as Krumbach and Kirchschlag to use local amenities as well as welfare and educational institutions.

There is no public transport service that covers the whole region. Vienna can be reached by car in roughly one hour. Residents in the region are thus highly dependent on motorized private transport. Regional buses only run along the B55 federal highway. One of the stations of the Aspangbahn train line can be reached by bus in roughly 20 to 40 minutes. However, these stations do not offer a frequent service (see Fig. 8). With tourism in mind, Austria's Bucklige Welt region is looking to develop services for e-bikes.

As a spa town, tourism is the dominant industry in the region. Several healthcare facilities are located at the outskirts of the municipality. The majority of jobs in Bad Schönau are either directly or indirectly reliant on tourism. The high number of overnight stays (over 200,000 per year, 550 per day) slightly surpasses the number of residents. Population growth in the municipality has been stagnant for a number of years and is now starting to fall in some areas (Kurgemeinde Bad Schönau 2020; Statistik Austria 2017, 2019).

Figure 8: Public transport accessibility in and around Bad Schönau

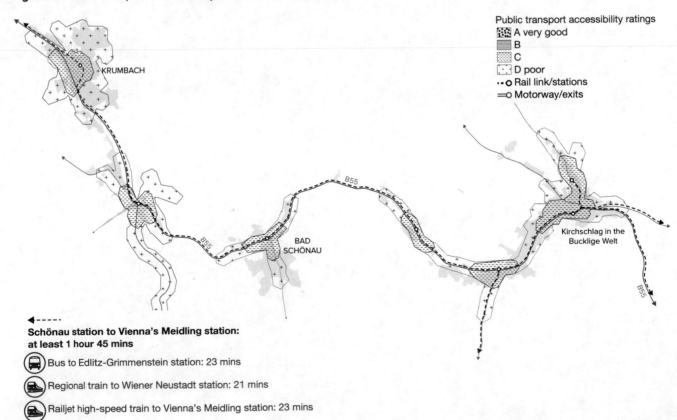

Public transport accessibility ratings
- A very good
- B
- C
- D poor
- ··○ Rail link/stations
- ⇒○ Motorway/exits

KRUMBACH

B55

BAD SCHÖNAU

Kirchschlag in the Bucklige Welt

◄ - - - - -
Schönau station to Vienna's Meidling station: at least 1 hour 45 mins

🚌 Bus to Edlitz-Grimmenstein station: 23 mins

🚆 Regional train to Wiener Neustadt station: 21 mins

🚅 Railjet high-speed train to Vienna's Meidling station: 23 mins

Graphic: Michael Gidam and Lucia Paulhart

TRANSFORMATION STAGE 1
MOBILITY IN THE "SHARED BUCKLIGE WELT REGION"

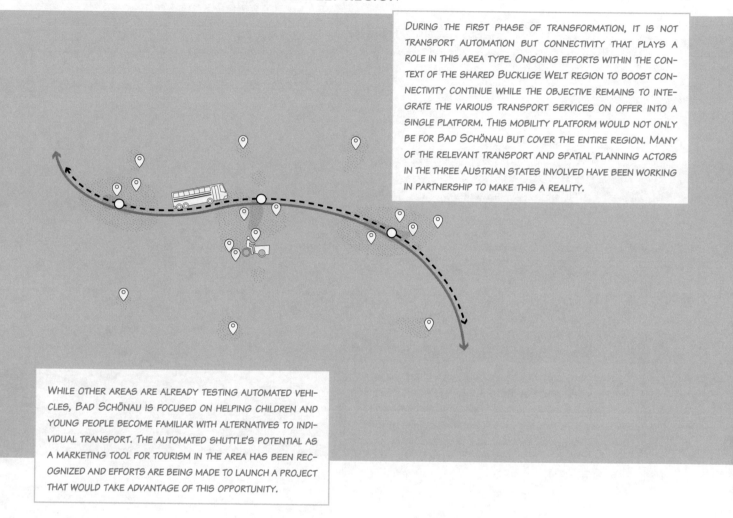

DURING THE FIRST PHASE OF TRANSFORMATION, IT IS NOT TRANSPORT AUTOMATION BUT CONNECTIVITY THAT PLAYS A ROLE IN THIS AREA TYPE. ONGOING EFFORTS WITHIN THE CONTEXT OF THE SHARED BUCKLIGE WELT REGION TO BOOST CONNECTIVITY CONTINUE WHILE THE OBJECTIVE REMAINS TO INTEGRATE THE VARIOUS TRANSPORT SERVICES ON OFFER INTO A SINGLE PLATFORM. THIS MOBILITY PLATFORM WOULD NOT ONLY BE FOR BAD SCHÖNAU BUT COVER THE ENTIRE REGION. MANY OF THE RELEVANT TRANSPORT AND SPATIAL PLANNING ACTORS IN THE THREE AUSTRIAN STATES INVOLVED HAVE BEEN WORKING IN PARTNERSHIP TO MAKE THIS A REALITY.

WHILE OTHER AREAS ARE ALREADY TESTING AUTOMATED VEHICLES, BAD SCHÖNAU IS FOCUSED ON HELPING CHILDREN AND YOUNG PEOPLE BECOME FAMILIAR WITH ALTERNATIVES TO INDIVIDUAL TRANSPORT. THE AUTOMATED SHUTTLE'S POTENTIAL AS A MARKETING TOOL FOR TOURISM IN THE AREA HAS BEEN RECOGNIZED AND EFFORTS ARE BEING MADE TO LAUNCH A PROJECT THAT WOULD TAKE ADVANTAGE OF THIS OPPORTUNITY.

1 PRINCIPLES OF THE BAD SCHÖNAU TRANSFORMATION

MOBILITY

CA SERVICES GO HAND IN
HAND WITH SOFT MOBILITY

TRANSFORMATION STAGE 2
AUTOMATED TOURIST MOBILITY

A MOBILITY CONCEPT IS DEVELOPED TOGETHER WITH OTHER MUNICIPALITIES IN THE BUCKLIGE WELT. AN INITIAL AUTOMATED MOBILITY PROJECT IS LAUNCHED IN BAD SCHÖNAU: A DIAL-A-BUS SERVICE WILL RUN BETWEEN HEALTH RESORTS AND OTHER TOURIST SITES IN THE TOWN CENTRE. TAILOR-MADE SERVICES WILL ALSO BE DEVELOPED FOR SPECIFIC TOURIST SEGMENTS.

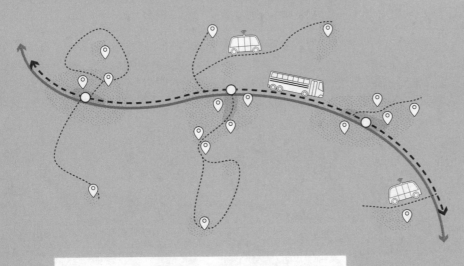

CYCLING WILL CONTINUE TO REMAIN POPULAR ON THE QUIET ROADS IN THE BUCKLIGE WELT BUT WILL NOW BE JOINED BY SHUTTLES, E.G. CARRYING PASSENGERS TO RESTAURANTS LOCATED AWAY FROM THE TOWN. THERE WILL ALSO BE IMPROVED LINKS TO THE ASPANGBAHN, WHICH WILL SEE MORE FREQUENT (AT THIS POINT, NON-AUTOMATED) SERVICES ALONG THE AREA'S SECONDARY ROAD.

2 PRINCIPLES OF THE BAD SCHÖNAU TRANSFORMATION

CA APPLICATION

AUTOMATED SHUTTLE BUSES
OPERATING ON ROUTES

MOBILITY

LINK CA SERVICES TO
OTHER SECTORS

MOBILITY

UPGRADE FEEDER ROADS
FOR CAV

TRANSFORMATION STAGE 3
CONNECTION TO REGIO-TRANSIT WIENER NEUSTADT

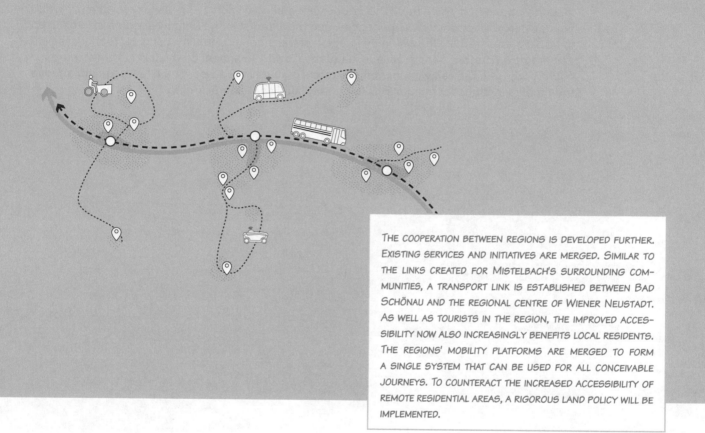

THE COOPERATION BETWEEN REGIONS IS DEVELOPED FURTHER. EXISTING SERVICES AND INITIATIVES ARE MERGED. SIMILAR TO THE LINKS CREATED FOR MISTELBACH'S SURROUNDING COMMUNITIES, A TRANSPORT LINK IS ESTABLISHED BETWEEN BAD SCHÖNAU AND THE REGIONAL CENTRE OF WIENER NEUSTADT. AS WELL AS TOURISTS IN THE REGION, THE IMPROVED ACCESSIBILITY NOW ALSO INCREASINGLY BENEFITS LOCAL RESIDENTS. THE REGIONS' MOBILITY PLATFORMS ARE MERGED TO FORM A SINGLE SYSTEM THAT CAN BE USED FOR ALL CONCEIVABLE JOURNEYS. TO COUNTERACT THE INCREASED ACCESSIBILITY OF REMOTE RESIDENTIAL AREAS, A RIGOROUS LAND POLICY WILL BE IMPLEMENTED.

3 PRINCIPLES OF THE BAD SCHÖNAU TRANSFORMATION

CA APPLICATION	MOBILITY	MOBILITY	SPACE

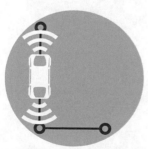

AUTOMATED SHUTTLE BUSES OPERATING ON ROUTES

SAFEGUARD ACCESS TO WIDER PUBLIC TRANSPORT HUBS

LINES MUST LINK CENTRAL FUNCTIONS

CA SERVICE CATCHMENT AREAS WITH HIGH DEVELOPMENT POTENTIAL

Discussion and reflection

The example of Bad Schönau, with its 500 residents, raises the question of where and at what stage it is wise to implement automated mobility services. This has only partly been answered with the acknowledgement that the first transformation stage could be implemented without any automation, but this may not necessarily be the case for other, larger tourist centres. In fact, the opposite may be true. This is because potential users, who visit for leisure or during holidays, tend to prefer new experiences, and this interest in the unfamiliar could also be used to boost automated driving within the context of transforming mobility. Local recreational activities are now far more popular as a result of the Covid-19 pandemic, which is why this approach could become more relevant, not only in terms of tourism but also, at a later stage, to improve the mobility options for populations living in similarly peripheral areas.

6. CONCLUSION

This explorative case study has shown that connected and automated vehicles – if they are to be instrumental in bringing about a mobility transformation – do not always have to be implemented according to the same rules or a general set of principles; instead, it is vital to understand the backdrop to, and motivating factors behind, the technology's application. This requires a bottom-up approach to the implementation of connected and automated vehicles. If these vehicles can be rolled out in a manner that addresses local problems and challenges, this will also increase the probability that the mobility system will be transformed sustainably.

The narratives playing out in the various area types and during the stages of transformation demonstrate that ambitious planning processes are required if a transformation towards a service-oriented, sustainable mobility model is ultimately to succeed in a wide range of spatial contexts. Our analyses give examples of specific planning requirements that will be essential in various transport and spatial scenarios.

Such a transformation requires the necessary patience to prepare and set in motion the desired change based on the technological requirements and (local) acceptance. Even if, for instance, automated commercial/internal mobility management or tourist mobility services initially appear to have hardly any quantifiable transformational effect, they could provide an opportunity to gain initial experience with service-oriented automated transport and thus form the basis for the wider application of such services as part of everyday mobility. Ultimately, our exploratory analysis of each of the four area types has resulted in a final transformational stage that is designed around a holistic mobility system and connected and automated vehicles that can be meaningfully implemented into the existing mobility offering (MaaS). At the end of this transdisciplinary process, we can pinpoint the following dynamics that are shared by all area types:

- Transformation stage 1: Problem-oriented, target group-specific and spatially confined pilot projects allow initial experiences with CAT to be gained.

- Transformation stage 2: CAT projects are met with widespread support and acceptance for the respective activity. As the technology continues to improve, services become more entrenched and spread to other areas and are also applied in other fields.

- Transformation stage 3: The services appeal to a wider range of target groups and can be expanded to serve new catchment areas. They integrate existing mobility systems such as public transport, active mobility and sharing services.

The initial steps to bring about each stage of these development plans are made in the respective area types by various groups of actors (large-scale industry and trade, the municipalities, the regions and the tourism industry). However, an intensive cooperation with policy planning actors will be required at least by transformation stage 3. This collaboration should ideally take place as early as possible. Preparations can be made for measures, such as the integration of a range of mobility services and/or providers into a single fare system and the creation of a shared mobility platform, without the need to wait for developments in vehicle technology.

All four case studies outlined here demonstrate that CAT cannot be seen as the silver bullet that can resolve every issue. Contrary to initial expectations, the technology will only have a positive effect in some areas. If CAT is to contribute to the transport/mobility transformation, automated mobility services must then be meaningfully combined with other public transport services as well as active forms of mobility. Finally, it is hoped that this case study will encourage the necessary plans (processes, decisions, analyses to identify potential for action) to be undertaken in a range of spatial contexts today. CAT will not be capable of initiating an ecologically sustainable transport and mobility revolution all by itself. Bringing major change to transportation seems to be an almost impossible task, especially in rural areas. Only if there is a shared interest among a wide range of actors to develop CAT not as an end in itself but as part of a high-quality public transport system that meaningfully complements active forms of mobility and optimally integrates a range of transport services, does there remain a pathway to revolutionizing transport. In this context, local actors should also be given greater and more comprehensive powers to develop new forms of cooperation with regard to new mobility and take subsequent steps together and decisively

REFERENCES

Backhaus, W., S. Rupprecht and D. Franco 2019. "Road vehicle automation in sustainable urban mobility planning – Practitioner Briefing". https://tinyurl.com/y32u3s7f (9/12/2020).

Beiker, S. 2015. "Einführungsszenarien für höhergradig automatisierte Straßenfahrzeuge", in *Autonomes Fahren. Technische, rechtliche und gesellschaftliche Aspekte*, ed. by M. Maurer, J. C. Gerdes, B. Lenz and H. Winner. Berlin/Heidelberg: Springer Vieweg, 197–218.

Buchholz, M., J. Strohbeck, A. M. Adaktylos, F. Vogl, G. Allmer, S. C. Barros and C. Ponchel 2020. "Enabling automated driving by ICT infrastructure: A reference architecture", *Proceedings of 8th Transport Research Arena TRA 2020*, Helsinki. https://arxiv.org/pdf/2003.05229.pdf (9/12/2020).

Canzler, W., A. Knie, L. Ruhrort and C. Scherf 2018. *Erloschene Liebe? Das Auto in der Verkehrswende. Soziologische Deutungen*. Bielefeld: transcript.

Dangschat, J. S. 2019. "Gesellschaftlicher Wandel und Mobilitätsverhalten. Die Verkehrswende tut Not!", in *Mobilität. Nachrichten der ARL* (49) 1, 8–11.

Dangschat, J. S. 2020a. "Raumwirksamkeit des individuellen hoch- und vollautomatisierten Fahrens", in *Mobilität, Erreichbarkeit, Raum – (selbst-)kritische Perspektiven aus Wissenschaft und Praxis*, ed. by A. Appel, J. Scheiner and M. Wilde. Wiesbaden: Springer VS, 103–122.

Dangschat, J. S. 2020b. "Nachhaltige Mobilität – Herausforderungen der Stadtentwicklung", in *Nachhaltige Industrie* 2/2020, 46–49.

Dangschat, J. S., and A. Stickler 2020. "Kritische Perspektiven auf eine automatisierte und vernetzte Mobilität", in *Jahrbuch StadtRegion 2019/2020*, ed. by C. Hannemann, F. Othengrafen, J. Pohlan, B. Schmidt-Lauber, R. Wehrhahn and S. Güntner. Wiesbaden: Springer VS, 53–74.

ERTRAC 2019 "Connected and Automated Driving Roadmap", ERTRAC Working Group "Connectivity and Automated Driving". Brussels. www.ertrac.org/uploads/documentsearch/id57/ERTRAC-CAD-Roadmap-2019.pdf (14/12/2020).

Environment Agency Austria 2019. "Klimaschutzbericht. Analyse der Treibhausgas-Emissionen bis 2017". Vienna. https://tinyurl.com/y4l5tujc (10/12/2020).

European Commission 2019. "Der europäische grüne Deal", COM(2019) 640 final. Brussels. https://ec.europa.eu/info/sites/info/files/european-green-deal-communication_de.pdf (1/9/2020).

European Commission 2020. "State of the Union: Commission raises climate ambition and proposes 55% cut in emissions by 2030". Download at https://ec.europa.eu/commission/presscorner/de-tail/en/IP_20_1599 (14/12/2020).

Görgl, P., J. Eder, E. Gruber and H. Fassmann 2017. "Monitoring der Siedlungsentwicklung in der Stadtregion+. Strategien zur räumlichen Entwicklung der Ostregion". Vienna: Planungsgemeinschaft Ost. https://tinyurl.com/y422s6zs (10/12/2020).

GVA (Gemeindeverband für Abgabeneinhebung und Umweltschutz) Mödling 2016. "Regionaler Leitplan Bezirk Mödling". https://tinyurl.com/y3nryplk (10/12/2020).

Harrouk, H. 2020. "BIG Designs Toyota Woven City, the World's First Urban Incubator", ArchDaily, 8/1/2020. https://tinyurl.com/yyo3v3fb (1/9/2020).

IEA (International Energy Agency) 2020. "Changes in transport behaviour during the Covid-19 crisis". Paris. https://tinyurl.com/y32cn6xa (1/9/2020).

IHK 2020. "Zukunftsfähigkeit von Gewerbegebieten: Bausteine und Best-Practice-Beispiele". Krefeld. https://tinyurl.com/y4awp6qh (10/12/2020).

IPCC (Intergovernmental Panel on Climate Change) 2018. "Global Warming of 1.5°C. An IPCC Special Report on the impacts of global warming of 1.5°C above pre-industrial levels and related global greenhouse gas emission pathways, in the context of strengthening the global response to the threat of climate change, sustainable development, and efforts to eradicate poverty", ed. by V. Masson-Delmotte, P. Zhai, H.-O. Pörtner, D. Roberts, J. Skea, P. R. Shukla, A. Pirani, W. Moufouma-Okia, C. Péan, R. Pidcock, S. Connors, J. B. R. Matthews, Y. Chen, X. Zhou, M. I. Gomis, E. Lonnoy, T. Maycock, M. Tignor and T. Waterfield. Download at www.ipcc.ch/sr15/download/ (10/12/2020).

Kirchengast, G., H. Kromp-Kolb, K. Steininger, S. Stagl, M. Kirchner, C. Ambach, J. Grohs, A. Gutsohn, J. Peisker and B. Strunk 2019. Referenzplan als Grundlage für einen wissenschaftlich fundierten und mit den Pariser Klimazielen in Einklang stehenden Nationalen Energie- und Klimaplan für Österreich (Ref-NEKP). Vienna: Verlag der ÖAW. https://tinyurl.com/yylaffec (10/12/2020).

Kugoth, J. 2020. "Gesetzentwurf: Wo Roboshuttles rollen sollen", Tagesspiegel Background, 11/5/2020. https://tinyurl.com/y2h7lwx9 (1/9/2020).

Kurgemeinde Bad Schönau 2020. "Öffentlicher Verkehr". https://tinyurl.com/y3anj5p3 (1/9/2020).

Kyriakidis, M., J. C. de Winter, N. Stanton, T. Bellet, B. van Arem, K. Brookhuis, M. H. Martens, K. Bengler, J. Andersson, N. Merat, N. Reed, M. Flament, M. Hagenzieker and R. Happee 2019. "A human factors perspective on automated driving", in Theoretical Issues in Ergonomics Science (20) 3, 223–249. DOI: 10.1080/1463922X.2017.1293187.

Lenz, B., and E. Fraedrich 2015. "Neue Mobilitätskonzepte und autonomes Fahren: Potenziale der Veränderung", in Autonomes Fahren. Technische, rechtliche und gesellschaftliche Aspekte, ed. by M. Maurer, J. C. Gerdes, B. Lenz and H. Winner. Berlin/Heidelberg: Springer Vieweg, 175–196.

Madadi, B., R. van Nes, M. Snelder and B. van Arem 2019. "Assessing the travel impacts of subnetworks for automated driving: An exploratory study", in Case Studies on Transport Policy (7) 1, 48–56.

Manders, T., and E. Klaassen 2019. "Unpacking the smart mobility concept in the Dutch context based on a text mining approach", in Sustainability (11) 23, 1–24.

Matthes, G., and C. Gertz 2014. "Raumtypen für Fragestellungen der handlungstheoretisch orientierten Personenverkehrsforschung", ECTL Working Paper 45. Hamburg: Hamburg University of Technology, Institute for Transport Planning and Logistics. https://tinyurl.com/y4plqe8m (10/12/2020).

McDonald, S. S. 2012. "Personal rapid transit and its development", in Encyclopedia of Sustainability Science and Technology. New York: Springer, 7777–7797. DOI: 10.1007/978-1-4419-0851-3_671.

Mitteregger, M., E. M. Bruck, A. Soteropoulos, A. Stickler, M. Berger, J. S. Dangschat, R. Scheuvens and I. Banerjee 2022. AVENUE21. Connected and Automated Driving: Prospects for Urban Europe, trans. M. Slater and N. Raafat. Berlin: Springer Vieweg. DOI: 10.1007/978-3-662-64140-8.

Mueller, A. S., J. B. Cicchino and D. S. Zuby 2020. What humanlike errors do autonomous vehicles need to avoid to maximize safety? Arlington: IHS.

ÖROK (Austrian Conference on Spatial Planning) 2017 "Für eine Stadtregionspolitik in Österreich. Ausgangslage, Empfehlungen & Beispiele", ÖROK recommendation No. 55. Vienna. https://tinyurl.com/ yyshmv5g (11/12/2020).

Pangbourne, K., M. N. Mladenović, D. Stead and D. Milaki 2020. "Questioning mobility as a service: Unanticipated implications for society and governance", in Transportation Research Part A: Policy and Practice 131, 35–49.

Perret, F., F. Bruns, L. Raymann, S. Hofmann, R. Fischer, C. Abegg, P. Haan, R. de Straumann, S. Heuel, M. Deublein and C. Willi 2017. Einsatz automatisierter Fahrzeuge im Alltag – Denkbare Anwendungen und Effekte in der Schweiz. Zurich: EBP and Basler Fonds.

Ratti, C., M. Bonino and M. Sun (eds.) 2020. "Eyes of the City", exhibition, UABB 2019, Shenzhen. https:// drive.google.com/file/d/1W9bfj7cWyr_RGeOst09pqOOKcgc5iw-u/view (1/9/2020).

Rudolph, F., T. Koska and C. Schneider 2017. "Verkehrswende für Deutschland: Der Weg zu CO2-freier Mobilität bis 2035". Wuppertal: Greenpeace. https://tinyurl.com/yxn2mfwq (10/12/2020).

Schwedes, O. 2017. Verkehr im Kapitalismus. Münster: Westfälisches Dampfboot.

Shladover, S. E. 2016. "The Truth about 'Self-Driving' Cars: They are coming, but not the way you may have been led to think", in *Scientific American* (314) 6, 52–57.

Shladover, S. E. 2018. "Connected and automated vehicle systems: Introduction and overview", in *Journal of Intelligent Transportation Systems* (22) 3, 190–200.

Sieverts, T. 2018. "Rurale Landschaften. Vom Aufheben des Ländlichen in der Stadt auf dem Wege in das Anthropozän", in *Rurbane Landschaften. Perspektiven des Ruralen in einer urbanisierten Welt*, ed. by S. Langner and M. Frölich-Kulik. Bielefeld: transcript, 31–47. DOI: 10.14361/9783839444283-003.

Soteropoulos, A., M. Berger and F. Ciari 2019. "Impacts of automated vehicles on travel behaviour and land use: an international review of modelling studies", in *Transport Reviews* (39) 1, 29–49.

Soteropoulos, A., M. Mitteregger, M. Berger and J. Zwirchmayr 2020. "Automated drivability: Toward an assessment of the spatial deployment of level 4 automated vehicles", in *Transportation Research Part A: Policy and Practice* 136, 64–84.

Stadtgemeinde Ebreichsdorf 2014. *Örtliches Entwicklungskonzept Ebreichsdorf (ÖEK)*. https://tinyurl.com/y6bjomy5 (11/12/2020).

Stadt Mistelbach 2014. *Örtliches Entwicklungskonzept Mistelbach (ÖEK)*. Download at www.mistelbach.at/politik-buergerservice/bauen-planen-raum/raum/oertliches-entwicklungskonzept/ (11/12/2020).

Statistik Austria 2017. "Abgestimmte Erwerbsstatistik 2018". Vienna. Download at www.statistik.at/web_de/statistiken/menschen_und_gesellschaft/bevoelkerung/volkszaehlungen_registerzaehlungen_abgestimmte_erwerbsstatistik/index.html. (1/9/2020).

Statistik Austria 2019. "Statistik des Bevölkerungsstandes". Vienna. Download at www.statistik.at/web_de/statistiken/menschen_und_gesellschaft/bevoelkerung/volkszaehlungen_registerzaehlungen_abgestimmte_erwerbsstatistik/bevoelkerungsstand/index.html (1/9/2020).

STRIA 2019. *Roadmap on Connected and Automated Transport: Road, Rail and Waterborne*. Brussels: European Commission.

SUM (Stadt-Umland-Management) 2020. "Wien und das Umland sind das Einzugsgebiet des SUM". www.stadt-umland.at/stadtregion/wien-und-das-umland.html (1/9/2020).

Taeihagh, A., and H. S. M. Lim 2019. "Governing autonomous vehicles: emerging responses for safety, liability, privacy, cybersecurity, and industry risks", in *Transport Reviews* (39) 1, 103–128.

The Austrian Federal Government 2020. "Aus Verantwortung für Österreich. Regierungsprogramm 2020–2024". Vienna. https://tinyurl.com/uvnle8f (1/9/2020).

Transport & Environment 2020. "How European transport can contribute to an EU -55% GHG emissions target in 2030". Brussels. https://tinyurl.com/y5rt2nmc (14/12/2020).

VCÖ (Verkehrsclub Österreich) 2016. "Schere zwischen Stadt und Land geht in der Mobilität zunehmend auseinander". Vienna. https://tinyurl.com/yyvevnmu (1/9/2020).

VCÖ 2019. "In Gemeinden und Regionen Mobilitätswende voranbringen", *Mobilität mit Zukunft* 1/2019. Vienna. https://tinyurl.com/y54wz84h (11/12/2020).

Wachenfeld, W., H. Winner, C. Gerdes, B. Lenz, M. Maurer, S. A. Beiker and T. Winkle 2015. "Use-cases des autonomen Fahrens", in *Autonomes Fahren. Technische, rechtliche und gesellschaftliche Aspekte*, ed. by M. Maurer, J. C. Gerdes, B. Lenz and H. Winner. Berlin/Heidelberg: Springer Vieweg, 9–37.

Wefering, F., S. Rupprecht, S. Bührmann and S. Böhler-Baedeker 2013. "Guidelines: Developing and Implementing a Sustainable Urban Mobility Plan". Brussels: European Commission.

Wittmayer, J., and K. Hölscher 2017. "Transformationsforschung. Definitionen, Ansätze, Methoden". Dessau-Roßlau: German Environment Agency. https://tinyurl.com/y3dstmst (10/12/2020).

APPENDIX

Figure A1: Participants in the focus groups

Daniela Allmeier	TU Wien, future.lab Research Center and Research Unit of Local Planning
Ian Banerjee	TU Wien, future.lab Research Center and Center of Sociology
Martin Berger	TU Wien, Transport System Planning Research Unit (MOVE)
Willem Brinkert	Zukunftsorte Munderfing
Emilia Bruck	TU Wien, future.lab Research Center and Research Unit of Local Planning
Jens S. Dangschat	TU Wien, Center for Sociology
Oliver Danninger	Lower Austrian State Parliament, head of regional planning and transport
Britta Fuchs	NÖ regional, Industrieviertel mobility management
Michael Gidam	TU Wien, Transport System Planning Research Unit (MOVE)
Peter Görgl	University of Vienna, spatial research and spatial planning
Erwin Hoffmann	Municipality of Mistelbach, department of planning and building inspection
Heinrich Humer	Municipality of Ebreichsdorf, local council, future working group
Mathias Mitteregger	TU Wien, future.lab Research Center
Marleen Roubik	Austrian Ministry of Climate Action, Environment, Energy, Mobility, Innovation and Technology
Martin Russ	AustriaTech, managing director
Rudolf Scheuvens	TU Wien, future.lab Research Center and Research Unit of Local Planning
Aggelos Soteropoulos	TU Wien, future.lab Research Center and Research Unit of Transportation System Planning (MOVE)
Andrea Stickler	TU Wien, future.lab Research Center and Center of Sociology
Johann Stixenberger	Zukunftsorte Waidhofen
Lucia Paulhart	TU Wien, future.lab Research Center
Renate Zuckerstätter-Semela	SUM-Nord, manager

Figure A2: Suitability of existing roads in Vienna South (area type A)

Automated Drivability
Suitability of roads for automated vehicles.

high suitability

low suitability

Graphic: Aggelos Soteropoulos

Figure A3: Line network for automated services in IZ NÖ-Süd (area type A)

Area assistant line network in transformation stage 1

Expansion by opening express lines (shown in green and blue) between the interfaces in transformation stage 2

Graphic: Stefan Bindreiter

Figure A4: Suitability of existing roads in the Mistelbach region (area type B)

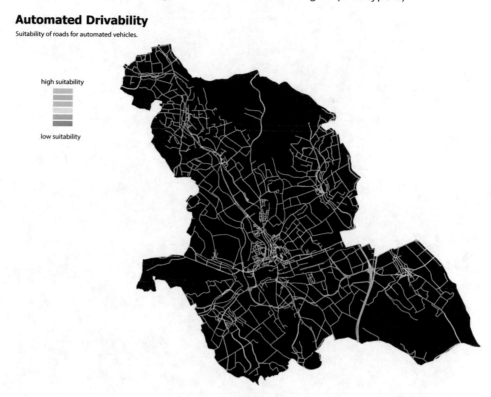

Automated Drivability
Suitability of roads for automated vehicles.

high suitability

low suitability

Graphic: Aggelos Soteropoulos

Figure A5: Changes in public transport accessibility in Ebreichsdorf (area type C) during the transformation stages

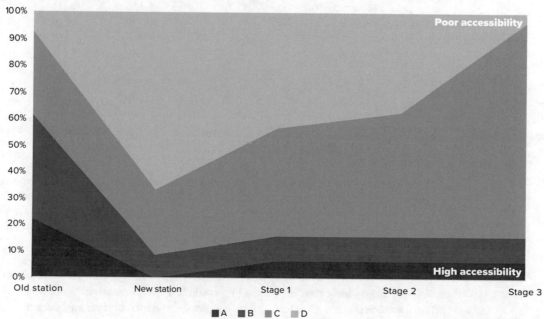

Source: the authors

15 Opportunities from past mistakes: Land potential en route to an automated mobility system

Mathias Mitteregger, Aggelos Soteropoulos

Mathias Mitteregger
TU Wien, future.lab Research Center

Aggelos Soteropoulos
TU Wien, future.lab Research Center and Research Unit Transportation System Planning (MOVE)

1. INTRODUCTION

After the initial hype, recent years have seen markedly more restrained assessments of how connected and automated vehicle technology will develop. A revolutionary scenario that sees a single IT company suddenly launch a successful automated driving system capable of handling every driving task as well as a human (Beiker 2015) is now widely considered unlikely, at least in the medium term. Instead of this presumed revolution, we are currently facing the challenge of a decades-long period of transition – a "Long Level 4" – during which only certain parts of the road network will be accessible to automated vehicles (Mitteregger et al. 2022, Soteropoulos et al. 2020).

However, researchers still anticipate an extensive urban transformation that will have impacts far beyond the transport sector. Past events have demonstrated that the transformation of connected technological systems (cf. Freeman/Perez 1988), such as the one currently taking shape as a result of the connection and automation of transport, will trigger a chain of reforms and revolutions that will impose considerable changes on societies and also have a visible impact on constructed spaces (cf. Headrick 1994, Schmitz 2001, Geels/Schot 2010). The spatial effects of this technological transformation will differ substantially between regions considered to be leading the way and those presumed to be lagging behind (Grübler 1992, Rogers 2003). Yet they will share one fundamental similarity: during the transition from one technology or transport system to another, the old system's need for land will vanish while its successor will place new demands on space (Grübler 1990). For instance, at the beginning of the railway era, land was acquired for the new transport infrastructure and a whole host of operational services by tearing down buildings in deprived parts of cities (Bruinsma et al. 2008) or by developing greenfield sites in suburban areas (Bellet 2009, Bertolini/Spit 2005).

A similar spatial development dynamic is also expected during the transition to a connected and automated transport system. In this article, we systematically examine the fundamental aspects of this shift while aiming to understand these elements quantitatively and qualitatively with the help of a case study that analyses Vienna's functional urban area (FUA). In Section 2, we present the framework of our analysis. To understand how land is directly used by various modes of transportation, we present a model that was developed based on product life cycle analyses. This model is illustrated using three types of transport (local public transport systems, individual transport and transport service providers, such as Uber and Lyft) and outlines the characteristic land use for each system's components. At the end of Section 2, existing sites used by transport service companies, who are frequently seen as the forerunners of automated mobility services, will be examined in greater detail. This analysis of urban planning characteristics and facilities will then be used to develop general spatial principles for automated mobility services and their potential need for land. In Section 3, this transport system land-use model will be applied to the Vienna FUA using a geographic information system (or GIS)-based analysis. Moreover, a data set from Herold, a company specializing in digital media and marketing services that publishes Austria's telephone directory, was used that locates and lists businesses in Austria according to their economic activity classification (otherwise known as their ÖNACE category). Here the components of individual transport are considered, while public transport components are explicitly excluded. In Section 4, we present the findings of our analysis of car-associated businesses, which examines their location and spatial characteristics, thus giving a regional dimension to the theory of sectoral reconstruction of the automotive industry. The final Section provides a discussion of these findings and a conclusion, as well as considering the extent to which existing spaces (rather than construction on greenfield sites) can be used for a potential future transformation.

2. THE PATH TO A CONNECTED, AUTOMATED AND SERVICE-ORIENTED TRANSPORT SYSTEM

Connected and automated mobility is usually described as a subset of shared mobility and Mobility as a Service (MaaS), or as the innovative system that could lead to a broad focus on service provision in the mobility sector (Shaheen/Cohen 2019). The transport sector is facing a considerable change which could see well-established opposites, such as individual vs. public transport and passenger vs. goods transport, become obsolete (Lenz/Fraedrich 2015; Mitteregger et al. 2022: 43). Previous studies on the possible spatial impacts of this change have notably considered land reclaimed from parked cars as the transformation spaces of tomorrow (Alessandrini et al. 2015, Heinrichs 2016, Stead/Vaddadi 2019). Numerous simulation studies on automated car- or ride-sharing services reach the conclusion that a reduction in private car use could lead to demand for parking spaces falling dramatically (by up to 80% or 90%; Soteropoulos et al. 2019).

While discussions concerning potential transformation spaces focus solely on car parks, the impact of the transformation of technological systems is underestimated. Firstly, these studies have access to limited knowledge as the decreasing need for parking spaces cannot be comprehensively ascertained with the models used; moreover, they are often based on a merely presumed relationship between (privately owned) vehicles that are replaced by automated car- and ride-sharing services and disappearing parking spaces (ibid.).[1] Secondly, the focus of these studies does not go far enough. How the economic effects of the transport revolution are assessed illustrates this, with the impacts extending far beyond parking space management. If owned mobility becomes situational and consumed, this will affect the entire automotive industry (Bormann et al. 2018) as well as the public transport sector (Sommer 2018), together with all businesses operating in these two areas. The possible spatial effects of such a transformation thus include not only parking spaces but potentially any space that is directly linked to the transport sector, whether it is used for manufacturing or sales, or even maintenance and servicing.

2.1 LAND DEMANDS OF VARIOUS MODES OF TRANSPORTATION

To ensure any potential for land transformation is fully taken into account, our analysis includes all economic sectors that are essential components of a coherent system to ensure the functioning of a single mode of transport (Fig. 1). For the observations presented here (as well as for the selected case study region), it shall be assumed – for reasons of simplification – that this transformation will have a lesser impact on companies building, maintaining, dismantling and recycling (road) infrastructure. This is because connected and automated driving systems will be able to make use of at least part of the existing road infrastructure (cf. the last two lines of Fig. 1 as well as STRIA 2019).

This general product life cycle was then used to create a more specific model for further analysis (see Fig. 2). Vehicle trading, a key sector, particularly for individual transport, was added alongside manufacturing, which plays only a marginal role in the region we are analysing. In the

---

1   Here we can see that the established methods and accepted analyses were designed for periods of "normal science" and thus do not adequately describe paradigm shifts (Kuhn 1967).

following table, which compares individual transport, local public transport and transport service providers, the key characteristics impacting land use already become clear. Here, "transport service providers" includes companies such as Uber, Lyft and Bolt (cf. Figs. 5 and 6) as well as specialist third-party providers offering services for drivers (e.g. Splend and Drover).

Figure 1: Diagram of the product life cycle of various modes of transportation

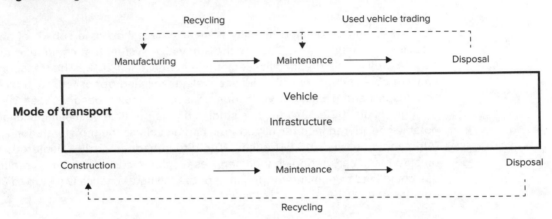

Source: the authors based on Spielmann/Scholz (2005: 86)

The comparison of transport modes shown in Figure 2 demonstrates that the components of a service-oriented form of mobility that is provided by either public or private transport operators will place different demands on space compared to those arising from individual transport, i.e. if there is wholesale, retail and used car trading, then it exists only in a radically altered form (cf. Bormann et al. 2018: 18). Current developments make clear the dilemmas facing vehicle traders in Europe (Eckl-Dorna 2020). The volume of transactions in individual transport after-sales services (spare parts trading, repairs and services, technical support, etc.), the added value of which is mostly generated locally, attests to the scale of the likely regional economic impacts (Cohen et al. 2006). In 2017, European turnover in this industry was calculated at €242 billion (McKinsey 2018) and it is predicted that the mobility transformation would see a completely new distribution of these sales.

Figure 2: Product life cycle of selected modes of transport and their components

Mode of transport	Local public transport	Individual transport (car)	Transport service provider
Manufacturing, vehicle trading	Vehicle manufacturers –	Vehicle manufacturers Wholesale, retail and used car trading	Vehicle manufacturers –
Operation Maintenance	Road/rail network, stops, train stations, depot, staff facilities, etc.	Road network, parking spaces, petrol stations, car washing facilities, garages, etc.	Road network, waiting bays, driver centres
Disposal	Scrap, waste management, recycling		

Source: the authors

2.2   LAND USE, COMPONENTS AND CHARACTERISTICS OF VARIOUS MODES OF TRANSPORTATION

Past events suggest that when one dominant mode of transport makes way for another, land conversions arise across all of the components listed in Figure 2. The various opportunities for development were exploited in different ways. The rise of the private car caused the railway network to shrink while depots and freight yards became obsolete. Workplaces and dormi-tory accommodation for train company staff also disappeared. Wherever the need arose, all of these sites became valuable resources for settlement development. Across Europe, new developments, urban megaprojects and even transport hubs for high-speed trains have taken shape where freight yards and depots once stood (Bertolini et al. 2012). Working-class housing estates were either torn down or adapted and procured for new uses (cf. Burghart/Hertenberger 2018). In rarer cases, instead of turning these areas into real estate, rail operators decided to ensure they were used as open spaces. In the following section, the spatial characteristics of different modes of transport are examined in detail based on their operational logic, with particular attention being paid to the recently emerging group of transport service providers.

2.3   LAND DEMANDS OF LOCAL PUBLIC TRANSPORT

Local public transport components are typically concentrated along the relevant infrastructure and in hybrid operational sites that fulfil functions such as maintenance, transport management and parking (Fig. 3). This concentration of functions is primarily a result of local transport operators needing to design internal processes that are as synergistic and efficient as possible. For instance, placing multiple functions at a single site makes it possible to utilize applications (e.g. spare parts storage for units operating on several lines) or general spaces within buildings (e.g. rooms for staff) more effectively. The spectrum of services offered by municipal infrastructure providers is proof that these synergies extend beyond the transport sector. The electricity generated by power stations for tram line, underground and commuter train services is still being fed into municipal power grids.

Figure 3: Erdberg underground station, Vienna

Photo: Vienna GIS

The inaccessibility of the railway network and local public transport operating sites usually means they become urban barriers that are gradually engulfed by growing towns and cities. At the same time, high passenger numbers at train stations and bus/tram stops boost their appeal for other functions, such as retail or office spaces, which has driven the formation of centres (Bertolini 2017).

2.4 LAND DEMANDS OF INDIVIDUAL TRANSPORT

Individual transport inverts the principles that apply to local public transport on multiple levels. Privately owned cars, for instance, have led to a rapid expansion of functions that were once concentrated and are now spread out along an unevenly dense road network across entire settlement structures. The car has thus contributed to the levelling up of spatial differences, a uniformity that is often viewed critically by urban planners and researchers, especially within the context of suburban space. That this – at least, initially – also meant social and spatial success is often overlooked in the relevant discourse (McLuhan 2003: 291–301).

The emotional weight attached to the car means that the components of individual transport (car dealerships, car washes, repair garages and petrol stations) serve not only to fulfil a certain function but are an expression of differentiation within the modern consumer society. This is because cars and car maintenance are closely linked to a sense of identity, especially in places where there are no, or only minimally appealing, public transport alternatives (Manderscheid 2014: 606f.). The pride associated with this vehicle is not shared collectively across nations or regions, as is the case with trains. The fact that businesses providing services linked to individual modes of transport are spread across entire settlement areas attests to a fragmented, mobile society, whose members are able to decipher the subtle differences between car brands and their accessories (Cresswell 2006, Bourdieu 1987).

Figure 4: Locations of businesses offering car-associated services along Triester Straße in Vienna

Photo: Vienna GIS

In settlement structures, the functions associated with the private car are mostly solitary or act as urban barriers. However, this is not just caused by each function's spatial requirements – be it a petrol station, a garage or a car dealership – but rather the combination with an excessive use of space caused by the demand for parking. From a historical point of view, this accounts for perhaps the most significant disruption to spatial planning to date: the car has reversed the relationship between a high concentration of people and centralized places that has been so pivotal throughout urban history. In the age of the car, roads with high volumes of traffic can no longer also be social spaces within the urban landscape as was the case, for instance, with main thoroughfares and boulevards (Marshall 2005: 5). According to this logic, attractors such as shopping or retail parks require motorways or wider bypasses, which intersect the space and remain isolated as they are surrounded by parking areas.

2.5   LOCATIONS USED BY NEW TRANSPORT SERVICE PROVIDERS

Rather than building their businesses around a taxi model, today's transport service providers, e.g. Uber, Lyft and Bolt, actually grew from ride-sharing schemes and car-sharing agencies that brought passengers and non-professional drivers together (Chan/Shaheen 2011). Initially, car and ride-sharing were primarily arranged through friends, acquaintances or colleagues. It was a set-up increasingly used in times when resources were scarce or to save money (for instance, after World War II in the US and during the 1970s oil crisis in Europe). This led to ride-sharing clubs, noticeboards and phone lists that gradually started to move online during the 1990s. The spaces dedicated to such activities, if they exist at all, are just as practically designed as the ride-sharing groups themselves: plain tarmac surfaces on motorway feeder roads, bypasses and lay-bys.

While modern-day ride-sharing providers corrupted this model of shared use by turning it into a money-making enterprise, some of the effects are now becoming apparent in other areas too. The success of ride-sharing providers is already having an impact on the flow of traffic and the modal split (SFCTA 2018) as well as becoming evident in the new demand for space at airports, train stations and shopping centres (Edelson 2017, Leiner/Adler 2019) as well as at the kerbside (cf. Chap. 8 by Bruck et al. in this volume). Moreover, little attention has so far been paid to the emergence of a new type of hybrid building that serves as a depot, a rest area and a co-working space (Greenblatt/Shaheen 2015, Epting 2019) and can offer some indication as to how much space automated mobility services might require.

Ride-sharing companies' global domination has resulted in the creation of operating sites where either the transport service provider or specialized companies offer services for drivers working in the sector and their vehicles. The table below compares four of these sites and their main characteristics (Fig. 5). In all of the examples analysed, similar features seem to have been the determining factor for the choice of location; the industry's history also appears relevant. The sites are located on the outskirts of urban zones or in suburban settlement areas close to motorway junctions. They are thus all situated outside of areas in which the demand is highest for their services (Anair et al. 2020). The location close to motorway exits suggests that relatively long journeys must be undertaken by drivers to reach their customers (to compensate for lower basic or rent costs). It is notable that each site offers good local transport links. This could be due to the fact that many drivers do not own their own vehicle but must rent a car from a transportation service provider or another company (Zwick 2018).

Figure 5: Urban planning features of transport providers' operating sites

	Lyft Driver Hub	Uber Greenlight Hub	Splend Member Support Center	Bolt Hub
Location	Windsor Park, Austin, TX	Bronx, New York, NY	Cricklewood, London	Chiswick, London
Surface area	2600 m²	2800 m²	2300 m²	N/A
Motorway	✓	✓	✓	✓
Site	Suburban	Periphery	Periphery	Periphery
Public transport	Bus	Bus, underground	Bus, commuter train	Bus, commuter train

Source: the authors

The characteristic features of the selected company sites reinforce this observation (Fig. 6). They clearly show how key services for non-professional drivers and vehicles become internalized. If we follow the logic of local public transportation, the aforementioned functions should become inherent to the mode of transport being provided. But in the hands of transport service providers, who follow the principles of the sharing economy, they become services that generate added value at the expense of drivers. The common practice of seeing employees as independent contractors (i.e. non-professional drivers are seen as contract workers, cf. Heller 2017) makes it possible to create a new hybrid form of working that follows the logic of both local public and individual transport, i.e. flexible spaces can be used for leasing businesses, training sessions, meetings or community events; one-to-one, sales or consultation meetings are conducted in smaller offices with prospective or existing drivers and also result in documented interdependence (Fig. 7 on the next page).

Figure 6: Facilities of mobility service provider hubs/centres

	Lyft Driver Center (Austin, TX)	Uber Greenlight Hub (Bronx New York, NY)	Splend Member Support Center (Cricklewood, London)	Bolt Hub (Chiswick, London)
Relaxation rooms	✓	✓	✓	✓
Sanitary facilities	✓	✓	✓	✓
Catering, kitchen, eating areas	✓	✓	✓	✓
Meeting rooms	✓	✓	✓	✓
Rooms for training and events	✓	✓	✓	✓
Help desk	✓	✓	✓	✓
Medical check-ups	N/A	✓	N/A	N/A
Insurance	✓	✓	N/A	N/A
Car rental	✓	✓	✓	N/A
Parking	✓	✓	✓	✓
Maintenance and repairs	✓	✓	✓	N/A

Source: Hu (2017), Charpentier (2020), Tucker (2019), Ongweso Jr./Koebler (2019), Splend (2019)

Aside from parking spaces that can be used not only by drivers but also for rented cars, numerous functions are offered for the operation and maintenance of vehicles (Fig. 6). Car repair and/or maintenance are conducted either by the driver (using tools or garages provided on-site) or by those operating the sites (Charpentier 2020). The cost is charged directly to the driver or drivers must become members to access this service. To ensure an appropriately high workload, the services on offer are sometimes made available to other customers.

The largest transport service providers have made clear that they are already preparing for a long transition to automated driving systems. It is expected that an extended period will begin during which automated mobility services and conventional services provided by a driver will be offered in tandem as part of a "hybrid network" (Chaum 2019, Sheikh 2018). In other words, every time a journey is requested, an algorithm will check whether current developments allow a route between the starting point and destination that is accessible to automated vehicles. If this is the case, a highly automated vehicle will be assigned the journey (unless the customer explicitly requests a conventional transport option). If the route is not appropriate for automated vehicles, the request will be passed on to a human driver. It is plausible that the number of functions assigned to drivers will gradually start to decrease compared to those that are assigned to vehicles. These sites could also potentially be used for new applications, e.g. they could become workplaces for dispatchers or personnel who monitor or control one or more highly automated vehicles (Yankelevich et al. 2018).

Figure 7: Advertisement on a bus in Chicago for legal specialists offering services specifically for the ride-sharing industry.

The maintenance and, here particularly, cleaning of highly automated vehicles will be seen as an extremely relevant challenge (Clements/Kockelman 2017: 13; Kucharczyk 2017). Service areas, such as the one in use at the Lyft Driver Hub, resemble production lines and demonstrate considerable future potential for automation. Companies that currently offer "predictive maintenance" in niche industries (such as heavy engineering and mining) could someday perform this service for transport service providers. Numerous actors (Bosch, Intel, Uptake) are already trying to gain a foothold on this market.

The urban-spatial effect of these spaces is likely to change as automation becomes more sophisticated. The need for parking spaces is still creating an integration challenge for urban planners, which is why these company sites currently feature the same characteristics as car-associated locations. However, as automation develops and algorithms take over the reins from drivers, parking spaces could also be used differently. For instance, they could only be available at off-peak times, although it remains uncertain whether this would still be possible at even more remote sites (as is the case for logistics companies) as vehicles currently have to travel far to pick up passengers. It is possible that other functions, such as those performed by dispatchers, will be outsourced globally. Such applications already exist in the mining sector (Frangoul 2018). The separation effect created within the streetscape could be amplified on the roads leading to and from sites in the medium term if the infrastructure needs to be upgraded for use with automated driving systems (Soteropoulos et al. 2020).

3. ANALYSING CAR-ASSOCIATED LAND CONVERSIONS WITHIN THE VIENNA CITY REGION

To provide a detailed analysis of the space development dynamic that could be initiated by the transformation to a connected, automated and service-orientated transport system, we looked at car-associated functions and businesses within the Vienna city region and considered their location and spatial characteristics. We established which car-associated businesses were present in the area using the model presented in Section 2 (Fig. 2). Our focus was solely on the transformation of existing spaces (Steen/Ryding 1993) and not any additional need for space in previously undeveloped areas (Jolliet/Crettaz 1996). We pinpointed potential transformation areas within the Vienna FUA and examined their individual characteristics. Figure 8 provides an overview of the methodical steps taken during analysis.

Figure 8: Steps for the analysis of car-associated land conversions in a city region

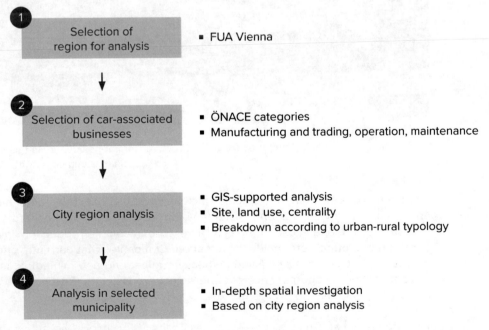

Source: the authors

3.1 CASE STUDY: THE VIENNA CITY REGION

The functional urban area of Vienna, as defined by the OECD (2020), was chosen for analysis. Beside the city of Vienna itself, the Vienna FUA encompasses additional municipalities with significantly different spatial characteristics. However, in terms of its settlement structure and integration, the FUA presents similar characteristics to other city regions in German-speaking countries and is thus an appropriate candidate for a case study investigating car-associated land conversions. The Vienna FUA is a useful example to demonstrate the sites and spatial characteristics of these potential transformation areas. It is also relevant (or even applicable) to other city regions.

In order to ensure the analysis accounted for a broad range of differentiation, Statistics Austria's "Urban-rural rypology" was used as a basis (Statistik Austria 2017). The city of Vienna and some surrounding municipalities fall under the "Urban centre (large)" category; however, given the Austrian capital's overall importance and the significant differences in comparison with nearby municipalities considered in the same category, the city itself was analysed separately. Figure 9 provides an overview of the Vienna FUA as well as the further classification of municipalities based on Statistics Austria's urban-rural typology.

Figure 9: Vienna FUA and area classifications according to Statistics Austria's "Urban-rural typology"

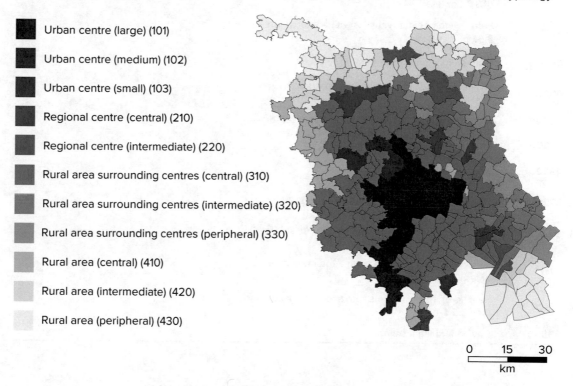

Urban centre (large) (101)

Urban centre (medium) (102)

Urban centre (small) (103)

Regional centre (central) (210)

Regional centre (intermediate) (220)

Rural area surrounding centres (central) (310)

Rural area surrounding centres (intermediate) (320)

Rural area surrounding centres (peripheral) (330)

Rural area (central) (410)

Rural area (intermediate) (420)

Rural area (peripheral) (430)

Source: the authors based on Statistics Austria (2017)

3.2   SELECTING CAR-ASSOCIATED BUSINESSES

Car-associated businesses were selected based on a data set provided by Herold, a database containing information on businesses and organizations throughout Austria. The ÖNACE categories were used to identify relevant business and/or organization categories, which were then assigned to one of three areas (manufacturing/trade, operation or maintenance) based on the model outlined in Figure 2. ÖNACE categories that have some association with car parts or accessories were added to the "maintenance" category. Businesses operating in the field of disposal or recycling were not considered at all. The ÖNACE categories used as well as their applicable area of activity are shown in Figure 10.

Figure 10: Selected ÖNACE categories

ÖNACE		Manufacturing and trade	Operation	Maintenance
29100	Manufacture of motor vehicles and motor vehicle engines	✓		
29200	Manufacture of bodies (coachwork) and trailers	✓		
29310	Manufacture of electrical and electronic equipment for motor vehicles			✓
29320	Manufacture of other parts and accessories for motor vehicles			✓
30990	Manufacture of other vehicles	✓		
33170	Maintenance and repair of motor vehicles			✓
45111	Wholesale trade of motor vehicles with a total weight of 3.5 t or less	✓		
45112	Retail trade of motor vehicles with a total weight of 3.5 t or less	✓		
45200	Maintenance and repair of motor vehicles			✓
45310	Wholesale trade of motor vehicle parts and accessories			✓
45320	Retail trade of motor vehicle parts and accessories			✓
47300	Retail sale of automotive fuel (petrol stations)		✓	
49320	Taxi operation		✓	
52211	Operation of multistorey and underground car parks		✓	
77111	Leasing of motor vehicles with a total weight of 3.5 t or less		✓	
77112	Renting of motor vehicles with a total weight of 3.5 t or less		✓	
77120	Renting of motor vehicles with a total weight of more than 3.5 t		✓	
85530	Driving and flight schools		✓	

Source: the authors

3.3 APPROACH TO THE CITY REGION ANALYSIS

At the city region level, the first step involved an observation of the current location of various car-associated functions, particularly in relation to other functions in the same locality. An analysis was also conducted to establish the type of area in which the businesses were located, i.e. in what way the land is used at these locations. For this analysis, land use data were obtained from OpenStreetMap (a free platform that collects and organizes freely usable geodata, which are provided in an open licence database for general use). Five categories were then established: residential, commercial/industrial, area with offices/business parks, area with retail businesses/shopping centres, and other (cf. Ramm 2019: 17; OpenStreetMap Wiki 2020).

The businesses' proximity to the nearest town centre was also analysed by calculating the distance between each site and the closest municipal administrative office (the relevant data were taken from the Herold data set). Due to the heterogeneous structure of its urban centres, the city of Vienna was excluded from this analysis.

The analysis of land used by businesses and functions, as well as their proximity to the nearest local centre, was further separated into the relevant spatial context as defined by Statistics Austria's urban-rural typology.

The results of the city region analysis were also used to carry out an in-depth spatial analysis in a selected municipality. This involved a detailed examination of car-associated land conversions in terms of their suitability for brownfield development. On the basis of the findings of our city region analysis, Vienna's 23rd district, Liesing, was selected.

4. FINDINGS OF AN ANALYSIS OF CAR-ASSOCIATED LAND CONVERSIONS WITHIN THE VIENNA CITY REGION

4.1 CITY REGION ANALYSIS

Categorizing businesses based on transportation system components and urban-rural typology

Within the Vienna FUA, a total of 3,082 car-associated business sites were identified. These were then categorized according to the table in Figure 2. The majority of businesses came under the headings of "Maintenance" (42%) and "Operation" (40%). Slightly less than a fifth of car-associated functions (18%) were classified as "Manufacturing and trade" (Fig. 11).

Figure 11: Car-associated businesses and their relevant category (manufacturing/trade, operation or maintenance) in %

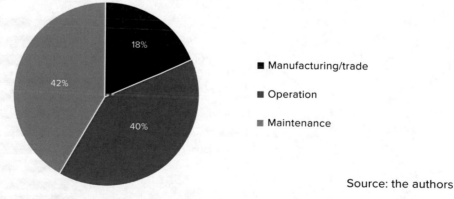

- ■ Manufacturing/trade
- ■ Operation
- ■ Maintenance

Source: the authors

If we examine the breakdown of businesses based on Statistics Austria's urban-rural typology (Fig. 12), we can see that the largest number of car-associated businesses are, as anticipated, located in Vienna (1,594 overall), followed by a total of 452 businesses in large urban centres, i.e. the ribbon development to the south of Vienna, as well as in the municipalities along the Danube situated to the north-west and south-east of the city. Overall there are 365 businesses located in municipalities classed as "Rural area surrounding centres (central)". The Vienna FUA is largely comprised of such municipalities. The number of businesses located in the subsequent three categories is relatively high. These are "Urban centre (small)" (150), "Urban centre (medium)" (131) and "Regional centre (central)" (104), which includes the municipalities of Wolkersdorf, Gänserndorf and Bruck an der Leitha. The latter are also the municipalities whose working populations comprise the highest percentage of commuters (Görgl et al. 2017: 181).

Figure 12: Number of car-associated businesses according to the urban-rural typology (n = 3,082)

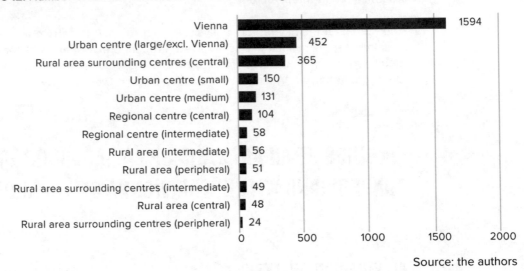

Source: the authors

The breakdown of businesses by sector and their classification according to the urban-rural typology demonstrates that a high percentage of companies providing maintenance services are located in "Rural area surrounding centres (peripheral)" (58%) and "Rural area (peripheral)" (51%). However, a significant number of businesses operating in manufacturing and trade appear to be present in centres or in areas surrounding centres, particularly in regional centres (29%; Fig. 13).

Figure 13: Breakdown of car-associated businesses by sector (manufacturing/trade, operation or maintenance) and urban-rural typology (n = 3,082)

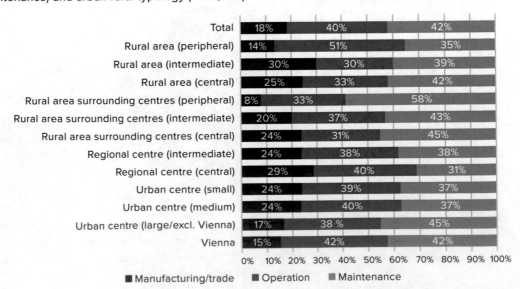

Source: the authors

Location of businesses within the city region

Figure 14 illustrates the location of businesses in the city region. The diagram shows the number of businesses and presents this figure in relation to all businesses and/or organizations located nearby. The areas that feature both a high number and a considerably high share of car-associated businesses, and are thus heavily influenced by this sector, are mainly located to the south of Vienna.

Figure 14: Number of car-associated businesses as a share of all businesses and/or organizations in the Vienna FUA

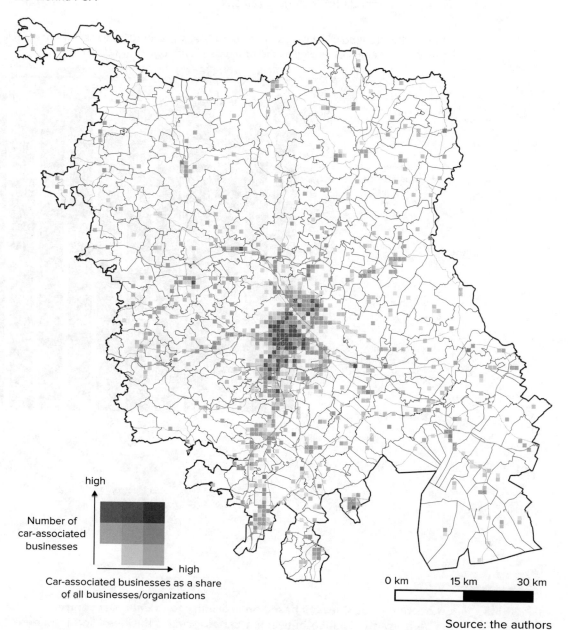

Source: the authors

Location of businesses according to land use

In terms of the location of businesses according to their land use (Fig. 15), it is clear that across the entire Vienna FUA, car-associated businesses are predominantly located in residential areas (67%) or in commercial/industrial areas (14%). By comparison, there are only a few businesses situated in areas with office blocks/business parks or with retail businesses/shopping centres (6% respectively). Regarding the urban-rural typology, the number of car-associated businesses in commercial/industrial estates is particularly high in central locations, especially in large urban centres (26%) as well as in centralized rural areas surrounding centres (17%). Residential areas with a higher share of businesses, on the other hand, tend to be located in smaller, peripheral municipalities, e.g. those that fall under the categories "Rural area (peripheral)" (90%) and "Rural area surrounding centres" (88%). It is interesting to note the high share of car-associated businesses located in areas with retail businesses or shopping centres in the "Regional centre (intermediate)" category (29%).

Figure 15: Breakdown of car-associated businesses according to sector (manufacturing/trade, operation or maintenance) as well as the urban-rural typology (n = 3,082)

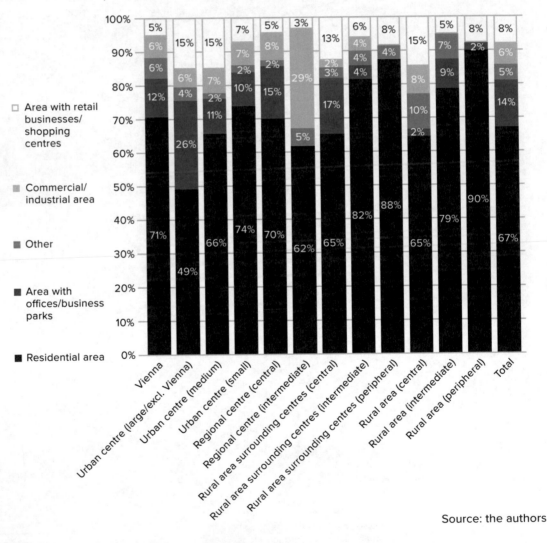

Source: the authors

Location of businesses based on proximity to the nearest centre

The average distance between a car-associated business and the nearest town centre is 1.3 kilometres (Fig. 16 on the next page). Particularly in peripheral areas, businesses are relatively close to the nearest town centre. This is especially true regarding businesses in the "Rural area (peripheral)" (695 m) and "Rural area surrounding centres (peripheral)" (861 m) categories. However, businesses tend to be located considerably further away from town centres in medium-sized (1,441 m) and small urban centres (1,496 m), presumably due to the settlement size.

Figure 16: Location of car-associated businesses based on proximity to town centres and urban-rural typology

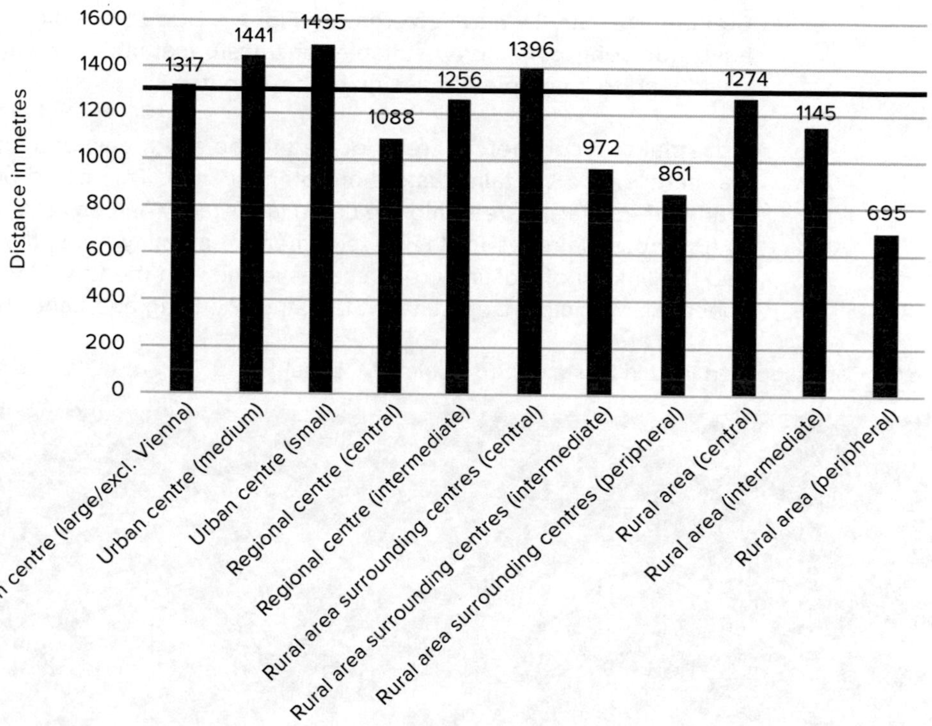

Distance in metres

Urban centre (large/excl. Vienna)	1317
Urban centre (medium)	1441
Urban centre (small)	1495
Regional centre (central)	1088
Regional centre (intermediate)	1256
Rural area surrounding centres (central)	1396
Rural area surrounding centres (intermediate)	972
Rural area surrounding centres (peripheral)	861
Rural area (central)	1274
Rural area (intermediate)	1145
Rural area (peripheral)	695

Source: the authors

4.2   IN-DEPTH ANALYSIS OF LIESING

Based on the findings of our city region analysis, Vienna's 23rd district, Liesing, was chosen for in-depth analysis. Our analysis of the number and share of car-associated businesses as a percentage of all businesses or organizations demonstrated that Liesing is an area where car-associated businesses play a considerable role. Moreover, Liesing's existing transport system and location also underline its relevance for this study: the district is located at the outskirts of Vienna and has transport connections both to the city's underground (specifically the U6 line) and commuter train network. The Südbahn (Southern Railway) and the A2 (E59) and A21 (E60) motorways, three large-capacity transport axes, pass through the district. Liesing has the highest motorization rate of any Viennese district, and its modal split has the highest share of car journeys (ÖIR 2015). Liesing's high affinity for cars is associated with its poor links to large-capacity local public transport. The district is growing rapidly, and there is high demand for property among buyers and renters alike (Lottes 2020). Its location at the urban periphery and its transport axes also make Liesing an attractive site for business.

This combination of factors could turn Liesing into a highly relevant target for automated mobility services (cf. Chap. 14 by Mitteregger et al. in this volume). Attractive services that complement existing local transport networks could become available on selected routes. The district's dynamic development in recent years has resulted in brownfield projects becoming an increasingly relevant issue. A lack of high-quality open spaces near residential areas is currently described as the key challenge. In addition, there are calls for the development of a more densely built local public transport network as well as the development and/or boosting of local district centres; such projects are already being funded (Municipal Department 21, District Planning and Land Use, City of Vienna 2015).

Car-associated land conversions could prove to be a valuable asset in achieving existing goals. Businesses from all sectors are concentrated along heavily used roads, such as Triester Straße, Breitenfurter Straße and Ketzergasse (Fig. 17). Due to the high volume of traffic, these roads have not been considered suitable for transformation; however, the high land use potential could lead to a complete redesign of these routes.

Industrial and commercial areas located to the south-east and further to the east of the historical centre also contain clusters of potential sites. This could be a step towards better integrating these large areas into the urban landscape. Across the entire district area, it is possible to find potential sites for the development of mobility hubs, the promotion of district centres and the linking of existing green spaces in line with the City of Vienna's plans to improve open space (cf. Municipal Department 21, District Planning and Land Use, City of Vienna 2015).

Figure 17: Car-associated land in Liesing (23rd district of Vienna)

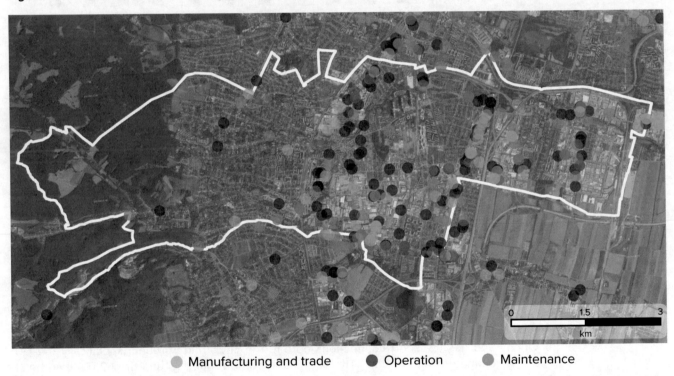

● Manufacturing and trade   ● Operation   ● Maintenance

Source: the authors based on basemap.at

5. CONCLUSION AND DISCUSSION

In Lewis Mumford's "The Highway and the City", which was first published in 1958 and then later revised, the author accuses his transport planning colleagues of historical amnesia, resulting in what he believed to be catastrophic urban planning and social impacts for the cities of his time.[2] As was the case during the railroad era, Mumford contended, urban space was once

2   "[They] lack both historic insight and social memory [and] accordingly, [...] have been repeating, with the audacity of confident ignorance, all the mistakes in urban planning committed by their predecessors who designed our railroads" (Mumford 1968, 94).

again being cleared and vast areas of land were once more being consumed by all types of traffic structure (Mumford 1968). Little attention was paid to Mumford's warning: the designs of North American and European cities attest to this. Land has been carved up and car-associated areas have generally been spread out across settlement structures. Now, at the beginning of the age of automated mobility, these areas have become the subject of debate. And as our mobility systems are on the cusp of revolutionary change, it is to be hoped that Mumford's call for historical insight will be heard as automated vehicles pose exactly the same type of risks. The rash allocation of land for use by each new mode of transportation also entails opportunities for future urban development, and this has been demonstrated by past events as well as current local development projects, e.g. at freight yards. These sites became valuable to local and district developers, allowing the creation of new spaces to reside, work and live within existing settlements. This resource has now been exhausted in Europe's growing cities. New developments have either already been built or are being used for other functions. While Europe's rail networks were being scaled back, train companies' property portfolios were growing.

The findings of this study could contribute to Mumford's call being adeptly put into action. Given the existing urban development objectives of European cities, regions and municipalities, which aim to not only create high-quality urban spaces but also make a tangible contribution to (at last) meeting defined climate targets, valuable spaces for development could arise during the transport system transformation. To avoid further soil sealing (and subsequently induced traffic), brownfield development, understood as "increasing residential and commercial units within a zoned area that has been largely developed, while simultaneously improving public space and adapting infrastructures" (Grams 2017), will play a key role. However, redeveloping existing sites has long been an obstacle difficult to negotiate with traditional planning tools. Securing and redeveloping current stock for a wide range of uses, including the option not to build on a site at all, is something rarely seen in today's planning practice.

It is to be assumed that the land use potentials examined in this article can never be fully realized during a systematic transformation of the transport system – from ownership right through to service provision – but sites could be partially and gradually repurposed over decades. Our study of the Vienna FUA and Liesing demonstrates that by employing a range of approaches adjusted to specific needs, existing development targets certainly could be met.

In realistic terms, local development currently takes place where the ownership structure allows, i.e. where a large owner, such as a rail company, has plans to develop connected segments of land. The planning mechanisms designed in the past will fall short in the years ahead. The findings of this analysis suggest that the age of the car, even as it reaches its potential end, could cause a final inversion of existing principles. Planners should prepare for land spread out diversely across settlement structures with an equally diverse ownership structure. However, throughout this study, it has become clear that these sites are predominantly found in residential areas, especially in smaller, rural urban areas, and that these spaces are mostly located close to town centres, especially in regional centres and rural areas. The scale and distribution of these spaces within the settlement structure, generally in car-friendly locations along main roads or motorways (as shown during our in-depth analysis), demonstrate the potential for a detailed accentuation in line with the understanding of brownfield development currently widely accepted among urban planners.

Transport service providers, but also local and regional mobility companies, could use parts of this land – at least that is what our analysis of existing sites suggests. Market leaders have set aside hundreds of millions of euros to this end (Hawkins 2018, Rapier/Wolverton 2019). However, the new automated mobility services model might, compared to individual transport, once again lead to a clustering of associated sites that could ultimately have a considerable impact on locally generated added value.

One of the impacts of previous studies that have focused on parking spaces and their relevance to connected and automated vehicles has been that the economic significance of the transformation to a connected, automated and service-oriented transport system long remained intangible (Clements/Kockelman 2017, Mitteregger et al. 2019). The disappearance of individual transport would lead to the vanishing of a key sector that generates added value at a local and regional level. If the sharing economy logic becomes increasingly established, it will result in precarious jobs that will almost certainly come under increasing pressure with the ongoing automation of the transport sector.

Rethinking streetscapes could be the key to driving a transformation that produces liveable spaces and helps to combat the climate crisis (Mitteregger et al. 2022). Liesing, in particular, a district that is intersected by one of the most important commuter axes for individual transport, Triester Straße, demonstrates areas of considerable potential. Clusters of car-associated businesses present some opportunities to shape urban space and these should not be ignored. If the transformation takes place as predicted, a serious error will have occurred – and Mumford would once again be proved right.

REFERENCES

Alessandrini, A., A. Campagna, P. Delle Site, F. Filippi and L. Persia 2015. "Automated vehicles and the rethinking of mobility and cities", in *Transportation Research Procedia* 5, 145–160.

Anair, D., J. Martin, M. C. de Moura and J. Goldman 2020. *Ride-Hailing's Climate Risks: Steering a Growing Industry Toward a Clean Transportation Future*. Cambridge, MA: Union of Concerned Scientists. https://tinyurl.com/tz5gwtt (15/4/2020).

Banham, R. 1971. *Los Angeles: The architecture of four ecologies*. Los Angeles: University of California Press.

Bauer, C. 2011. *7 Tools zur Innenentwicklung: Die Metron Dichtebox: der Siedlungsraum der Schweiz soll begrenzt und die Entwicklung nach Innen gelenkt werden*. Brugg: Metron. https://bit.ly/38sZRzj (4/5/2020).

Beiker, S. 2015. "Einfühungsszenarien für höhergradig automatisierte Straßenfahrzeuge", in *Autonomes Fahren. Technische, rechtliche und gesellschaftliche Aspekte*, ed. by M. Maurer, J. C. Gerdes, B. Lenz and H. Winner. Berlin/Heidelberg: Springer Vieweg, 197–218.

Bellet, C. 2009. "The introduction of the high speed rail and urban restructuring: the case of Spain", in *City Futures* 9, 4–6.

Bertolini, L. 2017. *Planning the Mobile Metropolis: Transport for People, Places and the Planet*. London: Palgrave/Red Grove Press.

Bertolini, L., C. Curtis and J. Renne 2012. "Station area projects in Europe and beyond: Towards transit oriented development?", in *Built Environment* (38) 1, 31–50.

Bertolini, L., and T. Spit 2005. *Cities on rails: The redevelopment of railway stations and their surroundings*. London: Routledge.

Bormann, R., P. Fink, H. Holzapfel, S. Rammler, T. Sauter-Servaes, H. Tiemann, T. Waschke and B. Weirauch 2018. "Die Zukunft der deutschen Automobilindustrie: Transformation by Disaster oder by Design?", WISO Diskurs 03/2018. Bonn: Friedrich-Ebert-Stiftung.

Bourdieu, P. 1987. *Die feinen Unterschiede*. Frankfurt am Main: Suhrkamp.

Braham, W. W., and J. A. Hale 2007. *Rethinking Technology: A Reader in Architectural Theory*. London: Routledge.

Bruinsma, F., E. Pels, H. Priemus, P. Rietveld and B. van Wee 2008. *Railway development: Impact on urban dynamics*. Amsterdam: Physica.

Burghart, W., and G. Hertenberger 2018. *Österreichs gefährdetes Kulturerbe: Vom Umgang mit dem Denkmalschutz: 70 Fallbeispiele*. Vienna: Schreybgasse.

Chan, N. D., and S. A. Shaheen 2012. "Ridesharing in North America: Past, present, and future", in *Transport Reviews* (32) 1, 93–112.

Charpentier, M. 2020. "Lyft Unveils 'One-Stop-Shop' Service Center For Drivers In Austin", KUT 90.5. Austin's NPR Station, 14/1/2020. https://tinyurl.com/qs2xrp2 (4/5/2020).

Chaum, M. 2019. "Uber Video Keynote", talk at the Swiss Mobility Arena, 16/9/2019.

Clements, L. M., and K. M. Kockelman 2017. "Economic Effects of Automated Vehicles", in *Transportation Research Record* (2606) 1, 106–114.

Cohen, M. A., N. Agrawal and V. Agrawal 2006. "Winning in the aftermarket", in *Harvard Business Review* (84) 5, 129–146.

Cresswell, T. 2006. *On the Move: Mobility in the Modern Western World*. New York: Routledge.

Eckl-Dorna, W. 2020. "So treibt der Dieselskandal Autohäuser in den Ruin", Der Spiegel, 20/2/2020. https://tinyurl.com/qvpk9fw (15/4/2020).

Edelson, Z. 2017. "Port Authority Bus Terminal to get total reset and other breaking news from annual RPA conference", *The Architect's Newspaper*, 21/4/2017. https://tinyurl.com/sx6dvks (15/4/2020).

Epting, S. 2019. "Automated vehicles and transportation justice", in *Philosophy & Technology* (32) 3, 389–403.

Frangoul, A. 2018. "How remote control centers are changing the way mining operations are carried out", *CNBC*, 4/7/2018. https://tinyurl.com/r2luk4m (15/4/2020).

Freeman, C., and C. Perez 1998. *Structural crises of adjustment: business cycles and technical change and economic theory*. London: Pinter.

Geels, F. W., and J. Schot 2010. "The dynamics of transitions: a socio-technical perspective", in *Transitions to Sustainable Development: New Directions in the Study of Long Term Transformative Change*, ed. by J. Grin, J. Rotmans and J. Schot. London: Routledge, 11–104.

Görgl, P., J. Eder, E. Gruber and H. Fassmann 2017. "Monitoring der Siedlungsentwicklung in der Stadtregion+. Strategien zur räumlichen Entwicklung der Ostregion", study commissioned by Planungsgemeinschaft Ost. https://tinyurl.com/trqqwdl (16/4/2020).

Grams, A. 2017. "Spielräume für Dichte. Der Innenentwicklungskompass als problemorientierte Methode für Verdichtung in kleinen und mittleren Gemeinden", IRL-Bericht 8. Zurich: vdf Hochschulverlag.

Greenblatt, J. B., and S. Shaheen 2015. "Automated Vehicles, On-Demand Mobility, and Environmental Impacts", in *Current Sustainable/Renewable Energy Reports* (2) 3, 74–81.

Grübler, A. 1990. *The rise and fall of infrastructures*. Heidelberg: Physica.

Grübler, A. 1992. "Technology and Global Change: Land Use, Past and Present", *IIASA Working Paper WP-92-002*, Laxenburg.

Hagemann, I. 2016. "Das gegenhegemoniale Moment der Demokratie. Gegenhegemoniale Projekte und demokratische Demokratie am Fallbeispiel der grünen Bewegung". University of Duisburg-Essen.

Hall, P. G., and K. Pain (eds.) 2006. *The polycentric metropolis: learning from mega-city regions in Europe*. London: Routledge.

Hawkings, A. J. 2018. "Lyft will spend $100 million on new driver support centers", *The Verge*, 23/5/2018. https://tinyurl.com/ycb94tw2 (16/4/2020).

Headrick, D. R. 1994. "Technological Change", in *The Earth as Transformed by Human Action*, ed. by R. Turner. Cambridge: Cambridge University Press, 55–68.

Heinrichs, D. 2016. "Autonomous driving and urban land use", in *Autonomous Driving. Technische, rechtliche und gesellschaftliche Aspekte*, ed. by M. Maurer, J. C. Gerdes, B. Lenz and H. Winner. Berlin/Heidelberg: Springer Vieweg, 213–231.

Heller, N. 2017. "Is the Gig Economy Working?", *The New Yorker*, 15/5/2017. https://tinyurl.com/utlvmct (15/4/2020).

Hu, W. 2017. "As Uber Woos More Drivers, Taxis Hit Back", *The New York Times*, 28/3/2017. https://tinyurl.com/vv8cqcm (15/4/2020).

Jolliet, O., and P. Crettaz 1996. "Critical surface-time 95. A life cycle impact assessment methodology including fate and exposure", EPFL – École polytechnique fédérale de Lausanne, 1996.

Kucharczyk, S. 2017. "How will maintenance change with the autonomous vehicle?", *ReadWrite*, 18/4/2017. https://tinyurl.com/td6s2hf (15/4/2020).

Kuhn, T. S. 1967. *The structure of scientific revolutions*. Chicago: The University of Chicago Press.

Leberstein, S. 2016. "Uber's car leasing program turns its drivers into modern-day sharecroppers", *Quartz*, 6/6/2016. https://tinyurl.com/yx7oecf5 (16/4/2020).

Leiner, C., and T. Adler 2019. "Transportation Network Companies (TNCs): Impacts to Airport Revenues and Operations – Reference Guide", *ACRP Research Report* 215.

Lenz, B., and E. Fraedrich 2015. "Neue Mobilitätskonzepte und autonomes Fahren: Potenziale der Veränderung", in *Autonomes Fahren. Technische, rechtliche und gesellschaftliche Aspekte*, ed. by M. Maurer, J. C. Gerdes, B. Lenz and H. Winner. Berlin/Heidelberg: Springer Vieweg, 175–195.

Lottes, G. J. 2020. "Der Wohnungsmarkt in Wien. Ein Marktbericht von Raiffeisen", Raiffeisenlandesbank Niederösterreich-Wien. https://tinyurl.com/tn55auo (16/4/2020).

Manderscheid, K. 2014. "The Movement Problem, the Car and Future Mobility Regimes: Automobility as Dispositif and Mode of Regulation", in *Mobilities* (9) 4, 604–626.

Marshall, S. 2005. *Street & Patterns*. London: Spon Press.

McKinsey 2018. "Ready for inspection – The Automotive Aftermarket in 2030". https://tinyurl.com/ss-b7ksk (15/4/2020).

McLuhan, M. 2003. *Understanding Media: The extension of man*. Corte Madera: Ginko Press.

Mitteregger, M., A. Soteropoulos, J. Bröthaler and F. Dorner 2019. "Shared, Automated, Electric: the Fiscal Effects of the 'Holy Trinity'", *Proceedings of the 24. REAL CORP*, International Conference on Urban Planning, Regional Development and Information Society. Karlsruhe.

Mitteregger, M., E. M. Bruck, A. Soteropoulos, A. Stickler, M. Berger, J. S. Dangschat, R. Scheuvens and I. Banerjee 2022. *AVENUE21. Connected and Automated Driving: Prospects for Urban Europe*, trans. M. Slater and N. Raafat. Berlin: Springer Vieweg. DOI: 10.1007/978-3-662-64140-8.

Mumford, L. 1968. "The Highway and the City", in *The Urban Prospect,* ed. by L. Mumford. New York: Harcourt, Brace & World, 92–107.

Municipal Department 21 (District Planning and Land Use), City of Vienna [Magistratsabteilung 21] 2015. "Perspektive Liesing. Ein Entwicklungskonzept für einen Stadtteil im Wachsen", *Projektzeitung #2*, 2/1/2015. https://tinyurl.com/tnjtguc (16/4/2020).

OECD 2020. "Functional urban areas by country". https://tinyurl.com/tawbkmz (15/4/2020).

OpenStreetMap Wiki 2020. "DE:Key:landuse". https://tinyurl.com/vo6prk9 (16/4/2020).

Ongweso Jr., E., and J. Koebler 2019. "Uber Office Has Nice Port-a-Potties for 'Employees Only,' Inferior Ones for Drivers", *VICE*, 5/12/2019. https://tinyurl.com/ufc2jxx (15/4/2020).

Pevsner, N. 1953. *The Buildings of England: Derbyshire*. Harmondsworth: Penguin.

Ramm, F. 2019. "OpenStreetMap Data in Layered GIS Format". https://tinyurl.com/yyukslna (16/4/2020).

Rapier, G., and T. Wolverton 2019. "Uber lost $5.2 billion in 3 months. Here's where all that money went", *Business Insider* 9/8/2019. https://tinyurl.com/yxxsm2qf (16/4/2020).

Rogers, E. M. 2003. *Diffusion of innovations*. New York: Free Press.

Schmitz, S. 2001. *Revolutionen der Erreichbarkeit. Gesellschaft, Raum und Verkehr im Wandel* (Stadtforschung aktuell, vol. 83). Wiesbaden: VS Verlag für Sozialwissenschaften.

SFCTA (San Francisco County Transportation Authority) 2018. "TNCs & Congestion. Final Report." https://tinyurl.com/y6w76ta6 (15/4/2020).

Shaheen, S., and A. Cohen 2019. "Shared ride services in North America: definitions, impacts, and the future of pooling", in *Transport Reviews* (39) 4, 427–442.

Sheikh, N. 2018. "Applying a Hybrid Network Approach to Deployment of Self-Driving Mobility Services", Automated Vehicle Symposium. https://tinyurl.com/uhwtb3k (15/4/2020).

Sieverts, T. 1998. *Zwischenstadt: Zwischen Ort und Welt, Raum und Zeit, Stadt und Land*, Bauwelt Fundamente 118. Wiesbaden: Vieweg+Teubner.

Sommer, C. 2018. "Neue Angebote für den ländlichen Raum", lecture at the "ZENTRALITÄTEN 4.0 – Mittelzentren im Zeitalter der Digitalisierung" symposium, 22/11/2018, Kassel.

Soteropoulos, A., M. Berger and F. Ciari 2019. "Impacts of automated vehicles in travel behaviour and land use: An international review of modelling studies", in *Transport Reviews* (39) 1, 29–49.

Soteropoulos, A., M. Mitteregger, M. Berger and J. Zwirchmayr 2020. "Automated drivability: Toward an assessment of the spatial deployment of level 4 automated vehicles", in *Transportation Research Part A: Policy and Practice* 136, 64–84.

Spielmann, M., and R. Scholz 2005. "Life Cycle Inventories of Transport Services: Background Data for Freight Transport", in *The International Journal of Life Cycle Assessment* (10) 1, 85–94.

Splend 2019. "We have a new Member Support Centre in London", press release, 17/4/2019. https://tinyurl.com/s3uwy97 (15/4/2020).

Statistik Austria 2017. "Urban-Rural-Typologie von Statistik Austria". https://tinyurl.com/rs5qj9s (16/4/2020).

Stead, D., and B. Vaddadi 2019. "Automated vehicles and how they may affect urban form: A review of recent scenario studies", in *Cities* 92, 125–133.

Steen, B., and S.-O. Ryding 1993. *The EPS enviro-accounting method. An application of environmental accounting for evaluation and valuation of environmental impact in product design.* APR: Stockholm.

STRIA 2019. *Roadmap on Connected and Automated Transport: Road, Rail and Waterborne.* Brussels: European Commission.

Tucker, C. 2019. "Uber-rival Bolt opens new hub in Chiswick, London, surging to 1.5 million London passengers in 6 months", *EU-Startups*, 25/12/2019. https://tinyurl.com/sbay9ul (15/4/2020).

Vlasic, B. & Isaac, M. 2016. "Uber Aims for an Edge in the Race for a Self-Driving Future", *The New York Times*, 18/8/2016. https://tinyurl.com/w928p8a (16/4/2020).

Yankelevich, A., R. V. Rikard, T. Kadylak, M. J. Hall, E. A. Mack, J. P. Verboncoeur and S. R. Cotton 2018. "Preparing the Workforce for Automated Vehicles". https://bit.ly/3qrzXC8 (15/4/2020).

Zwick, A. 2018. "Welcome to the Gig Economy: neoliberal industrial relations and the case of Uber", in *GeoJournal* (83) 4, 679–691.

PART IV
Governance

Jens S. Dangschat, Ian Banerjee, Andrea Stickler

The broad technological transformation and accompanying connection and automation of transport (CAT) present a major challenge to society and have the potential to trigger profound social change ("second industrial revolution"). How should politics and planning administrations, companies and civil society respond?

The new technologies will fundamentally change the labour market, development in city regions, transport and communication in everyday life. In turn, this will give rise to considerable opportunities for economic growth and competition as well as for improving the safety and efficiency of road transport. However, economic and political shifts on the macro level are also expected. A wide range of development dynamics will become apparent on the regional/local level, which will lead to new and deepened forms of social inequality. Which developments are desirable and which are not? How are the advantages and disadvantages of these technologically defined developments distributed and how can this imbalance be kept to a minimum? For example, can the introduction of CAT help to reduce the urban-rural divide?

Social science studies on digitalization in general and CAT in particular overwhelmingly come to the conclusion that the implementation of new socio-technical systems require an altered form of governance in order to not only avoid unwanted side effects but also support a necessary socio-ecological transformation. In this context, the setting of an adequate legislative and regulatory framework is just as necessary as a negotiation process between diverging interests, which can vary dramatically from place to place. Do innovative approaches to governance already exist?

Digitalization is completely redefining this area because it is not only new actors (the global IT industry, companies in the platform economy, start-ups, mobility providers, etc.) who need to be involved. As communication media are in a constant state of flux, steering access to these mostly global companies and their underlying business models will be challenging for local/regional or national policy and planning institutions.

In the chapter *New governance concepts for digitalization: Challenges and potentials*, Alexander Hamedinger discusses new governance concepts, which he applies to CAT for the first time. In this context, governance is not exclusively positive; rather, it depends which actors are involved in which constellation in which negotiation and decision-making processes. In Hamedinger's view, that new forms of governance are required is primarily due to four areas in which transformations have been taking place since the 1990s: in society itself, in the political administrative system, in the climate and above all in the field of technology, which has an accelerating impact on transformations in all other areas. Furthermore, with "reflexive governance", "tentative governance" and "regulative liberalism", the author presents three forms of steering and coordination that have been developed as a possible means to react to these new flexible challenges.

In the second article in this section of the book, entitled *How are automated vehicles steering spatial development in Switzerland?*, Fabienne Perret and Christof Abegg describe the impact of CAT on spatial development in Switzerland. They focus in particular on traditional and innovative models of public transport. Underpinning their findings are two comprehensive studies that address on the one hand the possible impacts and resulting courses of action for business and general policy and planning, and on the other the necessary conditions in terms of political, legal, economic, technological, societal and ethical aspects. Scenarios with different combinations of economic and political influence are developed and general large-scale impacts on Swiss settlement structure as well as detailed impacts on cities and agglomerations are discussed. Finally, a possible and necessary regulatory framework for the various levels of government, with particular importance attached to cities and agglomerations, is outlined and mooted. The article ends with a plea to cities and agglomerations "not to lose control of the steering of mobility and spatial development".

In her chapter *Learning from local transport revolution projects for connected and automated transport: New mobility services in Lower Austria from the perspective of discourse and hegemonic stability theory*, Andrea Stickler analyses new mobility offers in rural areas in Lower Austria and the extent to which they can not only contribute to the mobility revolution but also call the "automobility system" into question. With an analytical approach based on discourse and hegemonic stability theory, she considers offers of station-based electric car sharing, non-profit transport services and ring-and-ride taxis as possible use cases for new mobility systems. The author comes to the conclusion that the contribution new mobility systems are able to make to the transport revolution is currently rather limited due to conflicting intentions, uses and applications. In her final analysis, she emphasizes that the implementation of CAT must be explicitly and effectively contextualized within the transport revolution, which will be contested and requires a broader change in values in which ultimately automobility itself must also be questioned.

In his article *Technological transformations, radical social change and the challenge for governance in connection with the development of connected and automated transport*, Jens S. Dangschat interprets the development of CAT as a profound change in a socio-technical system. To this end he puts social development and the steering of policy and planning in the context of the second modernity with its more flexible structures, networks and plural developments. Subsequently, the author classifies the development of CAT from the perspective of sociotechnology before moving on to discuss the policies and steering of the technological transformation. However, the focus here is not on local/regional strategies but on the narratives of automation and unambiguity of the technological developments in the context of the dominant dispositif of automobility. To analyse the social impact of CAT, Dangschat proposes turning our attention to the narratives' political horizontal and vertical transfers with which "realities" and a lack of alternatives become established.

In the final chapter, *Data-driven urbanism, digital platforms and the planning of MaaS in times of deep uncertainty: What does it mean for CAVs?*, Ian Banerjee, Peraphan Jittrapirom and Jens S. Dangschat discuss the technological, economic and policy and planning context of CAT: data-driven urbanism, digital platforms and multiple mobility services (Mobility as a Service – MaaS) as well as their steering in a period of great uncertainty. In the context of the smart city debates, the technological and epistemological aspects of this computer-assisted development are discussed in terms of the handling of data. With the platform economy, new business models – often the brainchild of newcomers to the industry – are entering the arena whose motivations and objectives are materially different. When it comes to business and policy decisions, the exploration of MaaS is fraught with uncertainty. Finally, the three aspects of sociotechnological transformation are used to make steering proposals within the framework of new forms of governance in view of future CAT platforms.

16 New governance concepts for digitalization: Challenges and potentials

Alexander Hamedinger

Alexander Hamedinger
TU Wien, Research Unit Sociology (ISRA)

© The Author(s) 2023
M. Mitteregger et al. (eds.), *AVENUE21. Planning and Policy Considerations for an Age of Automated Mobility*, https://doi.org/10.1007/978-3-662-67004-0_16

1. INTRODUCTION

The term "governance" is generally understood to mean the steering and coordination of actions and its broadest definition encompasses various forms and their processes (cf. Gailing/Hamedinger 2019). They range from official, i.e. state-run, top-down steering and coordination to forms of self-steering without state actors (Mayntz/Scharpf 1995, Blatter 2007). This understanding of governance is grounded in social theory and differs from a narrower understanding that emphasizes the shift from state forms of steering and coordination. This shift, it is argued, is part of the transition from regimes of accumulation with increasing flexibilization. As part of the regulatory system, the state becomes modernized ("administrative modernization") and more entrepreneurial (Harvey 1989). At the same time, it becomes open to the involvement of non-state actors in political decision-making outside of the "classic" paths of liberal democracies. Public-private partnerships, networks, but also participatory processes in the sense of citizen participation are expressions of this transition from government to governance. Political and administrative actors expect that governance will bring about an improvement in the effectiveness and legitimation of steering and coordination.

However, advocates of critical planning theory and urbanism (including Yiftachel/Huxley 2000, Purcell 2009) object to the euphoria surrounding governance that can be observed in European urban development since the 1980s, saying that, among other things, participatory processes as forms of governance do not in fact lead to the anticipated democratization of society but in contrast can foster the reproduction of social inequality. They maintain that governance actually furthers the neoliberalization (deregulation and commodification policy) of politics, the economy and society. A similar interpretation is contributed to the debate by Swyngedouw (2013) when he speaks of the political order in the capitalist city today, which at bottom is post-political because the mitigation of deep-seated social conflicts is neglected and the possibilities for strong economic actors to influence policy are increased.

Furthermore, those who take this critical line point to the dominance of "technocratic elites" in recent forms of governance (e.g. in "policy networks") as well as the lack of transparency and democratic legitimation of steering and governing by networks or public-private partnerships (cf. Jessop 2000, Zürn 2009, Hamedinger 2013). In summary, therefore, what is criticized about governance – understood as a new form of steering and coordination in which actors from various social spheres participate – is the partial lack of a legitimate foundation. In addition, it is alleged that such forms of governance support neoliberalization and reproduce social inequality. Of course, empirical research would need to be conducted on such accusations for all forms of steering and coordination, including purely state-run forms, as well as for more recent governance concepts.

From today's perspective, this debate surrounding government and governance, deregulation and re-regulation, commodification and recommunalization, and the opening and closing of regulatory systems (Reckwitz 2019) is more topical than ever. This is related to the context of governance, which has changed fundamentally in Europe since the 1980s. Four aspects should be highlighted here:

1. Climate change, which demands an about-turn regarding both the aims of spatial, social and economic development ("degrowth") and the forms of governance ("new municipalism"). Just how difficult this is, is shown by Andrea Stickler (see Chap. 18 in this volume) using the example of policies related to the mobility transformation.

2. Social change, especially growing socio-economic inequalities that are the product of, among other things, structural effects on the labour market and changes in welfare state systems. In light of this, forms of steering and coordination are required that will promote and stabilize social cohesion and not reproduce social inequality.

3. The change in political-administrative systems and liberal democracies. For a long time the latter have been facing the challenge of inclusion given the increasingly heterogeneous nature of our societies. It is more and more challenging to reconcile such varied interests and represent this diversity. For instance, the proportion of the population that does not have the right to vote is rising rapidly in cities (in 2019 this affected 28% of the population of voting age in Vienna according to Verlic/Hammer 2019), and with it their inclusion in such systems is growing more difficult.

4. Technological change, which requires targeted governance in order to preclude any negative effects on the other named areas (the environment, society, democracy; see Chap. 19 by Dangschat in this volume). At the same time, digitalization is changing the nature of governance itself. Banerjee et al. (Chap. 20 in this volume) speak of "digital modes of governance" that would be possible as a result of large quantities of data, and claim that the regulation of data streams (their production and use) additionally requires a comprehensive and inclusive approach to governance.

What challenges do technological change – here primarily understood as digitalization – present for governance structures and processes? With which governance concepts should policymakers and administrators react to these challenges? These questions will be addressed briefly below.

2. THE CHALLENGES OF DIGITALIZATION

In principle, these challenges can be described in terms of the four dimensions that characterize governance structures and processes. The descriptions here are general as further detail about specific areas of digitalization (e.g. platforms like Airbnb or connected and automated transport, CAT) would require in-depth empirical analysis.

2.1 ACTORS, INSTITUTIONS AND ACTOR CONSTELLATIONS

Banerjee et al. (see Chap. 20 in this volume) speak of "commercial inflections" that need to be regulated as policymakers and administrators are increasingly dependent on privately operated platforms and the collection of data, which is often carried out by large data corporations (e.g. Microsoft, Google, Facebook). In their article Schulz and Dankert (2016) use the term "Governance by Things" to refer to the power shifts in the constellations of actors who had previously been pivotal to policymaking: they speak of a shift in power in favour of private companies that develop and implement codes and algorithms. On the city level precursors of the digitalization debate – which revolves around the role of technology in developing urban economies and societies – are often labelled with the term "smart city".

However, it becomes apparent at this point that the digital transformation entails a conflictual negotiation process between strong economic operators (primarily from the IT sector), municipal policymakers and administrators and, in some places, social movements (cf. Bauriedl/ Strüver 2018: 2) and that technology-centred and -deterministic perspectives are increasingly put under pressure to justify themselves. The breadth of today's spectrum of actors is a defining characteristic of "urban governance" (Cruz et al. 2019). Yet unlike the implications found in some critical discourses, municipal policymakers and administrators actively contribute to the coproduction of the smart city or are even the main drivers behind this development.

As such, government – the political administrative system as an actor in a smart, data-protected spatial development – once again takes centre stage of steering and coordination enterprises (see Chap. 19 by Dangschat in this volume). However, policymakers and administrators are often unable to contend with the rapid developments related to digitalization. Among other things, this concerns their knowledge of socio-technical conditions and the effects of digitalization (see Chap. 20 by Banerjee et al. in this volume), the legal possibilities to contain negative effects and the communication and decision-making processes within political administrative systems, which are inflexible and time-intensive for good reason. Furthermore, the vertical organizational structure of these systems impedes any rapid and interdepartmental reaction to new problems in urban development that may arise as a result of digitalization (e.g. in the case of Airbnb). Another challenge is touched on by Schulz and Dankert (2006): the development of codes and algorithms is based on explicit knowledge to solve problems. In contrast, political decision-making processes and our everyday routines are often based on tacit knowledge that cannot be verbalized. This form of knowledge is embedded in political cultures and social practices, with their respective social norms and values, and influences our behaviour. Using the example of automated driving, Schulz and Dankert (ibid.: 8) rightly state: "Test-drives with autonomous cars show that they are able to stick to traffic rules slavishly, but still cannot decide when to override rules reasonably [...]." Another associated aspect is the fact that political and planning cultures are bound to a specific time and place.

Finally, the question arises of who can in fact become an actor in the data-driven city. Banerjee et al. (Chap. 20) address the well-known problem of the digital divide, which refers to the unequal spatial distribution of digital infrastructures as well as the unequal competencies and economic resources of different social groups. Due to the speed of recent developments in digitalization, even "digital natives" could soon become "digital immigrants". Ultimately, they continue, the digital divide exacerbates social inequality and adds new dimensions to it. In summary, one gains the impression that the criticisms of governance voiced at the outset – power shifts in favour of private companies and technocratic elites, reproduction of social inequality – might be compounded by digitalization if there is no regulatory intervention.

2.2 LEGITIMATION

These changes in actor constellations are problematic when they lead to an aggravation of pre-existing legitimation crises in state systems and/or when the output of governance in the area of digitalization leads to negative developments (such as the aforementioned increase in social inequality). Schulz and Dankert (2006) do indeed see a legitimation problem, since unlike the development of laws, which are discussed politically and decided in the classical arenas of liberal democracies, there are no political debates when codes are developed and private companies do not have to prove their legitimacy politically.

2.3 INTERACTION PROCESSES

Digitalization could improve the processes of negotiating various actors' interests and of political decision-making (e.g. via interactive platforms, forms of e-democracy) and hence contribute to democratization. Cruz et al. (2019: 11) emphasize that digital technologies have changed the way in which cities are governed (e.g. via "city labs"). However, according to Blühdorn and Kalke (2020) the democratization process is often undermined and not strengthened by such technologies: the digital revolution is shifting the negotiation of conflicting and contradictory interests into "narrow discursive spaces" (ibid.: 11); a genuine discourse concerning "objective checks" does not occur in digital forums: "In a very similar way, digital filter bubbles and echo chambers also reduce complexity and block or prohibit thinking about certain topics, further fragmenting and polarizing society and suffocating democratic discourse" (ibid.: 11). The digital revolution or, taken one step further, the data-driven city incapacitates responsible citizens.

2.4   GOVERNANCE OUTPUT

Schulz and Dankert (2006: 2) explain their notion of a "Governance by Things" thus: "we consider that with the IoT [Internet of Things] the 'code is law' paradigm [...] might enter the physical world with all its consequences. We call this the 'Governance by Things'". In other words: it is claimed that the codes that define the way hard- and software function have a regulatory impact on concrete human behaviour (in addition to other regulatory forms like laws or social norms) in the physical world. Schulz and Dankert (ibid.) immediately point to the obvious challenge that goes hand in hand with this idea: governments are faced with the problem of how to influence the development of codes by private companies such that they correspond to legitimized political aims (e.g. climate protection, social cohesion, the common good). The question is: at what point do codes become a public issue and hence need to be publicly discussed? The demand made by Schulz and Dankert (ibid.: 12) is that the "Governance by Things" itself needs governance in order to "socialize" the codes.

3.   SEARCHING FOR NEW GOVERNANCE CONCEPTS

What governance concepts would be appropriate tools with which to respond to these serious challenges? Two approaches are discussed below: "reflexive governance" and "tentative governance".

3.1   REFLEXIVE GOVERNANCE

The considerations of Voß and Kemp (2006) concerning steering theory begin with the concept of sustainability and the question of which form of governance a sustainable transformation of the economy and society would require. For Voß and Kemp governance is an interaction process in which various actors with different – in many cases conflicting – interests meet. The result of such interaction processes includes specific policies or regulatory arrangements, in sum a specific kind of governance. The processes of societal problem-solving described by the term governance encompass collectively analysing and defining problems, setting targets and evaluating solutions as well as coordinating action strategies.

The authors argue that sustainability necessitates new ways for society to solve problems due to its cross-boundary character (spanning problems in society, the economy and environment; spanning steering sectors; spanning actors and institutions; spanning territories). They propose that reflexive governance is an appropriate way to react to these complex steering and coordination challenges. Reflexivity refers to the various forms of governance that are applied to solve certain social problems: "Reflexive governance thus implies that one calls into question the foundations of governance itself, that is, the concepts, practices and institutions by which societal development is governed, and that one envisions alternatives and reinvents and shapes those foundations" (ibid.: 4). Central to this is reflection on the cycle of problem creation and problem-solving, on the problems of steering and coordination themselves, their effects, associated uncertainties and ambivalences. At another point, this is also called "second-order governance" (ibid.: 7). Such reflection demands open and learning-oriented interaction processes as are already characteristic of transdisciplinary research or participative forms of decision-making.

This complexity of sustainability problems is related to the following:

1. Heterogeneity of elements in the socio-ecological transformation that exposes the limits of specialized expert knowledge, calls into question classic forms of knowledge production and requires newer forms as applied in transdisciplinary research.

2. Uncertainties associated with the impossibility of predicting with certainty the effects and dynamics of complex socio-ecological transformations. This calls for adaptive strategies and structures in order to learn from mistakes.

3. Path dependency: future developments are influenced by structures that have grown over time (e.g. value structures, institutional structures, problem-solving routines), which can only be changed slowly and with difficulty. However, it is not possible to perfectly predict development paths. With the aid of, among other things, scenario techniques or the participative modelling of policies, some alternative future developments could be anticipated and lock-ins avoided. To this is added the fact that the concept of sustainability cannot be objectively defined by science, but rather is always subject to the value judgements of various actors (from science, politics, planning, civil society, business). The targets of sustainability are therefore "moving". The consequence for Voß and Kemp (ibid: 15f.) is: "Sustainability is thus an ambiguous and moving target that can only be ascertained and followed through processes of iterative, participatory goal formulation." Ultimately, when implementing strategies for a sustainable transformation, various actors would have to be coordinated and allocated to various spatial levels (local, regional, etc.), among other things. In sum, Voß and Kemp (ibid.) identify demands on reflexive governance strategies: "integrated (transdisciplinary) knowledge production", "adaptivity of strategies and institutions", "anticipation of the long-term systemic effects of action strategies", "iterative, participatory goal formulation", "interactive strategy development".

This concept of reflexive governance has some potential in its approach to the above-mentioned challenges posed by digitalization:

- In such open, learning-oriented and inclusive interaction processes, "Governance by Things" could be widely discussed. This would involve questions regarding the specific aims of such governance and the search for the best form of governance for digitalization itself (i.e. where is state-led steering, co-governance, self-regulation or a hybrid needed?). Digitalization requires reflexive governance so that the aforementioned negative effects on society (social inequality), democracy (disenfranchisement) and the environment do not materialize.

- Knowledge about the conditions and effects of digitalization is distributed across various groups of actors, primarily from business but also in civil society (NGOs), politics and administration. A transdisciplinary production of knowledge that systematically integrates these various forms of knowledge and operating logics is hence a fundamental prerequisite for targeted steering and coordination. To develop strategies for dealing with digitalization, the keywords are therefore: interactive and inclusive. Any power imbalance would need to be deliberately addressed and redressed in the processes themselves.

- There is another characteristic of digitalization that speaks for a reflexive governance strategy: the speed of technological developments and the complexity of these developments' impacts (ignorance of the unintended consequences of digitalization and unpredictability regarding the context; cf. Chap. 19 by Dangschat in this volume). Adaptive learning strategies and the participative development of scenarios are undoubtedly suitable means to deal with these challenges. Institutional structures would need to become more adaptive and regular evaluations and monitoring of the effects of digitalization would have to form the basis of learning processes.

3.2 TENTATIVE GOVERNANCE

For Kuhlmann et al. (2019), the starting point for this governance concept are the uncertainties and unexpected risks that might arise as a result of emerging developments in science and technology. Furthermore, they are matters in which the power to influence governance solutions is shared among various actors. In order to deal with these context-specific conditions, Kuhlmann et al. (ibid.) consider a flexible and deliberative governance approach to be necessary that constitutes a good middle ground between providing the necessary stability on the one hand and flexibility on the other, i.e. "tentative governance": "We consider governance to be 'tentative' when it is designed, practiced, exercised or evolves as a dynamic process to manage interdependencies and contingencies in a non-finalizing way; it is prudent (e.g. involving trial and error, or learning processes in general) and preliminary (e.g. temporally limited) rather than assertive and persistent" (ibid.: 3). It is an extremely process-oriented approach to governance that is open, learning-oriented and adaptive and allows experimentation with various solutions to problems. Dangschat (Chap. 19 in this volume) calls for just such a version of governance when it comes to dealing with CAT.

According to Kuhlmann et al. (2019), governance encompasses various kinds of governing in various contexts, though they emphasize that hierarchical steering does not work as well now as it did in the past: coordination between various actors has become more important than steering via direct control mechanisms. However, both are needed, they say: tentative governance as an open and flexible approach to steering and coordination and "definitive forms of governance" (ibid.: 3) that are more stable and involve strong regulatory interventions by the state in order to make the achievement of certain steering goals possible. In addition, the authors differentiate between "intentional" and "incidental" effects of governance. A definitive form of governance could become tentative governance if unforeseeable adaptations become necessary in the steering process. Tentative governance can mostly be found in contexts in which "soft" (such as benchmarking) and "hard" forms of regulation coexist (the authors call these "hybrid arrangements"). Tentative governance is often tied to the "shadows of hierarchy", i.e. associated with hard forms of steering (such as legislation). The authors admit that tentative governance overlaps with other forms of governance already named in the literature, such as with reflexive and adaptive governance. However, Kuhlmann et al. (ibid.) claim that their concept of tentative governance is broader than the approach of reflexive governance since there is a reflection on both the cognitive foundations of governance itself and on the way it functions. They consider the majority of commonalities to exist with "experimentalist governance", which is based on flexibility, openness, learning and reversibility. An advantage of their concept of tentative governance resides in the possibility to apply it in various spatial, temporal and social contexts. Kuhlmann et al. view it as a heuristic approach, eventually revealing various ideal processes of production or change in the social order.

The concept of tentative governance also bears some potential regarding digitalization:

- As the authors themselves emphasize, the advantage of tentative governance consists in making clear that in the context of innovation and technological development, governance has to be conceptualized such that unrealistic hopes for steering are avoided. Uncertainties, complexity and dynamics require among other things such forms of governance.

- Even more than in the concept of reflexive governance, it is stressed that strongly process-oriented and open forms of governance need to be tied to other, stronger regulatory and hierarchical forms of steering ("shadow of hierarchy"). When it comes to dealing with the conditions and consequences of digitalization, this seems essential bearing in mind that some developments in the field of technology and innovation (e.g. platform urbanism) require a regulatory framework. A combination of hard and soft steering and coordination is possible with this concept.

- Tentative governance can be used in various contexts as a research strategy. The concept is helpful for identifying the context-specific challenges of digitalization (such as in different types of city).

4.   SUMMARY

At bottom, more empirical research on the correlation between (urban) governance and digitalization is necessary: "Advances in technology, the cost reduction of specialized hardware, and the open source and open data movements are redrafting the rules of the game for public services, community engagement, and urban entrepreneurship. [...] However, our understanding of the implications of these changes is still meager" (Cruz et al. 2019: 11). There is still insufficient empirical research on the impact that digitalization is having in various areas of urban development (mobility, energy, etc.) on governance systems that have evolved over time and how a governance of the governance of digitalization (in essence reflexive governance) can be established. For reflexive governance, however, an open discussion is imperative between "technology enthusiasts" who only see the advantages of the data-driven city and "technology critics" who (exclusively) fear an exacerbation of social inequalities and an erosion of democracy (cf. ibid.). Moreover, in various areas of urban development, digitalization must take into consideration the complexity and heterogeneity of cities. To achieve this, forms of governance are needed that are adapted to the respective context, as is required by the concept of tentative governance.

Reflexive and tentative governance therefore have some potential in terms of the governance of digitalization. However, there is a material difference in their starting points: whereas in the concept of reflexive governance the aim of steering and coordination is sustainability, tentative governance is about searching for new governance forms in the context of innovation and rapid technological change. The concept of tentative governance thus remains relatively neutral when it comes to the aim of steering. Kuhlmann et al. (2019) admit this themselves. From an analytical perspective this is understandable because it means it is possible to conceive of various – as they say – ideal types of producing social order. From the point of view of actors from politics, administration, business and civil society, however, this might be of little help as it would first oblige them to identify urban development goals and only in a second step would they be able to search for forms of governance that would enable them to realize these goals. Due to the multiple crises we currently face (cf. Brand 2014) – affecting the economy, society, the environment and our democracies – we undoubtedly need governance objectives that work towards socio-ecological transformation or sustainability.

A commonality and at the same time a criticism of both governance concepts is the way in which they largely neglect matters of legitimation. This is in effect an old accusation pointed at governance (see the introduction to this chapter), but it is not addressed in the concepts elaborated on here. That being said, questions concerning the legitimation of the learning- and process-oriented forms of governance stipulated by both concepts need to be asked because they would reveal which actors are excluded and who really has the right to play a part and have a voice in interactive strategy development and transdisciplinary knowledge production. At its heart, it is a matter of laying bare the democratic quality of such new governance concepts. Connecting this with classic arenas of democratic decision-making, as at least touched upon in the concept of tentative governance, will not suffice. New forms of legitimation must be found for both reflexive and tentative governance. Otherwise, democratic deficits and the legitimation crisis surrounding political administrative systems will be reproduced or even exacerbated.

Another critical point concerns the output of governance, above all the steering of various actors' behaviour via digitalization measures (such as via CAT in terms of mobility behaviour). Schulz and Dankert (2006) have already made clear the legal problems of steering via codes and algorithms, while Blühdorn and Kalke (2020) point to digitalization's negative effects on democratic politics. Matters of governance and democracy could and should be discussed using the example of digital "nudging", which means nudging behaviours in a desired direction by means of digital stimuli. Nudging in the area of consumer behaviour (e.g. energy consumption) raises questions regarding not only the digital surveillance of behaviour but also restrictions on rights to freedom. Is nudging a means of manipulation or does it increase the subject's freedom of choice, as Thaler and Sunstein (2019) claim? Have the aims of nudging been broadly and transparently negotiated in a democratic process? Are these aims visible to consumers? How are the private companies who conceive the selection design for behaviour control politically monitored?

Consequently, there are a number of questions to be answered both theoretically and empirically before blindly jumping on the bandwagon of governance concepts that seem good at first glance. But might there at least be a steering paradigm that could serve as a superordinate guiding principle for the governance of digitalization in future? Reckwitz (2019) outlines a possible middle course between the two dominant paradigms to date, the social corporatist paradigm and "apertist" liberalism, a regulatory and a dynamizing paradigm. The entanglement of politics and economics so characteristic of apertist liberalism (i.e. in the form of public-private partnerships), deregulatory measures and privatization as well as the increasing diversity and number of steering actors are surely the essence of the transition from government to governance. The negative effects of this liberalism are considerable; according to Reckwitz (cf. ibid.: 271ff.) the lack of social and state framing of markets has led among other things to an exacerbation of social inequalities, and a power shift in favour of international organizations, the legal system and tacitly strong economic actors has accelerated the erosion of democracy.

Another driver behind the loss of confidence in liberal democracies is digital media, Reckwitz continues, which shape public opinion and decision-making and with which traditional institutions are unable to keep pace. The author describes a new steering paradigm that is intended to supersede this neoliberalism: "regulative or embedding liberalism" (ibid.: 285). This involves adopting regulations in the interests of creating social order, reacts in a regulatory capacity to new social and cultural issues (such as with state regulation of markets and to protect individual rights) and retains a liberal foundation that uses the dynamics of markets and globalization and embeds them in a new framework. For governance, our understanding of the role of the state is significant: "In regulatory liberalism, the state is once again a more active state – though in a new incarnation that differs markedly from the steering national state of social corporatism [...]" (ibid.: 290). The description continues: "From the perspective of liberal governance, however, societal processes can only ever be steered indirectly, with the aid of incentives and deterrents, and it must be anticipated that these actions may have unintended consequences" (ibid.: 292).

The steering paradigm of regulative liberalism proposed by Reckwitz could prove well suited to the governance of digitalization as the negative effects of digitalization could be dealt with by means of social and ecological regulation without undermining entrepreneurial innovations and market dynamics, which are important for technological developments. However, if it is intended that steering and coordination should mainly function by means of incentives, then the same democratic reservations apply as with nudging. This is something that would need to be widely discussed by all relevant actors in the search for a new steering paradigm in the area of digitalization.

REFERENCES

Bauriedl, S., and A. Strüver (eds.) 2018. *Smart City – kritische Perspektiven auf die Digitalisierung in Städten*. Bielefeld: transcript.

Blatter, J. 2007. *Governance – theoretische Formen und historische Transformationen*. Baden-Baden: Nomos.

Blühdorn, I., and K. Kalke. 2020. "Entgrenzte Freiheit. Demokratisierung im ökologischen Notstand?", IGN Interventions, Feb. 2020, ed. by Institut für Gesellschaftswandel und Nachhaltigkeit (IGN), WU Wien. www.wu.ac.at/fileadmin/wu/d/i/ign/IGN_Interventions_Feb_2020.pdf (31/6/2020).

Brand, U. 2014. "Sozial-ökologische Transformation", in *Kurswechsel* 2/2014, 7–18.

Cruz, N. F. da, P. Rode and M. McQuarrie. 2019. "New Urban Governance: A Review of Current Themes and Future Priorities", in *Journal of Urban Affairs* (41) 1, 1–19.

Gailing, L., and A. Hamedinger 2019. "Neoinstitutionalismus und Governance", in *ARL Reader Planungstheorie*, ed. by T. Wiechmann, vol. 1. Berlin: Springer, 167–178.

Hamedinger, A. 2013. "Governance, Raum und soziale Kohäsion – Aspekte einer sozial kohäsiven stadtregionalen Governance", unpublished habilitation thesis, TU Wien.

Harvey, D. 1989. "From Managerialism to Entrepreneurialism: Formation of Urban Governance in Late Capitalism", in *Geografisker Annaler* 71 (B), 3–17.

Jessop, B. 2000. "Governance Failure", in *The New Politics of British Local Governance Stoker*, ed. by G. Stoker. London: MacMillan Press, 11–32.

Kuhlmann, S., P. Stegmaier and K. Konrad 2019: "The tentative governance of emerging science and technology: A conceptual introduction", in *Research Policy* (48) 5, 1091–1097.

Mayntz, R., and Scharpf, F. W. (eds.) 1995. *Gesellschaftliche Selbstregelung und Politische Steuerung*. Frankfurt am Main/New York: Campus.

Purcell, M. 2009. "Resisting Neoliberalization: Communicative Planning or Counter-Hegemonic Movements?", in *Planning Theory* (8) 2, 140–165.

Reckwitz, A. 2019. *Das Ende der Illusionen. Politik, Ökonomie und Kultur in der Spätmoderne*. Frankfurt am Main: Suhrkamp.

Schulz, W., and K. Dankert 2016. "'Governance by Things' as a challenge to regulation by law", in *Internet Policy Review* (5) 2, 1–20.

Swyngedouw, E. 2013. "Die postpolitische Stadt", in *sub\urban – Zeitschrift für kritische Stadtforschung* (1) 2, 141–158. DOI: 10.36900/suburban.v1i2.100.

Thaler, R. H., and C. R. Sunstein 2019. *Nudge. Wie man kluge Entscheidungen anstößt*, 14th ed. Berlin: Ullstein.

Verlic, M., and K. Hammer 2019. "Mind the Gap – Achtung demokratische Beteiligung". *A&W blog*, 31/10/2019. https://awblog.at/achtung-demokratische-beteiligung/ (9/9/2020).

Voß, J. P., and R. Kemp 2006. "Sustainability and Reflexive Governance: Introduction", in *Reflexive Governance for Sustainable Development*, ed. by J. P. Voß, D. Bauknecht and R. Kemp. Cheltenham: Edgar Elgar, 3–30.

Yiftachel, O., and M. Huxley 2000. "Debating Dominance and Relevance: Note on the 'Communicative Turn' in Planning Theory", in *Journal of Planning Education and Research* (24) 2, 907–913.

Zürn, M. 2009. "Governance in einer sich wandelnden Welt – eine Zwischenbilanz", in *Perspektiven der Governance-Forschung*, ed. by E. Grande and S. May. Baden-Baden: Nomos, 61–76.

17 How are automated vehicles driving spatial development in Switzerland?

Fabienne Perret, Christof Abegg

Fabienne Perret
ETH, Head of the Transportation Division, EBP Schweiz AG

Christof Abegg
University of Bern, Team Leader Urban and Regional Economic Development, EBP Schweiz AG

© The Author(s) 2023
M. Mitteregger et al. (eds.), *AVENUE21. Planning and Policy Considerations for an Age of Automated Mobility*, https://doi.org/10.1007/978-3-662-67004-0_17

1. A SWISS PERSPECTIVE

Automated transport will gradually reshape our definition of mobility, and the change it brings will hardly make a stop at Switzerland's borders. In recent years and with increasing intensity, researchers, policymakers and practitioners have been engaging with automated driving. In many respects, discussions mirror those in other countries, though they are much more strongly driven and shaped by public transport providers than by car manufacturers. Switzerland's unique institutional and spatial setting adds another layer of nuance. The country's complex federalist system requires all state levels to be in dialogue, making it imperative for the federal government, the cantons, cities and communities to align their policies. In addition, there is a dense network of urban spaces, agglomerations and urban and mountain regions — all sitting in a relatively small geographical radius.

This chapter is a synthesis of two transdisciplinary studies conducted by EBP Schweiz AG in conjunction with various partner institutions over the past four years in Switzerland.

The first multi-client study, carried out together with the Swiss Association of Cities, BaslerFonds, and additional partners, explored the use of automated vehicles in everyday life. Focused on identifying "Potential Applications and Effects in Switzerland", the project compiled a basic report (EBP, BaslerFonds, Swiss Association of Cities and Additional Partners 2017), various in-depth reports covering specific issues[1] (EBP, Basler-Fonds, Städteverband, and Additional Partners 2018a–2018f) as well as a synthesis document highlighting policy options for all partners (EBP, BaslerFonds, Städteverband and Additional Partners 2018g). The co-funding partners comprised representatives of the state at the city and canton levels, various public and private transport companies, as well as an insurance company and Switzerland's national traffic information authority.

The second study, conducted on behalf of the Swiss Foundation for Technology Assessment (TA-SWISS), was published under the title "Automatisiertes Fahren in der Schweiz: Das Steuer aus der Hand geben?" ("Hands on the wheel? Automated driving in Switzerland", available in German only) in February 2020 (Perret et al. 2020). In condensed form, it assesses the opportunities and risks of connected and automated driving for policymakers and society and specifies the conditions that must be created to pave the way for automated cars in Switzerland — including who should be responsible for creating them. Taking into view politics, legislation, economics, technology, society and ethics, the study delivers a final assessment and formulates recommendations for decision makers, and elected officials in particular. In addition, a short explanatory video was produced to make the project results accessible to a broader public.

The aim of these studies was not to provide comprehensive guidance to solve the technical and legal issues at hand. Instead, they describe the technology in sufficient detail to identify its

---

1   Transport technology, data and IT infrastructures, potential forms of shared transport (public transport and shared mobility services), impacts on road transport safety, freight transport and urban logistics, impacts on resources, the environment and the climate, as well as specific challenges faced by cities and agglomerations

main implications and distil a set of key planning and social considerations. In their respective assessments, both studies then derived from these insights a set of policy recommendations not only for stakeholders at all levels of government, but also for transport companies, the industry and their respective associations. As part of their scope, they also took into view the areas in which Switzerland, being a comparatively small nation, has it within its power to influence or shape these developments, and in which areas it will have to bend to international standards and regulations.

By summarizing and blending key findings from these two studies, this chapter will take on the form of a synthesis that simultaneously offers fresh takeaways. Building on an analysis of the opportunities and risks of automated driving (Section 2), the article will take into consideration three scenarios for the future use of automated vehicles in Switzerland (Section 3), including their spatial impacts (Section 4). Acknowledging the country's federalist structure, the chapter then outlines the need for action as well as policy options for the federal government, cities and agglomerations (Section 5). To conclude, Section 6 will show how automated driving in Switzerland has the potential to contribute to sustainable mobility if all levels of government along the entire policy cycle succeed in aligning their actions.

2. THE OPPORTUNITIES AND RISKS OF AUTOMATED DRIVING

Automated driving has the potential to fundamentally reshape our understanding of mobility and our passenger and freight transport systems. While semi-automated systems are already in operation and available on the Swiss market today, it is the arrival of the next levels of automation and the increasing connectivity of vehicles that will bring about significant change. Such highly and fully automated vehicles are currently being developed by car manufacturers, with some experts saying they will be market-ready by 2025, while others estimate it will be closer to 2050 until we see their breakthrough. In any case, the professional world is engaged in lively discussions seeking to forecast the possible ways in which connected and automated driving will impact our everyday lives.

From the vantage point of today it seems obvious that the possible uses of automated vehicles offer up countless opportunities. They can benefit groups such as children or seniors by giving them access to individual forms of mobility. They can improve road safety. Infrastructures can be operated more efficiently, thus creating additional capacities and increasing connectivity across Switzerland. CAVs can also help to enhance traffic management, reclaim urban space and create new types of public transport (PT) or shared mobility services (SMS). The latter are already being tested in pilot schemes in the form of automated shuttles, for instance in Sion, Fribourg, Meyrin, Neuhausen, Zug and Bern. The logistics and freight transport sectors, too, are likely to see efficiency gains and an increase in attractiveness once automated systems become operational.

On the other hand there are a number risks that can be anticipated, the largest being the potential increase in traffic and vehicle kilometres travelled. Fully automated vehicles could, for example, autonomously commute between people's homes and workplaces or schools, and deliver goods. Additional empty runs would translate into more vehicle kilometres travelled. The liberty to use travel time for other activities can reduce commuters' personal travel time costs, but these advantages can be turned into reinvestments in longer and more frequent trips, promoting further urban sprawl. In addition, the emergence of new competing services can challenge the crucial role played by public transport today. What is more, the increase in vehicle kilometres travelled can undermine the federal government's goals to ensure the efficiency of Switzerland's transport infrastructure. But here, further development is either impossible or has been ruled out by government policies.

For state actors it is crucial to gauge the opportunities and risks that automated driving systems will bring. In this context, one of the prerogatives is to define the role of the state and divide up responsibilities among the public and the private sectors. There is still time to lay out the route towards automation, but this window is likely to close soon given the predicted speed of technological advancement. These issues will affect a broad range of specialized areas: alongside technology and traffic-related questions, there are several challenges involving the economy, legislation, politics, society and ethics.

3. AUTOMATED VEHICLES: POTENTIAL TO ENTER THE SWISS MARKET

To highlight the potential future uses of automated vehicles, the TA-SWISS study developed three scenarios. Even if digitalization in the transport sector is advancing rapidly, its path forward is by no means preconfigured: future technological breakthroughs and policy decisions might well call into question path dependencies that are taken as a given today. In the present study, we are primarily interested in future scenarios with fully automated vehicles that allow drivers to transfer full control to autonomous systems.

3.1 THREE SCENARIOS: GENERAL ASSUMPTIONS AND SPECIFIC CHARACTERISTICS

All three scenarios share the underlying assumption that general megatrends (i.e. trends moving society towards a growing, ageing and increasingly individualized population) will impact the course of development. Each scenario, however, will handle one trend with priority, assigning it greater significance or assuming that it will gain critical momentum and determine broader changes. Two variables are key: one takes into account the spatial differences across Switzerland, while the other distinguishes between highly individualized and shared use of automated vehicles. Shared transport comprises all forms of transport in which a person uses a non-private vehicle which they share – at least temporarily – with additional passengers during their journey, i.e. carpooling or ride-sharing.

In the first scenario (highly individualized use), use of automated vehicles is highly individualized throughout all spaces, and development is market-driven and largely free from regulation. Car manufacturers as well as data suppliers and processors are the main drivers of this transformation. The state intervenes only where the market faces safety risks or bottlenecks, in these cases passing minimal regulation.

The second scenario sees the emergence of new shared mobility services in cities and agglomerations. In the introduction phase, uptake of automated vehicles is actively overseen by the public sector. By way of appropriate regulation, it creates market advantages for shared transport in densely populated areas, regulates data exchange and manages traffic flow. These interventions are designed to ensure that cities and agglomerations are not overwhelmed with traffic and to keep mobility attractive for all. Outside these spaces, where traffic is lighter, policymakers see no need for interventions and trust the market to drive development.

Scenario 3, which focuses on a strong uptake of shared mobility services throughout Switzerland, calls for very active and comprehensive intervention by the state. Operating shared transport services in regions and at times with low demand requires appropriate compensation models and consistent legislation. Creating the necessary nationwide reliable communications infrastructure will require significant investments. In this scenario, energy and environmental targets are a priority for policymakers.

3.1.1   Scenario 1: Highly individualized use

In scenario 1, passenger transport sees a shift in traffic away from public transport towards motorized private transport (MPT), triggered by the market-driven and individualized use of automated private vehicles. In addition, it forecasts a steep rise in the number of empty runs resulting, for example, from passenger pick-ups and drop-offs, as opposed to a drop in shipment sizes in freight transport, which effectively lead to a drop in the concentration of demand for both passenger and freight transport and an increase in vehicle movements in all spaces. Figure 1 shows which means of travel are used to meet the demand for passenger transport.

Especially in centres, demand is likely to exceed existing traffic capacities, leading to breakdowns. This in turn makes urban centres less accessible, resulting in increased pressure on agglomerations and peripheral areas. In effect, overall traffic performance in Switzerland is set to increase due to longer distances travelled as a result of spatial (re)distribution, which, unless halted by spatial planning interventions, promotes urban sprawl. For further details on this scenario, see the discussion of the EPB study below.

The low concentration of transport demand creates the need for additional infrastructure and vehicles and thus energy, leading to a spike in emissions. The state, having put manufacturers and data suppliers in charge of connecting vehicles with one another, is not forced to invest in complex public data infrastructures. Figure 4 below provides an overview of the impacts of all scenarios.

3.1.2   Scenario 2: New services in cities and agglomerations

With the emergence of new mobility services, shared transport takes on a key role in densely populated areas. This will likely lead to decreases in motorized private transport in these areas. In peripheral areas, by contrast, shared transport is of minor significance. Empty runs are more frequent, mainly due to pick-ups/drop-offs of passengers or goods. Thus, concentration will increase in cities and agglomerations but decrease in less densely populated areas. Despite these diverging developments, the overall number of vehicle movements across Switzerland is expected to remain unchanged. Figure 2 shows which means of travel are used to meet the demand for passenger transport. Freight transport is expected to follow a similar trend.

Centralized traffic management in cities and agglomerations pairs demand and supply. Accessibility is similar to current levels or has even improved slightly. Since travel times in automated vehicles can be used to engage in other activities, peripheral areas are now more attractive as residential locations. In effect, overall traffic performance in Switzerland increases due to longer travel distances – in scenario 2, however, this applies only for regions outside agglomerations, which is why the total increase is somewhat lower than in scenario 1.

The demand for vehicles and energy has not changed significantly for the time being. Only by ensuring high CAV penetration levels and motivating road users to shift from motorized private transport to shared mobility services in cities and agglomerations can demand for vehicles be cut. This would mean that fewer vehicles would be needed to satisfy demand for passenger movement. This in turn would cut energy consumption for vehicle production and operation, as well as expected emissions. With traffic now being managed in cities and agglomerations, existing infrastructure is used as efficiently as possible, taking away the need for large-scale infrastructure expansions. For these benefits to materialize, the public sector must ensure connectivity. Ideally, a significant rise in the use of shared taxis has slightly reduced the space required for parking in densely populated areas. In peripheral areas, local transport infrastructure expansions remain an option to handle increased transport demand. Figure 4 below provides an overview of the impacts of all scenarios.

3.1.3 Scenario 3: Shared transport services throughout Switzerland

The pivotal role played by shared transport is underlined by significant drops in individual mobility across all regions. Empty runs are minimized and limited to shared taxis. Traffic is highly concentrated, reducing the number of vehicle movements needed to accommodate transport demand. Figure 3 shows which means of travel are used to meet the demand for passenger transport. Freight transport is expected to follow a similar trend.

Centralized traffic control actively matches demand and supply. The lower concentration of shared taxis in peripheral areas may lead to waiting times, which impacts accessibility. In addition, the high costs of operating public transport services could make mobility more expensive. Overall traffic performance in Switzerland would drop.

Demand for vehicles is significantly reduced. Reduced traffic performance has entailed a cut in energy consumption for vehicle production and operation, as well as expected emissions. With traffic now being managed in all spaces, existing infrastructure is used as efficiently as possible, taking away the need for large-scale expansions. For this, the public sector must ensure connectivity. In all spaces, the widespread switch to shared transport has significantly reduced the space required for parking. Figure 4 on the next page provides an overview of the impacts of all scenarios.

Figure 1: Visualization of the impact of CAVs on transport in scenario 1. The number of vehicles shown represents the amount of traffic occurring in each scenario for a constant number of road users.

Source: Perret et al. (2020)

Figure 2: Visualization of the impact of CAVs on transport in scenario 2. The number of vehicles shown represents the amount of traffic occurring in each scenario for a constant number of road users.

Road users

MPT: occupied vehicles

MPT: empty runs

SMS (shuttles)

PT

City
Agglomeration
Peripheral areas

Source: Perret et al. (2020)

Figure 3: Visualization of the impact of CAVs on transport in scenario 3. The number of vehicles shown represents the amount of traffic occurring in each scenario for a constant number of road users.

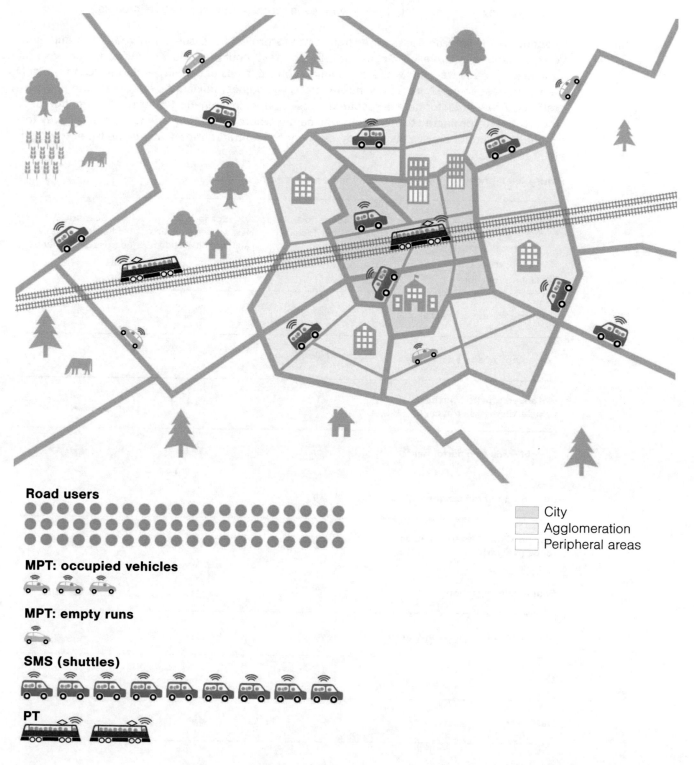

Road users

MPT: occupied vehicles

MPT: empty runs

SMS (shuttles)

PT

City
Agglomeration
Peripheral areas

Source: Perret et al. (2020)

3.2 THE SCENARIOS IN COMPARISON

Figure 4 below compares the three scenarios in terms of their qualitative impacts.

In scenario 1 freight and passenger transport are projected to see an increase in traffic and vehicle kilometres travelled. Scenario 3, by contrast, could lead to a significant decrease in vehicle kilometres travelled while maintaining current levels of mobility and logistics for freight transport. In scenario 2, increases in rural areas will be compensated by decreases in agglomerations. The key factor here is resource consumption: in scenario 1 energy consumption and land usage are projected to increase significantly, whereas scenario 3 would pave the way for a significant reduction as a result of increasingly concentrated trips. Data usage is set to rise in all three scenarios.

Figure 4: Overview of the qualitative impacts of all three scenarios

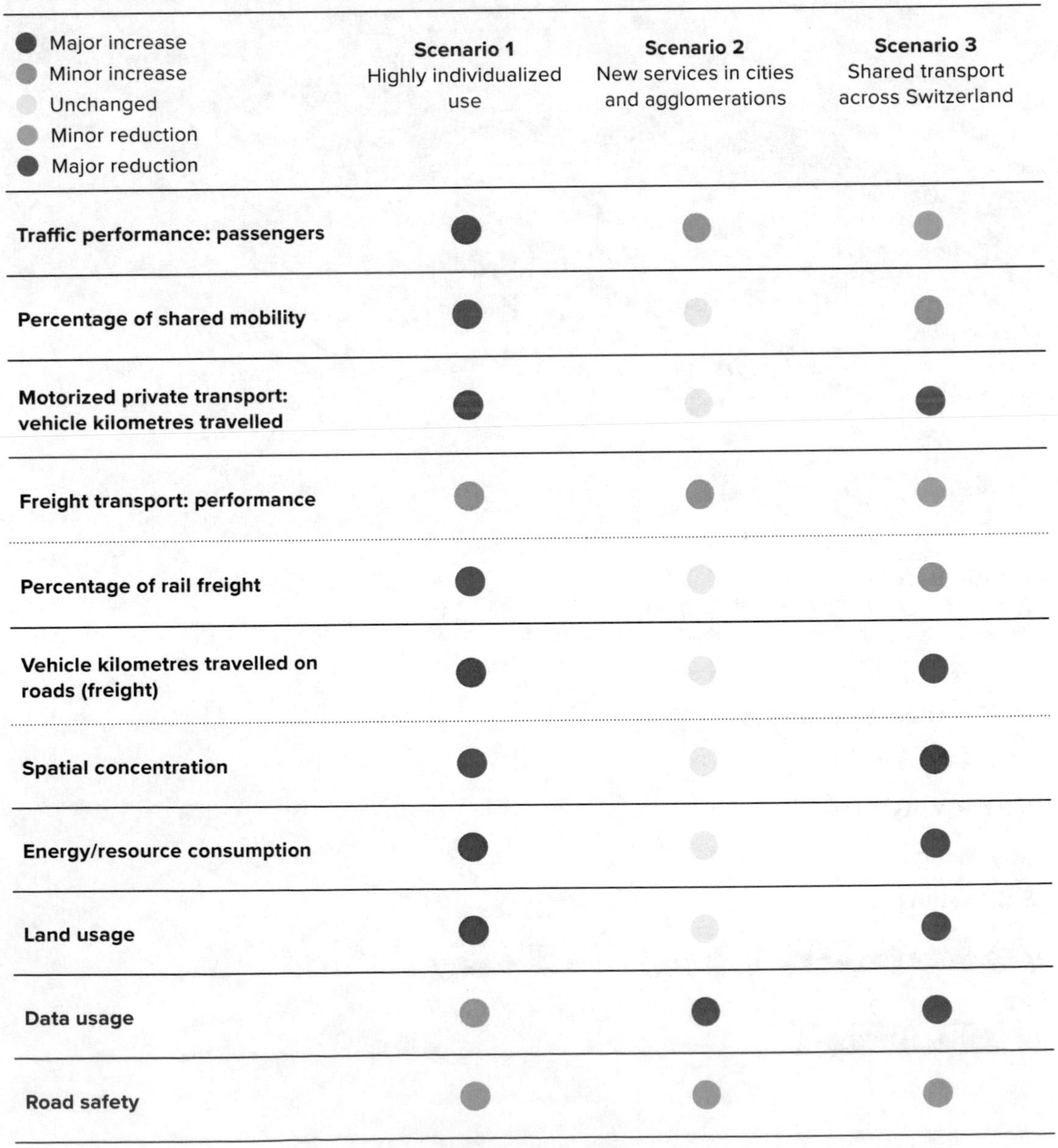

Legend	Scenario 1 Highly individualized use	Scenario 2 New services in cities and agglomerations	Scenario 3 Shared transport across Switzerland
● Major increase ● Minor increase ○ Unchanged ● Minor reduction ● Major reduction			
Traffic performance: passengers	Major increase	Minor increase	Minor reduction
Percentage of shared mobility	Major increase	Unchanged	Minor reduction
Motorized private transport: vehicle kilometres travelled	Major increase	Unchanged	Major reduction
Freight transport: performance	Minor increase	Minor increase	Minor reduction
Percentage of rail freight	Major increase	Unchanged	Minor reduction
Vehicle kilometres travelled on roads (freight)	Major increase	Unchanged	Major reduction
Spatial concentration	Major increase	Unchanged	Major reduction
Energy/resource consumption	Major increase	Unchanged	Major reduction
Land usage	Major increase	Unchanged	Major reduction
Data usage	Minor increase	Major increase	Major increase
Road safety	Minor reduction	Minor reduction	Minor reduction

Source: Perret et al. (2020)

4. IMPACTS ON SPATIAL DEVELOPMENT: REGIONAL AND NATIONWIDE EFFECTS

4.1 A POSSIBLE PATH FOR AUTOMATED DRIVING

While the TA-SWISS study presented three scenarios, the research project "Use of Automated Vehicles in Everyday Life – Potential Applications and Effects in Switzerland" took a different approach, devising a storyline on the use of automated vehicles in Switzerland – a path for the introduction of CAVs that experts consider feasible. However, it does not claim to represent the only way forward.

The storyline is defined as a sequence of states delineated by transitions with shifting preconditions (Fig. 5). Each subsequent stage is reached when all conditions are met, which include legal dimensions (approvals, insurance, traffic law), technology milestones, changes in infrastructure, and social acceptance. All in all, six stages were defined for the study.

Figure 5: The storyline illustrated

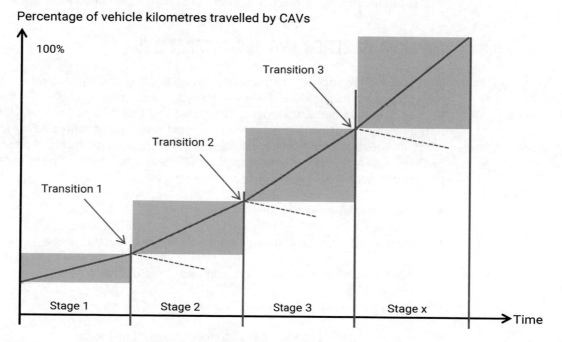

Percentage of vehicle kilometres travelled by CAVs

Source: EBP/BaslerFonds/Swiss Association of Cities and Additional Partners (2017)

The storyline reduces complexity in order to identify the salient aspects shaping actors' decisions. It thus provides a useful basis for further and more detailed studies. In addition, the coherent states distilled from the storyline succinctly summarize how automated driving will affect the various levels of government and spatial planning. The sub-project "Cities and Agglomerations" set out to analyse the impacts of the use of automated vehicles on spaces across Switzerland. Automated driving triggers effects on two spatial scales. The study focused on identifying local effects in densely populated cities and agglomerations. In addition, broader effects on the national level on settlement structures were also investigated.

4.2   NATIONWIDE IMPACTS ON SETTLEMENT STRUCTURES

The shift in availability of motorized private transport is likely to have profound impacts on Switzerland's spatial structure. Rural areas stand to benefit the most: increased capacities will translate into substantial gains in accessibility, and the use of travel time will make commuting to central spaces more attractive. Especially in rural areas with limited public transport services, people with restricted access to mobility, such as children or seniors, will experience improvements.

Cities and agglomerations, on the other hand, will be characterized by lower or even negative gains in accessibility as a result of breakdowns caused by the expected increase in traffic and limited capacities. Additional traffic and the high concentration of vehicles also threaten to undermine living conditions and the experience of public spaces in urban environments.

As a result of this relative shift in attractiveness, rural areas will be confronted with sustained or increasing sprawl. On the other hand, efficiency gains in public transport and new forms of shared mobility services can be expected to boost the attractiveness of urban areas and strengthen existing urbanization trends. These developments will vary with each scenario. Spatial effects will be determined by the interplay – or competition – between motorized private transport, public transport and shared mobility services. Taken by itself, automated mobility will not have a profound impact on spatial relations. Other factors such as demographic changes, digitalization, shifting value chains, or climate change also need to be taken into account.

4.3   IMPACTS IN CITIES AND AGGLOMERATIONS

In describing the impacts of automated driving in cities and agglomerations for each scenario, the study distinguishes the following types of traffic: moving traffic, stationary traffic, public transport, shared mobility services, pedestrian and bicycle traffic as well as freight traffic. The potential effects of automated driving in urban areas are determined by the function of the roads in question and must be assessed in relation to other uses and people's lived mobility experiences. Aiming to provide a spatially differentiated analysis, the study thus devised a typology distinguishing the following mobility spaces:

- main arteries in urban regions ("main arteries")

- main roads in urban and neighbourhood centres ("central streets")

- roads serving residential neighbourhoods ("residential streets")

- roads serving industrial areas ("industrial areas")

- multimodal transport hubs in settlement areas ("transport hubs").

Moving traffic
New services, potential empty runs, and new users (seniors, children) will contribute to an increase in vehicle kilometres travelled. The capacity of existing infrastructure could be boosted, provided that automation helps to reduce time gaps between vehicles. This gain will be disproportionally higher than the uptake of automated vehicles across the overall fleet. Some roads, such as main arteries, will register capacity gains as a result of automation, whereas capacity levels on others, like central streets, will remain more or less static, resulting in a redistribution or uneven distribution of additional traffic. All in all, additional traffic can be expected to overcompensate the increase in capacity, leading to noticeable shifts in all mobility areas. These

capacity effects may help to postpone or even render unnecessary individual infrastructure development projects.

Establishing segregated infrastructure for automated vehicles in settlement areas is hardly feasible due to a lack of space and a high concentration of junctions. As long as the number of human-driven vehicles remains significant and reference vehicles (e.g. public transport, emergency services, winter maintenance) remain similar in size, a reduction of lane widths is an unrealistic option.

On main arteries, the effects of automated driving will be predominantly positive. Overall, their streetscapes are less complex than, for example, in centres where they are shared by multiple road users. Main arteries do not provide direct access to many of the adjacent uses. This creates opportunities for traffic flow, which should improve as the share of CAVs increases, since these vehicles travel more densely and at more consistent speeds. Conversely, however, the rise in traffic, along with the fact that automated vehicles travel with reduced time gaps, will simultaneously reinforce segregation.

Dynamic lane assignments can deliver further improvements, increasing traffic capacities by taking advantage of the fact that commuter flow is often unidirectional. At junctions, traffic lights can be operated with greater complexity and responsiveness. This would benefit not only CAV traffic, but also improve the flow of motorized private transport at junctions. In cities and agglomerations, main arteries are therefore ideally suited to accommodating additional traffic, as automation on these roads can bring the greatest capacity gains. On the one hand, such capacity gains could end up restricting other road users, or even to them being excluded from access to such roads altogether. On the other, such gains can be used to expand public transport services. To implement dynamic lane assignments and traffic management systems at junctions, extensive organizational effort and a digital communication infrastructure are needed. In case these systems fail, there need to be fallback guidelines to ensure the functionality of the road network. This, in turn, would require ensuring long-term maintenance and operability of the necessary analogue equipment as well as operational flexibility.

In urban centres, roads not only serve transport functions, but also settlements. Therefore, it is crucial to take into consideration the effects of additional traffic, which are strongly influenced by the frequency of parking manoeuvres, the handling of freight, and expected empty runs. Potential capacity gains from automation are likely to be offset by increases in trips and freight handling. Additional traffic and the increased segregation it entails interfere with public transport, restrict large crossing areas for pedestrians and other users, interfere with bicycle traffic and reduce the quality of public spaces.

Stationary traffic
Parking spaces are becoming less of an issue as sharing and pooling services as well as the possibility for providers to run empty trips lower demand for (dispersed) parking. Adding to this is the fact that automated vehicles can be parked at remote locations after use, which means parking is no longer located-based. This trend, however, results in additional traffic: empty runs and trips to further-off parking amount to more vehicle kilometres travelled on the adjacent road network. In addition, on-demand and pooling services as well as privately owned automated vehicles will need designated off-street stopping areas for passenger pick-up and drop-off.

Central streets are characterized by a diversity of different uses and dense in- and outbound traffic, with car parks and parking spaces ensuring access to adjacent uses. If demand decreases, kerbside parking spaces can be given over to other uses and/or be redesignated as loading/unloading and passenger pick-up/drop-off areas. Parking can be relocated to new and repurposed parking areas, while "freed-up" parking spaces can be made available for a variety of alternative uses, such as new bicycle lanes, broader pavements or additional greenery to improve street design, depending on the specific context. In return, fully automated vehicles will need designated loading/unloading and passenger pick-up/drop-off zones, for which current legislation will need to be amended.

Residential neighbourhoods experience lower levels of in- and outbound traffic than central streets, with parking spaces being used to leave idle vehicles until they are needed again. Reduced individual vehicle ownership and consolidated parking in city and town centres will deliver two major benefits to residential neighbourhoods: streetscapes can be made more attractive for pedestrian and bicycle traffic, and communities have more space to enrich neighbourhood life. Although stopping on residential streets occupies additional space, it is limited to the streetscape or private driveways, requiring no additional regulation.

Public transport and shared mobility services

Conventional and consolidated public transport services increasingly running automated trains, trams and large buses will continue to serve as the backbone of the public transport system in cities and agglomerations. Automation, connected vehicles, and new (digital) mobility services will create considerable potential for rationalization, trailblazing new, cheaper and more attractive public transport services, in particular for offers operating with lower passenger volumes. This means that new shared mobility services will complement public transport, especially in peripheral areas, primarily using small to medium-sized vehicles running partially or fully demand-based trips. A crucial role will be played by new services operated by public or private mobility providers offering individual "front door pick-ups" – without pre-scheduled stops and with near-instant availability. The proliferation of responsive shared mobility services will increase the need for additional stopping areas, but it is currently difficult to assess whether accessibility standards should be optional or mandatory. Designated stopping areas and existing public transport stops can also be used by automated vehicles. However, whether or not such areas can be opened up to accommodate additional uses will primarily be determined by the frequency of these services.

Automation will also unlock new potential in multimodal transport hubs. Existing public transport lanes and stops can accommodate additional uses, especially during off-peak times, to make available new services and improve connecting services. Across the public transport network, automation can also take the pressure off traditional, heavily congested multimodal transport hubs by creating more tangential connections using on-demand services. Tram automation will also raise specific issues. Operations can be expected to switch to segregated tracks, which will require more space in relation to mixed traffic. At multimodal transport hubs, this can have (additional) impacts on permeability by restricting crossing pedestrian and bicycle traffic.

Pedestrian and bicycle traffic

Today, shared road space is successfully lived and promoted especially in urban centres and residential neighbourhoods. This primarily benefits pedestrian and bicycle traffic. As automation increases, however, safety and capacity considerations could lead to restrictions for pedestrians, limiting opportunities to cross streets in order to create dedicated lanes for automated vehicles. This would have adverse effects on mixed traffic environments, as automated driving would be prioritized over pedestrian and bicycle traffic. Interaction of automated vehicles with pedestrian and bicycle traffic as well as non-automated vehicles remains an issue yet to be resolved. Communication among road users is mainly based on gestures and visual acknowledgement, which automated vehicles cannot perform in the same way. Conversely, pedestrians and cyclists can also be expected to interact differently with automated vehicles.

The long transition phases with mixed traffic involving both automated and non-automated vehicles suggest that existing road marking, signal and steering systems will remain in use over a longer period. These extended mixed traffic scenarios may also mean that automated vehicles will operate responsively in mixed traffic areas, "playing it safe", so to speak, which could in turn absorb most of the expected benefits generated by increased capacity and smoother traffic flow. This will make it necessary to strike a balance between making optimum use of automation and ensuring pedestrian and bicycle traffic quality and safety.

The impacts of automated vehicles on pedestrian and bicycle traffic are likely to affect all mobility spaces. In the urban system as a whole, it will become more challenging to create or maintain continuous traffic networks for pedestrians and cyclists.

In urban and neighbourhood centres, solutions interfering with large crossing areas will negatively impact pedestrian and bicycle traffic quality. In such environments supporting a high diversity of uses, permeability and freedom of movement are key issues. On the other hand, large streets will benefit from the elimination of roadside parking. This freed-up space can be reassigned to serve a variety of other uses.

Pedestrian traffic is also a key factor in multimodal transport hubs, where pedestrian traffic flows must have space to spread out for users to be able to reach tightly timed connecting public transport services. The increasing uptake of shared mobility services will add more complexity to the status quo, making it more difficult to identify traffic flows. Visually and hearing-impaired people in particular experience difficulties accessing transport hubs if there is a lack of safe and easy-to-use crossing areas. This effectively limits the functionality of these facilities, which are so pivotal to ensuring seamless mobility. In the future, stricter standards should be applied to the design of traffic and waiting areas as well as access kerbs.

In residential neighbourhoods, automated vehicles can help to reduce travel speeds and ensure speed limits are observed, thus increasing safety for other road users. "Driverless pickups" of passengers or goods need to be viewed with caution. These can lead to additional traffic that congests public space and creates unforeseen safety risks. This raises the question whether such "pick-ups" should require permits or whether empty runs in such spaces should be prohibited per se.

On main arteries, managing automated and non-automated vehicles and pedestrian and bicycle traffic during the long transition phases remains a challenge. We need to find ways to integrate all road users into efficient and responsive traffic management systems, for example at traffic junctions. One of the key issues is to create the infrastructure needed to enable the various human and automated road users to communicate with each other.

In industrial areas in particular, the focus on automated (freight) traffic may risk impacting pedestrian and bicycle traffic, which should "clear the way" for commercial traffic wherever possible. Ultimately, this thinking may end up fully segregating different modes of transport, thus restricting or even eliminating permeability. Especially larger industrial developments that can easily be cut off from adjacent areas could become "forbidden cities". As a rule, however, industrial areas, too, must be safe environments for commuters and visitors, and must be served by various modes of transport providing attractive links to adjacent areas.

Freight transport
Freight transport brings additional traffic to cities and agglomerations. In both the business-to-consumer and business-to-business segments, we can expect to see the number of shipments increase, while their average size will shrink. The introduction of automated freight vehicles is likely to accelerate the increase in traffic, but its specific impact will depend on whether freight flows can be consolidated and vehicles are operated at capacity both when delivering and returning freight. Once driverless freight vehicles are a reality, goods will have

to be handled in clearly defined zones, and additional handling equipment will have to be easily available on demand. This will also mean that more tasks will be delegated to senders or recipients or other (automated) systems, such as delivery robots or parcel lockers.

Early adoption of automated driving could be seen in industrial areas, which make ideal test sites for a number of reasons. Firstly, the streetscapes are often less diverse, with less pedestrian and bicycle traffic. Also, they often have easy access to the motorway network, which reduces automated trips on lower-level roads to a minimum. Low traffic speeds also make navigation and steering easier for such vehicles. Secondly, commercial areas also create economic pressure to adopt automated driving, as businesses stand to directly benefit from efficiency and productivity gains resulting from the adoption of new logistics concepts.

Urban planning considerations or design requirements concerning the provision of specific infrastructure or road markings are likely to be assigned lower priority than in specifications for main streets in urban or town centres. Automated shuttles are also suitable for operation on access roads in industrial areas. Especially when planning comprehensive transformations of commercial areas or developing new estates, streetscapes can be designed from the outset to accommodate the needs of automated driving or smart logistics. The economic argument – increasing the efficiency of freight transport – could lead to a prioritization of businesses' access and design considerations over the needs of pedestrians, cyclists and other road users. Returning to automated freight handling areas, there remains a lack of guidance specifying the party responsible for their operation and management.

In inner cities, commercial uses make it essential to accommodate deliveries for retail stores, hotels and restaurants as well as courier, express delivery and parcel services throughout the day. This entails high levels of freight and logistics traffic as well as extensive demand for freight handling zones and stopping areas.

Impacts in residential neighbourhoods will largely depend on the type of business models that will come to populate the last-mile delivery market, and whether these will use pick-up and delivery points, automated parcel lockers or delivery robots.

4.4 SPATIAL IMPACTS: THE THREE SCENARIOS COMPARED

The overview in Figure 6 on the next page shows which mobility spaces in cities and agglomerations are likely to be most affected by individual modes of transport, without looking at whether these changes will be positive or negative. It also shows the impact of individual modes of transport on mobility spaces in cities and agglomerations. Overall, main arteries and industrial areas are affected least. Due to their functions and uses these are the most complex mobility spaces. Urban and neighbourhood centres, which are characterized by dense functions and uses that spur urban development, register a set of overlapping effects. It must be pointed out, however, that the functionality of urban transport systems is determined by the quality of the interaction of its respective subspaces and its various modes of transport.

Advances in technology and assessments of their potential impacts in cities and agglomerations remain fraught with uncertainty. It is also a matter of impossibility to determine purely the spatial impacts of the adoption of automated vehicles, as automated driving is only one of the modal, social and economic factors shaping the development of urban transport systems. Adding to this is the fact that for the decades to come, cities and agglomerations will continue to see transport networks being shared by vehicles with varying degrees of automation and, in the long term, also with non-automated vehicles and more road users.

Figure 6: Mobility spaces and the extent of change in modes of transport

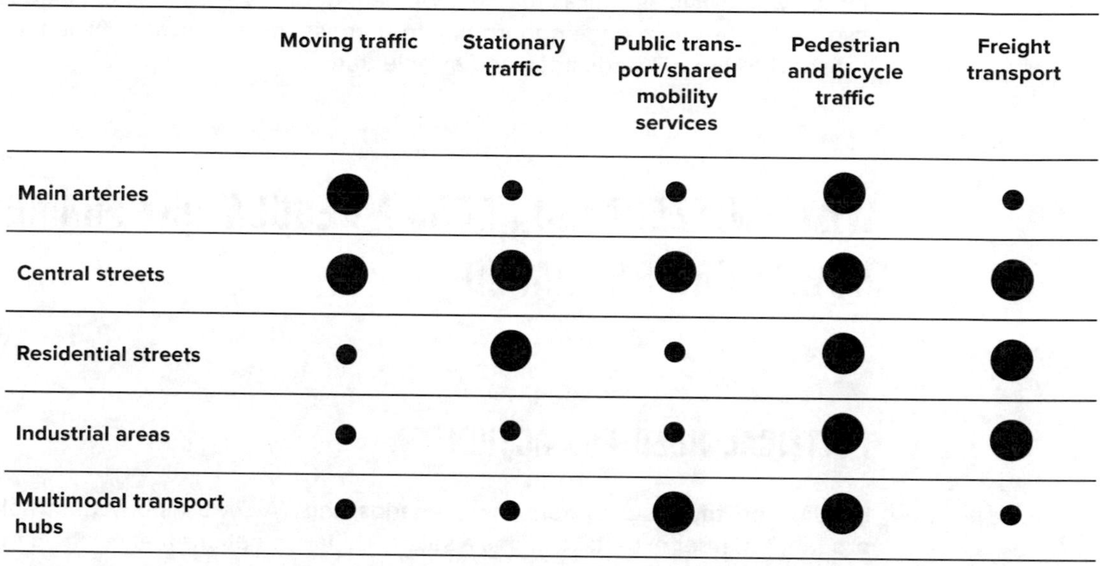

	Moving traffic	Stationary traffic	Public transport/shared mobility services	Pedestrian and bicycle traffic	Freight transport
Main arteries	● (large)	• (small)	• (small)	● (large)	• (small)
Central streets	● (large)	● (large)	● (large)	● (large)	● (large)
Residential streets	• (small)	● (large)	• (small)	● (large)	● (large)
Industrial areas	• (small)	• (small)	• (small)	● (large)	● (large)
Multimodal transport hubs	• (small)	• (small)	● (large)	● (large)	• (small)

Source: EBP/BaslerFonds/Swiss Association of Cities and Additional Partners (2018d)

By having briefly looked at these local, small-scale impacts we can also assess and evaluate the three scenarios laid out in the TA-SWISS study in terms of their transport policy and spatial policy implications.

Scenario 1, which assumes highly individualized use of automated vehicles, identifies two key challenges: a massive increase in congestion, or in other words, negative impacts on quality of life and public spaces in urban environments, as well as an increasing trend towards urban sprawl in rural areas. To what extent shifts in attractiveness spurred by the uptake of automated vehicles will affect large-scale spatial developments in Switzerland will primarily depend on regulatory decisions concerning automated driving as well as steering decisions relating to the transport sector and spatial planning. Here, automated driving can open up opportunities to optimize intermodal mobility chains, improve connectivity of different modes and means of transport, and integrate groups with limited mobility. All things considered, however, the scenario is at odds with the country's climate, transport and spatial development policies.

Scenario 3, in which automated vehicles are used collectively throughout Switzerland, raises questions concerning the profitability, i.e. the financial viability of nationwide shared services, bringing to the fore a number of transport, state and regional policy considerations: which public services does Switzerland as a federalist country want to fund, and what level of quality should they provide? And who will pay for them? From today's perspective and in light of the discussions that have taken place over the past decades, it seems unrealistic to assume that such a regulatory framework will be developed and receive funding. Even so, targeted shared services can bring improved access to mobility in peripheral areas, or at least help to support existing services.

Cities and agglomerations in particular are likely to view scenario 2 as the preferred, if not the only viable path forward. The majority of representatives of cities and agglomerations were sceptical when asked about the potential impacts of CAVs, as feedback gathered during project discussions shows. From their viewpoint, the long-term impacts of CAVs will largely depend on whether the advantages of automated driving can be used to strengthen cooperative transport. An increase in motorized private transport (relative to other modes of transport), on the other hand, is largely expected to cause negative impacts, with cities and agglomerations fearing competition between individualized and public transport, or restrictions affecting pedestrian and bicycle traffic as well as the development of public space. Simultaneously, these fears can also be framed as a huge opportunity to harness the significant efficiency gains in motorized transport to benefit pedestrian and bicycle traffic.

5. WHY SWITZERLAND NEEDS A REGULATORY FRAMEWORK, AND HOW IT CAN BE SHAPED

5.1 POLITICAL NEED FOR ACTION

Building on the insights from the scenarios, the TA-SWISS study then held a workshop that brought together experts to define a set of regulatory options that can help to steer expected advances in automated driving in Switzerland. The experts agreed that automated driving will arrive and that regulations are needed today, and not in 20 or 30 years. A laissez-faire approach would entail major shifts away from collective towards individual transport. Such a development could offset the efficiency gains that automated driving can bring and lead to more congestion and/or demands for infrastructure expansion.

However, experts had diverging opinions regarding the effects that regulatory measures should aim for, and there was no clear preference among the group for one of the outlined scenarios. This question has a normative dimension that must ultimately be aligned with a societal target. Switzerland's policies, however, are currently either lacking such targets, or their respective aims are at odds with each other, and it is unlikely that clarification will be achieved in the near future. There is an array of related issues that are both highly complex and yet currently too vague to be able to bring Switzerland closer to reaching societal and political consensus.

Next, the regulatory options developed by the group of experts in the workshop were discussed with stakeholders representing politics, administration, the economy and associations. They similarly agreed that the advent of automated driving will have a significant impact on the transport system, pointing primarily to the expected gains in efficiency. It also became clear, though, that automated driving can and must support efforts to make transport more sustainable on the whole.

Representatives agreed that it is crucial and critical to encourage lively societal and political debate to stake out regulatory options. Unlike experts, however, stakeholders already appear to have found common ground on key issues. The implicit consensus concerns the need to sustain Switzerland's transport system and the pivotal status of public transport, and the notion that Switzerland can and should play a pioneering role in further developing automated mobility. Moving forward, the needs of pedestrians and cyclists, too, should be taken into account, especially in densely populated areas. Building a greener and climate-resilient transport system is an overarching prerogative.

5.2 ROLE OF THE STATE AND SCOPE FOR ACTION

Two measures are needed to allow Level 3 automated driving systems on the road in Switzerland and lay the groundwork for further policy actions.

Firstly, in dialogue with its neighbours, Switzerland must prepare approval processes for semi-automated as well as highly and fully automated vehicles. This is to ensure that all Level 3 and higher automated vehicles available on the market can be used for passenger and freight transport operations. Here, the primary focus is on creating approval processes for automated systems and addressing liability issues. Next, and again in coordination with neighbouring countries, safety standards on handover times and other details must be defined. Further national-level regulations regarding connectivity, data exchange or specifications on energy-efficient drives must be examined in good time.

Secondly, the state should introduce mandatory training to educate drivers of (fully) automated vehicles, including options such as a "driving licence light", i.e. a certificate confirming eligibility to supervise highly automated Level 3 and 4 vehicles.

Beyond these two mandatory measures, government actions in response to the advent of connected and automated vehicles in Switzerland can take two distinct directions:

1. **"Enabling":** The state influences policies but generally remains in the background, limiting its role to creating a minimum-regulation environment to support private and public actors in creating market innovations.

2. **"Leading":** Building on clear political objectives, decision makers choose a "strong state" approach, actively adopting regulations that either restrict or delay the introduction of connected and automated transport systems, or proactively encourage these developments.

Discussions among study participants produced no majority preference for either of the outlined roles. Views diverged both among experts and stakeholders, depending on their political affiliation or on which particular aspect of connected and automated transport was being debated. Accordingly, both directions are seen as viable policy paths for Switzerland. Seeking to produce adequate responses to this breadth and uncertainty of opinion regarding the technological, legal and temporal development of connected and automated vehicles, we will assign each policy option to one of the two paths.

For its role as an **enabler**, influencing policies but generally remaining in the background, the state should consider the following options in addition to the mandatory measures outlined above:

■   Encourage societal and political discussion on the ethical and data protection-related aspects of the data generated by connected and automated vehicles. Address the risks and opportunities of data usage openly with civil society actors.

■   Produce in good time and for all levels of government a data policy that protects the interests of the public sector. Alongside ensuring data sovereignty and access rights, this includes defining the data the public sector needs to perform its tasks. Align policies with smart city strategies and Open Government Data (OGD) principles. In addition to complying with international data exchange standards, Switzerland can formulate stricter quality standards and specifications regarding the metadata contained in the exchanged data.

■   Update Switzerland's public transport act and related bye-laws. New providers should be given access to the passenger transport sector, which has been shielded by strict regulation in the past. Conversely, shared transport providers should be given greater flexibility to run services that complement existing public transport offers (for instance in the form of concessions that define minimum services for a designated area instead of timetabled or regular services – provided that this approach does not undermine the transport system's overall efficiency in serving all areas) or they could be partially exempt from meeting accessibility regulations.

- Ease approval processes and lift requirements to be met by licensed transport companies, allowing them to focus on research and development. Traffic law, liability regulations and provisions for businesses should be designed to help public and private transport providers run economically viable and appropriately financed services.

- Create attractive conditions to support the creation of an efficient mobile network. Switzerland currently has several privately operated networks. Building a state-operated "utility network" to accommodate private sector services could prove to be an interesting alternative that should be further explored.

- Build a "conditional open data" platform[2] that allows all actors to share their own as well as access third-party data (ASTRA 2018). This platform should also make available basic topographic and traffic data, as well as event, measurement and aggregated sensor data. Among other things, this will require enhanced technology systems for fully digitalized and automated workflows, and regulations ensuring that these systems are built to conform with emerging international standards.

On the other hand, the state can act as a **leader**, building on clear political objectives. To this end, decision makers should introduce the following measures and instruments in addition to the recommendations made so far:

- Launch a nationwide dialogue on the future of mobility in Switzerland, encouraging political representatives and researchers as well as civil society actors and business to join the conversation, and ensuring that debates also address the safety and desirability of technology-based systems.

- Define targets for Switzerland's mobility sector, taking into account the requirements of the state at the federal level, the cities, the cantons, society and the economy, as well as passenger and freight traffic. If national-level targets conflict with other policy areas, these must be addressed.

- Bringing together all levels of government and listening to business and civil society, develop ideas and proposals on how connected and automated driving in cities and agglomerations can help to strengthen shared forms of mobility and transport in particular. This will likely have to be backed by incentives for shared mobility services and require a new framework for the operation of public transport services, along with possible restrictions for motorized private transport. Solutions may include market-based instruments, obligations placed on concessions or approval processes (such as no empty runs), or information and persuasion.

- Consolidate traffic-steering competencies at the national level. For this, cities, cantons and the federal government must determine the body that is to take responsibility for traffic steering and how they plan to resolve overlapping competencies. Together with research-

---

2 "Conditional open data" means that all participating parties can access the available data free of charge for non-commercial use provided they agree to supply some of their own data in return. The expanded data set can then be accessed by the entire community and used for new applications, the only exception being "refined data", i.e. marketable information. It is up to the participating parties to draw the line between data and information (ASTRA 2018).

ers and transport companies, they will also have to identify the indicators used to steer traffic (occupation rates, minimum transport length, vehicle kilometres travelled, choice of destination, parking spaces, stops, etc.).

Summing up, it becomes clear that the need for regulation is highest if the state takes on a "leading" role. Without vision and strategy, a strong state will struggle to provide clear leadership and be unable to steer developments in a direction that aligns with its preferences. Concisely formulated targets and strategies are the outcome of political negotiation processes which may take some time to conclude in Switzerland. If, in turn, decision makers want the state to act as an "enabler", such visions and strategies can be developed iteratively. Such a policy involves greater intervention than a purely reactive approach but requires a less stringent debate on visions and strategies, which are needed if the state is to pursue a proactive leadership approach.

5.3   SPECIFIC POLICY OPTIONS FOR CITIES AND AGGLOMERATIONS

For cities and agglomerations, one of the past and future challenges is to make transport sustainable and city-friendly. For this reason, policies aiming to shape the use of automated vehicles also need to take into consideration their implications on the transport system as well as on spatial and urban planning. With this in mind, the BaslerFonds study formulates ten strategies for cities and agglomerations, each including a set of suggested measures.

- **Take an integrated and city-friendly approach to mobility planning:** How can automated driving contribute to a city's sustainable and city-friendly vision of mobility?

- **Test possibilities, share experiences, and learn:** Tests and pilot schemes help all actors to gain experiences and learn about how automated driving impacts the transport system and spatial development. By sharing knowledge, cities as well as cantons and federal-level actors can create synergies and gradually build knowledge.

- **Put cities and agglomerations' needs and concerns on the agenda:** Cities and agglomerations should actively follow current developments and uses of automated driving, share knowledge and experiences, and develop shared positions. On this basis, they can then put their needs and concerns on the agenda when negotiating with the federal government and the cantons.

- **Promote dialogue and build awareness:** Public discourse on all levels raises awareness of automated driving and prepares the ground for a broadly supported debate on its impacts and effects on transport and urban space.

- **Limit additional traffic:** Suitable measures must be put in place to minimize or effectively manage the additional traffic resulting from new mobility services and empty runs.

- **Strengthen shared mobility schemes:** In terms of land use, public transport and shared mobility services are the most efficient modes of transport. These services must be kept attractive, affordable and efficient, and policies must be put in place that influence the modal split and favour sustainable transport through attractive offers.

- **Promote intelligent traffic flow management and steering:** Higher penetration of automated vehicles and increased connectivity can help to manage traffic more intelligently and influence spatial and temporal peaks in traffic demand.

- **Optimize public and private use of land:** Repurpose vacant parking spaces to make urban space more attractive and develop concepts to provide parking, stopping and freight handling areas for automated passenger and freight transport.

- **Ensure all road users are safe:** Adopt targeted measures to minimize safety risks on shared road spaces, both during the transition towards automated driving and once full automation has arrived.

- **Make urban logistics city-friendly:** Cities and agglomerations can support and regulate private sector efforts to establish efficient freight delivery concepts.

6. STEERING MOBILITY AND SPATIAL DEVELOPMENT: KEEPING OUR HANDS ON THE WHEEL

Building on its three scenarios, the TA-SWISS study explored the potential paths that automated driving might take in Switzerland. The study's aim was not to promote a specific scenario, but to formulate policy options that can either be applied to all scenarios or highlight the specific roles the state can adopt depending on its preferred degree of intervention.

The partners involved in the BaslerFonds study, by contrast, clearly laid out the direction they would like cities and agglomerations to take in order to ensure that spaces and streetscapes remain attractive in the future. Their preference is to promote shared mobility services in urban spaces, as outlined in scenario 2 of the study. In order to ensure that sustainable mobility arrives in cities and agglomerations, it is not enough for the state to act as an enabler and trust the market to drive developments.

To harness this transformation, however, society first needs clearly defined targets for a future mobility strategy. Taking into consideration the policy cycle[3] (Fig. 7), such mobility strategies are ideally defined in advance of regulatory interventions, i.e. negotiated between the state at the federal, city and canton level and society and business. These strategies should be flexible enough to accommodate individual policies for each type of region, i.e. the state should lead with varying intensity while also addressing conflicting targets with other policy areas.

Only after achieving a consensus on these objectives should all levels of government proceed to review and define the instruments and regulations for each policy action area as described in this chapter. It goes without saying that this is not a purely linear process. Objectives and measures can and should also be developed iteratively, especially if decision makers want the state to act as an enabler. Such a policy involves greater intervention than a purely reactive approach but requires a less stringent debate on visions and strategies, which are needed if the state is to pursue a proactive leadership approach.

In addition, all levels of government need to stay up to date with current developments for them to launch appropriate and timely policy responses. Particular attention should be paid to regulations ratified by the EU, but also to legislation passed in other pioneering states, as well as general technological advances in connected and automated driving. Only by promptly

---

3   https://pb-tools.ch/politikzyklus/, www.staatsfragen.de/tag/politikzyklus/ (in German).

Figure 7: Policy cycle for the use of automated vehicles in Switzerland

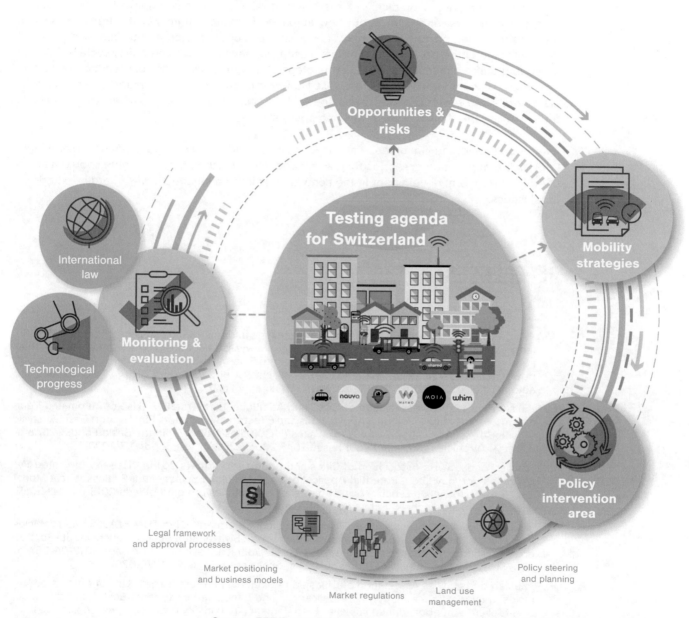

Source: EBP/BaslerFonds/Swiss Association of Cities and Additional Partners (2018g)

integrating insights from abroad into Swiss policy cycles can policy options be evaluated and, if necessary, effectively updated to ensure their alignment with the country's mobility strategies.

It needs to be communicated clearly that the negative impacts of automated driving can be effectively addressed or at least minimized. For this, the federal government, cantons, cities and communities must jointly develop suitable monitoring and controlling systems and test their applicability. Such a collaborative vein is also needed to create a regulatory framework to steer mobility and traffic management. If the effects of such a framework remain negligible, or if it fails to steer developments altogether, then the mobility strategies, the legal framework or the implemented instruments need to undergo review. In principle, all levels of government must remain open to amending their current approaches.

From today's perspective many of the future opportunities and challenges of automated driving are difficult to model or predict. Test systems and pilot schemes can help all actors to gather crucial experiences and learn about how automated driving will impact the transport system and spatial development. Thus, a key takeaway from both studies is that pilot schemes are pivotal for the entire policy cycle. For this reason, we have updated the policy cycle in Figure 7 by adding a fifth dimension. Now, the four steps of the policy cycle are embedded in a testing agenda to account for the broader context. However, because implementation is based on special permits and provisional approvals, the cycle is accelerated, thus providing critical insights for improving the design of the remaining policy stages.

On the one hand, the aim of these applications is to build a body of robust evidence and highlight viable application scenarios for Switzerland. On the other, these insights should also be used by the federal government in the context of its international activities in order to address legal issues.

REFERENCES

ASTRA (Swiss Federal Roads Office) 2018. "Bereitstellung und Austausch von Daten für das automatisierte Fahren im Strassenverkehr", 7/12/2018, Bern. www.astra.admin.ch/dam/astra/de/dokumente/abteilung_strassennetzeallgemein/bereitstellung-austausch-daten-automatisiertes-fahren.pdf.download.pdf/Bereitstellung%20und%20Austausch%20von%20Daten%20f%C3%BCr%20das%20automatisierte%20Fahren%20im%20Strassenverkehr.pdf (9/7/2020).

EBP, BaslerFonds, Swiss Association of Cities and Additional Partners 2017. "Use of Automated Vehicles in Everyday Life – Potential Applications and Effects in Switzerland. Final Report Basic Analysis (Phase A)", Zurich. https://www.ebp.ch/sites/default/files/project/uploads/BaslerFonds%20Automated%20Vehicles_Summary%20Phase%20A%20oct%202017_en_2.pdf (1/7/2020).

EBP, BaslerFonds, Swiss Association of Cities and Additional Partners 2018a. "Use of Automated Vehicles in Everyday Life – Potential Applications and Effects in Switzerland. Schlussbericht Modul 3a, Verkehrstechnik", Zurich. www.ebp.ch/sites/default/files/project/uploads/2018-04-04%20aFn_3a%20Verkehrstechnik_Schlussbericht_0.pdf (1/7/2020).

EBP, BaslerFonds, Swiss Association of Cities and Additional Partners 2018b. "Use of Automated Vehicles in Everyday Life – Potential Applications and Effects in Switzerland. Schlussbericht Modul 3b, Daten und IT-Infrastrukturen", Zurich. www.ebp.ch/sites/default/files/project/uploads/2018-04-05%20aFn_3b%20Daten%20und%20Infrastrukturen_Schlussbericht_0.pdf (1/7/2020).

EBP, BaslerFonds, Swiss Association of Cities and Additional Partners 2018c. "Use of Automated Vehicles in Everyday Life – Potential Applications and Effects in Switzerland. Schlussbericht Modul 3c, Mögliche Angebotsformen im kollektiven Verkehr (ÖV und ÖIV)", Zurich. www.ebp.ch/sites/default/files/project/uploads/2018-04-19%20aFn_3c%20Mögliche%20Angebotsformen%20im%20kollektiven%20Verkehr_Schlussbericht_0.pdf (1/7/2020).

EBP, BaslerFonds, Swiss Association of Cities and Additional Partners 2018d. "Use of Automated Vehicles in Everyday Life – Potential Applications and Effects in Switzerland. Schlussbericht Modul 3d, Städte und Agglomerationen", Zurich. www.ebp.ch/sites/default/files/project/uploads/2018-08-30%20 aFn_3d%20Städte-Agglomerationen%20Schlussbericht_1.pdf (1/7/2020).

EBP, BaslerFonds, Swiss Association of Cities and Additional Partners 2018e. "Use of Automated Vehicles in Everyday Life – Potential Applications and Effects in Switzerland. Schlussbericht Modul 3e, Ressourcen, Umwelt, Klima", Zurich. www.ebp.ch/sites/default/files/project/uploads/2018-04-09%20 aFn_3e%20Ressourcen%2C%20Umwelt%2C%20Klima_Schlussbericht_0.pdf (1/7/2020).

EBP, BaslerFonds, Swiss Association of Cities and Additional Partners 2018f. "Use of Automated Vehicles in Everyday Life – Potential Applications and Effects in Switzerland. Schlussbericht Modul 3f, Güterverkehr/City Logistik (Strasse)", Zurich. www.ebp.ch/sites/default/files/project/uploads/2018-03-28%20 aFn_3f%20Güterverkehr%20und%20Citylogistik_Schlussbericht_0.pdf (1/7/2020).

EBP, BaslerFonds, Swiss Association of Cities and Additional Partners 2018g. "Use of Automated Vehicles in Everyday Life – Potential Applications and Effects in Switzerland. Synthesis Document", Zurich. https://www.ebp.ch/sites/default/files/project/uploads/181113_Synthese_BaFoaFn_eng_def_5.pdf (1/7/2020).

EBP, Fonds für Verkehrssicherheit 2018. "Automatisiertes Fahren Auswirkungen auf die Strassenverkehrssicherheit. Schlussbericht vom 31. Mai 2018", Zurich. www.ebp.ch/sites/default/files/project/uploads/2018-05-31%20aFn_Verkehrssicherheit_Schlussbericht.pdf (9/7/2020).

Perret, F., T. Arnold, R. Fischer, P. de Haan and U. Haefeli 2020. *Automatisiertes Fahren in der Schweiz: Das Steuer aus der Hand geben?*, TA-SWISS 71/2020. Zurich: vdf.

18 Lessons from local transport transition projects for connected and automated transport

A discourse and hegemony theory-based assessment of new mobility services in Lower Austria

Andrea Stickler

Andrea Stickler
TU Wien, future.lab Research Center and Research Unit Sociology

© The Author(s) 2023
M. Mitteregger et al. (eds.), *AVENUE21. Planning and Policy Considerations for an Age of Automated Mobility*, https://doi.org/10.1007/978-3-662-67004-0_18

1. INTRODUCTION

In this chapter, we will consider selected projects relating to the transport or mobility transformation[1] in rural and suburban areas, examining them in the context of connected and automated driving. The hypothesis outlined in this study is that, while connected and automated transport can fundamentally change the existing "system of automobility" (Urry 2004), the changes in question have to be seen in the context of specific local challenges and contemporary conflicts around mobility systems. This analysis therefore focuses on local mobility projects that look beyond simply replacing automobility's fossil fuel combustion systems, instead aiming to achieve a broader-based transition via new mobility services such as public transport microsystems, non-profit lift services and car-sharing schemes.

The analysis draws on an empirical investigation in the region of Lower Austria and takes discourse and hegemony theory as its frame of reference. In contrast to other socio-technical theories of transformation, which place greater emphasis on technological niches and their interplay with established systems or regimes (Geels 2012, Loorbach et al. 2017, Kemp et al. 2012), the chosen lens of discourse and hegemony theory focuses more on existing power relations and the role of societal consent in transformation processes (Laclau/Mouffe 2000, Nonhoff 2006, Wullweber 2012). According to Antonio Gramsci's concept of hegemony, change – and this includes the change required to transform transport systems – can only be achieved if there is a certain level of consent or consensus among the population.

Taking Gramsci's premise as its starting point, this empirical research examines recent transport transformation projects offering alternatives to fossil fuel-powered, privately owned cars, specifically in rural and suburban areas of Lower Austria. These case studies allow us to empirically identify contemporary challenges, tensions and contradictions in the implementation of such projects and to recognize which groups or institutions are ready to embrace the transport transformation and what needs can be met by new mobility services.

Here, three types of project are assessed: electric car-sharing schemes, non-profit lift services and ring-and-ride taxis, the latter serving as an example of centralized micro public transport systems. This analysis is then used to ascertain which of the structures inherent in such new mobility services are effective under current conditions and which might, in conjunction with the connection and automation of transport, play a key role in future. As per the theoretical premises outlined above, it assumes that today's relatively stable discursive structures would not significantly shape future actions; new developments and breaks with the past can happen at any time, but the current structure makes certain outcomes more likely than others. By way of background, the next two sections first briefly examine the politics of the transport transformation and illuminate the underlying premises from discourse and hegemony theory.

---

1   Parts of this text are based on the dissertation project "Automobilität im Umbruch? Gegenwärtige Stabilisierungen oder Transformationen der automobilen Hegemonie", developed at the Institute of Spatial Planning's Research Unit of Sociology and TU Wien's future.lab. The terms transport transformation and mobility transformation are mostly used interchangeably in political discourse. For a more precise differentiation of the two terms, see Manderscheid in Chap. 4 of this publication.

2.   THE POLITICS OF THE TRANSPORT TRANSFORMATION

In the realm of climate, energy and environment policy, there has been an increasing focus on transport in recent years, with pressure for political action resulting in legally binding targets[2] at international, national and regional levels. In Austria, both the Mission 2030 national climate and energy strategy and the recently published national climate and energy plan (a plan the EU's governance regulation 2018/1999 obliges each member state to draw up) stress the need for targets and measures for the transport sector and emphasize the key importance of an environment- and innovation-friendly transport and mobility transformation[3] (Federal Ministry for Sustainability and Tourism/Federal Ministry for Transport, Innovation and Technology 2018, Federal Ministry for Sustainability and Tourism 2019), while local politicians (Österreichischer Städtebund 2015) are increasingly recognizing the importance of such a transition. The term transport or mobility transformation is often interpreted in very different ways. In general, though, it revolves around the problematization of greenhouse gas emissions from transport, and from fossil fuel-based automobility in particular, and around aims to reduce car traffic volumes or replace private car journeys with other modes of travel such as public transport, new mobility services, walking or cycling. In politics, this call for a transport and mobility transformation has now become an established policy demand, one that is increasingly gaining broad-based societal support.

The specific policies and strategies proposed to achieve such a transition, on the other hand, remain highly controversial. It is reasonable to assume that decisive action needs to be taken both in passenger and in goods transport; after all, with greenhouse gas emissions still rising, especially in urban centres and along through routes, transport is a particular problem area relative to other sectors (Federal Ministry for Sustainability and Tourism 2019: 6). The focus of such proposals is on trends such as the sharing economy, new mobility concepts – especially in rural areas – and on boosting public transport and promoting active travel. From a technological perspective, key roles are envisaged for electrification and digitization, while connected and automated transport, which combines both, is also expected to help facilitate the transport transformation (Federal Ministry for Transport, Innovation and Technology 2018).

Despite decades of demands for such a transition, it has failed to happen on any significant scale (Schwedes 2011: 14), and the car remains the dominant form of transport in many contexts (Manderscheid 2014). Moreover, the transport sector's greenhouse gas emissions have increased in recent years, as has the number of vehicles on Austria's roads (Statistics Austria 2020). This raises questions as to why so little progress has been made towards the transport transformation, despite political acceptance of its necessity. The next chapter uses perspectives from hegemony and discourse theory in order to examine the politics of the transport transformation in theoretical terms.

---

2   In Austria, these legally binding targets are primarily defined in the "transport sector targets" section of the country's climate protection act.

3   In line with the EU's Effort Sharing Regulation, Austria has pledged to achieve a 36% reduction in greenhouse gas emissions by 2030 compared to 2005 levels, a target also adopted by the provincial governments. The largest cuts are set to come in the transport and buildings sectors. Transport currently has the highest emissions of any sector, being responsible for some 46% of all emissions (excluding emissions trading). To hit that 2030 target, emissions need to be reduced by around 7.2 million tonnes of CO_2 equivalent to around 15.7 million tonnes. In 2017, greenhouse gas emissions for Austria's transport sector (excluding aviation) amounted to 23.6 million tonnes of CO_2 equivalent, which equates to a 73% increase since 1990, and almost two thirds of road transport's greenhouse gas emissions were caused by passenger transport.

3. PERSPECTIVES FROM DISCOURSE AND HEGEMONY THEORY

The notion of a transport transformation involves particular shifts in the hegemonic discourse around transport policies. The desire to see certain changes in the transport sector is driven primarily by environmental concerns, though safety considerations also play a role (see Chap. 10 by Mitteregger in this volume). The following theoretical appraisal of these shifts uses a discourse and hegemony theory-based frame of reference, applying this to the politics of the transport transformation. Thinking on how hegemony theory and concepts from discourse theory intersect often draws heavily on the work of Laclau and Mouffe (Laclau/Mouffe 2000; Laclau 1990, 2002), which was then developed and empirically operationalized by authors such as Nonhoff (2006), Vey (2015) and Wullweber (2012, 2014). The term hegemony refers to the predominance of certain patterns of articulation or social constructs. Hegemony is not fixed, however, but always in flux, resulting from a discursive practice that builds on and modifies given discursive structures (Nonhoff 2006: 137). Hegemonic practice forms part of the political realm's discursive structure and evolves within that realm. Demands for a transport transformation can be read as a hegemonic practice that seeks to reinterpret or alter the structures of automobility in a specific way.

Today's automobility can be seen as a relatively stable discursive formation,[4] albeit one that, given the ever-louder demands for a transport transition, is contested and in need of reform. While demands for a decarbonization-based transport transition have not so far resulted in a counterhegemony that might rival automobility, there have nonetheless been various political articulations that challenge or call into question the universal primacy of the automobile (Federal Ministry for Sustainability and Tourism/Federal Ministry for Transport, Innovation and Technology 2018, Federal Ministry for Sustainability and Tourism 2019). New technologies such as electric vehicles, shared mobility, Mobility as a Service or connected and automated transport aim to move on from fossil fuel-based automobility and represent shifts in the hegemonic discourse. The demand for a transport transformation is gaining traction as a transport policy idea and forming a key nodal point[5] that has an extremely broad range of connotations, bringing together various issues relating to the future of mobility.

In the contemporary debate around the transport transformation, state interventions play a key role. Recent state interventions aimed at bringing about such a transformation have, however, also resulted in the establishment of new forms of subordination. Electric vehicles, for instance, remain controversial within society (Brunnengräber/Haas 2020), as do structural disincentives to car use (e.g. via taxation). Even the meaning of the term transport transformation itself is

---

4   Socio-spatial and group-specific differences in car use play a key role. While the number of people owning a car is going down in many cities, car ownership continues to increase in rural areas (VCÖ 2015). The level of car use and the accompanying stability of automobility as a discursive formation is also linked to the quality of public transport services, while mobility researchers taking a milieu-based approach have highlighted the social differentiation associated with different mobility patterns and car usage (Dangschat 2017, Beck/Plöger 2008).

5   In discourse theory, this nodal point is described as an "empty signifier". The term "signifier" was coined by de Saussure, a semiotician who explored the relationship between language and reality. Signs consist of the signified (the thing being indicated, i.e. the concept) and the signifier (the thing doing the indicating; in the case of language, the "sound image"). The relationship between the signified and the signifier is essentially arbitrary, though given that the relationship is a product of society, not entirely random (Hagemann 2016: 16). The universalization of the specific that occurs in hegemonic projects renders the signifier empty, with the sign becoming detached from its meaning.

Figure 1: Transport transition as a nodal point

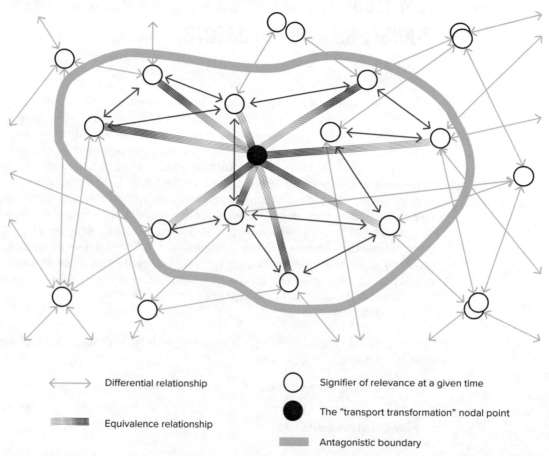

⟷ Differential relationship	⚪ Signifier of relevance at a given time
▬▬ Equivalence relationship	⚫ The "transport transformation" nodal point
	▬ Antagonistic boundary

Source: the author, adapted from Glasze (2008: 194)

contested. Environmental demands thus exist in multiple forms that "depend upon the manner in which the antagonism is discursively constituted" (Laclau/Mouffe 2000: 210). These demands can be anti-capitalist, anti-individualist and authoritarian but they can also be libertarian, social-ist and reactionary. The nature of such an environmental demand is variable rather than fixed from the start; the manner in which the demand is articulated is by no means preordained.

The use of the term transport transformation creates specific equivalent relationships with oth-er important terms, such as technological progress, innovation, economic growth, safety and sustainability or renewable energy sources. These terminological relationships (equivalence relationships) point up problems with today's (car-dominated) mobility systems, lending them a positive aspect (Wullweber 2014: 291). At the same time, the term transport transformation sets itself apart from other key concepts such as fossil fuels or urban sprawl, thus forming a dif-ferential relationship and antagonistic boundary. Various societal actors regard their interests as being tied up with the concept of a transport transformation. In this context, there is now a largely established narrative relating to the transformational processes around electric vehi-cles, car sharing and automation.

The hegemonic shift towards a transport transformation is thus being driven by a range of so-cietal forces. As a result, this transformation, which aims to tackle key societal challenges, has now been successfully associated with the public interest.

4. NEW MOBILITY SERVICES AS EXAMPLES OF TRANSPORT TRANSFORMATION PROJECTS

To ascertain what shifts or schisms the push for a transport transformation has brought about in the discourse around automobility, it is worth looking in detail at existing transport transformation projects. By analysing key parameters and evaluating interviews with a range of relevant actors, this section aims to show what general demands have arisen in connection with new mobility services. It also looks at factors affecting the implementation and consolidation of these projects, examining where challenges and problems remain. The projects selected were station-based electric car-sharing schemes, non-profit lift services and ring-and-ride taxi services. Based in suburban or rural areas, these projects all aim to provide alternatives to private car use. The 12 interviewees were project initiators, representatives of local authorities or commercial providers of new mobility schemes. In addition to these interviews, talks and seminars given by representatives of the province of Lower Austria or of its energy and environment agency were also analysed. It was, however, not possible to also gather broad-ranging input on users' perspectives and perceptions of these services in wider society – more research is thus needed here.

Figure 2 below gives an overview of the various new project types being offered within the context of mobility as a service.

Figure 2: Mobility schemes (Mobility as a Service – MaaS)

Fixed-route services

High-quality bus and rail services

Taxi systems

E.g. taxi voucher system

Ring-and-ride taxis with fixed pick-up points

Mobility as a Service (MaaS)

Ride-sharing

Sharing rides in a user's own car

Using professional drivers or in a car club

Car-Sharing

Sharing a privately owned vehicle (with neighbours, for instance)

Using EVs provided by a car club or mobility service provider

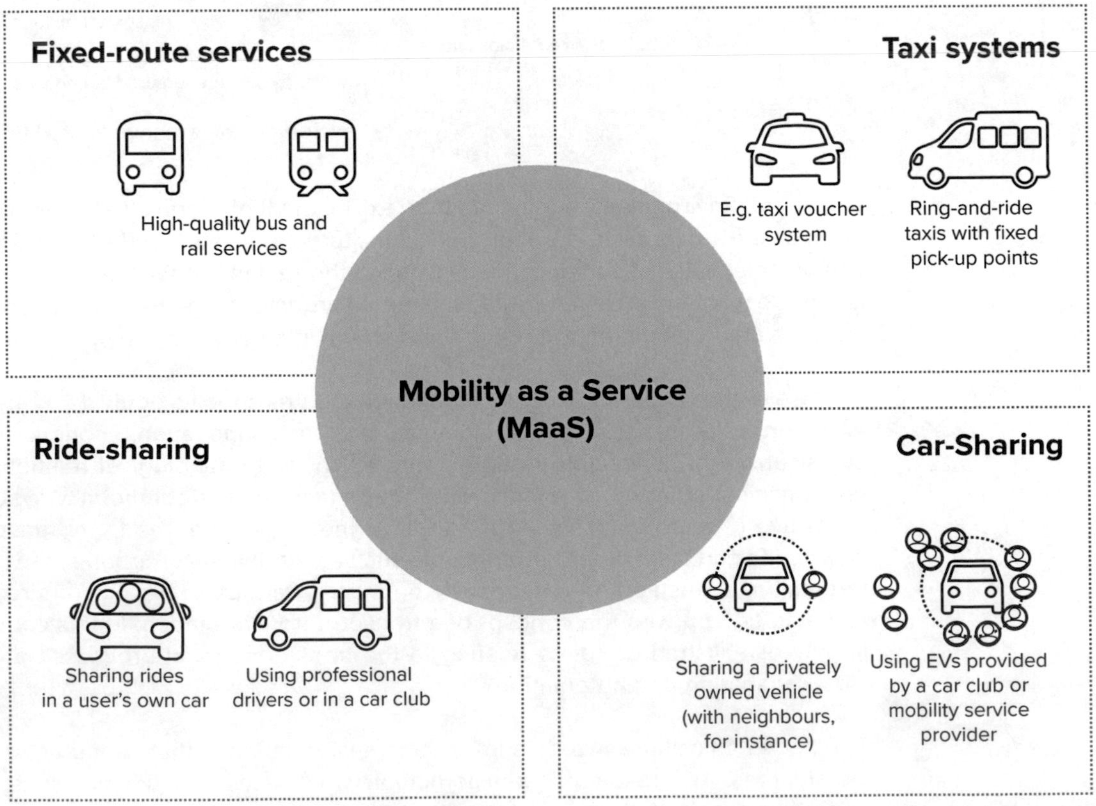

Source: the author, adapted from Danninger (2019)

Table 1: Comparison of selected mobility services

	Station-based electric car sharing	Non-profit lift services	Ring-and-ride taxi services
Primary operator	Municipality, company	Lift club	Municipality, region
Primary service type	Public sector, commercial (with public-sector support)	Private, non-commercial (with public-sector support)	Public sector (in conjunction with contracted commercial transport firms/platform operators)
Costs for users	Medium	Low	Medium
Responsibility for driving	Vehicles driven by the users themselves	Vehicles driven by volunteer drivers	Vehicles driven by professional drivers
Area covered	Vehicles accessed from fixed locations but can also be used beyond municipal boundaries	Services primarily operated within municipal boundaries	Operates alongside public transport, servicing predefined pick-up points

Source: the author

Table 1 above compares the three selected models – electric car sharing, non-profit lift services and ring-and-ride taxis – with regard to operator, service type, usage costs, responsibility for driving and area covered.

The aim in analysing these three different kinds of mobility service is to assess how transport transition projects can achieve a hegemonic shift in the relatively stable "system of automobility" (Urry 2004). The key question here is whether a large number of actors can be convinced that the implementation of such a project is essential to the public interest and persuaded to act accordingly.

4.1 STATION-BASED ELECTRIC CAR SHARING

In recent years, numerous electric car-sharing schemes have launched in suburban or rural areas; these are mostly locally based though some are regional. Electric car sharing allows a large number of users to try out and use an electric car in everyday situations. These schemes are often promoted as a way for households to avoid running a second vehicle. Compared to running a car of their own, the costs incurred by car-sharing users are relatively low. The innovative and sustainable nature of such projects provides an image boost for the operator – in rural areas, this is mostly the municipality. In Lower Austria, electric car-sharing schemes are mostly provided by a charity, municipality or professional operator, with the type of operator being key in determining whether the service is subject to commercial law.

For an electric car-sharing scheme to be economically viable, it needs 20 to 30 users per vehicle. Schemes can be initiated by private individuals, companies or municipalities. Users generally pay an annual charge of €100 to €300, plus a rate per kilometre (€0.10 to €0.20) and/or per hour (€1 to €5). Cars are generally booked via an app or some other online booking system. Caruso and Ibiola are among the platforms using such a booking system. The booking platform displays the availability of the scheme's vehicles and the requisite charging time between bookings. Lower Austria is considered a model province in the field of electric car sharing. More than 120 EVs are available in car sharing schemes across over 90 municipalities. Increasingly, such schemes are being operated in conjunction with EV-based lift services (Komarek 2019).

Figure 3: Electric car-sharing schemes in Lower Austria

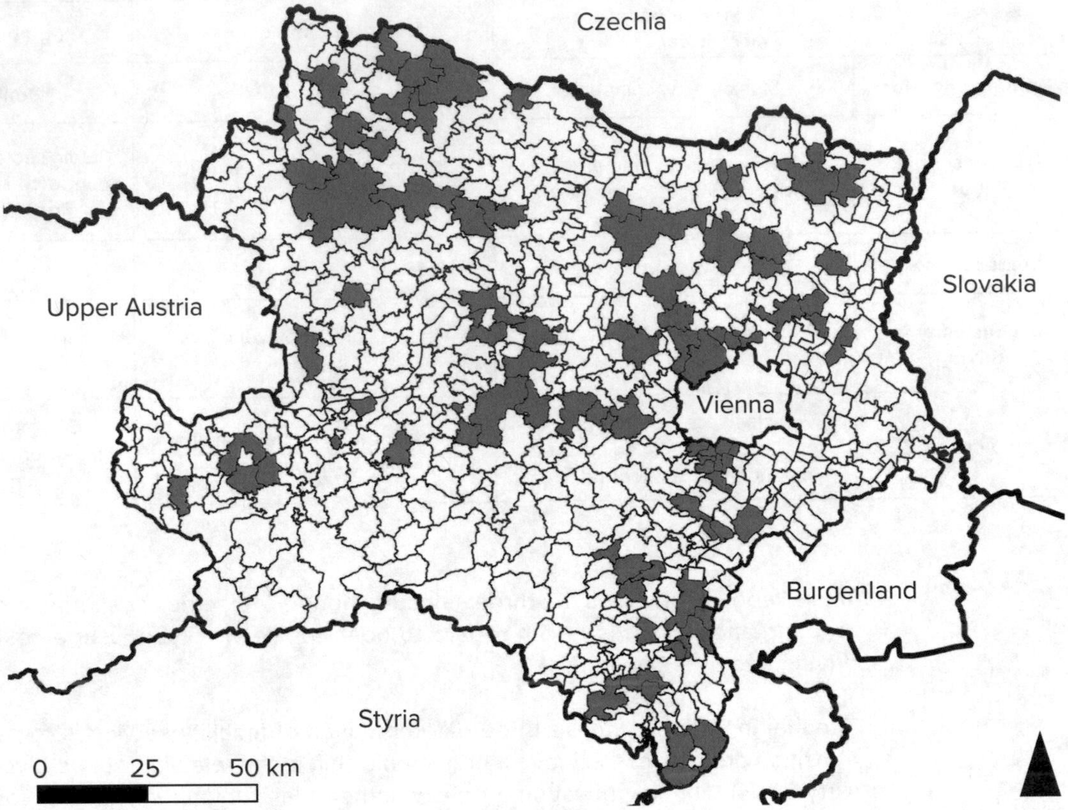

Czechia

Slovakia

Upper Austria

Vienna

Burgenland

Styria

0   25   50 km

Source: the author, adapted from eNu data (2019)

It became clear from examining and talking about electric car-sharing schemes that, in rural areas, these tend to require strong political impetus, especially in the initial phase. They need someone involved in policy and planning to take up the issue and champion it in public. This is how an interviewee who had initiated an electric car-sharing scheme put it:

> *"In my opinion, what it takes to get these things off the ground is an organizer, i.e. a central contact person, ideally a political representative, who is prepared to take up the cause and sell it to people as a great thing."* (Interview A)

These new mobility schemes are often linked to a desire among local politicians to be innovative and attract attention. Municipalities celebrate their new mobility schemes and gain plaudits for them, for instance by entering the "Clevermobil" competition, via which the province of Lower Austria recognizes particularly innovative projects. This desire to be innovative can also be seen in the numerous networking meetings and seminars relating to new mobility services, at which there is a certain competitiveness and rivalry between political representatives when it comes to their municipality's innovations. Moreover, the implementation of new mobility services is strongly reliant on local commitment among a municipality's officials and representatives. Politicians or planners at higher levels of government are less likely to be involved in driving new electric car-sharing schemes, though they do sometimes provide additional support. Several interviewees referred to the need for commitment at municipality level. Responding to the question of how successful electric car sharing currently is, one car-sharing project's director stated:

"That depends a lot on municipalities' commitment and on how strongly they back it. Some municipalities are very active, and their schemes are going very well." (Interview B)

That requisite degree of commitment at municipal level, however, often meets with criticism, exposing initiators to significant personal risk, particularly in rural areas. Another interviewee who championed the creation of a new electric car-sharing scheme (Interview E) describes the implementation of electric car sharing as a "constant battle". The benefits of such schemes often go unrecognized and unacknowledged, and experience has shown that, despite the efforts of local political representatives, there is often limited public interest in the use of electric car sharing, with many pilot projects not being taken forward as a result. For instance, several municipalities conducted surveys to assess public interest in electric car sharing, but did not receive a sufficiently broad response. Interviewees also emphasized the importance of choosing the "right" moment to launch such a scheme, with two municipal representatives describing their unsuccessful launches as follows:

"After I'd been to a talk at Lower Austria's energy agency in 2016 [...], we just gave it a go [...], publicizing it in the local newspaper. Only one person signed up though, so then we just dropped the idea." (Interview H)

"We've put so much time into a potential electric car-sharing scheme already, organizing information events that generated only very limited interest. I've invested countless hours in it, but it all came to nothing. Practically no one turned up to the information events. Maybe we went for it too early and were a bit ahead of our time." (Interview D)

The municipal representatives interviewed for this research described how expressions of interest were not forthcoming after postal or media communications on potential electric car-sharing schemes. That engendered a feeling of resignation among political representatives and a critical attitude towards electric car-sharing projects that were in operation elsewhere. Some sceptical municipal representatives also argued that electric car-sharing schemes being presented as success stories needed to be examined critically as the users they were attracting did not always correspond to the desired target group (car drivers switching to a sharing model). The project initiators interviewed here also raised long-term provision and continuity of usage as challenges for new mobility services such as electric car sharing. Electric car sharing can even lead to users eventually buying an EV of their own, after which they then hardly use the car-sharing scheme any more.

"Continuity of use is another challenge. What happens if people pull out because they've bought a vehicle themselves, for example? The system is then undermined." (Interview G)

If it serves as a temporary solution until users buy a vehicle of their own, then car sharing can potentially further consolidate automobility's hegemony, even if more of the individual vehicles end up being electrically powered.

Interest in car sharing remains particularly limited in areas where most people own a car and are already heavily dependent on automobility. Interviewees repeatedly expressed scepticism regarding schemes celebrated as success stories by municipal representatives. Due to the high degree of rivalry around innovation in municipalities, many schemes are given an overwhelmingly positive spin, with any problems swept under the carpet. One municipal representative gave the following critical account of such spin:

"Many of the schemes only work because the municipality itself is 'obliged' to use it. That means all municipal politicians and officials are obliged to use the service, which is not the idea at all. These schemes are kidding themselves with their claims of success because they're not persuading people to switch to new mobility solutions. It needs to be driven much more by popular demand — with people saying to politicians 'Hey, let's do that, we want this!', but that isn't happening." (Interview D)

According to this interviewee's account, car sharing only works if it is also regularly used by municipal officials, meaning it fails in its actual aim of persuading people to ditch individual car ownership in favour of sharing cars with other users. The interviewee also describes the lack of popular demand for the introduction of electric car sharing. Despite these experiences and the scepticism around electric car sharing, such schemes have become increasingly prevalent in Lower Austria, though they differ greatly in how they are organized. The integration of these schemes into an overarching mobility platform is where observers see the greatest potential for further development, but such a move is hindered by their heterogeneous providers and organizational structures. Evidently, there is not always popular interest in electric car-sharing schemes; often they are not a response to broad-based subordinate demands for automobility to be transformed. On the other hand, they represent a very specific expression of opposition to automobility's hegemony, one driven primarily by state intervention.

Figure 4: Lift services in Lower Austria

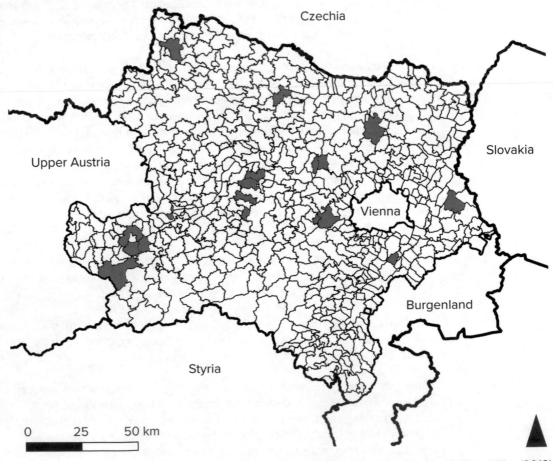

Source: the author, adapted from Wels-Hiller (2019)

4.2 NON-PROFIT LIFT SERVICES

Another niche examined in this chapter on practical transport transition projects is non-profit lift services. Using volunteer drivers, these schemes, in which volunteers drive members of a club from A to B by electric car, can be classed as decommodified mobility services. This model grew out of the community bus service, a citizen-initiated form of local transport launched in Lower Austria in 2011. The aim of non-profit lift services is to better integrate less mobile individuals into community life and to boost communication and interaction between different generations. In addition, it is hoped that lift services will reduce the need for existing pick-up/drop-off services for limited-mobility individuals (such as the elderly or children), while lift services are sometimes also associated with a boost to the local economy as they can be used for local shopping trips. Unlike electric car sharing, a lift service does not require the electric vehicle to be located in the immediate vicinity of the user's place of residence. Such services allow a large number of people to gain their first experiences of electric vehicles, something the provincial government of Lower Austria regards as an important objective, hoping to thereby raise awareness of electromobility. In the same context, lift services have also been discussed as a potential substitute for a household's second car. Their costs to users are relatively low and they can also complement public transport networks (Komarek 2019).

These services are mostly provided by an organized club, albeit with the backing of the municipality. They are non-profit-making and charge accordingly, though their status in commercial law needs to be checked with the district commission. Those transporting people for money generally require a licence (be it a public transport licence or a taxi or car rental licence). To simplify the legal situation, standardized statutes have been developed and made available by Lower Austria's energy and environment agency, while service providers are encouraged to work with taxi firms and driving schools in order to limit competition for trade and assess drivers' competence. Having enough motivated drivers is said to be the key factor in whether such a service is successful, with 20 to 30 thought to be the minimum number required. Another critical step is defining the service's operating times and planning drivers' shifts accordingly. Non-profit lift services are only allowed to carry officially registered club members. The latter pay monthly or annual membership fees. Mostly, lift clubs distinguish between regular and associate (accompanying) members. Schemes can charge flat-rate membership fees, though there is also the additional option of charging for individual journey costs. This is done via booking, billing and admin tools such as Emilio, Tullnerbacher or similar. In some cases, services are subsidized via Lower Austria's local public transport funding scheme (Komarek 2019).

Non-profit lift services are often run in combination with electric car-sharing schemes. In 2019, there were around 25 such services registered in Lower Austria, primarily in rural areas (ibid.). The province's energy and environment agency provides support for EV-based lift services, particularly in the launch phase, helping with rough cost calculations and advice on implementation, arranging a promotional evening event and providing standardized statutes as well as a communications package for residents.

As these lift services are organized as clubs, setting one up requires a high degree of local commitment and personal contacts with potential drivers. One interviewee involved in local politics who had experience of initiating a lift service describes the launch process as follows:

> "It was a very stressful process, I spent days driving around and looking in coffee shops for people who might be interested. By just approaching and asking people, I found 25 members who were prepared to join up on the spot. Personal contact was very important; after that word got around." (Interview C)

The process was thus heavily reliant on social media and on making personal contact. It is important for there to be a central contact person, particularly in the early set-up phase, and to decide which groups can be targeted: some interviewees cited younger pensioners as a key target group, along with young families keen to volunteer and make contacts in their new area. Another interviewee involved in local politics who had experience of setting up a lift service describes the starting point as follows:

> "We have a lot of newcomers in our municipal area – young families who have completed their studies and want to move back to the countryside. We're blessed with an extremely good location here [a municipality on the outskirts of Vienna]. It's just 30 minutes to Vienna, and the same to St. Pölten. But that also means many of those who move to the area hardly have any extended family here. We started off with a ring-and-ride taxi service, but in practice hardly anybody used it. Then after a bit of brainstorming, we came up with the idea of volunteer lift services, using electric vehicles of course and relying on volunteer drivers." (Interview F)

Setbacks were a recurring theme for many of those interviewed for this study. The launch of new mobility services did not always go well from the start. While the interviewee in Interview F reported that the launch of a ring-and-ride taxi service was initially unsuccessful, the lift service enjoyed greater success thanks to the specific social group targeted in that municipality and their particular needs. With many residents hardly having any extended family in the area, the willingness to volunteer and get involved in such clubs was high. A lift service not only serves to provide physical transportation, it also boosts the sense of community among members. The social contact afforded by the lift club thus performs a useful social function. This aspect was emphasized by multiple interviewees:

> "As I mentioned already, the social aspect of it all is key. Because it's so challenging, many older drivers view it as a way of keeping active in old age, because you have to think a lot and because it's important for older people to still have a purpose." (Interview F)

> "The people doing most of the driving are younger pensioners; they then meet up to clean the car, do maintenance on it, and so on. And the drivers tell me they find it very fulfilling, so in that sense our scheme is like a mini social project." (Interview H)

Those who describe working for a non-profit lift service as fulfilling are primarily older people. The way the service is marketed or framed plays a key role. Some lift services emphasize not only the contribution they make to sustainable mobility but also their economic benefits. One interviewee stressed that this broad-based framing is actually a must if the service is to be successful (Interview F). Desire for media coverage among local politicians can also play a key role in the setting up of a lift service. Both municipal representatives and lift club directors celebrate their role in such mobility services, earning commendations from higher political authorities for the innovative nature of their schemes. The interviewee describes this media coverage as follows:

> "It's also important to have the mayor on board, because then it'll be in the newspaper. That in itself brings in a lot of people who want to do some good, and then a lot of others who follow their lead." (Interview F)

Various interviewees emphasized that, in order to promote the service better, reach more people and gain broader popular support, it helps to have municipal representatives involved in the lift club. Often the mayor also serves as club chair, which sends a positive message to the public and can help to boost the scheme. Promotion in local or regional print media also plays an important role in growing membership.

Judging by the interviews, dissuading households from running a second car and better integrating individuals with limited mobility are the primary reasons for establishing lift services. Achieving a far-reaching transformation of hegemonic automobility, on the other hand, is generally not among the immediate aims. When it comes to their driver pool, lift services attract a particular section of the population (primarily younger pensioners), while usage is only open to registered lift club members. Lift services thus remain a very specific solution. There does, though, seem to be a higher level of interest among the population than is the case with electric car sharing, though broader demands for far-reaching changes to the hegemony of the car are rarely voiced in conjunction with such services.

4.3 RING-AND-RIDE TAXIS

The third project type examined as part of this analysis is ring-and-ride taxi services. Ring-and-ride taxis and micro public transport systems in general serve to supplement regular public transport systems. They help to ensure broader coverage beyond fixed-route services (in metropolitan areas, on highly frequented routes and at peak travel times). On less frequented routes and at off-peak times, it is almost impossible to provide standard public transport cost-effectively, so local, demand-led mobility services are a useful way of filling gaps.

In Lower Austria, ring-and-ride taxi services are operated by taxi or car rental firms. Journeys can be booked for predetermined departure times, with passengers collected from set pick-up points, from where they are taken directly to specific drop-off points. Unlike conventional public transport operators, ring-and-ride taxi operators do not require a special licence. Passengers pay the standard fare for public transport plus a "convenience supplement". Use of ring-and-ride taxis is not limited to set public transport routes within the service area (Gausterer 2019).

The advantage of ring-and-ride taxis is that they enable local public transport to be provided more cost-effectively and allow standard schedules to be supplemented at times of low demand. As they are not tied to fixed routes, ring-and-ride taxis can service not just specific stops but also entire geographical areas, meaning target group-specific offerings can be created. The requirement to prebook by phone and the slightly higher fares, meanwhile, are cited as disadvantages.

When planning a ring-and-ride taxi service, the target group, operating times and service area all need to be defined. The specifics of the service have to be planned in conjunction with the municipality and the local transport authority. The municipality then calculates the cost of employing local taxi firms and draws up the necessary contracts. Often it will work with a mobility platform provider (such as ISTmobil). Initiators can apply for joint funding from federal, provincial and municipal authorities via Lower Austria's local public transport funding scheme (Land Niederösterreich 2020). Another benefit of ring-and-ride taxis is that, unlike lift services, they do not operate in a legal grey area. If need be, journeys can cross municipal boundaries as long as this is covered contractually. In addition, this kind of service is not reliant on the commitment of volunteer drivers, nor does a dedicated members' club need to be set up (Gausterer 2019). In contrast to volunteer lift services, ring-and-ride taxis are thus available to anyone and not just to members.

Interviewees stressed dedicated oversight and effective marketing as key to launching such a scheme successfully. Ring-and-ride taxi services benefit from clearer statutory regulation than lift services and are mostly set up in conjunction with local taxi firms and bus companies. Interviewees described the launch process as very time-consuming and stressful:

> *"Getting to that point was a very protracted process, requiring a great deal of patience and a lot of meetings. I see it as a process that's still ongoing, because there is regular need for improvements and we are constantly fine-tuning the concept." (Interview I)*

Ring-and-ride taxis operate on specific public transport routes, servicing particular stops. Deliberate efforts are made to avoid competing with fixed-route public transport, however, with taxis picking up only at selected stops. One interviewee describes the local ring-and-ride taxi's service area as follows:

> *"We've now got 634 pick-up points, some of which are in neighbouring municipalities because it makes more sense that way. We only go from stop to stop and we don't want to compete with public transport systems, which is why we put a lot of emphasis on ensuring the service feeds into the public transport system." (Interview I)*

Figure 5: Ring-and-ride taxis in Lower Austria

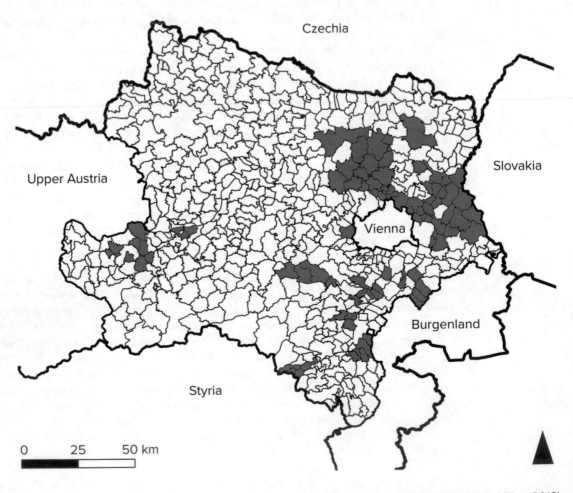

Source: the author, adapted from Wels-Hiller (2019)

Setting up a ring-and-ride taxi service is a long-term process in which pick-up points have to be regularly renegotiated. Launching and operating such a system comes at a high cost to the municipality[6] and also requires a great deal of legal, organizational and financial support from policymakers and planners at higher levels of government. Ring-and-ride taxi services generally operate between various fixed pick-up points. In order to meet the objective of integrating limited-mobility individuals, however, certain exceptions are considered:

> "We generally don't operate door to door; limited-mobility individuals can apply for home pick-up however, but this has to be officially registered." (Interview I)

As long as the service is sufficiently attractive and suitably promoted, ring-and-ride taxis tend to be well used (Interview I). The extent to which they persuade people to switch to public transport cannot yet be precisely gauged, though the limitations predefined stops and routes place on their potential use represent a clear disadvantage in comparison with owning a car or using lift services.

5. HEGEMONIC SHIFTS WITH NEW MOBILITY SERVICES

New mobility services represent congruent but often very specific solutions (with regard to their target groups, operating times and service area) that fit into the existing transport transformation discourse. The innovations they represent act primarily within the "system of automobility" and do not therefore call that system into question. On the contrary: they seek to reshape it via new supplementary services (making it more sustainable or socially integrative) and thus to consolidate it. In many cases, there is neither a widespread desire for automobility to be renounced nor the acceptance within society that new mobility services should be introduced on a broad scale. Adding new facets to the existing mobility system, such services are thus not so much actively supported as passively tolerated.

As new mobility services are introduced, certain tensions become apparent. These include competitive relationships between new services and established firms, though state-run schemes make efforts to avoid these. Shoring up traditional sectors such as the taxi trade and public transport, on the other hand, creates its own tensions when it comes to the introduction of new mobility services. Another issue is the highly specific nature of these new mobility services, which vary greatly in how they are organized and have thus far not really lent themselves to integration within a common platform.

It is clear from the above analysis that these new mobility schemes do not generally call automobility's dominant status into question. Instead, they add supplementary mobility services that, at best, might allow households to do without a second or third car and individuals without a driving licence or vehicle to get out and about by car. In addition, such new mobility services are often not rooted in people's thoughts and actions, but brought about via state interventions as authorities compete to demonstrate innovation. As a result, these would-be transport transformation projects should be seen as political rather than as hegemonic. There is, after all, no attempt here to create services with the kind of universal meaning and validity required for

---

6   The costs to the municipality are mostly charged in proportion to the number of inhabitants.

large sections of society to move away from today's high level of dependence on automobility. Antagonistic demands for more far-reaching change, as voiced by certain action groups, such as the campaign network Bürgerinitiative Verkehrswende Niederösterreich or the cycling lobby group Radlobby Niederösterreich, have been sidelined in the discourse. To a certain extent, transport transformation projects can thus help to consolidate automobility, even if the modes of transport used become more differentiated.

Nonetheless, hegemonic shifts can arise at any time as a consequence of new mobility services. The problematization of automobility arising from the transport transition debate opens up space for more universal calls for automobility's transformation, providing scope for projects that could challenge it to establish themselves and develop over the long term. Ultimately, the need for a transport transformation is now rarely questioned, with critical voices focusing merely on how and not whether it should happen. Despite its relative stability, automobility is thus by no means immutable; in fact, it is only as a discursive and dynamic phenomenon that it can maintain its predominance. The potential consequences of these insights for connected and automated transport are examined in the final section of this chapter.

6. CONSEQUENCES FOR CONNECTED AND AUTOMATED TRANSPORT

The study *AVENUE21. Connected and Automated Driving: Prospects for Urban Europe* presented a very ambivalent picture of the spatial and social impacts of connected and automated transport, which differ greatly according to the underlying policy and planning scenarios (Mitteregger et al. 2022: 99–140). The hopes and expectations associated with automated driving also vary (see Chap. 19 by Dangschat in this volume for more details), while the *AVENUE21* study provided a comprehensive overview of hopeful and sceptical expectations regarding automated transport (ibid: 33–46). The main positive expectations were a reduction in the number of accidents, efficient traffic flow management, enhanced regulation of vehicle speed and easier identification of available parking spaces, a drop in energy consumption and a decrease in the number of vehicles (and hence an increase in the available road space), the enhancement of intermodality, greater social inclusion for limited-mobility groups, stress-free driving and the ability to use travel time for other activities. On the other hand, there is much scepticism regarding assumptions around reduced traffic volumes and decarbonization, with doubts raised in particular by the enhanced convenience of CAVs and the increased attractivity of outlying residential areas, the potential increase in traffic levels due to empty runs and the integration of new road users previously unable to travel by car (Dangschat 2019). In addition, the impacts of automated driving could undermine the objectives of sustainable urban development (compact cities, cities of short distances, ecomobility). Many of the positive assumptions can only be realized once there is a high prevalence of automated vehicles and are particularly open to question if there is a lengthy period of mixed-traffic flows (Mitteregger et al. 2022).

For the positive expectations regarding the technology's environmental and socially integrative potential to be fulfilled, the synergetic effects of automation, electromobility and sharing need to be exploited (ibid.). While automated vehicles are for the most part still being tested in controlled environments, the mobility services analysed here have already been operational for some time. In some cases, these projects already combine electromobility and sharing, while automation could be added in future. The spatial, economic, environmental and social impacts of automated driving are still highly uncertain (ibid.), which makes insights and conclusions derived from existing transport transition projects all the more useful. Given this general uncertainty around automation, it is worth examining three aspects of the empirical case study analysis and considering them in the context of connected and automated transport:

- New mobility services in rural and suburban areas have to date been very specific (with regard to target groups, operating times and service area) and have not always gone hand in hand with broad-based, popular demands to move away from automobility and towards a more holistic transport system. In a similar way, it is possible that automated vehicles and mobility services could be integrated into pre-existing structures without any politicization of demands for a transport transition. In addition, the nature of such a transport transition remains very much open to interpretation, which can lead to internal tensions. If the aim is for new mobility services to form part of a broad-based transport transformation, then they need to pose much more of a challenge to the current system of automobility. Only then will such mobility services establish themselves as hegemonic and be capable of breaking our high levels of car dependency and solving the problems that individual car ownership brings. If new connected and automated transport services aimed at combining electro-mobility, vehicle sharing and automation are to be implemented on a large scale, they therefore need to go hand in hand with widespread demands for the current system of automobility to be transformed and they need to enjoy broad-based support within society.

- Contemporary projects providing new mobility services differ greatly in terms of how they are organized and structured – they rely on strong local commitment, involve complex coordination between disparate actors and feature varying organization types, operational models or target groups. Bringing these diverse schemes together in one hegemonic transport transition programme is a major challenge, as is integrating them under one "Mobility as a Service" umbrella. The contemporary projects analysed here suggest that the introduction of connected and automated vehicles will be contested and require extensive coordination and negotiation processes. We should not therefore assume it will be a linear, conflict-free or top-down process. From a hegemony theory perspective, achieving the kind of change required for a transport transition requires popular consent, with the people also being granted the power to influence what form such projects take.

- In addition, implementing collective, shared or public transport systems requires a more broad-based value shift. Encounters with new mobility solutions can act as a trigger for this value shift: lift services, ring-and-ride taxis, car sharing, or testing environments for connected and automated driving can help to convey the advantages of these systems over today's automobility (in terms of cost, maintenance, space utilization, emissions, etc.) and to foster a value shift. In that regard, today's new mobility projects could also strengthen forces that are antagonistic towards hegemonic automobility. While non-profit lift services are likely to suffer as a result of automation, potentially losing their social function, ring-and-ride taxis are likely to benefit, thanks to a lowering of operating costs. Car sharing could become increasingly automated and potentially move from a station-based to a free-floating model. If automation allows for driverless operation in certain environments, we could see increasing convergence between car sharing and taxi systems or lift services. Connected and automated vehicles could then positively impact this value shift towards a broader acceptance of shared transport systems, in part at least because of the general fascination exerted by technological innovations.

REFERENCES

Beck, S., and W. Plöger 2008. "Lebensstile und Mobilität", in *Stadtentwicklung. Lebensstile und Mobilität*. Bundesverband für Wohnen und Stadtentwicklung, vhw FW 1, 48–51. https://tinyurl.com/y82e6zmn (23/4/2020).

Brunnengräber, A. and Haas, T. 2020. *Baustelle Elektromobilität. Sozialwissenschaftliche Perspektiven auf die Transformation der (Auto-)Mobilität*. Berlin: Transcript.

Dangschat, J. S. 2017. "Wie bewegen sich die (Im-)Mobilen? Ein Beitrag zur Weiterentwicklung der Mobilitätsgenese", in *Verkehr und Mobilität zwischen Alltagspraxis und Planungstheorie. Ökologische und soziale Perspektiven*, ed. by M. Wilde, M. Gather, C. Neiberger and J. Scheiner. Wiesbaden: Springer VS, 25–52.

Dangschat, J. S. 2019. "Automatisierte und vernetzte Fahrzeuge – Trojanische Pferde der Digitalisierung?", in *Infrastruktur und Mobilität in Zeiten des Klimawandels, Jahrbuch Raumplanung* vol. 6, ed. by M. Berger, J. Forster, M. Getzner and P. Hirschler. Vienna: Neuer Wissenschaftlicher Verlag, 11–28.

Danninger, O. 2019. "Dekarbonisierung der Mobilität in NÖ", talk held as part of the 5th e-mobility conference "e-mobil in niederösterreich" in St. Pölten. www.ecoplus.at/media/14895/1_danninger_dekarbonisierung.pdf (30/1/2020).

eNu (Energie- und Umweltagentur des Landes NÖ) 2019. "Projekte in Niederösterreich", www.umweltgemeinde.at/e-carsharing-in-niederoesterreich (31/1/2020).

Federal Ministry for Sustainability and Tourism 2019. "Integrated National Energy and Climate Plan for Austria, 2021–2030", 18/12/2019. Vienna. https://ec.europa.eu/energy/sites/ener/files/documents/at_final_necp_main_en.pdf (23/5/2022).

Federal Ministry for Sustainability and Tourism and Federal Ministry for Transport, Innovation and Technology 2018. "#mission 2030: Austrian Climate and Energy Strategy". https://gruenstattgrau.at/wp-content/uploads/2020/10/mission2030_oe_climatestrategy_ua.pdf (23/5/2022).

Federal Ministry for Transport, Innovation and Technology 2018. "Austrian Action Programme on Automated Mobility 2019–2022". Vienna. https://www.bmk.gv.at/dam/jcr:56570b3f-9b2a-42b7-838c-4a201a501ef3/action_automated_mobility_2019-2022_ua.pdf (23/5/2022).

Gausterer, F. 2019. "Projektabwicklung und Empfehlungen für Trägerorganisationen", talk held as part of the seminar "E-Fahrtendienst in NÖ" organized by eNu (Energie- und Umweltagentur des Landes NÖ), 6/11/2019, St. Pölten.

Geels, F. W. 2012. "A socio-technical analysis of low-carbon transitions: introducing the multi-level perspective into transport studies", in *Journal of Transport Geography* 24, 471–482.

Glasze, G. 2008. "Vorschläge zur Operationalisierung der Diskurstheorie von Laclau und Mouffe in einer Triangulation von lexikometrischen und interpretativen Methoden", in *Historical Social Research* (33) 1, 185–223.

Hagemann, I. 2016. "Das gegenhegemoniale Moment der Demokratie. Gegenhegemoniale Projekte und demokratische Demokratie am Fallbeispiel der grünen Bewegung", dissertation, Universität Duisburg-Essen.

Kemp, R., F. W. Geels and G. Dudley 2012. "Introduction", in *Automobility in Transition? A Socio-Technical Analysis of Sustainable Transport*, ed. by F. W. Geels, R. Kemp, G. Dudley and G. Lyons. New York/London: Routledge, 3–28.

Komarek, M. 2019. "e-Mobilität & e-Fahrtendienst in NÖ", talk held as part of the seminar "E-Fahrtendienst in NÖ" organized by eNu (Energie- und Umweltagentur des Landes NÖ), 6/11/2019, St. Pölten.

Laclau, E. 1990. *New Reflections on the Revolution of our Time*. London: Verso.

Laclau, E. 2002. "Was haben leere Signifikanten mit Politik zu tun?", in *Emanzipation und Differenz*, ed. by E. Laclau. Vienna: Turia & Kant, 65–78.

Laclau, E. and C. Mouffe 2000. *Hegemonie und radikale Demokratie. Zur Dekonstruktion des Marxismus*. Vienna: excerpts.

Land Niederösterreich 2020. "NÖ Nahverkehrsfinanzierungsprogramm (NÖ NVFP)", http://www.noe.gv.at/noe/OeffentlicherVerkehr/Foerd_NOE_NVFP.html (1/2/2020).

Loorbach, D., N. Frantzekaki and F. Avelino 2017. "Sustainability Transitions Research: Transforming Science and Practice for Societal Change", in *Annual Review of Environment and Resources* (42) 1, 599–626.

Manderscheid, K. 2014. "Formierung und Wandel hegemonialer Mobilitätsdispositive: Automobile Subjekte und urbane Nomaden", in *Zeitschrift für Diskursforschung* (2) 1, 5–31.

Mitteregger, M., E. M. Bruck, A. Soteropoulos, A. Stickler, M. Berger, J. S. Dangschat, R. Scheuvens and I. Banerjee 2022. *AVENUE21. Connected and Automated Driving: Prospects for Urban Europe*, trans. M. Slater and N. Raafat. Berlin: Springer Vieweg. DOI: 10.1007/978-3-662-64140-8.

Nonhoff, M. 2006. *Politischer Diskurs und Hegemonie. Das Projekt "Soziale Marktwirtschaft"*. Bielefeld: transcript.

Österreichischer Städtebund 2015. "Positionspapier des Städtebundes zum bundesweiten Handlungsbedarf im Bereich des städtischen/stadtregionalen ÖPNV", 18/3/2015.

Schwedes, Oliver 2011. "Statt einer Einleitung", in *Verkehrspolitik. Eine interdisziplinäre Einführung*, ed. by O. Schwedes. Wiesbaden: VS Verlag, 13–36.

Statistics Austria 2020. "Pressemitteilung: 12.165-005/20. Pkw-Neuzulassungen". www.statistik.at/web_de/presse/122440.html (15/1/2020).

Urry, John 2004. "The 'System' of Automobility", in *Theory, Culture & Society* (21) 4–5, 25–39.

VCÖ (Verkehrsclub Österreich) 2015. "Beim Autobesitz geht Schere zwischen Stadt und Land immer weiter auseinander". www.vcoe.at/presse/presseaussendungen/detail/vcoe-beim-autobesitz-geht-schere-zwischen-stadt-und-land-immer-weiter-auseinander (26/3/2020).

Vey, J. 2015. *Gegen-hegemoniale Perspektiven. Analyse linker Krisenproteste in Deutschland 2009/2010*. Hamburg: VSA. https://tinyurl.com/ya5vz95f (23/4/2020).

Wels-Hiller, S. 2019. "NÖ Gemeindebus-Modell. Grundlagen und Fördermöglichkeiten." Talk given as part of the seminar "E-Fahrtendienst in NÖ" organized by eNu (Energie- und Umweltagentur des Landes NÖ), 6/11/2019, St. Pölten.

Wullweber, J. 2012. "Konturen eines politischen Analyserahmens – Hegemonie, Diskurs und Antagonismus", in *Diskurs und Hegemonie. Gesellschaftskritische Perspektiven*, ed. by I. Dzudzek, C. Kunze and J. Wullweber. Bielefeld: transcript, 29–58.

Wullweber, J. 2014. "Leere Signifikanten, hegemoniale Projekte und internationale Innovations- und Nanotechnologiepolitik", in *Diskursforschung in den Internationalen Beziehungen*, ed. by E. Herschinger and J. Renner. Baden-Baden: Nomos, 270–306.

19 Connected and automated transport in the socio-technical transition

Jens S. Dangschat

Jens S. Dangschat
TU Wien, Research Unit Sociology (ISRA)

© The Author(s) 2023
M. Mitteregger et al. (eds.), *AVENUE21. Planning and Policy Considerations for an Age of Automated Mobility*, https://doi.org/10.1007/978-3-662-67004-0_19

"[...] we live in a world that not just is changing, it is metamorphosing. Metamorphosis implies a much more radical transformation in which the old certainties of modern society are falling away and something quite new is emerging" (Beck 2016: 3).

1. INTRODUCTION

Alongside questions of ethics, legislation, safety, financing, environmental friendliness and climate protection, the development of connected and automated transport (CAT) is primarily discussed as a technological challenge in the context of broad and diverse digitalization. By contrast, the social impacts or even the societal embedding of the technics associated with CAT developments still play a vastly subordinate role in the contemporary scientific discourse surrounding CAT and are occasionally portrayed as being highly uncertain.

This lack of research on CAT in the social sciences is all the more remarkable considering that the current changes in society are very closely tied to processes of globalization, cosmopolitization, acceleration (cf. Rosa 2015) and digitalization, and it can be assumed that these processes will have a significant impact on the use of automated vehicles. Although predicting the future is not traditionally one of sociology's core competences, generally accepted diagnoses of the present made by the social sciences should not be ignored but considered in terms of their relation to CAT. Since Beck's publication on the "risk society" (Beck 1992) at the latest, there has been discussion within the social sciences of the "detraditionalized modes of living" in industrial societies, the "destandardization of labor" and the "individualization of lifestyles and ways of life" within "reflexive modernization" (ibid.; cf. also the concept of "liquid modernity" by Baumann 2000). In his theory of the "real-time society", Weyer proceeds on the assumption that in the

"[...] future society [...] traditional concepts will no longer be effective because the boundaries between planning and action, between autonomy and control, but also between steering and self-steering will increasingly blur" (Weyer 2019: 11).

On the basis of these considerations, Kesselring developed – with many cross-references to Urry's "mobilities turn" (cf. Urry 2000, 2007, 2009; Sheller/Urry 2006, 2016) – the concept of "reflexive mobilities" (cf. Bonß/Kesselring 2001; Kesselring 2008, 2020). According to this, mobility ought to be understood among other things "[...] as an inconsistent, contradictory and ambivalent principle of modernity" (Kesselring 2020: 162), which is in marked contrast to the notions of traditional transport planning and steering as well as the classic sciences of technics and engineering, in which deviations from the linearity of rational logic are interpreted as examples of the rebound effect (cf. Santarius 2012). With the "second modernity" or rather "reflexive modernity" approach, aspects like insecurity, ambivalence and plurality – which are relevant for human activities in addition to rationality – are also taken into consideration.

A principal branch of theorizing and research in the social sciences has focused on the parallelism of technological, technical and societal processes. For example, Saint-Simon (1814) attributes the triumph over feudalism to the strengthened middle classes, which "predicate their self-confidence on economic, technical and scientific successes" (Häußling 2014: 13). The sociological question would therefore be "how social coexistence, social norms and structures as well as social change function as a result of the incorporation of technics" (ibid.: 13). Even Marx (1867/1990) attributes social upheavals like the restructuring of the class system or socio-cultural change ("consciousness") and new patterns of behaviour, among other things, to the mechanization of the world of work. Schumpeter (1942) sees as the driver of the business cycle the interplay between more and less innovative entrepreneurs, in the context of which technical and economic progress is key to the success of products, companies and national economies. Building on Schumpeter's concept of the "creative destroyer", Christensen (1997) introduced the concept of "disruptive technology".

Digitalization and the associated connection and automation of transport are mostly categorized as a "disruptive technological change" by large international consultancy firms (cf. also Jonuschat et al. 2016). However, when the changes associated with these technologies – some of which are yet to be developed – and in particular their manifold applications are so fundamental, then it is imperative that the social significance of these developments is analysed. Yet it is especially in periods of considerable social change that analyses in the social sciences deliver varied interpretations. In order to make policy and planning-related but also business decisions that are geared towards the common good, business stability and sustainable development, it is necessary to recognize the structure, the differentiation and the dynamics of present-day societies.[1]

Firstly, that requires us to agree on the extent and dynamics of social change. That implies taking into account the main drivers of social change and evaluating their impact (see section 2). How sociology has addressed the influence of technological and technical developments is shown in section 3. A critique of the widely discussed approach of Frank W. Geels (2004) to analyse the macro-meso-micro correlation between socio-technical innovations can be found in section 4. Following this, it is argued that it is essential to critically analyse the political creation of notions concerning technical progress in general and the connection and automation of transport in particular (see section 5). Finally, we turn our attention to "digital modernity" (cf. Canzler/Knie 2016) and consider the role that CAT will play in its refiguration (see section 6).

2. TRANSFORMATION OF (MODERN) SOCIETY

"In order to comprehend the social force of digital transformation, it makes sense to think in terms of larger social contexts" (Weyer 2019: 10).

Within the social sciences there is no doubt that at least modern societies and those in emerging economies are currently in a phase of rapid transformation that is affecting institutions,

---

1   When they analyse society, policymakers and public administrations rely above all on official statistics. However, they provide entirely inadequate data when it comes to social differentiation because the important socio-cultural dimension is completely lacking and the socio-economic dimension is only described superficially. "Flying blind" like this is not only problematic due to a lack of differentiation but because it also contributes to a continuation of thinking and acting in the outmoded categories of first modernity (cf. Dangschat 2015). Furthermore, a science that is dependent on these data and their spatialization in administrative units is significantly disadvantaged, which ultimately has a negative impact on the expertise coming from the social sciences that is applied in political, planning and business consultancy.

organizations, structures, attitudes and behaviours (cf. Sennett 1998). In this context the nation state is becoming less important as a regulatory power that steers the economy and seeks social balance,[2] while companies with increasing interdependencies and dependencies connect on a global scale and develop their own dynamics.

The socio-economic, socio-demographic, socio-cultural and socio-spatial dimensions of social structures change and diverge. Societies are undergoing a change in their values with increasingly conflicting positions and sectional interests, and – in light of the growing antithesis between time and material resources – with lifestyles converging globally in line with those of the educated urban middle classes (cf. Dangschat 2020). This value change in particular is giving rise to "multiple modernities" (cf. Eisenstadt 2006).

The specialist literature cites as causes of these developments terms such as globalization (cf. Sassen 2001, Wallerstein 2004), acceleration (cf. Rosa 2015), individualization (cf. Beck 1995), flexibility (cf. Sennett 1998), digitalization (cf. Elliott 2018), network society (cf. Castells 2010) and mobility (cf. Urry 2000), which have subsequently become part of their general interpretation in society via the media discourse.

In German-language research on social inequality, there are three main trends concerning how to categorize societal processes (at least in post-industrial countries) in terms of forms of social de- and restructuring (cf. Dangschat 2020):

- as a consolidation and revival of existing class structures but with a different outward appearance; the neoliberal global economy has revealed new class relations not only between the Global North and the Global South but above all within nation states (cf. Dangschat 1998);

- as a "disembedding" from traditional attachments (origins, education, social situation, understanding of gender roles, etc.) at the same time as a "re-embedding" in new forms of socio-cultural differences (social milieus, lifestyles; cf. Vester et al. 2001); or

- as a "disembedding" from traditional attachments, but without the opportunity to permanently fall back on new collective orientations and instead having to constantly redefine and reaffirm oneself (transition from first to second modernity; cf. Beck et al. 2001).

These three archetypal approaches are in opposition with one another as "schools" and differ in the sense that, viewed in terms of a criticism of capitalism, they emphasize the altered outward appearances (from the erosion of the welfare state to individual consumption patterns), stress the new levels of communitization and socialization or consider societies to be in the process of a fundamental transition, i.e. they differ primarily in their assessment of how deep social changes are or will be.

This chapter follows the third approach, because it explicitly assumes that the technological and technical transformations that will foreseeably come to pass in the context of digitalization will have a massive influence on societies. Furthermore, in the theory of the transition to second

---

2   Yet at the beginning of the Covid-19 pandemic, there was a clear countermovement with a substantial increase in trust in the nation state and the ruling parties in most European democracies even though (or because) they had greatly encroached on citizens' fundamental rights. However, the easing of these restrictions shows that this countermovement was short-lived; not only have sectional interests again come to the fore, but the space for various conspiracy theories and fake news has grown further.

modernity – also known as "reflexive modernity" (cf. Beck et al. 1996, Beck et al. 2001) – the concept of mobility is considered part of a very wide-ranging context that will define the course of modernity and that can once again focus our attention on the social integration of a new mobility technology like CAT (cf. Bonß/Kesselring 2001, Kesselring 2020).

Into this broader understanding are added on the one hand the thoughts of Urry and Sheller on the "mobilities turn", which have had a profound impact on sociology (cf. Urry 2000, 2007, 2009; Scheller/Urry 2006, 2016; Sheller 2011). On the other, the "space of flows" discourse, i.e. the reorganization of spatial arrangements in light of technologically induced "real times" (cf. Castells 1989), is incorporated, as is that of "(socio-)spheres", i.e. the dissolution of nation states' significance (cf. Albrow 1998), of "scapes", i.e. "deterritorialized" landscapes (cf. Appadurai 1996, Urry 2003), of "fixities" and "motion" (cf. Brenner 1998), as well as the view that modern societies are network societies (cf. Castells 2010).

2.1 THE TRANSITION FROM FIRST TO SECOND MODERNITY

Consequently, the development of modernity is closely connected to technological development and mobility. Over the course of modernity, there have been constant – sometimes step-by-step – changes in:

- the structure of societies (differentiation of classes via the division of labour and the dominance of (urban) middle classes),

- industrial production (Fordism on the basis of assembly line production, Taylorism),

- technologies (as a consequence of the dominance of key technologies as depicted in Kondratiev waves), and

- political processes (democratization, forms of the welfare state in the course of Keynesianism), bureaucracy and the education system (compulsory schooling for all).

In many sciences and in the arts, there was growing scepticism in the 1970s regarding the assumption of modernization processes' linearity and purposiveness. The "end of modernism" was declared and replaced by various "post-models": postmodernism, post-Fordism, post-industrial, postcolonialism, post-structuralism, post-growth, etc. There has also been speak of "turns" like the "cultural turn", the "communicative turn" or the "mobility turn". Whereas the "post" terms merely suggest that something has come to an end and it is disputed in its further discussion how the current and conceivable futures are to be understood, the "turn" approaches assume a change in thinking, mostly as a rejection of dominant linear and logical ideas. For example, Lash (1999) is of the opinion that the "new modernity" coincides with an altered form of rationality, which challenges the positivist viewpoint in particular.

At variance with these dissolutions and repositionings, the disciplinary mainstream of sociology insists on the theory of an ongoing modernization. The sociologist Ulrich Beck has repeatedly and persistently argued against this mainstream: he advocates studying social inequality "beyond status and class" (Beck 1983) and sees modern society as a "risk society" (Beck 1992) that can only be overcome by heading "towards a new modernity" (subtitle). Only later did he refer to this approach as "second modernity" or "reflexive modernity" (Beck et al. 1996, 2001).

In his estimation, it is due to the logic of (traditional) modernity that technological developments, capitalist logic and the disregard for the increasingly problematic climate and ecological crises are leading to more and more industrial catastrophes that not only can no longer be stopped with the possibilities or entrenched strategies of traditional modernity but are in fact intensifying.[3] By contrast, what is needed is a new reflectiveness, new forms of political negotiation (cf. Hajer/Wagenaar 2003) and trust in processes rather than in structures and institutions.

In addition, Beck criticizes his colleagues for persisting in thinking in terms of nation state-based societies. The global networking of flows of production, trade and above all financial capital undermines traditional industrialized nations' decision-making and steering power in favour of transnational corporations (cf. Beck 1997). The consequences are a weakening of the welfare state, social disintegration processes, the erosion of many familiar social relationships and the growing "release" from accustomed social contexts on the individual level, which Beck summarizes in his "individualization theory" (cf. Beck 1995).

Beck even sees the sciences as undergoing change: in his judgement, in second modernity they are not only consulted for problem-solving but are at the same time seen as the cause of problems, because scientific analyses are deemed to have questionable sides due to, among other things, a vast torrent of unrelated detailed results in the course of their practical realization. On the one hand this leads to an uncertainty caused by partly contradictory results, on the other people from all areas of society are able to cherry-pick "their truths" from inconsistent scientific findings – the discussion surrounding the speed and way in which we emerge from Covid-19 lockdowns is a good example.

Beck's collaboration with Lash and Giddens on "reflexive modernity" (cf. Beck et al. 1996) and the parallel works by Sennett (1998) led to a very fruitful cooperation between the four scientists at the London School of Economics (LSE) for several years. Together with his colleagues, Beck formulated the very far-reaching hope that humankind would succeed in shaping their future "reasonably" with the aid of this theory, and that on the basis of an analysis of current (global) problems approaches for their improvement would be developed.

In his reflection on the significance of mobility in the first and second modernity, Kesselring compared the differences between the two as follows (see Overview 1 on the next page).

This comparison makes it clear that there is tension when it comes to rolling out CAT: the technological development of the vehicles and infrastructures is just as firmly established in the first modernity mindset as political strategy papers and engineering publications (clear structures, rationality, certainty, predictability, etc.). However, they are already – and will be even more so in future – facing a social context that corresponds to that described by second modernity: discontinuity, transitive structuring, uncertainty and risk (see section 6).

2.2   SIGNIFICANCE OF (AUTO-)MOBILITY IN MODERN SOCIETIES

In this discourse, mobility is viewed as a basic principle of modernity (cf. Kesselring 2020: 161). According to Sennett (1977), unlimited individual mobility is regarded as a "fundamental right" for all citizens in modernity. In the 1970s "Free driving for free citizens" was not only a slogan by ADAC, Germany's biggest motoring association, but also their political demand. A range of processes that go hand in hand with and shape modernity like rationality, individuality, globalization, acceleration, competition and freedom of movement have facilitated the "automobilization" of modern societies.

---

3   At the time, Beck had in mind the chemical factory explosion in Bhopal, India, in 1984 and its devastating consequences. Moreover, his publication was released in May 1986, only a few days after the explosion of the atomic power plant in Chernobyl.

Overview 1: Criteria for differentiating between first and second modernity

	First modernity	Second modernity
Reaction to ambivalence	purification	pluralism
Optimal solution	"one best way" solution	"multiple best way" solution
Categories of order	structures, rules, consistency	networks, "scapes", "flows"
Security	security and certainty	risk and uncertainty
Scientific unambiguity	foreseeability and predictability, growing stability	unpredictability, growing volatility
Structure and dynamics	continuity and development	discontinuity and change
Orientation	target-oriented	process-oriented
Scale	national order	cosmopolitan contingency
Connections	enduring, stable connections	connectivity as a problem and project
Order	enduring, stability-oriented (national) structures and order	temporary, transitive (transnational) structuring and disorder
Borders	fixed borders and maintenance of borders	flexible borders and flexible border management

Source: the author after Kesselring (2020: 178)

Automobility is more than just the use of cars; it is a complex, self-reinforcing socio-material system of technological and cultural processes, policies, norms and practices. Modern societies are consequently irrevocably bound in time and space to the use of fossil fuel-powered cars, their long-lasting infrastructures and suburban settlement structures, to the politics of supporting traditional industries, and to the cultural expectations and experiences of the organization of everyday life.

According to Urry (2004: 26f.), automobility consists of the interaction of six components that constitute the "specific character of dominance":

1. the car as a vehicle manufactured by the "iconic" companies of the most important sector of industrialization in the 20th century

2. that is – after housing – the most important consumer good to act as a status symbol for its user or owner.

3. Through its technical and social links with the industries, a powerful complex of infrastructures, repair and leisure companies emerges that is produced via urban development and urban planning.

4. The car embodies the global form of "quasi-private" mobility, which shapes work and private life and thereby dominates and suppresses other forms of mobility.

5. It defines a "good life" culture and appropriate middle-class mobility.

6. As a result of the materials, energy and space used to produce the vehicles, the costs and space of the roads and infrastructures, and the emissions and consequential health and social costs, the car is the largest consumer of natural resources.

A similar argument is made by Canzler and Knie (2019):

> "[The] [...] car became part of everyday life; more and more, it also structured the way people organized their lives and opened up new, unforeseen potentials. Where to work? Where to live? With the car, there were suddenly many more options. The outskirts and rural regions were tapped as settlement areas, commuting to one's place of work was presented as normal. Participation in society no longer depended on where one lived; the car made it possible to access new areas in one's private life, too; covering distances became affordable for most people in society" (Canzler/Knie 2019: 14).

The automated vehicle, which is presented as smart, clean, safe and socially inclusive, is intended to counteract this negative image that is emerging of cars. Replacing a conventional vehicle with an automated one is per se not a contribution to more sustainable mobility. Even if the development is related to a "propulsion transition", there is the risk that the mobility transformation will be thwarted by CAT (cf. Manderscheid 2020).

In addition to safety and reliability, first modernity transportation research is primarily dominated by the rationality of covering distances quickly and affordably. Transport is divided into categories of "performance", which is contrasted with the "performance capacity" of infrastructure. If forecasts suggest that the performance capacity might be limited due to an increased volume of traffic in the course of motorized private and goods transport, it is proposed that the infrastructure be expanded proactively. The design of the streetscape has largely bowed to this principle to the disadvantage of other uses.

Due to the growing interest in this field shifting towards mobility research, the linearity of engineering and technics-dominated transportation research has been challenged, with the focus now being directed at the actors involved. Among approaches influenced by psychology and sociology, the application side is dominant and explains on the one hand "irrational" action and on the other the mobility of different social groups (cf. Scheiner 2009; Dangschat 2013, 2017b).

Ambivalences arise from the very different mobility behaviours of social groups due to subjectively motivated optimization. Thus in second modernity mobility is no longer purely rational but varied, no longer linear but rather non-linear (cf. Kesselring 2020: 172–175). However, it is less possible than ever before to translate the plurality associated with this into a uniform transformation process (towards sustainable mobility). In this context Schneidewind et al. (2018: 11) speak of a necessary special (transformative) "literacy", i.e. an ability to understand the different motivations and courses of action and steer them by means of policies and planning. If "reflexive modernization" proceeds unexpectedly, unnoticed and unintentionally, then rational planning reaches its limits (cf. Kesselring 2020: 172).

In their reflections on technological transformations, Sovacool and Axsen (2018) explored how electrification, sharing and automation will change the transport system and hence also automobility. They classify the eight aspects of automobility ("frames") into a four-field system of individual vs. social significance and functional vs. symbolic effects (see Overview 2).

Overview 2: Impacts of connected and automated vehicles on aspects of automobility

Frame	Type	Autonomous vehicles (SAE Levels 4 and 5)[4]
cocooning and fortressing	private-functional	strengthened
mobile digital offices	private-functional	strengthened
expression of gender identity	private-symbolic	maintained
expression of class and wealth	private-symbolic	maintained, perhaps strengthened
environmental stewardship	societal-functional	weakened, as it leads to increased energy use
suburbanizing	societal-functional	strengthened, as it leads to longer commute distance
oil independence	societal-symbolic	independent*
innovativeness	societal-symbolic	strengthened

*The author's assessment is at variance with that of Sovacool/Axsen.
Source: the author based on Sovacool/Axsen (2018: 740f.)

This overview makes it clear that CAT is more likely to strengthen than weaken automobility. Sovacool and Axsen (2018: 740) also estimate the strength of the evidence underlying their assessments. I judge both environmental responsibility and oil dependency somewhat differently from these two authors (see the frames marked with an asterisk in Overview 2). Furthermore, they consider a series of their estimates to be not or insufficiently certain, such as use as a digital office and as a symbol of wealth and social status.

The momentum of CAT's technological and technical development will therefore hardly contribute of its own accord to mitigating the climate crisis, consumption of land and resources or the social symbolism of automobility (cf. Mitteregger et al. 2022a: 148ff). In contrast: as a result of the "reinvention" of the car as safe, smart, innovative and clean, automobility will be strengthened. As the aims of a sustainable transport development are highly likely to be undermined by connection and automation (cf. Dangschat 2017a, 2019; Milakis et al. 2017; Dangschat/Stickler 2020), it is the task of policymakers and planners to put in place in good time the basic conditions for the development and roll-out of connected and automated vehicles and the associated infrastructure.

---

4   The standards of SAE International describe levels of automation; according to this, the SAE Level 4 standard refers to highly and SAE Level 5 to fully automated vehicles.

3. SOCIOTECHNOLOGY'S VIEW OF TECHNOLOGICAL AND SOCIAL TRANSFORMATIONS[5]

> *"Technics does not [have] a significance per se [...], but rather [...] the significance [is] negotiated more or less divisively between the stakeholders"* (Häußling 2014: 240).

The introduction of connected and automated transport is considered a fundamental, partly disruptive technological development that will strongly influence the transport system, mobility, the use of space and other social developments (cf. Braun et al. 2019). Sociotechnology[6] in particular has explored the interrelationship of technological/technical and social developments. In this special branch of sociology, the terms *Technik* ("technics") and *Technologie* ("technology") are often used synonymously, especially in the international discourse. In this article, the term "technics" is used to describe processes with which scientific insights are made practically usable.[7] Conversely, "technology" is the science of technics and uses a basis of scientific, technical and engineering insights to examine the way in which resources and materials are transformed into production processes and finished products.

Rammert (2007: 18) differentiates three basic and distinct perspectives of sociotechnology that have developed over recent decades:

- *Technological determinism:*[8] Technics is the decisive factor in a broadly understood social development (societal consequences of technics or rather technical development; see section 3.1).

- *Social constructivism:* Social constructs and institutional, economic and cultural institutionalizations define the emergence and the form of technics (see section 3.2.2).

- *Interactionist perspective (pragmatism) or associationist perspective* (actor-network theory; see section 3.3): Technics and society form a unit in which each permeates the other.

The economist Möhrle (2018) defines as the sociological understanding of technology the

> *"systematic subsumption and integration of individual technics into a technique aimed at specific targets and for specific purposes, including social technologies, e.g. a process of conflict settlement. Modern technologies*

---

5   A comprehensive overview the various approaches and questions is provided by Weyer (2008) and Häußling (2014: 129–354).

6   In German, sociotechnology is called "Techniksoziologie" or the "sociology of technics".

7   In German, another, primarily colloquial, use of the word "Technik" is for certain established approaches, methods and skills for which the word "technique" is used in English (breathing technique, speaking technique, massage technique, painting technique, etc.); this use of the term is not what is meant when it is used in this article.

8   "Technikdeterminismus" or "technical determinism" in German.

define and shape social relationships and social change to a high degree; therefore, they cannot be viewed independently of society and have to be judged on the basis of whether they are socially acceptable".

This definition reveals a clear practical application of technological determinism, as is dominant in neoclassical economics. However, current approaches analytically draw a distinction between the two areas but assume a mutual permeation and influence in terms of their dynamics (see section 3.3). When it comes to connected and automated vehicles (up to and including SAE Level 4), that means analysing and shaping the human-machine interfaces on the individual level of application. For the production of the technologies on the macro level, however, the technological, economic, scientific and governmental narratives are just as significant as state regulatory systems and marketing strategies (see sections 5 and 6).

In order to be able to carry out a well-founded analysis of the interrelationship and extent of permeation of technological and social developments, it is advisable to be guided by socio-technological theories and methods (see sections 3.2 and 3.3). Häußling (2014: 16) defines the subject matter of sociotechnology as follows:

> *"Sociotechnology concentrates on the interrelationships between real technics, process technics and technologies on the one hand and society and the social on the other. The focus can be on both the application context and the production context of technics. The former context is about technics' appropriation processes [...]. Negotiation processes in the development of technology are at the fore in the latter context."*

When it was stated above that social science studies on CAT are conducted considerably less frequently than technological studies, this definition makes it clear that almost all social science studies on CAT thus far have concerned above all (potential) application or appropriation processes (for exceptions cf. Dangschat/Stickler 2020; Stickler 2020b or rather Manderscheid 2012, 2014, 2020, who however studies primarily the impact of technological change on automobility).

In the following description of different approaches, attitudes to the application of technics (see section 3.1) and to the production of technics (see section 3.2) are differentiated in the first step. In section 3.3 the currently dominant approach of "science and technology studies" and here in particular the theory of large technical systems and actor-network theory are described.

3.1 APPROACHES TO THE APPLICATION OF TECHNICS (TECHNOLOGICAL DETERMINISM)

With the approaches to the application of technics, a perspective is adopted according to which technologies or rather technical systems have an impact on society (e.g. on workflows, health, mobility, communication, but also on regional and social inequalities). If the opinion is held that a technological change materially influences social change, one speaks of "technological determinism". Prominent exponents are Schumpeter and Ogburn; the technology assessment[9] approach can be traced back to the latter (cf. Häußling 2014: 14).

There are very distinct typologies with which to analyse technical development and its interrelationship with social processes (for an overview cf. Häußling 2014: 11–86). Rammert (2003:

---

9   "Technikfolgenabschätzung" in German or "technics assessment".

296) distinguishes five stages over the course of history during which the human-machine interfaces have shifted in favour of robots; artificial intelligence (AI) is deemed highly significant in the fifth stage:

- *passive technics* (tools),

- *active technics* (machines),

- *reactive technics* (cybernetic mechanisms as combinations of machines and sensor technology – according to this automated vehicles of SAE Level 2 should be classed as cybernetic technics),

- *interactive technics* (multi-agent systems in which a solution is sought via mutual agreement in a manner fitting for the situation and the system steered accordingly – according to this automated vehicles of SAE Levels 3 and 4 should be classed as interactive technics), and

- *transactive technics* (intelligent systems in which one's own and others' actions are integrated into a total action in order to be able to independently reflect on and change given aim-means ratios – according to this automated vehicles of SAE Level 5 should be classed as transactive technics).

Foerster (1993: 357) also uses the term "machine" for input-output relations[10] that go beyond purely technological applications. According to him, there are two manifestations:

- *trivial machines*, i.e. unmistakable input-output relationships, whereby a certain stimulus makes a clear and predictable impact; that can apply to other "plannable" processes that take place "logically" and "rationally";

- *non-trivial machines*, i.e. ambiguous input-output relationships, in which the "machine" reacts to the input to achieve an output that it perceives to be optimized; these are "autopoietic machines" or in fact people whose behaviour (output) cannot be analysed with mechanical thinking by the observers; this category includes the development of AI, for example.

This differentiation is important to the extent that the concepts of humankind in classic sciences of technics and engineering differ significantly from those in the dominant social sciences and humanities. Whereas in the former the trivial machine concept of humankind is dominant (homo oeconomicus), the latter tend to assume that which Foerster (1993) referred to as "non-trivial machines" (situation-specific reflexivity). This is particularly apparent in traditional technology assessment and rebound research, because there humans' actions that scientists consider not to be rational are classed as "rebound behaviour". The interpretations and actions are in fact optimized actions from the actor's point of view in light of their perceptions and values, which in turn are based on their socialization experiences.

Another differentiation refers to the position in the context of a business-technology cycle (such as Schumpeter's business cycle or Kondratiev waves) or rather product life cycle. Welge (1992: 270) draws a distinction between

---

10　"The expression 'machine' refers in this context to the well-defined functional characteristics of an abstract entity, and not primarily to a system of cogs, buttons and levers, even though such systems can realize these abstract functional entities" (Foerster 1993: 357).

- *Basic technologies* (technologies in the maturity stage of the product life cycle or rather as fundamental technologies in a business-technology cycle),

- *Key technologies* (technologies that drive market growth) and *pacemaker technologies* (innovative or disruptive technologies that trigger a new cycle).

With connected and automated vehicles, basic technologies would be all those in operation in vehicles classed as SAE Levels 1 and 2. Examples of key technologies are those that are the backbone of the innovative sensor systems, the efficiency increases and the decarbonization of propulsion systems, the acceleration of charging times for electric vehicles and apps with which multimodal mobility is organized (MaaS – Mobility as a Service). Pacemaker technologies are currently still in the development phase – they lie in the areas of new fuels, further increases in the efficiency of computer performance, but most of all in the further development of artificial intelligence to steer autopoietic systems.

In order to cover the spectrum of questions posed by sociotechnology, Weyer (2008: 11) proposes four dimensions; his approach is clearly dominated by technological determinism. However, in Overview 3 the "social structures of technics" aspect is omitted because Weyer exclusively covers the micro-level range of application; it is therefore integrated in the area "consequences of technics". There is no separate area for a socially differentiated perspective because it is also necessary for the emergence and shaping of technics.

Overview 3: Fields of analysis in sociotechnology

Fields of analysis	Questions
Emergence of technics	• How do new technologies emerge? • How do new technologies become established and how do existing technologies persist? • What social processes (power, interest, expectations) shape the emergence and implementation of new technologies and technics?
Consequences of technics	• What impacts do processes of technicization have on socio-economic, socio-demographic and socio-cultural structures and processes? • What impacts do processes of technicization have on work processes, mobility and communication (manufacturer-user interface)? • What importance do social factors have for the functioning of technical systems (human-machine interface)?
Shaping/steering technics	• With what means is the development of technologies and technics steered (power, interests)? • Who plays what role (policymakers, planners, lawmakers, companies, start-ups in niche areas, science, media, civil society)?

Source: the author based on Weyer (2008: 11) and Tarmann (2018: 27f.)

If these fields of analysis are applied to CAT, it becomes clear that scientific interests in the conjectured (!) consequences of CAT outweigh by far any other interests. The field of shaping and steering technics is primarily dealt with in strictly legal terms, with the various development interests being addressed rather formally and oriented towards institutional systems. Power relations or rather IT companies' considerable interest in the disruptive development of CAT, which is less oriented towards transport itself and mobility and more interested in collecting personal on-trip data, are hardly even mentioned (cf. Dangschat 2017a, 2019). Social science studies on the emergence of CAT technics thus far remain the exception. Initial considerations in this regard can be found in Manderscheid (2014), Dangschat (2019), Freudendal-Pedersen et al. (2019) and Dangschat and Stickler (2020).

3.2 APPROACHES TO THE PRODUCTION OF TECHNICS

Technics are not suddenly there and do not grow on trees. They are based on inventions (innovations), have to be integrated into systems and find a market. Interests and constellations of leadership and power are associated with all three of these steps. That means that all technics become established within a by no means linear or rational negotiation process in a specific social context (desires, expectations, reservations). The what and how, with what dynamics and to whose advantage, are thus always tied to a certain power relation and a constellation of interests in a "sphere of influence"[11] and in a certain period.

3.2.1 DEMAND-PULL VERSUS TECHNOLOGY-PUSH THEORIES

There are two opposing schools of thought when it comes to the possible drivers of a socio-technical transformation: the theory of demand ("demand-pull") and the theory of the pressure for technological development ("technology-push"; cf. Stefano et al. 2012):

- With the demand-pull theory, it is assumed that demand influences the prices of applied technologies and hence determines the dynamics of market penetration (cf. Schmookler 1966).

- With the technology (technics)-push theory,[12] it is assumed that the possibility to use existing technics not only means that innovative goods and services are perceived to be preferable, but that companies and regional authorities support above all scientific and technical developments (cf. Schumpeter 1942).

A policy that follows the demand-pull theory entails the risk – as is the trend in some industries (cf. Hoppmann 2015) – that mature (superannuated) technics are supported for longer than

---

11 Whereas the connection between technological and social developments was long seen as mostly limited to a region or nation state, the developments in the context of digitalization irrefutably show territorially unlimited connections. That becomes clear with second modernity approaches (cf. Beck et al. 2001; Bonß/Kesselring 2001) and the organization of power and interest constellations in scales (cf. Brenner 2019).

12 Especially in Anglo-Saxon literature, this approach is referred to as "technology push". On the one hand, that is due to the different linguistic usage of the term "technology" in English compared to German and on the other it depends on the era, i.e. on fashions in academia. Here, an attempt is made to differentiate between the two terms in line with the definition formulated above.

necessary (cf. Hoppmann et al. 2013).[13] Furthermore, when demand is subsidized by national funds, it cannot be ruled out that products by foreign manufacturers are also being bought. State funding of technology would, however, provide targeted support for innovations in that country (cf. Peters et al. 2012).

For a long time – often too long – policies and industries pursue successful developments based on outdated basic technics. Since the 1970s in particular, this state of affairs has become apparent in the old industrial areas of North America and Europe (steel and coal, shipbuilding, leather industry, etc.): here the past success of steering by the state and of the industrial products (1950s to the early 1970s) was a major obstacle to timely and systematic political and technical change (on the north-south divide in German, cf. Friedrichs et al. 1986). This led to path dependencies that not only involved doubling-down on basic technics, but also to a culture of elites involving the so-called "steel barons" that ultimately became anchored in certain structures and mentalities (lock-in effects) due to (party) political steering and company developments.

3.2.2 SOCIO-CONSTRUCTIVIST APPROACHES

Socio-constructivism can be traced back to the research by Berger and Luckmann (1969). According to them, it is not the "facts" that are relevant, but the interpretations of social structures, processes and things that create "reality" (cf. Knorr-Cetina 1989). Socio-psychological findings show that people do not act on the basis of objective facts but always in light of their own experiences of socialization and values (habits) as well as within their time and financial resources (social situation) in line with their perceptions and assessments (cognitions; cf. Bamberg 2004). For this reason, Foerster considers people and social groups to be "non-trivial machines" (see above).

That "realities" are socially constructed affects both the production (technical knowledge is socially constructed) and the application of technics (subjective interpretation of technics). The way technics are dealt with in their production and application leads to them being integrated into the respective activity as "problem-solvers". A frequently overlooked role is played in this by science itself, since "[...] both science and technology are socially constructed cultures" (Pinch/Bijker 1987: 21).

That there are social differences in interpretations is evident in various aspects that are relevant in the context of CAT. On a global scale, assessments of the safety of vehicles and of the benefits of connectivity, concern about tests being conducted where respondents live and their willingness to spend more for CAT technologies differ greatly (see Table 1 on the next page).

Yet even within Europe there are differences between nation states (see Table 2 on the next page): especially in Germany and Austria, misgivings about data security are considerable and few respondents saw a positive aspect to improved connectivity.

Such results from international surveys, which are carried out almost exclusively by large international consultancy firms, have an enormous impact on the media and politics. From a social science perspective, however, they are almost entirely meaningless or rather they confirm again and again well-known general tendencies. On the one hand, such data are heavily de-

---

13   A good example of this is the debate around compensation for losses in the automotive industry caused by the Covid-19 pandemic: should subsidies only be available for cars with post-fossil fuel propulsion systems or also for those with "modern combustion engines"?

Table 1: Acceptance of aspects of connected and automated transport by country (in %)

	GER	USA	JPN	KOR	IND	CHN
Autonomous vehicles are not safe.	45	48	47	46	58	35
Increased vehicle connectivity is beneficial.	36	46	49	56	76	80
Concern about tests on public roads where they live	46	51	41	48	57	32
Percentage of consumers unwilling to pay more for CAT technologies	41	34	30	11	8	7

Source: Deloitte (2019)

Table 2: Acceptance of aspects of connected and automated transport in Europe (in %)

	AUT	GER	FRA	ITA	GBR	BEL	ESP
Autonomous vehicles are not safe.	42	45	38	25	49	50	33
Increased vehicle connectivity is beneficial.	33	36	42	63	49	54	55
Concern about tests on public roads where they live	42	46	40	39	52	40	48
Percentage of consumers unwilling to pay more than €400 for safety improvements	64	71	76	69	68	72	64
Percentage of consumers unwilling to pay more than €400 for better infotainment	86	84	84	79	76	88	78
Percentage of consumers unwilling to pay more than €400 for improved autonomy	66	67	70	59	61	70	62
Percentage of consumers unwilling to pay more than €400 for connectivity improvements	76	79	79	74	70	80	69
Increased vehicle connectivity is beneficial.	33	36	42	63	49	54	55
Concern about biometric data being captured and shared with external parties	62	62	54	36	54	44	49

Source: Deloitte (2020)

pendent on the quality of the surveys: were they truly representative of the respective country (socially and spatially differentiated)? On the other, the results are only presented in terms of averages and no information is given about how assessments may differ between social groups or settlement structures (e.g. urban/rural). Above all, such studies ensure that consultancy firms are commissioned with more surveys; they are much less about gaining true insights into people's concerns about and openness towards new technologies.

Another approach comes from psychology, which explains how "people in general" (in the sense of a basic pattern of human behaviour) make decisions. In a secondary analysis of over 75 international studies on the acceptance of automated vehicles that were published between 2013 and 2019, Jing et al. (2020) identified the major factors that influence the acceptance of CAT. They are on the one hand personality traits (trust, discernible benefit of the technology, ease of use and social control) and on the other characteristics and impacts of the automated vehicles themselves (driving safety, financial factors as general characteristics, and data security, personal safety, meeting mobility needs, environmental friendliness, driver comfort, identification with the vehicle and the assessment of travel time as other characteristics).

Yet here, too, there are two inherent weaknesses: first, statements from different technical, cultural, spatial and policy and planning contexts are taken together and analysed as "human acceptance" without any reflection on problems of transferability or context dependency. That is important to the extent that most studies are based on fully automated vehicles categorized as SAE Level 5, i.e. on vehicles that do not even exist yet. Second, there is no differentiation between social groups or types of settlement structure, which would be essential in light of increasing social differentiation.

3.3 INTEGRATIVE APPROACHES

The aim of "integrative" approaches is to overcome both technological and social determinism and to view the development of technology/technics as a "seamless web", i.e. as directly linked to societal – scientific, economic, political, socio-economic and socio-cultural – conditions as well as their development. Maintaining a critical distance from one-sided viewpoints, various approaches – some clearly distinct from one another – have developed since the 1980s (cf. Häußling 2014: 226–278). A considerable proportion of them are based on a constructivist perspective (cf. Berger/Luckmann 1969). According to this, there is no "objective reality", but rather people produce their own "subjective realities" with the aid of their sensory organs, with their individual cognitive skills and interpretation patterns, that then guide their actions. Within constructivism, however, some points of view are more radical than others (cf. Knorr-Cetina 1989).

According to this, technics is not neutral, not "suddenly there", knows no one inventor, is always connected to a time and place (i.e. a social context comprising power relations and constellations of interests) and should always be viewed dynamically. Consequently, technical products are always the result of negotiation processes among very different actors, with market penetration being just one indicator. Even prior to prototypes, images of (un-)desirable futures are developed, logics of practical constraints are formulated and arguments and institutions are excluded from the discourse.

It bears mentioning that the role of science in the construction of "facts" is addressed in only a few approaches. Exceptions include the sociology of knowledge approaches of the Strong Programme (cf. Bloor 1999) and the Empirical Programme of Relativism by Collins (1981). In contrast to the American sociologist Robert K. Merton, who views scientific findings as unambiguous, the Strong Programme sees even these as a social construct. This makes it necessary to consider the constellations of interest behind the establishment of scientific "truths".

Collins (1981) criticizes the approaches in which scientists only analyse "the world out there" and deny their own role. He considers there to be three stages to negotiation processes within science. What is interesting is that he also attributes an "interpretive flexibility" to systems of scientific examination, which is particularly relevant for the first stage: the communication and influence of the results (reputation, rhetorical skills). In the second stage, one viewpoint gains acceptance, among other things because alternative perspectives have been abandoned.[14] In the third stage, these "findings" become the subject of political decisions and social belief.[15]

Yet here, too, there is no differentiation between the natural and technical sciences on the one hand or between the social sciences and the humanities on the other; after all, strong closure processes still exist between them (for example, concerning the acceptance of different methods or the importance of mathematical logic).

Science and Technology Studies (STS)

Due to increasingly dynamic and far-reaching social change, science and technology studies (STS)[16] emerged from the late 1980s as a new perspective in sociotechnology and as a very interdisciplinary field of research (cf. Bijker et al. 1987). By incorporating the philosophy of science, the history of technology[17] and sociotechnology, qualitative methods were applied in addition to quantitative ones (ethnographic methods, interviews, discourse analysis, etc.; cf. Beck et al. 2012: 11). The focus on the conditions under which technologies and technics are produced was accompanied by analyses of steering and governance.

The central socio-constructivist argument states that both the sciences and technologies/technics are socially constructed cultures (cf. Pinch/Bijker 1987: 21). As such, doubt is cast on the idea of findings being divorced from time or place – as the natural sciences suggest – and even rationality is seen as "relative to a surrounding culture and relative to the farragos in which it can be claimed that the rational can be found and in which the rational must prevail" (Beck, S., et al. 2012: 224).

Currently, attention is focused on the new challenges posed by AI and its impacts on human-machine interfaces, which will be significant for Industry 4.0 and the Internet of Things (IoT) in particular. Ultimately, however, the speed and depth of technological change is not least a political matter.

Large Technical Systems (LTS)

The historian of technology Hughes (1987) addresses large technical systems (LTS) from a socio-constructivist perspective and within the STS approach.[18] In order to be functional, technical systems comprise a "seamless web" of linked elements like physical artefacts, organizations, natural resources, scientific elements and legislations. A technical system is deemed "large" if it is distributed over a large area, if its impact is long-lasting, if it comprises complex technologies and technics that need to be coordinated, and if it has a high level of connectivity (is part of a network) – all these aspects unequivocally apply to the system of connected and automated transport. According to Hughes (ibid.: 57–77), a large technical system becomes established over the course of seven stages (see Overview 4).

---

14 For instance, by only backing pure electric vehicles as the only post-fossil fuel type of propulsion system.

15 The current analyses of the Covid-19 pandemic and the measures to control it are a good example of the relevance of this approach.

16 The abbreviation STS is now also understood to mean science, technology and society studies.

17 "History of technics" in German.

18 The T in LTS alternates between "technical" and "technological" in Hughes' various texts on the subject. For the sake of consistency with the differentiation between the two terms outlined above, this article exclusively uses "large technical systems".

As the development of connected and automated driving is only in its infancy, it is important from a social science perspective to turn our attention to its "system builders". Who are they? How do they act? Which stakeholders prevail when with which arguments? Furthermore, it is interesting how the "technological style" of automated mobility develops and becomes established. How do the discourses of "technology transfer" combine with those of acceptance? Moreover, it is important for policy and planning decisions how the system builders are dealt with and how it is ensured that their interests do not jeopardize the aims of sustainable development (cf. Rupprecht et al. 2018).

Overview 4: Seven stages of the development of large technical systems (LTS; based on Hughes 1987)

	Stage	Description
1	Invention	socio-technical development, mostly as a disruptive invention
2	Development	system builders translate the invention into a future-fit technical development in due consideration of economic, ecological, political and social aspects
3	Innovation	the technical development is converted into marketable products/systems (production, marketing, distribution, services)
4	Technology transfer	adaption of the marketable products/systems to various time- and space-dependent contexts
5	Technological style	experiences with the application of the products/systems that evolve into mainstream applications on the one hand and are used in their gradual further development on the other
6	Growth, competition and consolidation	As a disruptive innovation, it has to gain acceptance in the face of established technical systems and their lock-in effects.
7	Momentum	phase of self-sustaining products/systems being firmly established on the market

Source: the author based on Häußling (2014: 246–248) and Tarmann (2018: 38f.)

Actor-network theory (ANT)

Actor-network theory (ANT) is an approach in the social sciences that has been developed within science and technology studies (STS) since the 1980s in order to research and explain scientific and technical innovations. It is considered the most prominent theory within STS, but is also controversial because it deems materiality to play a key role in socio-technical constellations and because doubt was cast over whether it is a theory or in fact just a (descriptive) method (cf. Latour 1996, Gad/Bruun Jensen 2010). ANT was subsequently developed into a comprehensive sociological theory and research method. It has become established as an independent position within sociology between technological and social determinism.

Significant theoretical contributions to actor-network theory come primarily from Callon (1999), Law (2006) and Latour (2007) and mostly emerged in the context of STS. While early research mainly analysed the way sciences and technologies were produced and functioned, ANT later worked through the basic concepts of sociology.

According to ANT, people never act as solitary individuals but always depend on other actors or entities. It is for this reason that the theory describes actor-*networks*. In addition, social, technical and natural objects are viewed as being caused by neither exclusively natural or technical nor exclusively social factors, i.e. there is an attempt to overcome both technological and social determinism.

The central idea of actor-network theory is that society is comprised of various elements and is arranged like a network. According to Law (2006: 431), the key finding of ANT is that "the social" only exists in the form of structured networks that consist of heterogeneous materials. These networks encompass not only social actors but also material things like technical artefacts or discursive concepts (cf. Peuker 2010).

That means that the connected and automated mobility of the future should be understood neither solely via its stakeholders nor solely via technological inventions or narratives on CAT in political strategies, the media and advertising. Instead, it is necessary to analyse the mutual permeations and the respective temporally and spatially bound networks in their entirety.

The main object of investigation in ANT is the coming together of diverse elements into more or less coherent actors. Of central interest are associations that arise between distinct (heterogeneous) entities – for example the reciprocal relationship between a technical development, changed priorities in research funding and political strategies for the propulsion, transportation and mobility transformation. Consequently, ANT reveals connections that are just as material (between things) as they are semiotic (between concepts).

With actor-network theory, an attempt is made to explain how material-semiotic networks are formed in order to act as a whole (e.g. automated transport is both a network and an actor, and for some purposes it acts as a single entity). ANT views explicit strategies that serve to integrate various elements in a network so that they appear as a coherent whole to the outside world as being a part of this.

However, the networks are not inherently coherent and can involve conflicts because they incorporate various interests and various narratives are produced (e.g. conflicts may arise between car manufacturers and the state registration offices or the population may have difficulty accepting automated vehicles; cf. Jing et al. 2020). The networks are always the product of negotiations between interests of different importance. Figure 1 illustrates the different approaches of sociotechnology as a tree structure.

Figure 1: Analytical approaches of sociotechnology

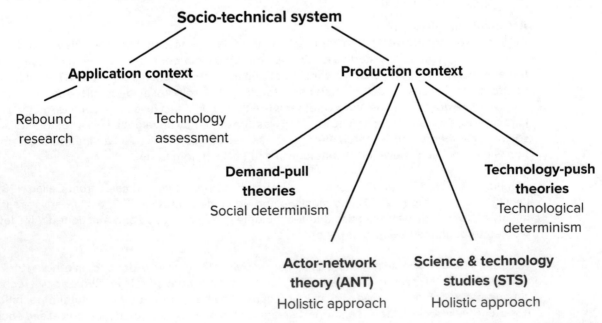

Source: the author

4. THE APPROACH OF GEELS – AND WHY HE DOES NOT GO FAR ENOUGH

"[Technological transitions] do not only involve changes in technology, but also changes in user practices, regulation, industrial networks, infrastructure, and symbolic meaning or culture" (Geels 2002: 1257).

The Netherlandish STS researcher Frank W. Geels explores technological innovations in the context of sociotechnological transitions. He dissociates himself from those sociotechnological approaches that claim that technology has no characteristics, no power and "does nothing". In various research contexts, he took the concept of the "technological regimes" of Nelson and Winter (1982) and developed it into a socio-technical regime by expanding the set of actors and paying greater attention to their skills, resources and interests.

Kanger et al. (2019: 48) also distance themselves from approaches that focus purely on the application of technics and technologies. As a rule they are, the authors argue, only focused on individual groups, most of which can be divided only into "early adopters" or "laggards". Furthermore, these models are static and ignore the reciprocity of production and application, i.e. co-creation. Moreover, it is at least implicitly assumed that change proceeds in a linear fashion and there is a disregard for the development processes and conflicts that have a formative influence on political culture, moral values, policy measures and the behaviour of those who use the socio-technical system (cf. Geels 2004).

"Technological diffusion" is understood in this approach

"[...] as a process of co-construction of the entire configuration of socio-technical systems, including the focal technology and the environments in which it will function" (Kanger et al. 2019: 48).

The authors proceed in a typically system-theoretical manner to the extent that they differentiate between the (sub-)system ("technology") and an environment in which technics is produced and applied. Based on this, they judge their approach to have three advantages (cf. ibid.: 48f.):

1. It is not just the users of the technologies who are taken into account, but also other social groups.

2. It raises awareness that the characteristics of socio-technical systems are not known at the outset, but that markets, their steering, infrastructures, actors, preferences and symbols only evolve in the course of market penetration.

3. It thus becomes clear that socio-technical processes can take different trajectories because the directionality and intensity of innovations are determined by the struggles between the system builders.

Specifically, Kanger and his co-authors assume five relevant dimensions of "societal embedding" (ibid.: 49f.) regarding the diffusion of innovations:

1. *Embedding in user environments* (in the broader sense): Here, the authors include more than just car drivers: they also consider other road users, the automobile trade, the me-

dia, police, policymakers and planners in order to incorporate the new technologies in professional and private everyday routines.[19] That requires "cognitive work", "symbolic work" and "practical work".[20]

2. *Embedding in the business environment*: This area addresses the development of industries and new business models, supply and distribution chains and repair shops. What is interesting here is the focus on innovations by new market entrants that cause the demise of traditional structures (they refer here to Schumpeter).

3. *Cultural embedding*: The articulation of positive discourses, narratives and visions creates cultural legitimacy and encourages societal acceptance of the new technics. Negative discourses about traditional technics and positive discourses about new technics influence consumer preferences and political support. Consequently, cultural embedding is a contradictory and contested area because it is where different interests clash.

4. *Regulatory embedding*: Here, it is above all the role of the state (policymakers and planners) that takes centre stage which (helps) shape the production, the market and the use of technics by setting new standards and introducing regulations.

5. *Embedding in the transnational community*: Through mutual learning, international embedding encourages the exchange of experience, consultation and negotiation. A consensus about standards supports a shared view of technologies.[21]

However, Kanger et al. (2019) overlook the fact their five dimensions of embedding are not equal. From a purely functional perspective, the steering framework provided by policymakers (and planners) creates a series of intersections with the world of business regarding funding and regulation. The three other areas of embedding are on entirely different levels.

The transnational level is not a separate dimension, but rather is merely a shift in scale to the international level (cf. the scales approach of Brenner 2019). The field of cultural embedding describes the type of struggle, the arguments and the "weapons" with which the "battle is fought" on each of the levels and between the various levels. This also includes regulatory embedding, which is likewise shaped by various cultures and power constellations.

---

19 Without explicitly saying so, the authors follow the approach of "actor-centred institutionalism" (cf. Mayntz/Scharpf 1995), according to which people's actions are always part of actor constellations. Their aim and interests comply with the external guidelines of the respective institutional context and their own permanent activity orientation as well as with situational motivations for action. Häußling (2014: 250) emphasizes the importance of this approach for large technical systems' ability to steer technology policy, i.e. the development of technics itself and the steering of infrastructure systems.

20 It is interesting that the role of (interpretative) science is ignored by the authors here.

21 When describing this level, Kanger et al. (2019) refer to the perspective of a historical transnationalism, of automobility and sustainable developments. In my view, this description is one-sided, showing only the positive aspects, and is therefore unrealistic because it is especially in the race to economize technical innovation that nation states and businesses are in fierce competition with one another and are more likely to isolate themselves or at best cooperate in order to make the competition more positive. Even the (misappropriated) realm of the sciences is hardly known for its efforts for consensus and cooperation.

Finally, the user dimension cannot be separated from the other dimensions because according to their understanding of their own approach the process of technical development is influenced by very diverse actors – in politics, in companies, on the local, national and international level. The mutual influence always and exclusively occurs in the context of cultural embedding, which is bound to a time and a place.

Building on various studies from his different research contexts in the multilevel perspective on technology transitions (MLP; cf. Kemp 1994, Rip/Kemp 1998), Geels (2002: 1260–1263) developed a multilevel approach that he applied in various publications to the transformation of different technologies and technics, refining his model in the process (cf. Geels 2004, 2005, 2006, 2011, 2012; Geels et al. 2012). On the landscape level, Geels and Schot (2007: 404) differentiate between four types of change based on the dynamics of four aspects (frequency, amplitude, speed and extent): regular change (slow and incremental), specific shock (sudden and dramatic change in few dimensions that leads to a return to the original state), disruptive change (intermittent, slow changes that only affect a few dimensions[22]) and "avalanche change" (the extent, speed and reach of the change are considerable).

These considerations centre on the socio-technical regime on the meso level, which consists of seven elements: technology, infrastructures, industrial networks and business strategies, technical and scientific knowledge, sectoral policies, markets and practices of the users and symbolic meaning (see Fig. 2). However, these regimes are integrated in a specific territorial context (which might be an industrial region but it is mostly understood to mean a nation state) and in a (global) landscape of (slow) developments (macro level).

According to Kemp et al. (2001), the micro level is home to technological niches, which are viewed as "incubation rooms" for "radical novelties" and from which technical innovations proceed.

In reaction to various criticisms of his approach, Geels (2011) specified his model. He now concedes that the "socio-technical" landscapes on the macro level function as an "exogenous context" and that this can indeed be the origin of the pressure on the socio-technical regime on the meso level that makes novelties possible. He continues to view the socio-technical regime as being in a "dynamically stable" equilibrium that also influences the niches (through expectations of their innovations and being incorporated in networks[23]).

Applying this model to CAT, it is doubtful whether the role of the niches and the "landscape developments" were assessed correctly. For the first driving assistant systems like ABS, technological innovations that are bottom-up from the niches may still have been relevant, but it is primarily the top-down effect of the global landscape's development focus that is responsible for the pressure to develop highly and fully automated vehicles (cf. Berkhout et al. 2004: 62). From here, globalization, competition between the major car manufacturers for market leadership on submarkets with different dynamics, companies new to the industry that come from the IT sector, political influence on (national) research funding in the context of locational competition, combating the climate crisis and the growing criticism of automobility have an impact

---

22   I find the term "disruptive" inappropriate for this type of change because in the general understanding of technics research it is interpreted as meaning a fundamental, very sudden change – i.e. more that which they refer to as an "avalanche".

23   Above all, the policy of funding transnational, national and regional public institutions as well as businesses funding university and non-university research – which go unmentioned here – are part of this.

Figure 2: A dynamic multilevel perspective of technological transformations

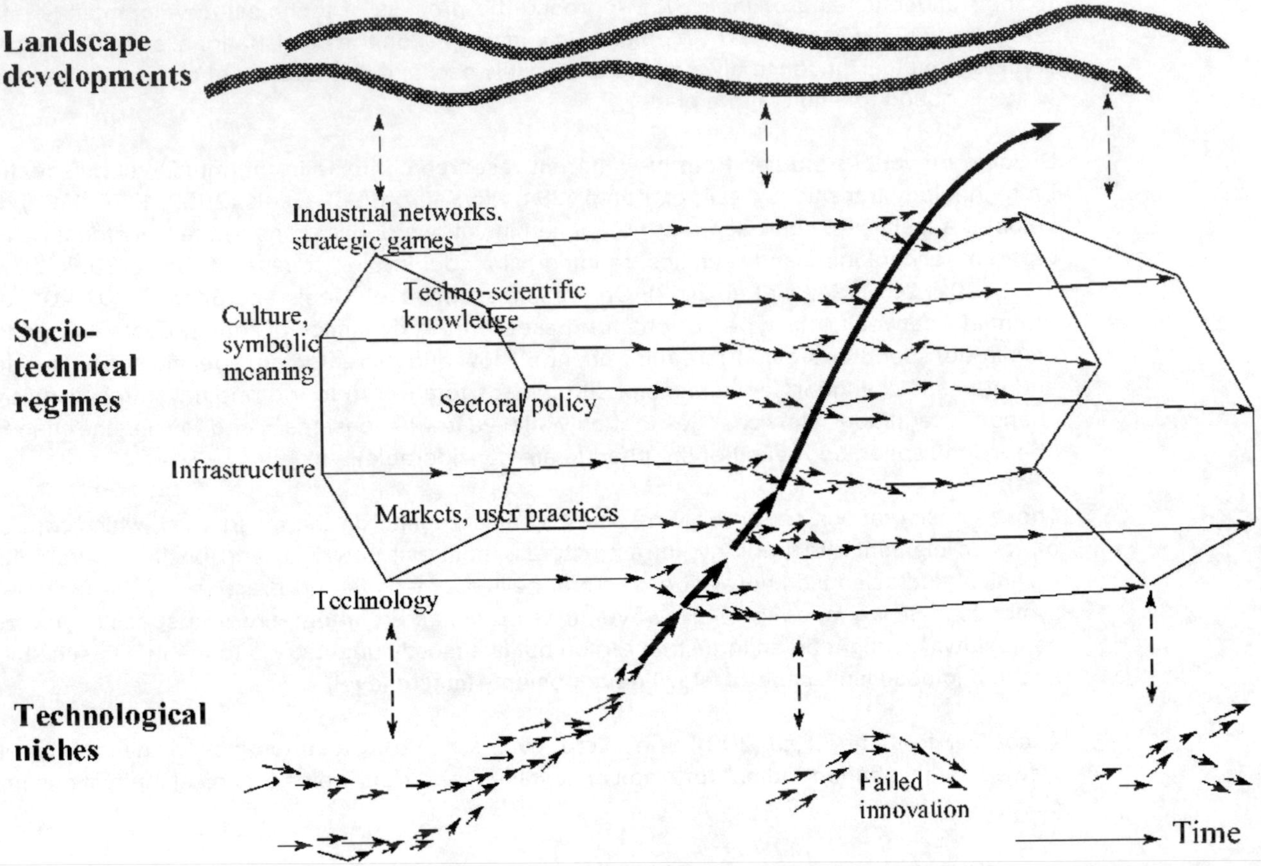

Landscape developments

Socio-technical regimes

Industrial networks, strategic games

Techno-scientific knowledge

Culture, symbolic meaning

Sectoral policy

Infrastructure

Markets, user practices

Technology

Technological niches

Failed innovation

Time

Source: Geels (2002: 1263)

on the socio-technical regime. That does not mean that individual technical developments, for example in the areas of sensor technology, data management or networking, take place in start-ups and fab labs – which however are rapidly integrated in the big companies in the IT sector, the supply industry or car manufacturing and hence become part of the socio-technical regime and/or the landscape.

Genus and Coles (2008) criticize that Geels' concept is too formal and that the actions – especially of political steering – are not sufficiently elaborated. They propose taking into greater account the approaches of the social construction of technology (SCOT), actor-network theory (ANT) and constructive technology assessment (CTA), all of which emerged in the context of socio-constructivist sociotechnology, because they specifically make allowances for stakeholders' interests and motivations (see section 3.3).

For this reason, the next section uses the "policy mobilities" and "mobility transfers" approaches to actively direct readers' attention to aspects of power and interest, above all on the macro level. Using these approaches, an analysis is conducted of how narratives from the "global landscape" – with their diverse regional and cultural contexts – are adopted in policies and planning and how they can take on entirely new dynamics. Especially on the level of cultural embedding, narratives are generated that are intended to increase the acceptance of connected and automated vehicles. With this discourse, the intention is on the one hand to ensure that automobility is not questioned and on the other to entrench a positive example of digitalization (cf. Dangschat 2019).

5. "MOBILITY TRANSFERS" AND "POLICY MOBILITIES": WHY THE CONTEXT AND CREATION OF IMAGES IS SO IMPORTANT FOR CRITICAL ANALYSIS

"[T]he target groups and appliers of scientific results in politics, business and the public [become] active coproducers in the social process of knowledge definition. [...] This is a development of great ambivalence." (Beck 1992: 157, emphasis in the original)

In contrast to the dominant sociotechnological approaches to the application of technologies and technics as well as the bottom-up approaches to technical innovations, current multilevel analyses in urbanism and regional science primarily explore structures, processes, interests and power structures on the (global) macro level and the creation of narratives around future development (cf. Farias/Bender 2010). Starting from a criticism of neoliberal trends towards the independent development of large metropolises in locational competition below the level of nation states, policy transfer (PT) research asked how policies (neoliberal vs deliberative), administrative organizations ("new public management"), aims and strategies are transferred from one political system to another (cf. Dolowitz/March 2000). Proceeding from still strongly structuralist positions, it was analysed how and why a policy transfer, for example, functions between the local/regional and nation state level (cf. Evans/Davies 1999). Mostly, policy transfers in a country or in a rather specific cultural area (of the "First World") were studied. It was primarily the factors that facilitate or hinder a transfer that were identified. In reaction to what was initially an exclusively positivist view of rational decisions, subjective processes in policy transfers were increasingly taken into consideration within the PT approach as well.

In the PT approach, there are two opposing (ideal typical) positions:

- that of convergence due to the adoption of (Western) values, market logics and the state of democracies, ultimately also due to the harmonization efforts of international associations of states, and

- that of divergence due to different path dependencies, power constellations, discourses, learning processes and political disputes.

At present it is generally assumed that these two trends are concurrent. The probability of convergence is higher the more a realignment is incorporated in a "great narrative". That also appears to apply to the global implementation of CAT. On the one hand there is countless evidence of automobility rapidly spreading or stabilizing in industrialized countries and emerging economies in accordance with convergence. However, implementation and acceptance will occur differently – not as a matter of principle, but with different dynamics.

Urban policy mobilities (UPM) research dissociates itself from the positions of PT and parallels with the theory of the network society, of second modernity, with the "mobility turn", with actor-network theory and with assemblage theory are explicitly made (cf. McCann/Ward 2012, Künkel 2015: 8f.; see section 2.1). According to UPM research permanent administrative levels become ever more permeable and become "scales" (cf. Brenner 1998). Due to certain mixtures of technical and administrative practices, new spaces are tapped and become comprehensible as a result of territories being "deciphered" and "recoded" ("assemblage"; cf. Deleuze/Guattari 1992: 699).

Especially with the increasing competition between metropolises, a new politics of scale is being pursued, whereby institutions and actors from various levels are part of a strategic network (cf. Brenner 2019: 206–233). The set of observed actors is thereby expanded significantly beyond politics and administration.

In addition, the focus of the analysis is moved from organizational structures towards discourses and power constellations by monitoring how ideas, concepts and strategies have developed via scales and over time (cf. Peck/Theodore 2010). Elements of ethnographic political research are also applied in this context (cf. Wiesner 2003). UPM research is interested in more than just the descriptive "how" of structures, in the question of "how, why, where and with what effects policies are mobilised, circulated, learned, reformulated and reassembled" (McCann/Ward 2012: 326). Ultimately, therefore, it is about how strategies are established by which system builders with what arguments and how global policy discourses are "broken down" into other scales.

In this context, science plays an important role as a mediator. With "vehicular ideas" it produces a new category of arguments within the discourses (cf. McLennan 2004; on the construction of mobility, cf. Mincke 2016). Building on a diagnosis of contemporary challenges, "problem-solving suggestions" are introduced to the discourse that are intended to bring a system into a desirable situation.[24] Peck (2012) explains this using the discourse on the "creative city", with which the shortcomings of political and administrative steering can be remedied or a sustainable development can be enabled. This viewpoint is relevant for the analysis of CAT to the extent that it is most notably presented by nation state and EU policies as the solution to manifold transport and mobility problems as well as environmental and equal opportunity issues.

In light of the current and foreseeable societal significance of digitalization and the connected and automated mobility embedded therein, it seems sensible to analyse the socio-technical transformation of transport on the basis of a multilevel model with the approaches of policy transfer (PT) and policy mobilities (UPM) in particular – such an analysis is currently lacking in the international literature.

6. DIGITALIZATION AND AUTOMATION AS DRIVERS OF THE SOCIO-TECHNICAL TRANSITION?

"The great transformation describes a massive ecological, technological, economic, institutional and cultural transition process at the beginning of the 21st century. [... That requires] a special (transformative) literacy, i.e. a competence with which to understand this dimension and its interaction, and the skill to translate this understanding into contributions to a sustainable development" (Schneidewind et al. 2018: 11).

Since the 1980s, globalization has been a "great narrative" that is predominantly presented as an opportunity. According to an argument that can be heard all over the world, if a nation state

---

24   A good example of this are the works by Richard Florida (2002) on the "creative class", which he championed as a prerequisite for competitive city development and which was adopted all over the world (cf. Dzudzek 2016: 95 ff.).

or business wants to defend or expand its own economic position, it should face up to the challenge proactively, rapidly and systematically. As a result of the network society, globalization has led to an increasingly accelerated and further-reaching mobility of people, goods, finances, information, technologies, ideas, cultures and ideologies, which in turn has relativized societies, local ties and the importance of nation states (cf. Beck 2007, Endres et al. 2016).

Over the course of the ongoing Covid-19 pandemic and the resulting lockdowns in almost all industrialized countries, very strict curfews were imposed that shut down industry and trade and interrupted supra-regional supply chains. There were supply shortages for intermediate goods and wares, especially for personal protective equipment and active pharmaceutical ingredients. That in turn not only completely changed transport and mobility at least temporarily, but also led to a(nother) critical reflection on globalization due to increased awareness of strong supra-regional dependencies.

Since the millennium, the globalization narrative has been dominated by automation via technology-based digitalization , as a result of which competition has emerged from smart cities in which it is no longer production and typical services that take centre stage but communication, mobility and data management. Consequently, funding is required above all for technologies and the key industries that are based on them in order to develop life sciences, artificial intelligence, the Internet of Things (IoT) and Industry 4.0, etc. (see Chap. 20 by Banerjee et al. in this volume). The lessons learned when dealing with the coronavirus crisis and the associated experience gained regarding working from home, teleconferences and distance learning will lead to another accelerated digitalization of gainful work, education, learning and everyday communication.

Especially in China, South Korea and Singapore, it is assumed that CAT is important for dealing with the pandemic because self-driving vehicles can be used to safely surveil and potentially disinfect public spaces or rapidly and safely transport medical goods. Furthermore, due to the non-existent risk of infection, nursing robots are very well suited to treating Covid-19 patients. Prototypes developed by start-ups are already being used there. By contrast, the car manufacturers and ancillary industries in Europe and North America are cutting back their investments in developing automation due to the financial difficulties caused by the pandemic.

The multifaceted and far-reaching technological transformation of digitalization in general and the (further) development of CAT in particular are happening at the same time as societies are undergoing sea changes – ultimately also as a consequence of globalization, digitalization and climate change. Deepening socio-economic inequalities coincide with blurred boundaries ("Entgrenzungen") and with a release from traditional structures providing the freedom and obligation to organize one's own life (cf. Hitzler 1984, Beck 1995). This is accompanied by raised expectations of steering by policymakers and planners (improving safety on the basis of a growing number of incalculable factors). At the same time, nation states are losing their steering power and their will to steer, which is leading to insecurities, especially in the middle classes who are at risk of downward social mobility.

This situation coincides with the discourse on the inevitability of CAT. Yet the application of technics require specialist knowledge and a high level of competence. This concerns the way in which technics can be applied, exert control and hence become a source of power (cf. Mackenzie/Wajcman 1985). Just as with automobility, the application of new technologies and technics has thus far always led to the exacerbation of existing social inequalities and/or the addition of new ones (cf. Cudworth et al. 2013, Urry 2012, Dangschat 2019). In the context of increasing electronic communication in general and the connection and automation of transport and the offer of new mobility services in particular, the exclusionary access to these systems via digital channels (i.e. the digital divide) plays a socially selective and marginalizing role that has as yet been paid little attention (see Chap. 20 by Banerjee et al. in this volume).

Alongside the repeatedly communicated potential advantages of CAT, especially in the form of highly and fully automated vehicles, it must also be permitted to ask how the idea of automated driving arose. And: how did it happen that digitalization, automation and connection came to be presented as "inevitable", as without any alternative? In the discourse surrounding CAT, local conditions, policy and planning alternatives or the reserved scepticism among the population play at most a subordinate role. Yet it should not be a matter of once again adapting settlement structures, public spaces and the distribution of locations to technological requirements, but rather implementing CAT in a spatially and socially differentiated way such that it is able to support the necessary transformations in propulsion, transport and mobility (cf. Rupprecht et al. 2018, Dangschat 2019, Manderscheid 2020 and Chap. 4 by Manderscheid in this volume).

That would mean taking a more careful approach to the discussion and differentiating

- according to the level of connection and automation, especially for the heterogeneous and high-risk Long Level 4 (cf. Mitteregger et al. 2022b),

- according to the environment in which CAT should be used (operational design domain – ODD; cf. Mitteregger et al. 2022b: 60–64), which includes in particular the possibility of preferably using rural regions and small towns as test areas alongside highways with good performance capacity (see Chap. 6 by Soteropoulos et al. in this volume), and

- according to social selectivity, which includes differences in purchasing power and attitudes to topics directly and indirectly related to CAT (automobility, familiarity with technics, environmental and climate protection, sharing, etc.) and the widened digital divide (cf. Dangschat 2019).

The complex of connected and automated transport can certainly be thought of as a "large technical system" (see section 4; cf. also Tarmann 2018) which is currently transitioning to the development phase in Europe; at most, technical subsystems in the gradually improved driving assistant systems are currently in the innovation phase (stage 3).[25]

The sciences of technics and engineering as well as politics on the EU and most national levels present connected and automated transport as a "safe" technical system (cf. Stickler 2020a) with which the number of accidents will be significantly reduced. Thinking of it as a deterministic mitigator of social problems corresponds to a political self-image of "traditional modernity". Even if managing the first phase of the Covid-19 pandemic led to increased trust in the political style of the "strong nation state", the current "normalization" development is revealing marked conflicts of interest, a contested role for scientific expertise and the strengthening of the influence of "new media", including conspiracy theories and fake news, i.e. it appears that once again second modernity is becoming important as the context of steering by policymakers and planners.

Moreover, it has become apparent in this development that the traditional concepts of policy and planning behaviour are no longer effective and that new negotiation process with additional actors and across various scales are therefore needed. In addition, it is to be expected that being guided by the inherent logic of the natural sciences will be a cause of contention for the new actors and especially the users of connected and automated vehicles. This concerns situations

---

25 Tarmann (2018: 40) considers "autonomous driving" to be in the phase of "technological transfer". I do not share this assessment because for one thing it is not certain whether other inventions will be necessary (phase 1), and for another the economic, ecological, political and social embedding (phase 2) is thus far – with the occasional exception in South East Asia and California – completely uncertain.

that Foerster (1993) calls "non-trivial machines" because perceptions, decisions and processes increasingly take place according to an "inherent logic". Here, not only are the producers and decision makers in the realms of policy and planning faced with new tasks, but the sciences too must be aware of their new role – which must amount to more than simply handing over responsibility via the questionable co-creation of an urban mobility lab.

If we are to believe Kesselring (2020) on mobility in second modernity, then in future the implementation context of CAT will be even more and increasingly characterized by unpredictability and discontinuity (see Fig. 1; cf. also Läpple 2011). There will continue to be hardly any unambiguity; instead, a range of "best-way" solutions will be developed in parallel – such as to shape the propulsion transformation, or sharing solutions to shape the mobility transformation or even CAT itself (cf. Shladover 2016). For policy and planning solutions, that means an element of risk, an increased necessity for a process-oriented approach and the involvement very different actors. Furthermore, the institutional actors on the nation state level are not the only steerers, since on the one hand multinational political decisions will be prescribed as a framework (primarily by the EU), and on the other the research and development constellations are more and more international and elude state steering, with technological innovations and their implementation frequently being based on "niche actors" (cf. Kanger et al. 2019).

What is overlooked with the technologically oriented approach, i.e. with the dominant logic of first modernity, is that neither policy and planning strategies nor the everyday world of citizens is deterministically organized – in contrast to mathematical logic (cf. Häußling 2014: 218) – rather, what is important is with which interests and with what power the respective institutional actors act within a "corridor of action" (cf. Mayntz/Scharpf 1995). Ambivalent individual actions must therefore always be considered in the context of contradictory system dynamics, temporal flexibilizations and spatial fragmentation (cf. Schwedes 2013: 284).

Consequently, it is not sufficient to describe the potential technological and technical capabilities of vehicles, infrastructures and their networks and to anticipate possible impacts on transport and spatial development with the aid of scenarios. Rather, according to Cresswell (2006: 3f.) it mainly comes down to how CAT is thematized, interpreted and presented in various ways as an imminent and inevitable fact. In this context, technical feasibility, the framing of policy and planning as well as research policy, the advertising and marketing of future mobility and associated images of futuristic vehicles and of (urban) landscapes[26] as well as interpretations from (social) science of contemporary societies and future trends play a significant role in the formation of ideologies about CAT.

This analytical division into "objective facts" and the "construction of realities" cannot be maintained in the Foucauldian sense of the "archaeological" approach (cf. Foucault 1972) and should instead address the issue – related to the reflections of Frello (2008) on CAT – of how it happens that the future of mobility (and hence the solution to all transport problems) is exclusively determined by highly and fully automated transport. While Sheller and Urry (2006: 211) demand that we monitor mobility discourses and mobility practices, Cresswell (2010) points to the importance of power constellations from which the discourses and actions on mobility emerge and are simultaneously produced and established. That means that it is – and should be even more so in future – above all the task of social science research to address not just the description and critical classification of the (potential) consequences of CAT but above all the context in which CAT emerges and its steering potential.

---

26   In most renderings of future urban mobility, vibrant uses of public spaces by other modes of transport are generally ignored, and in every case a deaf ear is turned to advice about social problems or social inequalities.

REFERENCES

Albrow, M. 1998. *Abschied vom Nationalstaat*. Frankfurt am Main: Suhrkamp.

Appadurai, A. 1996. *Modernity at Large. Cultural Dimensions of Globalization*. Minneapolis, MN: University of Minnesota Press.

Bamberg, S. 2004. "Sozialpsychologische Handlungstheorien in der Mobilitätsforschung: Neue theoretische Entwicklungen und praktische Konsequenzen", in *Verkehrsgenese – Entstehung von Verkehr sowie Potenziale und Grenzen der Gestaltung einer nachhaltigen Mobilität*, ed. by H. Dalkmann, M. Lanzendorf and J. Scheiner. Mannheim: MetaGIS Infosysteme, 51–70.

Baumann, Z. 2000. *Liquid Modernity*. Cambridge: Polity Press.

Beck, S., J. Niewöhner and E. Sørensen 2012. *Science and Technology Studies. Eine sozialanthropologische Einführung*. Bielefeld: transcript.

Beck, U. 1983. "Jenseits von Klasse und Stand? Soziale Ungleichheiten, gesellschaftliche Individualisierungsprozesse und die Entstehung neuer sozialer Formationen und Identitäten", in *Soziale Ungleichheiten*, ed. by R. Kreckel. Göttingen: Schwartz, 35–74.

Beck, U. 1992 (1986). *Risk Society: Towards a New Modernity*. Trans. Mark Ritter. London/Newbury Park/New Delhi: SAGE Publications.

Beck, U. 1995. "Die 'Individualisierungsdebatte'", in *Soziologie in Deutschland. Entwicklung, Institutionalisierung und Berufsfelder, theoretische Kontroversen*, ed. by B. Schäfers. Opladen: Leske + Budrich, 185–198.

Beck, U. 1997. *Was ist Globalisierung? Irrtümer des Globalismus – Antworten auf Globalisierung*. Frankfurt am Main: Suhrkamp.

Beck, U. 2007. *Weltrisikogesellschaft. Auf der Suche nach der verlorenen Sicherheit*. Frankfurt am Main: Suhrkamp.

Beck, U. 2016. *The Metamorphosis of the World: How Climate Change Is Transforming Our Concept of the World*. Cambridge/Malden, MA: Polity.

Beck, U., W. Bonß and C. Lau 2001. "Theorie reflexiver Modernisierung – Fragestellungen, Hypothesen, Forschungsprogramme", in *Die Modernisierung der Moderne*, ed. by U. Beck and W. Bonß. Frankfurt am Main: Suhrkamp, 11–59.

Beck, U., W. Bonß and C. Lau 2004. "Entgrenzung erzwingt Entscheidung: Was ist neu an der Theorie reflexiver Modernisierung?", in *Entgrenzung und Entscheidung*, ed. by U. Beck and C. Lau. Frankfurt am Main: Suhrkamp, 13–61.

Beck, U., A. Giddens and S. Lash 1996. *Reflexive Modernisierung. Eine Kontroverse*. Frankfurt am Main: Suhrkamp.

Berger, P. L., and T. Luckmann 1969. *Die gesellschaftliche Konstruktion der Wirklichkeit. Eine Theorie der Wissenssoziologie*. Frankfurt am Main: Fischer.

Berkhout, F., A. Smith and A. Stirling 2004. "Socio-technical regimes and transition contexts", in *System Innovation and the Transition to Sustainability: Theory, Evidence and Policy*, ed. by B. Elzen, F. W. Geels and K. Green. Cheltenham: Edward Elgar, 48–75.

Bijker, W., T. P. Hughes and T. J. Pinch (eds.) 1987. *The Social Construction of Technological Systems: New Directions in the Sociology and History of Technology*. Cambridge, MA: MIT Press.

Bloor, D. 1999. "The Strong Programme in the Sociology of Science", in *Scientific Inquiry*, ed. by R. Klee. New York: Oxford University Press.

Bonß, W., and S. Kesselring 2001. "Mobilität am Übergang von der Ersten zur Zweiten Moderne", in *Die Modernisierung der Moderne*, ed. by U. Beck and W. Bonß. Frankfurt am Main: Suhrkamp, 177–190.

Braun, S., S. Schatzinger, C. Schaufler, C.-M. Rutka and N. Fanderl 2019. "Autonomes Fahren im Kontext der Stadt von Morgen (AFKOS)". Stuttgart: Fraunhofer IAO. http://publica.fraunhofer.de/eprints/urn_nbn_de_0011-n-5436689.pdf (23/6/2020).

Brenner, N. 1998. "Between fixity and motion: accumulation, territorial organization and the historical geography of spatial scales", in *Environment and Planning D: Society and Space* 16, 459–481.

Brenner, N. 2019. *New Urban Spaces: Urban Theory and the Scale Question*. Oxford: Oxford University Press.

Callon, M. 1999. "Actor-network theory – the market test", in *Actor Network Theory and After*, ed. by J. Law and J. Hassard. Oxford/Malden, MA: Blackwell, 181–195.

Canzler, W., and A. Knie 2016. "Mobility in the age of digital modernity: Why the private car is losing its significance, intermodal transport is winning and why digitalisation is the key", in *Applied Mobilities* (1) 1, 56–67.

Canzler, W., and A. Knie 2019. "Autodämmerung: Experimentierräume für die Verkehrswende", strategy paper. Berlin: Heinrich-Böll-Stiftung. DOI: 10.25530/03552.4.

Castells, M. 1989. *The Informational City: Information Technology, Economic Restructuring, and the Urban-Regional Process*. Oxford/Cambridge: Blackwell.

Castells, M. 2010/1996. *The Rise of the Network Society*, part 1 of the trilogy *The Information Age: Economy, Society, and Culture*, vol. 1. 2nd ed. with new preface. Chichester: Wiley-Blackwell.

Christensen, C. M. 1997. *The Innovator's Dilemma: When New Technologies Cause Great Firms to Fail*. Boston: Harvard Business School Press.

Collins, H. 1981. "Stages in the Empirical Programme of Relativism", in *Social Studies of Science* 11, 3–10.

Cresswell, T. 2006. *On the Move. Mobility in the Modern Western World*. New York/London: Routledge.

Cresswell, T. 2010. "Towards a Politics of Mobility", in *Environment and Planning D: Society and Space* (28) 1, 17–31.

Cudworth, E., P. Senker and K. Walker 2013. "Introduction: Contested Futures: Technology, Inequality and Progress", in *Technology, Society and Inequality. New Horizons of Contested Futures*, ed. by E. Cudworth, P. Senker and K. Walker. New York: Peter Lang, 1–16.

Dangschat, J. S. 1998. "Klassenstrukturen im Nach-Fordismus", in *Alte Ungleichheiten – Neue Spaltungen, Sozialstrukturanalyse 11*, ed. by P. A. Berger and M. Vester. Opladen: Leske + Budrich, 49–88.

Dangschat, J. S. 2013. "Eine raumbezogene Handlungstheorie zur Erklärung und zum Verstehen von Mobilitätsdifferenzen", in *Mobilitäten und Immobilitäten. Menschen – Ideen – Dinge – Kulturen – Kapital, Blaue Reihe – Dortmunder Beiträge zur Raumplanung 142*, ed. by J. Scheiner, H.-H. Blotevogel, S. Frank, C. Holz-Rau and N. Schuster. Essen: Klartext, 91–104.

Dangschat, J. S. 2015. "Gesellschaftliche Vielfalt – Heraus- oder Überforderung der Raumplanung?", in: *Jahrbuch Raumplanung 2015. Department für Raumplanung*, vol. 3, ed. by J. S. Dangschat, M. Getzner, M. Haslinger and S. Zech. Vienna/Graz: Neuer Wissenschaftlicher Verlag, 15–38.

Dangschat, J. S. 2017a. "Automatisierter Verkehr – was kommt da auf uns zu?", in *Zeitschrift für Politische Wissenschaft* 27, 493–507. DOI: 10.1007/s41358-017-0118-8.

Dangschat, J. S. 2017b. "Wie bewegen sich die (Im-)Mobilen? Ein Beitrag zur Weiterentwicklung der Mobilitätsgenese", in *Verkehr und Mobilität zwischen Alltagspraxis und Planungstheorie. Ökologische und soziale Perspektiven*, ed. by M. Wilde, M. Gather, C. Neiberger and J. Scheiner. Wiesbaden: Springer, 25–52. DOI: 10.1007/978-3-658-13701-4_3.

Dangschat, J. S. 2019. "Automatisierte und vernetzte Fahrzeuge – Trojanische Pferde der Digitalisierung?", in *Infrastruktur und Mobilität in Zeiten des Klimawandels, Jahrbuch Raumplanung*, vol. 6, ed. by M. Berger, J. Forster, M. Getzner and P. Hirschler. Vienna: Neuer Wissenschaftlicher Verlag, 11–28.

Dangschat, J. S. 2020. "Gesellschaftlicher Wandel, Raumbezug und Mobilität", in *Wechselwirkungen von Mobilität und Raumentwicklung im Kontext des gesellschaftlichen Wandels, Forschungsberichte der ARL 14*, ed. by U. Reutter, C. Holz-Rau, J. Albrecht and M. Hülz. Hannover: Akademie für Raumentwicklung in der Leibniz-Gemeinschaft, 32–75.

Dangschat, J. S., and A. Stickler 2020. "Kritische Perspektiven auf eine automatisierte und vernetzte Mobilität", in *Schwerpunkt: Digitale Transformation, Jahrbuch StadtRegion 2019/2020*, ed. by C. Hannemann, F. Othengrafen, J. Pohlan, B. Schmidt-Lauber, R. Wehrhahn, S. Güntner. Wiesbaden: Springer VS, 53–74. DOI: 10.1007/978-3-658-30750-9.

Deleuze, G., and F. Guattari 1992. *Tausend Plateaus – Kapitalismus und Schizophrenie II*. Berlin: Merve.

Deloitte 2019. "2020 Global Automotive Consumer Study: Is consumer interest in advanced automotive technologies on the move? Global focus countries". https://tinyurl.com/ybrtrlfu (30/4/2019).

Deloitte 2020. "2020 Global Automotive Consumer Study: Is consumer interest in advanced automotive technologies on the move? Europe". https://tinyurl.com/y87563bo (23/6/2020).

Dolowitz, D. P., and D. Marsh 2000. "Learning from abroad. The role of policy transfer in contemporary policy-making", in: *Governance* (13) 1, 5–23.

Dzudzek, I. 2016. *Kreativpolitik. Über die Machteffekte einer neuen Regierungsform des Städtischen.* Bielefeld: transcript.

Eisenstadt, S. N. 2006. "Die Vielfalt der Moderne: Ein Blick zurück auf die ersten Überlegungen zu den 'Multiple Modernities'", Themenportal Europäische Geschichte. www.europa.clio-online.de/2006/Article=113 (3/4/2020).

Elliott, A. 2018. *The Culture of AI. Everyday Life and the Digital Revolution.* Milton: Routledge.

Endres, M., K. Manderscheid and C. Mincke 2016. "Discourses and ideologies of mobility: an introduction", in *The Mobilities Paradigm. Discourses and Ideologies*, ed. by M. Endres, K. Manderscheid and C. Mincke. Milton Park: Routledge, 1–7.

Evans, M., and J. Davies 1999. "Understanding policy transfer. A multi-level, multi-disciplinary perspective", in *Public Administration* (77) 2, 361–385.

Farias, I., and T. Bender (eds.) 2010. *Urban Assemblages: How Actor-Network Theory changes urban studies.* London/New York: Routledge.

Florida, R. 2002. *The Rise of the Creative Class. And How It's Transforming Work, Leisure, Community and Everyday Life.* New York: Basic Books.

Foerster, H. von 1993. "Mit den Augen des andern", in *Wissen und Gewissen. Versuch einer Brücke*, ed. by S. J. Schmidt. Frankfurt am Main: Suhrkamp, 350–363.

Foucault, M. 1972. *Archaeology of Knowledge and the Discourse on Language.* New York: Pantheon.

Frello, B. 2008. "Towards a Discursive Analytics of Movement: On the Making and Unmaking of Movement as an Object of Knowledge", in *Mobilities* (3) 1, 25–50.

Freudendal-Pedersen, M., S. Kesselring and E. Servou 2019. "What Is Smart for the Future City? Mobilities and Automation", in *Sustainability* 11, 221. DOI: 10.3390/su11010221.

Friedrichs, J., H. Häußermann and W. Siebel 1986. *Süd-Nord-Gefälle in der Bundesrepublik? Sozialwissenschaftliche Analysen.* Opladen: Westdeutscher Verlag.

Gad, C., and C. Bruun Jensen 2010. "On the Consequences of Post-ANT", in *Science, Technology & Human Values* (35) 1, 55–80.

Geels, F. W. 2002. "Technological transitions as evolutionary reconfiguration processes: a multi-level perspective and a case-study", in *Research Policy* (31) 8/9, 1257–1274.

Geels, F. W. 2004. "From sectoral systems of innovation to socio-technical systems: Insights about dynamics and change from sociology and institutional theory", in *Research Policy* (33) 6/7, 897–920.

Geels, F. W. 2005. *Technological Transitions and System Innovations: A Co-Evolutionary and Socio-Technical Analysis.* Chaltenham: Edward Elgar.

Geels, F. W. 2006. "Multi-Level Perspective on System Innovation: Relevance for Industrial Transformation", in *Understanding Industrial Transformation. Views from Different Disciplines*, ed. by X. Olsthoorn and A. J. Wieczorek. Dordrecht: Springer, 163–186.

Geels, F. W. 2011. "The multi-level perspective on sustainability transitions: Responses to seven criticisms", in *Environmental Innovation and Societal Transitions* (1) 1, 24–40. DOI: 10.1016/j.eist.2011.02.002.

Geels, F. W. 2012. "A socio-technical analysis of low-carbon transitions: introducing the multi-level perspective into transport studies", *in Journal of Transport Geography* 24, 471–482. DOI: 10.1016/j.jtrangeo.2012.01.021.

Geels, F. W., R. Kemp, G. Dudley and G. Lyons (eds.) 2012. *Automobility in Transition? A Socio-Technical Analysis of Sustainable Transport.* London: Routledge.

Geels, F. W., and J. W. Schot 2007. "Typology of sociotechnical transition pathways: Refinements and elaborations of the multi-level perspective", in *Research Policy* (36) 3, 399–417.

Genus, A., and A.- M. Coles 2008. "Rethinking the multi-level perspective of technological transitions", in *Research Policy* (37) 9, 1436–1445. DOI: 10.1016/j.respol.2008.05.006.

Häußling, R. 2014. *Techniksoziologie.* Baden-Baden: Nomos.

Hajer, M. A., and H. Wagenaar (eds.) 2003. *Deliberative Policy Analysis: Understanding Governance in the Network Society.* Cambridge: Cambridge University Press.

Hitzler, R. 1984. "Sinnbasteln: zur subjektiven Aneignung von Lebensstilen", in *Das symbolische Kapital der Lebensstile. Zur Kultursoziologie der Moderne nach Pierre Bourdieu*, ed. by I. Mörth and G. Fröhlich. Frankfurt am Main/New York: Campus, 75–92.

Hoppmann, J. 2015. "The Role of Deployment Policies in Fostering Innovation for Clean Energy Technologies: Insights from the Solar Photovoltaic Industry", in *Business & Society. A Journal of Interdisciplinary Exploration* (54) 4, 540–558.

Hoppmann, J., M. Peters, M. Schneider and V. H. Hoffmann 2013. "The Two Faces of Market Support: How Deployment Policies Affect Technological Exploration and Exploitation in the Solar Photovoltaic Industry", *Research Policy* (42) 4, 989–1003. DOI: 10.1016/j.respol.2013.01.002.

Hughes, T. P. 1987. "The Evolution of Large Technological Systems", in *The Social Construction of Technical Systems: New Directions in the Sociology and History of Technology*, ed. by W. E. Bijker, T. P. Hughes and T. Pinch. Cambridge, MA/London: MIT Press, 51–82.

Jing, P., G. Xu, Y. Chen, Y. Shi and F. Zhan 2020. "The Determinants behind the Acceptance of Autonomous Vehicles: A Systematic Review", in *Sustainability* 12, 1719. DOI: 10.3390/su12051719.

Jonuschat, H., A. Knie and L. Ruhrort 2016. "Zukunftsfenster in eine disruptive Mobilität. Teil 1: Mobilität in einer vernetzten Welt". Berlin: Innovationszentrum für Mobilität und gesellschaftlichen Wandel (InnoZ). https://tinyurl.com/yc3zwee3 (22/6/2020).

Kanger, L., F. W. Geels, B. Sovacool and J. Schot 2019. "Technological diffusion as a process of societal embedding: Lessons from historical automobile transitions for future electric mobility", in *Transportation Research Part D: Transport and Environment* 71 (June), 47–66.

Kemp, R. 1994. "Technology and the transition to environmental sustainability: the problem of technological regime shifts", in *Futures* (26) 10, 1023–1046.

Kemp, R., A. Rip and J. W. Schot 2001. "Constructing transition paths through the management of niches", in *Path Dependence and Creation*, ed. by R. Garud and P. Karnøe. Mahwah, NJ/ London: Lawrence Erlbaum, 269–299.

Kesselring, S. 2008. "The mobile risk society", in *Tracing Mobilities: Towards a Cosmopolitan Perspective*, ed. by W. Canzler, V. Kaufmann and S. Kesselring. Aldershot/Burlington, VT: Ashgate, 77–102.

Kesselring, S. 2020. "Reflexive Mobilitäten", in *Das Risiko – Gedanken über und ins Ungewisse. Interdisziplinäre Aushandlungen des Risikophänomens im Lichte der Reflexiven Moderne. Eine Festschrift für Wolfgang Bonß*, ed. by H. Pelizäus and L. Nieder. Wiesbaden: Springer VS, 155–193.

Knorr-Cetina, K. D. 1989. "Spielarten des Konstruktivismus. Einige Notizen und Anmerkungen", in *Soziale Welt* 40 1/2, 86–96.

Künkel, J. 2015. "Urban policy mobilies versus policy transfer", in *sub/urban. zeitschrift für kritische Stadtforschung* (3) 1, 7–24.

Läpple, D. 2011. "Das Jahrhundert der Städte und die Diversität städtischer Entwicklungsmuster", in: *Stadt und Urbanität*, ed. by M. Messling, D. Läpple and J. Trabant. Berlin: Kadmos, 34–64.

Lash, S. 1999. *Another Modernity: A Different Rationality*. Oxford: Blackwell.

Latour, B. 1996. "On actor-network theory", in *Soziale Welt* 47, 374.

Latour, B. 2007. *Eine neue Soziologie für eine neue Gesellschaft. Einführung in die Akteur-Netzwerk-Theorie*. Frankfurt am Main: Suhrkamp. Original: *Reassembling the Social: An Introduction to Actor-Network-Theory*, Oxford: Oxford University Press, 2005.

Law, J. 2006. "Notizen zur Akteur-Netzwerk-Theorie: Ordnung, Strategie und Heterogenität", in *ANThology. Ein einführendes Handbuch zur Akteur-Netzwerk-Theorie*, ed. by A. Bellinger and D. J. Krieger. Bielefeld: transcript.

Mackenzie, D., and L. Wajcman 1985. "Introductory essay: the social shaping of technology", in *The Social Shaping of Technology*, ed. by D. Mackenzie and L. Wajcman. Milton Keynes: Open University Press.

Manderscheid, K. 2012. "Automobilität als raumkonstituierendes Dispositiv der Moderne", in *Die Ordnung der Räume. Geographische Forschung im Anschluss an Michel Foucault*, ed. by H. Füller and B. Michel. Münster: Westfälisches Dampfboot, 145–178.

Manderscheid, K. 2014. "Formierung und Wandel hegemonialer Mobilitätsdispositive: Automobile Subjekte und urban Nomaden", in *Zeitschrift für Diskursforschung* (2) 1, 5–31.

Manderscheid, K. 2020. "Antriebs-, Verkehrs- oder Mobilitätswende? Zur Elektrifizierung des Auto-mobilitätsdispositivs", in *Baustelle Elektromobilität. Sozialwissenschaftliche Perspektiven auf die Transformation der (Auto-)Mobilität*, ed. by A. Brunnengräber and T. Haas. Bielefeld: transcript, 37–68.

Marx, Karl 1867/1990. *Capital: Critique of Political Economy*, vol. I. Trans. Ben Fowkes. London: Penguin in association with New Left Review.

Mayntz, R., and F. W. Scharpf 1995. "Der Ansatz des akteurzentrierten Institutionalismus", in *Gesellschaftliche Selbstregelung und Steuerung*, ed. by R. Mayntz and F. W. Scharpf. Frankfurt am Main: Campus, 39–72.

McCann, E., and K. Ward 2012. "Policy assemblages, mobilities and mutations: Towards a multi-disciplinary conversation", in *Policy Studies* (34) 1, 2–18. DOI: 10.1111/j.1478-9302.2012.00276.x.

McLennan, G. 2004. "Travelling with vehicular ideas: The case of the third way", in *Economy and Society* (33) 4, 484–499.

Milakis, D., B. van Arem and B. van Wee 2017. "Policy and society related implications of automated driving: a review of literature and directions for future research", in *Journal of Intelligent Transportation Systems* 21 (4), 324–348. DOI: 10.1080/15472450.2017.1291351.

Mincke, C. 2016. "From mobility to its ideology: when mobility becomes an imperative", in *The Mobilities Paradigm. Discourses and Ideologies*, ed. by M. Endres, K. Manderscheid and C. Mincke. Milton Park: Routledge, 11–33.

Mitteregger, M., E. M. Bruck, A. Soteropoulos, A. Stickler, M. Berger, J. S. Dangschat, R. Scheuvens and I. Banerjee 2022a (2020). *AVENUE21. Connected and Automated Driving: Prospects for Urban Europe*. Berlin: Springer Vieweg. DOI: 10.1007/978-3-662-64140-8.

Mitteregger, M., E. M. Bruck, A. Soteropoulos, A. Stickler, M. Berger, J. S. Dangschat, R. Scheuvens and I. Banerjee 2021b (2020). "Connected and automated transport in the Long Level 4: Settlement development, transport policy and planning during the transition period", in *AVENUE21. Connected and Automated Driving: Prospects for Urban Europe*, ed. by M. Mitteregger, E. M. Bruck, A. Soteropoulos, A. Stickler, M. Berger, J. S. Dangschat, R. Scheuvens and I. Banerjee. Berlin: Springer Vieweg, 57–98.

Möhrle, M. G. 2018. "Technologie". https://wirtschaftslexikon.gabler.de/definition/technologie-48435/version-271688 (24/3/2020).

Nelson, R. R., and S. G. Winter 1982. *An Evolutionary Theory of Economic Change*. Cambridge, MA: Bellknap.

Peck, J. 2012. "Recreative City: Amsterdam, vehicular ideas, and the adaptive spaces of creative policy", in *International Journal of Urban and Regional Research* (36) 3, 462–485.

Peck, J., and N. Theodore 2010. "Mobilizing Policy: Models, methods, and mutations", in *Geoforum* (41) 2, 169–174.

Peters, M., M. Schneider, T. Griesshaber and V. H. Hoffmann 2012. "The impact of technology-push and demand-pull policies on technical change: Does the locus of policies matter?", in *Research Policy* (41) 8, 1296–1308.

Peuker, B. 2010. "Akteur-Netzwerk-Theorie (ANT)", in *Handbuch Netzwerkforschung*, ed. by C. Stegbauer and R. Häußling. Wiesbaden: VS Verlag für Sozialwissenschaften, 325–335.

Pinch, T. J., and W. E. Bijker 1987. "The Social Construction of Facts and Artifacts: Or How the Sociology of Science and the Sociology of Technology Might Benefit Each Other", in *The Social Construction of Technical Systems: New Directions in the Sociology and History of Technology*, ed. by W. E. Bijker, T. P. Hughes and T. Pinch. Cambridge, MA/London: MIT Press: 17–50.

Rammert, W. 2003. "Technik in Aktion: Verteiltes Handeln in soziotechnischen Konstellationen", in *Autonome Maschinen*, ed. by T. Christaller and J. Wehner. Wiesbaden: VS Verlag für Sozialwissenschaften, 289–315.

Rammert, W. 2007. *Technik-Handeln-Wissen. Zu einer pragmatischen Technik- und Sozialtheorie*. Wiesbaden: Springer.

Rip, A., and R. Kemp 1998. "Technological change", in *Human Choice and Climate Change 2*, ed. by S. Rayner and E. L. Malone. Columbus, OH: Battelle, 327–399.

Rosa, H. 2015 (2005). *Social Acceleration: A New Theory of Modernity*. Trans. Jonathan Trejo-Mathys. New York, NY: Columbia University Press.

Rupprecht, S., S. Buckley, P. Crist and J. Lappin 2018. "'AV-Ready' Cities or 'City-Ready' AVs?", in *Road Vehicle Automation 4*, ed. by G. Meyer and S. Beiker. Heidelberg et al.: Springer International Publishing, 223–233.

Saint-Simon, H. de 1814. De la réorganisation de la société européenne. Paris: Delonay.

Santarius, T. 2012. "Der Rebound-Effekt: Über die unerwünschten Folgen der erwünschten Energieeffizienz", *Impulse zur WachstumsWende*. Wuppertal: Wuppertal Institut für Klima, Umwelt, Energie.

Sassen, S. 2001. *The Global City: New York, London, Tokyo*. Princeton, NJ: Princeton University Press.

Scheiner, J. 2009. *Sozialer Wandel, Raum und Mobilität. Empirische Untersuchungen zur Subjektivierung der Verkehrsnachfrage*. Wiesbaden: VS Verlag für Sozialwissenschaften.

Schmookler, J. 1966. *Invention and Economic Growth*. Cambridge, MA: Harvard University Press.

Schneidewind, U. et al. 2018. *Die Große Transformation. Eine Einführung in die Kunst gesellschaftlichen Wandels*. Frankfurt am Main: Fischer.

Schumpeter, J. A. 1942. *Capitalism, socialism and democracy*. New York/London: Harper.

Schwedes, O. 2013. "Räumliche Mobilität im gesellschaftlichen Transformationsprozess – Ein Resümee", in *Räumliche Mobilität in der Zweiten Moderne. Freiheit und Zwang bei der Standortwahl und Verkehrsverhalten*, ed. by O. Schwedes. Münster: Lit, 273–287.

Sennett, R. 1977. *The Fall of Public Man*. London/Boston: Faber & Faber.

Sennett, R. 1998. *Der flexible Mensch*. Die Kultur des neuen Kapitalismus. Munich: Siedler.

Sheller, M. 2011. "Mobility", in *Sociopedia.isa, 1–12*. DOI: 10.1177/205684601163.

Sheller, M., and J. Urry 2006. "The New Mobilities Paradigm", in *Environment and Planning A: Economy and Space* (38) 2, 207–226.

Sheller, M., and J. Urry 2016. "Mobilizing the new mobilities paradigm", in *Applied Mobilities* (1) 1, 10–25.

Shladover, S. E. 2016. "The Truth about 'Self-Driving' Cars: They are coming, but not the way you may have been led to think", in *Scientific American* (314) 6, 52–57.

Sovacool, B. K., and J. Axsen 2018. "Functional, symbolic and societal frames for automobility: Implications for sustainability transitions", in *Transportation Research Part A: Policy and Practice* 118, 730–746. https://tinyurl.com/yae5rufr (23/6/2020).

Stefano, G. di, A. Gambardella and G. Verona 2012. "Technology Push and Demand Pull Perspectives in Innovation Studies: Current Findings and Future Research Directions", in *Research Policy* (41) 8, 1283–1295.

Stickler, A. 2020a. "Automatisiertes und vernetztes Fahren als Zukunftsperspektive für Europa? Eine Diskursanalyse der gegenwärtigen europäischen Politik", in *Baustelle Elektromobilität. Sozialwissenschaftliche Perspektiven auf die Transformation der (Auto-)Mobilität*, ed. by A. Brunnengräber and T. Haas. Bielefeld: transcript, 93–116.

Stickler, A. 2020b. "Automobilität im Umbruch? Gegenwärtige Stabilisierungen oder Transformation der automobilen Hegemonie", unpublished PhD dissertation, TU Wien.

Tarmann, P. 2018. "Autonomes Fahren: Chancen und Risikowahrnehmungen. Eine qualitative Analyse der gesellschaftlichen Akzeptanz selbstfahrender Autos", unpublished master's thesis, Karl-Franzens-Universität Graz.

Urry, J. 2000. *Sociology beyond Societies: Mobilities of the Twenty-First Century*. London: Routledge.

Urry, J. 2003. *Global Complexity*. Cambridge: Polity.

Urry, J. 2004. "The 'System' of Automobility", in *Theory, Culture & Society* (21) 4/5, 25–39.

Urry, J. 2007. *Mobilities*. Cambridge: Polity.

Urry, J. 2009. "Mobilities and Social Theory", in *The New Blackwell Companion to Social Theory*, ed. by Bryan S. Turner. Hoboken, NJ: Wiley-Blackwell, 477–495.

Urry, J. 2012. "Social networks, mobile lives and social inequalities", in *Journal of Transport Geography* 12, 24–30.

Vester, M., P. von Oertzen, H. Geiling, T. Hermann and D. Müller 2001. *Soziale Milieus im gesellschaftlichen Strukturwandel. Zwischen Integration und Ausgrenzung*. Frankfurt am Main: Suhrkamp.

Wallerstein, I. 2004. *World-System Analysis: An Introduction*. Durham: Duke University Press.

Welge, M. K. 1992. *Planung: Prozesse – Strategien – Maßnahmen*. Wiesbaden: Springer Gabler.

Weyer, J. 2008. *Techniksoziologie. Genese, Gestaltung und Steuerung sozio-technischer Systeme.* Weinheim/Munich: Juventa.

Weyer, J. 2019. *Die Echtzeitgesellschaft. Wie smarte Technik unser Leben steuert.* Frankfurt am Main: Campus.

Wiesner, A. 2003. "Ethnografische Politikforschung", in *Politik als Lernprozess. Wissenszentrierte Ansätze der Politikanalyse*, ed. by M. L. Maier, F. Nullmeier, T. Pritzlaff and A. Wiesner. Wiesbaden: VS Verlag für Sozialwissenschaften, 141–166. DOI: 10.1007/978-3-663-11061-3.

20 Data-driven urbanism, digital platforms and the planning of MaaS in times of deep uncertainty: What does it mean for CAVs?

Ian Banerjee, Peraphan Jittrapirom, Jens S. Dangschat

Peraphan Jittrapirom
Nijmegen School of Management, Radboud University and Global Carbon Project (GCP),
National Institute for Environmental Studies (NIES)

Ian Banerjee
TU Wien, Research Unit Sociology (ISRA)

Jens S. Dangschat
TU Wien, Research Unit Sociology (ISRA)

© The Author(s) 2023
M. Mitteregger et al. (eds.), *AVENUE21. Planning and Policy Considerations for an
Age of Automated Mobility*, https://doi.org/10.1007/978-3-662-67004-0_20

1. INTRODUCTION

This paper offers a critical review of three coevolving socio-technical paradigms: (a) "data-driven urbanism", (b) digital platforms and (c) "Mobility-as-a-Service" (MaaS). It explores the complex relationship unfolding between data-driven cities and digital platforms, while drawing on MaaS as a case to discuss the challenges of implementing mobility services via digital platforms. Inferences are drawn from the ongoing debate accompanying these three paradigms to identify potential criteria for the design of socially accountable governance models for the deployment of connected and automated vehicles (CAVs). Grounded in current trends, the research builds on the assumption that CAV-based transport services will be offered through digital platforms of some sort in the future. Also based on this assumption, we believe that if national governments and municipalities are interested in deploying CAVs, they will be well advised to learn from the experience gained from the practices of these three emerging paradigms.

Complying with the overall aspiration of this publication, we are eager to move away from understanding CAVs as a game-changing techno-economic novelty and towards understanding them as a socio-technical innovation that might help us to design sustainable, inclusive and participatory human environments. To achieve this, we argue it is indispensable to understand the geography, architecture and "logic" of the emerging cyber-physical landscape produced by the interweaving of physical and digital formations and the practices of data-driven cities, of which CAVs will potentially be constitutive elements (see Fig. 1).

The first section of this paper reflects on the technological, conceptual and epistemological shifts taking place in the discourse and practices of data-driven urbanism by highlighting the growing influence of big data and the computational understanding of the city. The second section looks at how online platforms are making use of big data and ingenious digital mechanisms to create ever-new forms of commercial services. It looks at the key challenges that surround their emerging practices. The third section looks at the challenge of planning MaaS under deep uncertainty. Apart from its technological feasibility, MaaS is associated with various uncertainties concerning issues such as network effects and questions around how it will affect the overall transportation system, future demand and the willingness of crucial stakeholders to cooperate (Jittrapirom et al. 2017a). Finally, the fourth section draws inferences from an analysis of the debate unfolding around the three paradigms mentioned above, and makes recommendations for the regulation, governance and design of potential CAV platforms.

1.1 BACKGROUND

The growing capacity to capture and intelligently process vast amounts of data lies at the centre of the data-driven city paradigm (Kitchin 2014, Leszczynski 2016). The intensifying relationship between new data regimes and city organization is increasingly leading to the emergence of practices in urban governance that primarily engage in the collection, management and commodification of voluminous amounts of varied, dynamic and interoperable data. Hopes placed on the data-driven city, popularly called the "smart city", revolve largely around the narratives of efficiency, economic opportunity, safety and sustainability. All aspects of contemporary life in the modern world are affected by the shift towards this data-based reorganization of activities and services – from mobility to education, from health to administration, from retail to energy.

An essential driver of this shift is the promise of securitization and enhanced quality of public life through increasingly centralized decision-making processes and service provision with the help of artificial intelligence (AI) and automation (see Chap. 10 by Mitteregger in this volume). The digitized notion of urban governance is based largely on new forms of predictive analytics, enabled first by the technological feasibility of capturing large amounts of socio-spatial data drawn from surveillance cameras and other sensors built in the city's infrastructures, and second by the possibility of deriving meaningful patterns of prognostic material from them. In addition to the collection of *infrastructural* data, increasing amounts of *personal* data are being captured by digital platforms. These entities use automated mechanisms to extract large volumes of specified personal data from the depths of digital space and provide users of this same space with increasing varieties of value-added services. The intelligent processing of real-time personal data of various formats is now making it possible to create lenses of what is sometimes called "sentiment analytics" (Leszczynski 2016, Bassoo et al. 2017). These are automated learning algorithms deployed to analyse and predict various patterns of human behaviour. The interoperability between these highly heterogeneous types of collected data and the convergence of different types of analytics has opened up unprecedented possibilities of new forms of organization but also control of everyday life. This has also resuscitated the notion of a "computational understanding" of the city – an idea that had already entered the urban imagination back in the 1950s (Hall 2002, Shelton et al. 2015).

The rapid proliferation of trans-local digital platforms and their high degree of penetration into all spheres of society are making them more and more indispensable for the functioning of contemporary cities. Their capacity to replace pre-digital, offline organizations have turned them into powerful agents of structural change. However, while they are benefiting society in innumerable ways, the untransparent practices of their globally operating commercial inflections have recently led to intense legal disputes (see 3.3) across continental borders. Regulators, particularly in Europe, are confronted with a plethora of new challenges associated with their social impact, citizens' acceptance, technological complexities, legal ambivalences and geopolitical consequences. Since the public sector and governments are increasingly reliant on corporate platforms, it is crucial to rethink the principles on which the design of these systems is based if we have the desire to sustain democratic values and guarantee an equal playing field for all. We assert that designing governance models in the digital age will to a great degree be about governing the practices of data-driven cities and digital platforms. This paper identifies the main challenges facing the regulation of these formations and the prospects of designing governance models that will engender benefits to the whole of society. The key challenges identified in this review are the question of how to instigate a *broad societal discourse* on the topic and how to enshrine *public values* into the elusive architecture of these digital formations.

1.2 THE FOURFOLD PROCESS: MENTAL MAP OF THE DATA-DRIVEN CITY

It is fairly evident by now that digital platforms will play a crucial part in the creation of new types of digital infrastructures that will shape the future of cities. A growing volume of literature on "platform urbanism" (Barns 2014, Bratton 2016, Langley/Leyshon 2017, Srnicek 2017, Artioli 2018, Söderström/Mermet 2020) indicates that there is increasing academic interest in the dynamics that are emerging between the socio-technical formation of digital platforms and the urban.

Drawing from this literature, we outline a mental map of the main underlying dynamics producing the cyber-physical landscapes of the data-driven city. It depicts a fourfold movement comprising (a) the collection of unprecedented amounts of infrastructural and personal data captured in physical and digital spaces, (b) the storage of collected data, (c) the processing of collected data to derive meaning and to simultaneously create economic or societal value out of them, and (d) feeding back the created values into the materialities of everyday life in the form of services.

This process is largely orchestrated by digital platforms (see 3.3.2) with the help of digital information systems, automated algorithms and AI. MaaS, CAVs, the Internet of Things (IoT), etc. are seen as constitutive elements in the incessant movement of this fourfold process.

Figure 1: Mental map of the production of cyber-physical landscapes in the data-driven city (fourfold movement). Physical objects and digital systems merge here to produce new realities and imaginaries of the urban. Digital platforms play a pivotal role in this process, while MaaS, CAVs, IoT, etc. are seen as its constitutive elements.

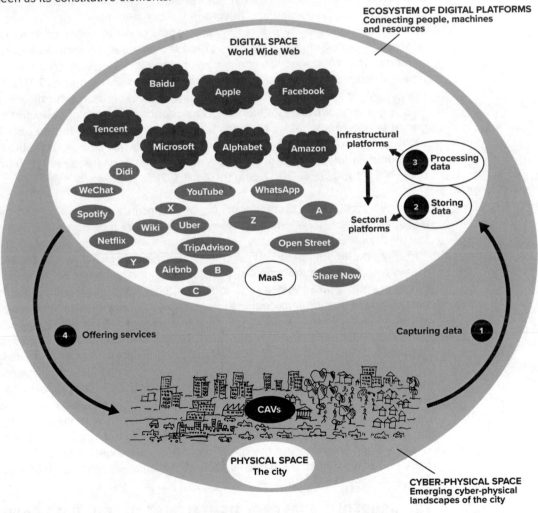

Source: Ian Banerjee

2. DATA-DRIVEN URBANISM

2.1 "CODE/SPACES": BUILDING BLOCKS OF THE CYBER-PHYSICAL LANDSCAPE

Two highly complex, contingent and open systems – cities and digital information systems – are currently being interwoven (see Fig. 1) to create a novel and vast landscape of cyber-physical materialities (Townsend 2013). Dodge and Kitchin (2004) describe spaces wherein software and the spatiality of everyday life become mutually constituted as "code/spaces". For example, the connected car is not just a car, it is a code/space embedded in large networks of various informational exchanges enabled by codes and algorithms – networks that Manuel Castells famously called "spaces of flow" (Castells 1996). The ontologies of hybrid forms of single code/spaces, connected with thousands of other code/spaces, are challenging our very understanding of objects, urban processes and human agency. The amalgamation of digital and physical spaces has reached a size and degree of sophistication that makes the invisible architecture of the World Wide Web as important as the visible architecture of the physical world – indeed in some cases, it seems, it has become more important.

2.2 DIGITALIZATION AND BIG DATA

To encapsulate the new discourse emerging around the broad notion of data-driven urbanism, it is helpful to look at the emergence of two important terms: big data and digitalization. Today technological change is mostly associated with digitalization, a process that has arguably brought about the most sweeping transformative changes to society since the industrial revolution. Central to all digitalization processes lies the technological capacity to capture and convert all data into binary information and to process them to generate economic or social values. These units of binary data are emerging as the new constitutive elements of all significant flows and activities of human civilization. Big data, a term popularized by John Mashey in the 1990s (Kitchin/Gavin 2016), refers to very large or complex sets of data that can be captured, stored and analysed through data-processing mechanisms (Kitchin/Gavin 2016).

The term often refers to the traces created in digital space from the activities, transactions and movements of millions of users and objects. Small data, in contrast, is the data collected from sources such as questionnaires, city audits, ethnographies, etc. Literature on the features of big data commonly refers to the three "Vs":

1. *volume*: consisting of petabytes or more of data;

2. *velocity*: consisting of data created in real time or near to it; and

3. *variety*: consisting of structured or unstructured data, temporally and spatially referenced. Also, the data collected need to be *fine-grained* in resolution and *relational* in nature; that is, capable of conjoining different types of data sets. They are also supposed to be *flexible*, meaning extendable and scalable at a later point (Kitchin 2014: 3).

2.3 THE DATA-DRIVEN CITY

Broadly speaking, there are two related interpretations of the data-driven city (Kitchin 2014: 1). The first vision pertains to a better management of the city from a largely technological and technocratic perspective and the second alludes to exploiting cities as information economies, wherein urban policies are directed mostly towards deploying digital infrastructures and services to boost digital entrepreneurship and the activities of the knowledge-based economy (Coletta et al. 2017).

As cities are revealing themselves to be the key sites for digital transformation and societal experimentation in the 21st century (Glaeser 2011, Offenhuber/Ratti 2014, Shelton et al. 2015), data is becoming both "the modus operandi and raison d'être" (Shelton et al. 2015: 16) of contemporary modes of urban governance. The real-time collection of large amounts of inexpensive data about urban activities and processes is also believed to enable new forms of adaptive management and digital modes of governance (see Chap. 16 by Hamedinger in this volume). As mentioned earlier, the push towards a data-based understanding of the city is largely framed by economic and political narratives of an efficient, cost-effective, sustainable, competitive, productive, open and transparent city (Kitchin 2014). It also entails the belief that a widescale, fine-grained and real-time grasp of the city will lead to an enhanced "control of urbanity" (Kitchin 2014: 3).

According to Kitchin (2014), there are three main capture sources for urban data:

a. directed data: these are traditional forms of data collected through CCTV, fingerprints, iris scans, etc.;

b. automated data: these are collected from the measuring points of meteorological sensors, automated forms of surveillance, GPS, built-in sensors in road infrastructure, public transport systems, RFID chips attached to rubbish bins, etc.;

c. volunteered data: these are data "gifted" by users from their interactions across social media, such as Facebook and Twitter, or user-generated data in mobile devices contributing to a common system such as OpenStreetMap or the project EMOTIVE (see below).

The information-capturing notion of the IoT has further opened up vast application possibilities in virtually all technologically based processes in everyday life. The IoT-based embedding of software (codes) into electronic gadgets of all kinds, transforms them from "dumb" to "smart" (Kitchin 2014: 4). Once the objects become smart, they are no longer independent entities but knots in global informational networks transmitting information that can subsequently be exploited for commercial use or for securitization achieved through various methods of surveillance.

As indicated above, an example of one of the most important applications of big data in the data-driven city is the securitization of public life with the help of sentiment analysis (Leszczynski 2016, Bassoo et al. 2017). Anticipatory security calculus designed to identify and divert risks of human behaviour by using volunteered data combined with automated data is expected to deliver unprecedented ways of predicting of human behaviour. A number of pre-emptive algorithmic calculi are being tested in different countries.

One experiment, called EMOTIVE – Extracting the Meaning of Terse Information in a Geo-Visualization of Emotion, is currently taking place in the UK. It is designed to monitor and map the emotional charge of online traffic and "shared atmospheres of affect" as expressed through individualized contributions to the social media platform Twitter (Leszczynski 2016: 1699). The aim of the project, as its website declares, is to predict and monitor selected events for the "benefit of social security" and to "safeguard the public from potentially harmful events" (EMOTIVE 2020). Subjected to such background analysis, the general public may legitimately question what checks and balances are in place to ensure that their private data is not being exploited and to determine and regularly reassess what is deemed "harmful" to whom, by whom.

This technology, however, bears a considerable amount of risk, as it may exacerbate data expressing prejudices against social groups, places and socio-spatial constellations. Additionally, the data will not be representative as normal situations will necessarily be under-represented. Moreover, if a local government does have this information: what will it do with it? Hire more police? Enforce shutdowns during certain periods? Or declare a neighbourhood a no-go area for tourists?

2.4 THE COMPUTATIONAL UNDERSTANDING OF THE CITY AND ITS COMPLICATIONS

The alluring promises of efficiency, safety and economic opportunities achieved through the use of big data and predictive analytics are increasingly leading to a computational understanding of the city (Shelton et al. 2015). This paradigm purports to comprehend urban phenomenon through a lens of digitalized data. It is inadvertently changing the culture and content of the urban debate that has evolved over the last hundred years around issues like public space, inequality, justice, public participation, etc. The explosion of digitized data and the role of cities as the main sites of their production, and the way these data are being used to reimagine the urban life of today and tomorrow, are moving to the centre stage of the urbanism debate (Batty 2012, Townsend 2013). Cities embracing digitalization and ICT as their main development strategy have been labelled "wired cities", "cyber cities", "intelligent cities", "smart cities", etc. (Kitchin 2014). "Smart city" is undoubtedly the most popular term in use today – a term, and the prerogatives of which, academics have exhaustively critiqued, pointing to issues such as growing corporatization of city government, violation of privacy rights, cyberattacks, etc. (see Hollands 2008, Greenfield 2013, Townsend 2013, Kitchen 2014, Söderström et al. 2014).

At the core of the computational understanding of the city is the belief that all functions of a city can be measured and monitored, and all malfunctions can be treated as technical problems. Even complex social situations can be disassembled into neatly defined problems that can be solved, or at least optimized, through computation (Kitchin 2014). This position is based on a staunch belief in the linear and logical manageability of all societal and socio-spatial problems, a position that may prove to be deeply inadequate to tackle the challenges of the future, as shown by the arguments of a view known as "second modernity" or "reflexive modernization" (Beck et al. 1996). This view points out the ambivalent, risk-prone unpredictabilities and insecurities, marked by flows instead of structures, and sees connectivity both as a problem and as a project (see Chap. 19 by Dangschat in this volume).

Capturing especially the data of real-time phenomena has created the impression that it is possible to manage and fix a situation even while it is still unfolding. The computational understanding of the city is by no means new; its historic precursors go back to the "quantitative revolution" started by geographers and planners in the 1950s. Ever since, think tanks, corporations and also academics have tried to tilt the discourse in urbanism from an ideographic and critical approach to a more rational, scientific and depoliticized attitude (Hall 2002, Shelton et al. 2015). What has substantiated the ideas of the geographers of the 1950s, is the enormous improvement in the computational power of computer chips and technologies to efficiently capture and handle vast amounts of data.

The computational view of the world seems to be the resurrection of a 200-year-old idea sprung out of the then burgeoning discipline of classical mechanics. It was assumed that if Laplace's demon, a figure popularized in the early 19th-century sciences, knew the location and momentum of every atom in the universe, it could precisely predict the future of any given object in it. The technocratic view of the city implies, like Laplace's demon, that we can only understand its processes fully if we have enough data; and good governance is only possible if we subsequently employ an evidence-based, algorithmically processed mechanism that can process this data. This method alone is believed to ensure a rational, logical and impartial decision-making process.

2.5   REGULATING THE DATA-DRIVEN CITY

Almost imperceptibly, data have ascended to become the most treasured resource and asset of the 21st century. While in the year 2000 there were three tech companies among the top ten most valuable corporations in the world, today, seven of them are data or tech companies – five American and two Chinese (Fengler/Gill 2019). The rise in the value of data and consequently in the power of those who possess them, has also brought the necessity to regulate them. This power and value have led to the flows and ownership of data becoming the most important subject of regulation – in addition, of course, to the long-existing flows of finance. Ongoing legal battles show how the regulation of data flows is still a globally fragmented endeavour (Aridi/Petrovčič 2020).

This regulation is shaped by the views of two opposing groups: the anti-interventionists and the pro-interventionists (Bostoen 2018). The former argue that intervention in digital markets should be kept to a minimum, while the latter essentially believe meaningful principles for platform regulation must be put in place. However, both groups believe competition laws have to be rewritten, with diverging beliefs about how to apply the new rules and with varying strictness. Big political struggles are likely to be fought in the future around the regulation of data flows – something already concentrated in the hands of a small number of corporate titans.

The Regulatory Policy Division of the OECD (2019) and the ten principles proposed by the House of Lords in the UK (2019) point to some of the key issues and broad challenges concerning the development of frameworks and regulatory standards for this new world.

1. **Transboundary challenge:** Given the digital economy's clear cross-border effects, solutions limited to the domestic domain will no longer suffice; international regulatory cooperation is needed to avoid arbitrage, protect consumer rights and promote interoperability across regulatory frameworks and enforcement, whilst creating a favourable environment for the digital economy to evolve.

2. **Lack of knowledge:** Most national legislatures are unaware of the socio-technical finesses of the evolving platform ecosystems. They are still based on pre-digital and pre-networked systems of governance. For example, "filter bubbles" or "personalization" are not part of the common legal discourse (Dijck et al. 2018: 157).

3. **Problem of speed:** Given the level of technical expertise involved, the uncertainty surrounding digital developments, and the overwhelming pace of digital transformation, governments need to actively engage a broad range of stakeholders, invest in foresight and horizon scanning, initiate regulatory impact assessments early in the policymaking process, and carry out post-implementation reviews.

4. **A whole-of-government approach:** This calls for increased dialogue and coherence between governmental bodies in order to meet institutional challenges and the cross-jurisdictional nature of the task. This may require the bringing together of key relevant players and the preparation of specific institutional responses by establishing thematic platforms. Governments also need to create a broader public debate by involving the civic sector.

What further complicates the creation of broader regulatory frameworks for the data-driven city, is the simultaneous emergence and convergence of new technologies that revolve around big data, artificial intelligence, robotics, cloud computing, IoT, 5G, along with new types of investment models (ITU 2020). It is becoming more and more evident that there cannot be a single framework of regulation: authorities have to perpetually navigate the complex and fast-moving digital landscapes of infrastructures, protocols, standards and "user services" (House of Lords 2019: 11). Also, what makes regulation in liberal democracies particularly difficult at this critical juncture, is the need for strong political will to engrave democratic values into the underlying architecture of digital technologies and social practices.

Regulating the digital world is a gargantuan task that will need the involvement of many countries, actors and negotiations on many different levels. Some observers in Europe say new regulatory authorities should be designed at the EU level (OECD 2019), while others say it would be more fruitful if national authorities simply collaborated more intensely. A number of national and international organizations, such as the UN-based International Telecommunication Union (ITU 2020), the Regulatory Policy Division of the OECD (2019), the Digital Charter of the UK (Digital Charter 2018) and Canada's Directive on Regulation, are in fact coming together to work on the topic in more concerted ways. How to safeguard citizens and institutions in the data-driven world without hampering innovation and competition, has become the key challenge for regulators. It is becoming evident that it will not be about more regulation but about a different approach to regulation (see Section 3.3).

2.6 GOVERNANCE: EXAMPLES OF ALTERNATIVE APPROACHES

Another critical question that arises in view of the immense influence of the data-driven city and the computational understanding of the urban is: how is it impacting the mindset of planners, administrators, politicians and citizens? If we say regulation is about controlling digital transformation, and governance is about finding consensus on socially shared visions for the future of people and places reached through complicated negotiation processes between the state, market and civil society, then it is evident that we must find new participatory processes of enforcement of regulations in the cyber-physical city.

Besides an overriding number of examples of the digitized city being technocratically and entrepreneurially exploited, we also see an increasing number of alternative approaches (Townsend 2013, Banerjee 2014, Banerjee/Fischer-Schreiber 2015). For example, in the conceptualizations of the cyber-physical city based on "open data", we see some promising prospects of democratic and inclusive development. This is a movement fighting to make big data a public prerogative and a common good. Many governments around the world have started to release various kinds of administrative and operational data using various kinds of open-data models (Ferro/Osella 2013, Leszczynski 2016). For example, the city of Santander in Spain created the app SmartSantanderRA to provide real-time information to citizens about around 2,700 places, such as libraries, public buses, bike rental services, etc. Together with the City of London, University College London (UCL) has developed London Dashboard, a data visualization app that tracks the city's performance in twelve areas, such as jobs, transport, etc. It communicates to citizens live feeds of real-time data. Another example is Dublinked in Dublin; this platform provides operational data from four authorities in Dublin in an open format, encouraging the creation of apps providing services of social value by using these data. The increasing number of such examples shows how different the practices of producing the data-driven city can be – owing to the initiatives of single actors or to locally specific conditions of culture, politics or governance. They also show that such progressive local initiatives can take place even within prevailing conservative ideologies.

3. DIGITAL PLATFORMS

3.1 RISE OF THE PLATFORM SOCIETY

This section looks at how data-based "transactional technologies" are becoming essential constituents of the data-driven city and how the challenge of regulating them is becoming the key question of shaping cities and society at large.

The idea of matching supply and demand by creatively processing of torrents of digitized data has led to the rise of one of the most disruptive socio-technical formations of the 21st century: the digital platform. Knowingly or unknowingly, digital platforms have become integral parts of our everyday lives. We make use of them through connectivity services offered by social media giants such as Facebook, mobility services by tech start-ups such as Uber, accommodation renting through Airbnb, educational services by Coursera, health services by PatientsLikeMe, rating apps like Tripadvisor, etc. Digital platforms position themselves as key intermediaries in the provision of services in all domains of society. Almost imperceptibly, they are reorganizing and reconfiguring communication, entertainment, mobility, travel, work, government and increasingly politics, by gradually embedding themselves in everyday human life.

Even though the term "digital platform", or simply "platform", is widely used by public authorities and the media, a workable definition for it is lacking. In a seminal publication on the "platform society", Van Dijk, Poell and de Waal (2018) define digital platforms broadly as a "programmable digital architecture designed to organize interactions between users – not just end users but also corporate entities and public bodies. It is geared towards the systematic collection, algorithmic processing, circulation and monetization of user data" (Dijck et al. 2018: 4). From a more market-centred viewpoint, the European Commission defines them as "an undertaking operating in two (or multi)-sided markets, which uses the internet to enable interactions between two or more distinct but interdependent groups of users so as to generate value for at least one of the groups" (European Commission 2015).

Platforms can take the shape of search engines, marketplaces, social media platforms, gaming platforms, content-sharing platforms, etc. An important delineation to be drawn between platforms is that some are designed to be essentially collaborative and not-for-profit and some operate with overt commercial interest (Stowel/Vergote 2016). Even though both rely on matching supply and demand, and both use technology to reduce transaction costs for their users, there are differences in their internal logic. There is a difference, for example, between wanting to enable modest sharing of, say, harvests from private gardens with neighbours or knowledge within a scientific community, and wanting to conquer global markets with the backing of billions of dollars of venture capital.

It is becoming more and more evident that online platforms are pervading all sectors of private and public life and transforming the fundamental organizational structures of society. The term "platform society" denotes the inextricable relationship growing between such online platforms, societal structures and individual behaviour. They are increasingly "penetrating the heart of societies – affecting institutions, economic transactions, and social and cultural practices [...]" (Dijk et al. 2018: 2). Automated forms of user transactions are replacing a growing number of offline organizations, as demonstrated, for example, in Estonia's much-acclaimed showpiece, the public sector platform e-Estonia (Priisalu/Ottis 2017), or in the United Kingdom's push towards creating a "government-as-a-platform" (Dijk et al. 2018). These examples show how far the platform's impact has penetrated the domain of government.

It is astonishing to see how a very small number of players could constitute the epicentre of the "platform revolution" (Parket et al. 2016). Located on the West Coast of the USA, merely five tech companies from the region – Alphabet/Google, Facebook, Apple, Amazon and Microsoft – have captured the digital spaces of the whole of North America, Europe and large parts of Asia and Australia. The only other country to have developed platforms of a comparable scale is China, with Tencent and Baidu. While the "Big Five" dominate the global market, there is a rapidly growing number of small actors shaping the local universes of platforms. Governments, businesses, entrepreneurs, universities, NGOs, cooperatives: all are contributing to producing a new, highly complex, interconnected landscape of platform practices.

3.2 FROM "SMART URBANISM" TO "PLATFORM URBANISM"

Even though the boundary between the (ill-defined) concepts of smart urbanism and platform urbanism is fuzzy, Söderström and Mermet (2020) have recently identified three features that distinguish platform urbanism from smart urbanism. They relate to the former's (a) materiality, (b) impact on everyday life, and (c) actual effect on how cities work and change. Materiality pertains to how platform processes are manifested in things like Uber cars or Airbnb rooms. This differs significantly from how surveillance cameras and sensors create impressive "smart city control rooms", such as the often-cited and illustrated room built by IBM in Rio de Janeiro. Barns (2018) makes the following remark about the second difference: "[p]latform urbanism enacted daily as we commute, transact, love, post, listen, tweet or chat, deeply implicates the everyday urban encounter" (Barns 2018: 6).

While smart urbanism is associated with extracting data, such as measuring, tracking and tracing people, platform urbanism is interactive, intimate and deeply engaged with everyday life. For these platforms, we gladly volunteer to "like" places, foods or hotels. While the smart city remains somewhat shrouded in techno-utopian or dystopian imagery, platform urbanism conveys tangible, intimate and everyday experiences of the city through its technology. Concerning the third difference, Leszczynski (2020) says that while smart city projects address municipalities, private companies or sometimes civil society, digital platforms directly target individual customers and "by reaching into the pockets of urbanites, [they] express a potential for individualized influence unprecedented by 'smart' infrastructure-urban configurations" (Leszczynski 2020: 5). She further says that platform urbanism should be considered a "reconfiguration, diversification and intensification" or "extension" of the smart city (Leszczynski 2020: 5).

These three differences may very well be the reason for the remarkable success of the sectoral platforms that offer services via apps (see 3.2.2). For example, in 2014, Apple's infrastructure platform iOS had 365 million users who downloaded 800,000 complementary apps, created by 200,000 firms – a process over which Apple has little ownership (Tiwana 2014). From 192 billion in 2018, the total number of app downloads increased to 204 billion in 2019 (Clement 2020). Almost unnoticed, these small waves of millions of digital apps are forming a giant tsunami of services, ready to crash onto the shores of the contemporary physical city in unprecedented ways.

3.3 THE PROBLEM OF PLATFORMS AND THE CHALLENGE OF THEIR REGULATION

Admittedly, platforms have greatly improved many of our lives in various ways. However, while modest and peer-to-peer platforms hardly cause disapproval, the practices of commercially operating corporate giants have generated considerable public outcry and calls for appropriate regulation of their elusive practices. Despite the very high degree of user acceptance, intrinsic flaws in the mechanisms of commercially operating global platforms have led to various confrontations between private and public interests – which have subsequently triggered a substantial debate around the question of how to regulate these new types of entities. Legal battles and backlashes against big tech companies culminated in 2017, making it a turning point for global platform politics. For example, Alphabet/YouTube faced a strike from major advertisers like The Guardian, Starbucks and Walmart; Alphabet was fined US$2.7 billion by European antitrust officials; German competition authorities took Facebook to court for their incomprehensible consumer agreements; the European Court ruled Uber to be a "transportation company" and not the "connective platform" of a "tech company"; and Equifax reported that they had lost 136 million social security numbers to thieves in the digital space. This was a year before the EU member states adopted the General Data Protection Regulation (GDPR), a novel attempt to protect personal data, albeit not specifically for the digital space.

There are three general ways in which regulation of the cyber-physical world is enforced (House of Lords 2019): (a) "regulation", (b) "co-regulation", and (c) "self-regulation". "Regulation" is about enforcing rules for specific types of activities. This is carried out by independent bodies with the power to monitor and enforce rules. "Co-regulation" takes place when digital enterprises set rules by themselves on a voluntary basis. This may be driven by the need to create trust, by corporate social responsibility or by other business interests. "Self-regulation" occurs when a regulatory body delegates responsibility for enforcing rules to an industry body. To understand the challenge of regulating digital platforms in a specific way, it is helpful to understand an interconnected range of issues stemming largely from three of their foundational features: their *legal engineering*, their *disruptive business models* and the technological finesses of their *digital architecture*.

3.3.1   LEGAL ENGINEERING AND DISRUPTIVE BUSINESS MODELS

Researchers have shown that the so-called platform revolution (Parker et al. 2016) was made possible largely by the platforms' capacity to accumulate vast amounts of personal data, create unprecedented degrees of connectivity and generate economic value below the radar of existing regulatory institutions (Dijk et al. 2018). The success of platforms is based on the premise that "they offer personalized services and contribute to innovation and economic growth, while efficiently bypassing incumbent organizations, cumbersome regulations, and unnecessary expenses" (Dijk et al. 2018: 9). This legal engineering is not merely accidental – it is believed to be the core of platforms' success: "Digital platforms obviously challenge the law, and this is a key feature and consequence of their operations. They like to show how the law is out-of-date with the new economy, and they even appear alien to the law. Indeed, they tend to negate the territorial aspect of the (State) law" (Stowel/Vergote 2016: 4).

For these sophisticated operators harbouring global ambitions with monopolistic tendencies, rules applicable on national territory appear an anachronism in the digital age, which poses the challenges of "reflexive modernization", where government is no longer limited to state boundaries but has to develop cross-border governance across scales (Brenner 1998, 2019). This conjuncture has led to intensifying tensions of a geopolitical nature between governments, corporations and civil society. Considering the differences in how the market and jurisdiction are viewed on either side of the Atlantic is crucial when designing competition policies for platforms. The discourse in the US is far from any form of consensus: while former presidential candidate Elizabeth Warren is talking about ways to break up the monopolistic power of tech giants, others are suggesting some of the existing antitrust doctrines should be relaxed. Like the European Union, the US courts are also looking for new ways to address the anticompetitive practices of the tech giants.

Globally operating platforms are fuelled by data, organized through algorithms, formalized through ownership relations, driven by business models and governed through user agreements (Dijk et al. 2018: 9). To understand the premises upon which the foundations of the corporate platforms' techno-legal and techno-commercial strategies are built, it is necessary to examine the functional building blocks of their *digital architecture* (structures) and their *mechanisms of processing data and performing transactions* (processes).

3.3.2   DIGITAL ARCHITECTURE: SOURCE OF TECHNO-LEGAL FUZZINESS

Much of the elusiveness of online platforms' operations stems from the intricate digital architecture of platforms defined mainly by: (1) algorithms, (2) infrastructural platforms, (3) sectoral platforms, and (4) ecosystems of platforms.

1. **Algorithms:** Algorithms are an essential technological ingredient comprising a set of automated instructions that transform input data into a desired output. They have the capacity to produce, reproduce and learn. Platform operators are currently moving from rule-based algorithms to machine learning-based algorithms driven by AI; this makes decision processes even more opaque for regulators to understand.

2. **Infrastructural platforms:** Van Dijk et al. (2018) distinguish two principal types of platform architecture: infrastructural and sectoral. The infrastructural platforms, mostly owned by the Big Five, are the foundational platforms upon which other platforms can be built. Like gatekeepers, they manage, process, store and channel the data that flows through them. They can be search engines, browsers, data services, data analytics services, video hosting platforms (like YouTube), geospatial services (like Google Maps), etc. For example, Netflix runs on the infrastructural platform Amazon Web Services, Spotify on Google Cloud (Dijk et al. 2018). They form the heart of the "ecosystem of platforms".

3. **Sectoral platforms:** The second type of platform are called sectoral platforms. These platforms serve particular sectors, such as transportation, news, education, food, health, finance, hospitality, etc. They can be owned by diverse organizations. Two well-known sectoral platforms are Airbnb and Uber. Both have claimed to be merely "connective platforms", that is, connecting dormant resources and people. The innumerable applications running on these sectoral platforms are popularly called "apps" (see Fig. 1 and Section 3.2). Potential platforms for CAVs will be built on such sectoral platforms.

4. **Ecosystems of platforms:** What has made platforms so powerful is the technological innovation of application programming interfaces (APIs). With the help of APIs, platforms can allow controlled access to their data by third parties – information on which they can build new applications (Helmond 2015). This important ingredient of interlocking functionality created the possibility of establishing large "ecosystems of platforms". For example, Google's search and advertising services can be coupled to its educational platforms; Facebook can produce news content, etc. They are also described as "assemblages of networked platforms" (Stowel/Vergote 2016, Dijk et al. 2018, House of Lords 2019). An important feature of these assemblages is that they enable corporate platforms to partner with a wide variety of non-profit and public players. The Western assemblage of networked platforms and their infrastructural services are almost entirely controlled by the Big Five in the USA. They are central to the ecosystems' overall design and data flow distribution. The exchange of information, goods and services with global outreach would be unthinkable without these platforms.

These four above-mentioned aspects explain a major part of the fuzziness that makes regulation so problematic. In a way, it can be seen as a fuzziness of identity. Most regulation systems in Europe depend on the division between infrastructure and sectors. Correspondingly, the task of regulating online platforms is traditionally compartmentalized, meaning each level of government assumes a limited range of regulatory categories. To avoid the costs of liabilities and responsibilities, platforms blur or deliberately obfuscate these categories. Airbnb does not identify itself as a hotel business, and Uber does not consider itself a taxi business; instead, both see themselves as connectivity platforms. In the case of Uber, their legal battle with legacy taxi firms revolves around the indistinction regarding whether Uber is a sectoral platform (a transportation company) or a platform that merely connects people. When data flows cannot be confined to one single sector, such as transport or health, then platform providers can evade national legislations and "elude the radars of public scrutiny" (Dijk et al. 2018: 158). This conflict was at the heart of the European court case mentioned above until the European Court of Justice finally ruled Uber to be a transportation company (Dijk et al. 2018).

Figure 2: Schematic diagram of the ecosystem of digital platforms. New values (services) are created by allowing sectoral digital platforms to build upon each other through the interlocking functionality of application programming interfaces (APIs).

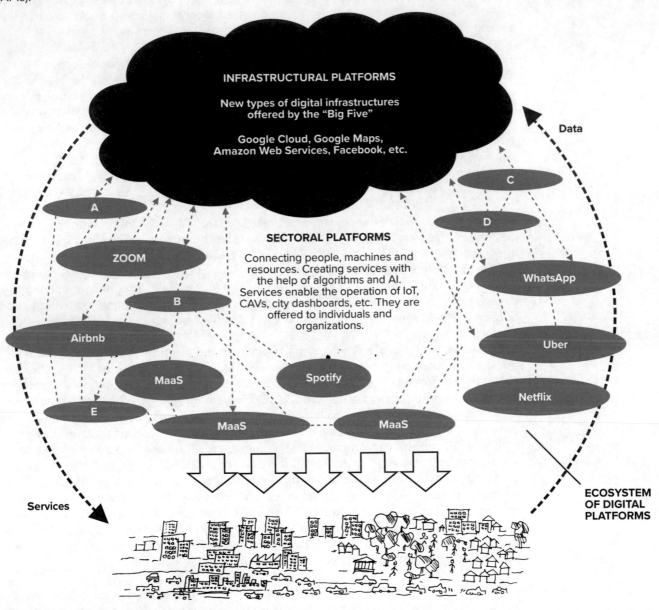

Source: Ian Banerjee

Governments will have to understand these mechanisms and immaterial dynamics of the ecosystem of platforms in order to define the material responsibilities of their online services.

Airbnb can help individuals to make some extra money and tourists to pay less – but who is going to clean the streets when the tourists have left? Students can consume "free" courses on Coursera – but who pays the teachers who prepared the courses? The economic success of all these platforms depends on private and public investments, largely paid for by taxation, i.e. transportation relies on infrastructure like motorways and railways as well as maintenance and cleaning companies. Without those hard infrastructures, neither Uber nor Airbnb will work. This argument lies at the heart of the current legal contentions.

3.3.3 MECHANISMS OF TRANSACTIONS: SOURCE OF SOCIO-TECHNICAL CONFLICTS

Platforms are not merely connectors between actors, they can also steer how they connect users with each other – and they can also shape social norms. Through a number of social plug-ins, such as the "like" button, or through rating, following, sharing, etc., sentiments, interests and opinions can be tracked and tilted by platforms in certain directions: steering users' behaviour, and potentially paving the way for new economic exchanges.

The mechanisms of processing data and conducting transactions are embedded in the architecture of platforms and are comparable to the "genes" of their structures: they carry their system's codes of reproduction. To design governance models for platforms, it is essential to understand the conflicts arising from the effects of these mechanisms. Van Dijk et al. (2018) have identified three of them as the main forces shaping the dynamics of the ecosystem; they are: (1) datafication, (2) selection, and (3) commodification.

1. Datafication is about collecting vast amounts of user data and subsequently aggregating, disaggregating, unbundling and rebundling them into digital products. Datafication is potentially an excellent means to contribute to general well-being. It can improve traffic management, solve health problems, etc. The availability and accessibility of (real-time) data shared through open standards could contribute greatly to societal benefits. Aggregated data sets may not only hold economic value but also public value. However, datafication has become one of the most contested issues in the field. Its vastness and complexity have been made possible by the tech giants' holy grail of high technology. The problem lies in the fact that these operators usually capture, store and resell data without compensating the public. This monetization takes place well outside the reach of regulators, and users usually lose control over their data after having accepted the platform's terms of service – hereafter, they become a proprietary asset. The movement of such large amounts of data (enabled by a handful of companies) makes it almost impossible to trace them and therefore there is little public accountability. How to make data traceable, their flows transparent, and bestow users with a better sense of ownership, are at the heart of the legal discourse around platforms.

2. Selection involves the mechanisms of filtering, ranking and personalization of data. Again, it is not intrinsically faulty; personalized healthcare, for example, can be very beneficial; but the very same mechanism can become detrimental for citizens if breaches in privacy lead to high rates for insurance for i.e. disabled persons.

3. Commodification is a mechanism that transforms data into economic value. In simple terms, this mechanism connects producers to customers, and importantly, advertisers to content. A myriad of monetization schemes shapes the "notoriously untransparent" practices of commodification (Dijk et al. 2018: 144). What complicates this simple mechanism is that corporate platforms often partner with non-profit and public players, and in so doing "render fluid the distinction between for-profit and non-profit, private and public, free and paid for, infrastructural and sectoral, complementors and connectors – and between global, national and local markets" (Dijk et al. 2018: 144). The accountability discussion raises the question of governance: who is accountable to whom? Who has ownership over what (Annany/Crawford 2016)? However, even here, commodification could in principle also create public values, though the ubiquitous public value-based design of platforms is still a far-fetched ideal.

In all mechanisms we see the repeated need for new taxonomies of governance or regulation that will make transactions more transparent and accountable. Essential for the governance of these mechanisms is to rethink the principles according to which these socio-technical systems are designed. It requires clarity about the values needed to inform the online systems (see Section 3.4).

3.3.4 POSSIBLE APPROACHES TO REGULATION: RULES-BASED OR PRINCIPLES-BASED?

All socio-economic activities take place within certain regulatory frameworks. These new types of commercial operators have exploited a momentary regulatory gap, which is currently being closed. The two main approaches to regulation are rules-based or principles-based (Stowel/Vergote 2016, Dijk et al. 2018, House of Lords 2019). Principles-based regulation focuses on outcomes, whereas rules-based regulation prescribes the format of compliance (House of Lords 2019: 14). An example of the latter are the data protection principles that underlie the General Data Protection Regulation (GDPR). These principles subsequently led to the GDPR's guidelines, codes of practice and certification: a principles-based approach can help to establish a common understanding for addressing issues that cut across sectors and can provide a common framework for regulators, executive bodies, policymakers and lawmakers (House of Lords 2019: 14). Also, the rapid pace of technological development makes a principles-based approach seem more appropriate, because it creates a certain flexibility by setting out the standards and expectations of service providers.

A general principle that is gaining interest is the outcome-oriented principle of "public value" (for more, see 5.2). The House of Lords (2019) has broken down the general public value-centric approach into ten subgroups: (1) parity of equivalent outcomes online and offline, (2) accountability, (3) transparency, (4) openness, (5) ethical design, (6) privacy, (7) recognition of childhood, (8) respect for human rights and equality rights, (9) education and awareness-raising, and (10) democratic accountability, proportionality and an evidence-based approach (for more, see House of Lords 2019: 15). The aim of working with these principles is to help the industry, regulators, government and users work towards a common goal of making the internet a better environment that is beneficial to all.

A crucial topic for the governance of the platform society is the challenge of the digital divide. There are three main issues here. Firstly, concerning infrastructure, the world is far from having ubiquitous presence of the internet – even in highly industrialized countries many peripheral areas are still undersupplied. Secondly, concerning individuals, not only is the question of purchasing devices to use Web 2.0 relevant, but so too are the competencies needed to manoeuvre in the cyber-physical space. When this fact is mentioned, if at all, it is usually labelled as the problem of the elderly, which it is therefore self-evidently assumed will fade away over time. This approach is to be questioned, as the progress in technological transformation, particularly in communication, is so rapid and complex that even the "digital natives" will sooner or later struggle to keep up with the speed of change and themselves turn into "digital immigrants". Thirdly, the different use of the digital world does not depend on age or gender but on social position. The digital divide results on the one hand in the intensification of existing social inequalities and on the other in the creation of new forms of inequality – in extreme cases it leads to groups' exclusion from digital societies (Rudolph 2019). This may happen with growing frequency, particularly if daily life is increasingly organized via apps (e.g. for access to mobility).

3.4 DIGITAL TRANSPORT PLATFORMS

3.4.1 DATAFIED MARKETPLACES, PRICING AND TRUST

From the perspective of digital platforms, urban transport platforms can be seen as datafied marketplaces (Dijk et al. 2018), where data extracted from various infrastructural and personal sources are matched with specific mechanisms to meet demand with supply, while lowering transactional costs (Dijk et al. 2018). Datafied marketplaces are increasingly being created by both the private and public sector – leading to the convergence of all modes of transport (see Section 4 on MaaS). However, the playing field is far from equal, as the amounts of data owned by tech companies vastly outweigh the data streams collected or owned by the public sector. The unwillingness to share their data may lead to a potential conflict of interest between commercial operators and the common good of a well-functioning, integrated transit system.

Two important aspects of the datafied marketplace of digital transport platforms are pricing and the organization of trust. Pricing plays a central role in the transport business; however, it remains unclear whose interests are served: is it to increase net gain for the platform? Is it to optimize travel and waiting times for all passengers? Or is it to optimize travel for premium customers at the cost of other groups (Dijk et al. 2018: 81)? Trust is another key element for the success of transport platforms. Platforms vitally need trust and a good reputation to operate: an increasing number of users want platforms to assume more responsibility. Normative rating apparatuses used by companies like Uber are well known. For CAVs, the trust systems will have to be engineered in different ways, for example by showing the degree of geographic coverage of a transport network company (TNC), or maybe how well they are serving structurally underserved areas.

3.4.2 GOVERNANCE: AN EXAMPLE OF AN ALTERNATIVE APPROACH

As many scholars have often pointed out, sharing models (including those for CAVs) will not automatically make transit systems more efficient for all social groups. In fact, the opposite will be the case if it is not governed with a comprehensive view. For example, ride-sharing services in New York have led to a fall in ridership of the metro since their introduction in 2016; consequently, the average speed in central Manhattan fell by 12% from 2010 to 8.1 miles per hour in 2017 (Fitzsimmons/Hu 2017).

Case study: São Paulo – Government as the central actor
An interesting example of designing a comprehensive approach towards platformization of an entire urban transportation system can be witnessed in the megacity of São Paulo – a city notorious for being segregated with starkly unequal access to services. The city management of São Paulo is currently recalibrating the commodification of the urban mobility system and testing a novel transport concept. In a new proposed law, the local government will sell "credits" to all network service providers that want to make use of the city's infrastructure. For each kilometre driven, a TNC will have to buy mobility credits, which will be auctioned off on a pay-as-you-drive model. These credits will be priced dynamically; for example, they will be made cheaper for providers offering public transport to the underserved, or for providing transport for disabled persons, to stimulate providers to serve a particular group of customers. They will also be used to promote certain labour market policies – for example, by reserving a certain amount of credits for female drivers.

This case study shows how public values can be incorporated into the conceptualization of a platform's mechanism of commodification. The goal is to maximize the transport system's efficiency from the point of view of public values rather than profit alone (Dijk et al. 2018). Findings from this large-scale experiment could be valuable for the widescale deployment of CAVs in the future.

4. THE CASE OF MAAS: PLANNING IN TIMES OF DEEP UNCERTAINTY

4.1 MAAS: BENEFITS AND CHALLENGES

Digitalization in the transport sector has brought about substantial changes to the incumbent transport industry. In particular, the way digital platforms are applied have significantly revolutionized mobility services: they provide new channels for travellers and service providers to interact and enable new business models and mobility services to emerge. Consequentially, the urban transport system landscape in cities around the world is transforming in ways that were unimaginable even a decade ago.

At the centre of this development is a new transport concept called "Mobility as a Service" (MaaS). MaaS is a transport concept that combines different transport services to provide travellers with a transport solution via a single interface. The available transport services are bundled together and offered to users in exchange for a monthly payment or a pay-as-you-go tariff, similar to mobile phone services. It provides a demand-orientated mobility service that can be tailored and customized to meet users' requirements (for reviews of MaaS, see Kamargianni et al. 2016, Jittrapirom et al. 2017, Pangbourne et al. 2018).

Arguably, the concept of providing mobility as a service even predates the automobile. With horse-drawn carriages, taxis, vehicle renting and sharing services, travellers could already access and benefit from these modes of transport without having to own them. Also, efforts to bring together different modes of transport by amalgamating vehicles, schedules and payment under the notion of integrated transport has been ongoing for decades. Smart mobility concepts, which aim to implement the IoT in the transport system to enhance convenience, accessibility and better management of individual mobility services, are already prevalent and have become part of daily life in several cities. So, what is new and unique about MaaS?

The novelties that set it apart are twofold. Firstly, the principal focus of MaaS is in meeting travellers' needs. MaaS seeks to combine existing advanced technologies to provide a layer of information exchange that can seamlessly integrate these different modes of transport to provide a mobility service that caters to users' needs (Finger et al. 2015). This focus is more explicit in MaaS than in the other previous transport concepts. In MaaS, each traveller is theoretically offered a range of mobility options, which are selected and optimized by a computer algorithm to best fit their needs. The provision of a single platform that combines all available modes will also ease accessibility and call attention to less familiar or new modes as well as the interchangeability between different modes.

Secondly, MaaS can bring together the positive attributes and characteristics of existing mobility concepts (e.g. sustainable, active and integrated mobility) to better manage the overall operation of urban transport systems (Wong et al. 2019). For instance, MaaS can contribute to the better use of the available capacity by enabling supply and demand to match better, and allowing travellers to switch spontaneously from, say, congested systems of transport to those with more available capacity. Also, MaaS can deter travellers from private vehicle use with its "access instead of ownership" paradigm, and potentially offer sustainable and seamless mobility services.

These innovations can combine to offer an effective solution to address localized urban transport challenges (e.g. congestion, air quality, noise) and respond to the global challenge of the climate emergency. MaaS can also offer several short-term perks (e.g. convenience, health benefits and other incentives) required to encourage travellers to use a more sustainable and environmentally friendly mode through its personalization and integration with other sectors (e.g. retail).

Despite the many potential benefits, there are a number of challenging issues around the planning of MaaS. For example, it is still questionable whether the aforementioned perks are sufficient to alter travellers' behaviour or whether it would need additional infrastructure to induce change – as the choice of mode is often argued to be rooted "in the human physiological structure" (Knoflacher 2007: 395). Moreover, "MaaS has considerable potential for deception" (Pangbourne et al. 2019: 13), as the narratives around MaaS are largely dominated by private businesses that usually see it solely as a means to achieve higher levels of efficiency and convenience in the transport sector. While these arguments may hold, efforts must be made to bring about a more balanced discussion that examines other aspects, such as equity, sustainability and environmental impacts.

The role of the public sector will be absolutely crucial in ensuring the involvement of other stakeholders, such as the civic sector, and in realizing the societal benefits associated with Mobility as a Service (Jittrapirom et al. 2018b). Although several public sector activities have been ongoing for a decade (see Box 1 below), more can be done to help the public to understand MaaS and encourage them to engage with the discourses that are currently still limited to tech companies, public service providers, academia and governmental organizations. The limited public involvement may stem from the novelty of MaaS as a concept; however, only a broader engagement of citizens can turn MaaS into a valuable constituent of future urban landscapes.

4.2 PILOT PROJECTS AND SCHEMES OF MAAS: GOVERNMENT AS A CENTRAL ACTOR

The potential prospects of MaaS in solving urban transport problems have attracted the interest of public policymakers from around the world. The European Commission, for example, has been very active in stimulating activities associated with MaaS (see Box 1). At the member state level – particularly in Austria, Finland, Netherlands, Finland and the UK – there are lively discussions and initiatives around MaaS (Polis 2017).

Box 1: A selection of activities related to MaaS with government as the central actor

The European Commission declared 2018 to be the "Year of Multimodality" to emphasize the importance of multimodality for EU transport systems and dedicated three key thematic areas to MaaS: digitalization, support for multimodal (physical and digital) infrastructure and innovation, and legislative framework to protect passenger rights during multimodal journeys. The Commission also started to provide funding in 2018 to support the advancement of MaaS through several financial mechanisms, such as Horizon 2020, Multiannual Financial Framework (MFF) and the new Framework Programme for Research and Innovation (FP9). Additionally, it also organized Digital Transport Days and the Transport Cybersecurity Conference for the first time in 2019.

In the Netherlands, MaaS has attracted keen interest from various parties. It is seen as a stepping stone towards building a sustainable and smart city, strongly driven by the promise of efficiency gained through new transport systems and the potential to create new business opportunities. In 2016 the MaaSifest task force was established, bringing together experts from various sectors to formulate action plans to accelerate the adoption of MaaS (MaaSifest n.d.). The Dutch central government and the Ministry for Infrastructure and Water Management have also shown interest in MaaS as an alternative approach to solving transport congestion alongside investment in infrastructure and public transportation. Several initiatives have been launched to support this ambition and to accelerate its realization. They include the implementation of projects, such as SCRIPTS in 2016 and the initiation of market consultations on MaaS in 2017. A tendering process in 2018 followed the latter for pilot projects in seven regions – Amsterdam, Eindhoven, Limburg, northern provinces, Rotterdam, Twente and Utrecht – several of these pilots are underway as of 2020.

In Finland, the active governmental support for MaaS is led by the Finnish Transport Agency, which published a report on MaaS in 2015. In 2016 the Helsinki Regional Transport Authority (HSL) agreed on a cooperation with Whim, a MaaS platform, to sell its single-journey ticket through its platform (HSL 2016). This agreement was the first of its kind in a real-world operational setting. The following year, the Finnish government published a new act on transport services, with the aim of creating the preconditions for the digitalization of transport services and new business models. The act consists of three components: (1) the provision on the interoperability of data and information systems (came into force on 1st January 2018); (2) provisions for air, maritime and rail markets and the opening up of data and regulation of transport registers; and (3) a reflection on the act's objectives of creating digital services. This act was adopted as of 1st July 2018 (Vayla 2018).

The government of the United Kingdom, headed by the Department of Transport, commissioned a study on MaaS to examine its feasibility for London in 2015 (Kamargianni et al. 2015). The study found that the application of MaaS would bring potential benefits to transport operators, and that it would outweigh the potential costs. Subsequently, the government called for a roundtable discussion on MaaS in 2017 (Governement Office for Science 2017) and thereafter launched a formal inquiry into the subject (Parliament 2017). Currently, there are several MaaS research projects running in the UK (e.g. MaaS4EU and Pro-MaaS by MaaS Lab). This also includes the operation of the Whim service (see Finland) in the Midlands region of the UK.

In Japan, the Ministry of Land, Infrastructure, Transport and Tourism (MLIT) and the Ministry of Economy, Trade and Industry (METI) launched a Smart Mobility Challenge Promotion council in 2019. The commission brought together local authorities, private business and other related institutions to discuss the potentials and opportunities of new mobility services in Japan. Later that year, the council called for a public tender and selected 19 advanced MaaS pilot projects with a total value of ¥3.1 billion (approx. €25 million). The 19 projects can be classified into MaaS for suburban, rural and depopulated areas, and also for tourism. Several of them have been in operation since late 2019 and are expected to continue till the end of 2020.

4.3 PLANNING MAAS DURING DEEP UNCERTAINTY

The pilot projects conducted in different countries have provided the first indications of the necessary preconditions for implementing MaaS on a large scale, as well as the quantification of possible impacts of MaaS on the performance of the transport system in general. However, these indications are still too limited in number and often too case-specific to allow broad generalizations. There is still a high degree of uncertainty surrounding MaaS as to how it should be implemented, and what the real benefits would be from its large-scale rollout (Jittrapirom et al. 2018a).

The level of uncertainty is high for several reasons. Firstly, knowledge about this novel transport concept is still limited. Several of the issues shaping the ongoing debate revolve around the ambiguities underlying the precise definition of the service, its overall effect on the urban transport system, and the uncertainties regarding users' and stakeholders' acceptance. The second dimension concerns the complexity of the urban transport system, which stems from the interconnectivity and interoperability of the entities within the urban system (Kölbl et al. 2008). The complex entanglements between these different entities make it difficult to predict future behaviours resulting from any single intervention (Pojani/Stead 2015, Jittrapirom et al. 2017). Thirdly, the evaluation of the outcome of interventions by decision makers can be uncertain. Although these outcomes may be forecasted with some certainty, the inherent subjectivity of their evaluation can be influenced by contingent factors, such as public mood at the time of measurement. Finally, uncertainties associated with external forces play a significant role. Specific forces, such as demographic development, can be forecasted with some accuracy using past data, whereas other forces, such as national economic development or other surprise events (i.e. a black swan), are more difficult to predict. Researchers are addressing these uncertainties by adapting appropriate planning techniques, for instance, robust decision-making (RDM) and Dynamic Adaptive Policy Pathways (DAPP; see Haasnoot et al. 2019, Lempert 2019, Walker et al. 2019).

Dynamic Adaptive Policy Pathways

A promising alternative to the orthodox planning technique that has been suggested in view of developing MaaS is Dynamic Adaptive Policymaking (DAP), which deals with uncertainties by assuming that they can be predicted with some accuracy using statistical or scenario-based approaches. The DAP approach first allows policies to be developed for novel concepts using available information and then focuses on reducing uncertainties during their implementation through monitoring and adaptation processes. It helps policymakers to deal with deep uncertainties regarding contested opinions and mental models of those involved in the planning process when those policy makers have no previous experience to draw upon (see Jittrapirom et al. 2018a for how DAP is applied to MaaS planning). However, there are several drawbacks of DAP. As a "predict and act" method, it is based on explorations of past experiences that may be limited in dealing with the implementation of MaaS. For instance, its development process often involves a limited group of experts, potentially restricting the perspectives and the comprehensiveness of the plan. Also, there are challenges in implementing the approach in practice, which includes dealing with complex and contested issues and establishing the trigger points for the complex system. Moreover, it needs to take into account the implications of institutions and governance (Bosomworth et al. 2017). Finally, DAP requires additional techniques to calculate the costs involved in shifting between different plans of the adaptive pathways (Haasnoot et al. 2019). Some studies have sought to address these weaknesses by combining DAP with computational simulation (Hamarat et al. 2013) and some by involving groups of different experts (Pas et al. 2012). Yet another study combines DAP with the Delphi method, an anonymized way to systematically capture the opinion of a group of experts (Jittrapirom et al. 2018b).

5. WHAT DOES THIS MEAN FOR CAVS?

5.1 COMPREHENSIVE PLANNING

As researchers have repeatedly pointed out, if unregulated, CAVs and other new mobility solutions – such as car-sharing, ride-sharing and ride-hailing services – are expected to make cars even more appealing and hence possibly draw passengers away from (existing) public transport systems (EU Science Hub 2019). Therefore, how to design a comprehensive transport system will remain the key challenge for mobility management. If a country or a city administration is willing to deploy CAVs while simultaneously subscribing to progressive policies of making transport cleaner and more equitable than its car-centred present, then it will have to engage with research and the political questions of planning that address questions revolving around (a) how to minimize the risks of implementing new mobility services by instigating dynamic forms of adaptive policymaking, (b) how to steer digitalization in a democratically, socially and ecologically sustainable direction, and (c) how to create a value-centric and public interest-oriented platform society.

5.2 PUBLIC VALUE-CENTRIC DESIGN: THREE CRITICAL QUESTIONS

As already hypothesized, in view of current trends it is highly likely that transport services for CAVs will be offered via platforms or platform ecosystems of some sort. It can be assumed that the conflicts arising from these future platforms will be very similar to those existing today. For example, despite the obvious differences, the ongoing legal battles fought between legacy taxi firms and Uber are likely to be no different from those potentially fought with CAV platforms if they threaten the order of existing labour markets, hide under the radar of sectoral legislation or undermine other aspects of public interest. Given the situation that the "implementation of platforms in society triggers a fierce discussion about private benefit and corporate gains versus public interests and collective benefits", and that "many platforms have grown surprisingly influential before a real debate about public values and common goods could get started" (Dijk et al. 2018: 3), the question that appears to lie at the heart of political commitments and the design of future CAV platforms is how they can incorporate public values and benefit the common good. This will mean that governments and regulators will first have to comprehend the intricate ways in which public values are contested or eroded through the practices of platforms, and then find a broad consensus about the principles that could shape their regulation (Stowel/Vergote 2016). Scholars such as van Dijk et al. (2018) and governmental bodies like the House of Lords (2019) have suggested conceptualizing the principle of a "public value-centric design" of the platform society as a whole. The question lying at its core is: in what kind of society do we want to live in the digital age? This seemingly lofty and philosophical inquiry quickly becomes tangible when coders actually start designing the platform's algorithms. Consciously or unconsciously, they build values into their designs – be they corporate, commercial or in the public interest.

The first question that arises in this context is: what are public values and which of these values do specific societies want to uphold (see also Section 3.3.4)? Moore (1995) describes public value as the value that an organization contributes to society to benefit the common good. As Bozeman (2007) says, the common good can be translated into a number of propositions that are achieved through collective participation in forming a shared set of norms and values. Public values may include privacy, safety, security, accuracy, etc. They may pertain to broader societal effects, such as fairness, accessibility, affordability, inclusiveness, democratic control or accountability. Also, public values are site- and case-specific, i.e. the values upheld in rural Japan may be very different from those in rural Germany. Public values are also ideologically variable and defining them is not the sole privilege of the public sector. Depending on the specificities of the country, this may imply the simple act of rereading and reemphasizing the values already inscribed in the respective nation's constitution. A challenge that would remain for governance is how public values can be advocated and the terms for their implementation negotiated. It is quite evident that there will be no globally universal recipes for their negotiation. Articulating which values are contested by whom and in which context may help to shape the current platform ecosystem in ways that will make it more responsive to public concerns. As mentioned in the section on data-driven urbanism, the governance of platforms will need a whole-of-government approach and the involvement of all relevant stakeholders in society.

The second question that follows is: who will be responsible for anchoring these public values in the platforms and which institutions could oversee the regulation? Besides the need for collaborative action among transnational actors to regulate the global ecosystems of infrastructure platforms, it will be local actors who will play an essential role in keeping an eye on the private sector. An effective way to meet this challenge might be to turn governments into central actors in the platform society. They could themselves become developers of comprehensive platforms centred on public values and collective goals, such as aspired to by the local government of São Paulo (see 3.3.2). Overseeing the regulation can perhaps also be undertaken by independent third-party institutions like civic watchdogs installed by civil society.

The third question is: how can public values be anchored in platforms? The main concern of a value-centric design of the platform society is how to retool the current platform ecosystem by tilting its underlying mechanisms towards "societal valorization" (Dijk et al. 2018:146). This broadly entails redefining the meanings assigned to the technology, implying that the action needed will be as much technological as sociopolitical. Since the 1980s schools of thought – such as science and technology studies (STS), actor-network theory (ANT) and social construction of technology (SCOT) – have pointed to the effects of broader societal factors in the development of science and technology (Pinter 2008). Contrary to technological determinism, they have shown with different theoretical underpinnings how technology is shaped by complex interactions between numerous vectors of interest and trajectories of change. In this coevolutionary process, the pace of digitalization – mainly driven by corporate players with a technocratic view of the world – is creating an immense lag in the formulation of complementary rationalities and alternative narratives. Efforts to deploy the digital products in a democratic direction will need to broadly anchor democratically endorsed narratives that simultaneously denote technological and ethical elements – such as "ethical technology", "urban commoning", "open data", "open-source ecologies" – in the mindsets of regulators, politicians, stakeholders and society at large. The technological anchoring and algorithmic translation will follow, or rather, coevolve with the wider context of sociocultural, political and environment discourse (Pinter 2008). It will be the result of a multi-institutional endeavour involving the nexus of public discourse, policy measures and technological fixes.

6. CONCLUSION

This paper has explicated the systemic interlinkages currently unfolding between the emerging practices of the data-driven city and digital platforms. It took MaaS as a case study to exemplify the insecurities arising in the planning of mobility services offered through a digital platform.

The choice of these foci for this review was based on the assumption that CAV-based services will be offered via digital platforms of some sort in the future and that this will take place in a data-driven city. The article has shown how the urban is being reframed by the conceptualization of the city as a crucible of data that have to be extracted in every way possible (data-driven city). The main tenets of this belief are based on the conviction that all urban processes can and should be measured and monitored, and all urban malfunctions can be treated as technical problems. This understanding of determinism and linear logic has been questioned since the failures of the urban modelling of the 1970s, with the critique of Forrester's model (Gray et al. 1972) and, among others, the classification of Lowrin models (Li/Gong 2016). Also, increasing social diversities are excluded from these models and the general understanding of human behaviour is reduced to rational behaviour only ("homo oeconomicus").

Further, this paper has shown how the convergence of the physical (spatialities) and the digital (codes) are creating a new kind of cyber-physical reality that is profoundly transforming the academic and managerial imaginaries of the urban. It has been argued that in order to create (socially responsible) mobility services with CAVs, it will be essential to understand the dynamics of these emerging hybrid landscapes. The authors sketched out a mental map of a fourfold process which, according to them, depicts the incessant production of the cyber-physical landscape in the data-driven city. This fourfold movement constitutes (a) the extraction of data from cyber-physical space, (b) the storage of data, (c) the analysis and construction of meaning out of these data and their simultaneous transformation into value in the form of services, and (d) these services augmenting the materialities of everyday life. The socio-technical formations of digital platforms play a key generative role in the production of these cyber-physical landscapes, while MaaS or CAVs play constitutive roles.

Moreover, this paper has demonstrated that digital platforms are shaping the process of digital transformation in profound ways. They are rapidly transforming the entire field of urban services – from news to education, from entertainment to health. Mobility services will play a key role in this newly evolving real-virtual landscape, including services offered around CAVs. It is evident by now that the city of the future will increasingly be built on digital platforms and apps. After effecting radical changes in the business sector, online platforms are now steadily challenging the structures, organizations and institutions of the state (government); the relationality of power flowing through the networks of state, market and civil society (governance); and also the techniques, procedures, programmes and strategies of state and non-state agencies that are shaping citizens' conduct (governmentality; Pieterse 2008, Jessop 2004).

While acknowledging the weight of the global framing conditions of politics, technology, the economy and societal change, the main conclusion of this paper is that the outcome of the digitalization and platformization of public services is neither entirely predetermined, nor is the future locked in by solely entrepreneurial logic or the spectre of surveillance. It is still possible for national governments and societies to envision a digital social order that reflects a democratic and equitable direction based on public interest, and to implement these novelties in adaptive manners that enable quick feedbacks of the lessons learnt. This can be achieved if the relevant actors (a) precisely understand the legal engineering and intricate mechanisms of commercially operating platforms, and (b) strive to incorporate public values and collective interests into the architecture of these digital formations, practices and routines. Promising experiments in the field substantiate this possibility. Public values include the fight against the many forms of digital divide (see 3.3.4).

Policy recommendations will have to move with care and reflexivity, because cutting through the vicissitudes of the emerging cyber-physical world means taking paths as yet untrodden. Seeing the current platform ecosystem being predicated on an architecture shaped overwhelmingly by economic values and corporate interests, and given the complexities of the presumably highly heterogeneous landscape of future service providers operating with diverse logics and interests, it is realistic to conceptualize a regulatory framework that clearly defines principles while leaving space for local adaption. The coordinates that anchor the authors' recommendations revolve around the principle of public values. Depending on the political specificities of the country, this may imply the simple act of rereading and reemphasizing the values already inscribed in the respective nation's constitution. The question lying at the centre of political commitment and the critical design of future platforms is: how can platforms incorporate locally negotiated public values and benefit the common good? The main lesson learned from the global discourse around digital platforms is that while platforms may enhance personalized benefits and economic gain, they simultaneously put pressure on collective means and public services. The fact that the activities of platforms do not automatically translate into public benefits is the main issue of contention. Public values are at the centre of a struggle over a more equitable and democratic platformization of society as a whole, of which mobility, health, education and other aspects of urbanism are subsets. From this it can be inferred that the key concerns for new governance models for CAV platforms will revolve around the following questions: (a) how will online platforms penetrate a specific transport sector, (b) how are the sectoral platforms embedded in the platform ecosystem as a whole, (c) which site-specific public values are identified, (d) how is their implementation negotiated, and (e) who will oversee their regulation? From the perspective of transport planning, a comprehensive approach and dynamic forms of adaptive policymaking and dynamic adaptive planning can address the looming danger that the deployment of personalized transport systems, such as MaaS, or potential future CAV platforms may bring and maximize the risk of decreasing inclusive public transport services or reduced affordability.

Governments have always negotiated with commercial parties to design regulatory frameworks. To build a trustworthy global platform ecosystem, they will have to distribute responsibilities among the market, state and civil society (Dijk et al. 2018). They will need a multi-stakeholder, multi-sectoral and, essentially, a multinational approach that will require a rethinking of social contracts based on equality and solidarity on a global scale. If countries want to protect their democratic values, build more equal societies, and if they believe that only the free flow of information can build a healthy society, then creating governance models for a responsible platform society will have to be a key agenda for governments in the 21st century.

This review made it evident that the myriad manifestations of the digital revolution will not only radically change the imaginaries of the city, but also the foundations of the socio-technical order of human civilization at large. In times of unprecedented corporate power, the digital space – initially conceptualized as a commons – shows signs of becoming both a splintered arena with an unequal distribution of power, resources and services, and a real-virtual place for geopolitical, ideological, technological and legal conflicts – carried out mainly between the USA, the European Union and China. In the maze of competing critical perspectives on the potential of an equitable future for the internet, views mainly oscillate between two poles: those with a pessimistic view based on the belief that on every conceivable front of digital change lingers the superior cunning of a system that is merely reinventing and perpetuating the existing conditions of inequality and exploitation; the other, more optimistic view is based on the belief that after a phase of initial disorientation and legal vacuum when it comes to regulatory mechanisms, rational policy agendas will incrementally stipulate the conditions of the digital world and make it – and by extension the real-virtual space it governs – a more equitable place.

While this paper was being completed, the world was being ravaged by Covid-19. One consequence of this unprecedented magnitude of disruption in the modern world was unquestionably the rapid acceleration of digitalization. While the world was rapidly moving its activities into the virtual world, the debate on basic human rights flared up, as country after country in the democratic world attempted to suspend citizens' rights to privacy by deploying digital platforms to track and trace infected persons. In rare cases, civil society was involved in the process of designing these apps. At the end of the crisis, studies will show which countries could both respond aptly to the crisis and inscribe democratic values into their spontaneous digital responses. In late March 2020, while the virus was in full swing, a UK delegate to the UN's International Telecommunications Union (ITU) said: "Below the surface, there is a huge battle going over what the internet will look like" (FT 2020). It will be of utmost importance to bring this battle to the surface of public discourse and create a broad and constructive debate on the future of the cyber-physical world. The stakes are high: do we want a greener and fairer world or do we accept a more unequal one? Do we want an urbanized world of openness and transparency or a world of surveillance and authoritarian control?

REFERENCES

Annany, M. and K. Crawford 2016 (2018). "Seeing without knowing: Limitations of the transparency ideal and its application to algorithmic accountability", in *New Media & Society* (20) 3, https://doi.org/10.1177/1461444816676645.

Aridi, A. and U. Petrovčič 2020. "How to regulate Big Tech?", in the 'Future Development' blog of the Brookings Institution and the World Bank, 13/2/2020, https://www.brookings.edu/blog/future- development/2020/02/13/how-to-regulate-big-tech/ (6/7/2020).

Artioli, F. 2018. "Digital platforms and cities: a literature review for urban research", in *Cities are back in town*. Working Paper (2018) 01, 1–34.

Banerjee, I. 2014. "Smart Cities: A contested marketplace for large corporations and small communities", in *Österreichische Ingenieur- und Architekten-Zeitschrift* (159) 1, 53–59.

Banerjee I. and I. Fischer-Schreiber 2015. *Digital. Communities 2004–2014. Selected Projects from Prix Ars Electronica*. Linz: ARS Electronica. https://www.aec.at/prix/en/kategorien/digital-communities/ (6/7/2020).

Barns, S. 2018. "Smart cities and urban data platforms: Designing interfaces for smart governance", in *City, Culture and Society* 12, 5–12. https://doi.org/10.1016/j.ccs.2017.09.006.

Batty, M. 2012. "Smart Cities, Big Data", in *Environment and Planning B: Planning and Design* (39), 191–193. https://doi.org/10.1068/b3902ed.

Bassoo V., V. Ramnarain-Seetohul, V. Hurbungs, T. P. Fowdur and Y. Beeharry 2018. "Big Data Analytics for Smart Cities," in *Internet of Things and Big Data Analytics Toward Next-Generation Intelligence*, ed. by N. Dey, A. Hassanien, C. Bhatt, A. Ashour and A. Satapathy, Studies in Big Data, vol. 30. Cham: Springer. https://doi.org/10.1007/978-3-319-60435-0_15.

Beck, U., A. Giddens and S. Lash 1996. *Reflexive Modernisierung. Eine Kontroverse*. Frankfurt am Main: Suhrkamp.

Bostoen, F. 2018. "Neutrality, fairness or freedom? Principles for platform regulation", in *Internet Policy Review* 7 (1). https://doi.org/10.14763/2018.1.785.

Bozeman, B. 2007. *Public Values and Public Interest: Counterbalancing Economic Individualism*. Washington, DC: Georgetown University Press.

Bratton, B. H. 2015. *The Stack: On Software and Sovereignty*. Cambridge, MA/London: MIT Press.

Brenner, N. 1998. "Between Fixity and Motion: Accumulation, Territorial Organization and the Historical Geography of Spatial Scales", in *Environment and Planning D: Society and Space* (16) 4, 459–481. https://doi.org/10.1068/d160459.

Brenner, N. 2019. *New Urban Spaces: Urban Theory and the Scale Question*. Oxford: Oxford University Press.

Castells, M. 1996. *The Rise of the Network Society*. Oxford: Wiley-Blackwell. https://doi.org/10.1002/9781444319514.

Chang, S. K. J., H. Y. Chen and H. C. Chen 2019. "Mobility as a service policy planning, deployments and trials in Taiwan", in *IATSS Research* 43 (4), 210–218. https://doi.org/10.1016/j.iatssr.2019.11.007.

Clement, J. 2020. "Number of mobile app downloads worldwide from 2016 to 2019", in *Statista*. https://www.statista.com/statistics/271644/worldwide-free-and-paid-mobile-app-store-downloads/ (6/7/2020).

Commons Select Committee 2017. "Mobility as a Service: Committee explores transformative potential", in *News from Parliament – UK Parliament*, 14/11/2017. https://www.parliament.uk/business/committees/committees-a-z/commons-select/transport-committee/news-parliament-2017/mobility-as-a-service-launch-17-19/ (23/1/2020).

Coletta C., L. Heaph, S.-Y. Perng and L. Waller 2017. "Data-driven Cities? Digital Urbanism and Its Proxies: Introduction", in *Tecnoscienza – Italian Journal of Science and Technology Studies 8* (2), 5–18.

Digital Charter UK 2018. "Digital Charter: A response to the opportunities and challenges arising from new technologies", policy paper. https://www.gov.uk/government/publications/digital-charter (6/7/2020).

Dijck, J. van, T. Poell and M. de Waal 2018. *The Platform Society: Public Values in a Connective World.* New York, NY: Oxford University Press.

EMOTIVE 2020. Homepage. http://emotive.lboro.ac.uk (15/1/2020).

EU Science Hub 2019. The Future of Road Transport – Implications of automated, connected, low-carbon and shared mobility. https://ec.europa.eu/jrc/en/facts4eufuture/future-of-road-transport (6/7/2020). http://doi.org/10.2760/668964.

European Commission 2015. "Consultation on Regulatory environment for platforms, online intermediaries, data and cloud computing and the collaborative economy", 24/9/2015, 5. https://cn-numerique.fr/files/uploads/2015/11/PositionCNNum_ConsultationonplatformsEUCommission.pdf (6/7/2020).

Fengler, W. and I. Gill 2019. "A new alphabet for Europe: Algorithms, big data, and the computer chip", in Future Development blog of the Brookings Institution and the World Bank, 18/4/2019, https://www.brookings.edu/blog/future-development/2019/04/18/a-new-alphabet-for-europe-algorithms-big-data-and-the-computer-chip/ (6/7/2020).

Financial Times 2020. "China and Huawei propose reinvention of the internet", in *Financial Times*, 27/3/2020. https://www.ft.com/content/c78be2cf-a1a1-40b1-8ab7-904d7095e0f2 (6/7/2020).

Finger, M., N. Bert and D. Kupfer 2015. "Mobility-as-a-Service: from the Helsinki experiment to a European model?", in *FSR Transport* (2015) 01. https://doi.org/10.2870/07981.

Getzner M., J. Kadi, A. Krisch and L. Plank 2018. "Plattform-Ökonomien: Kennzeichen, Wirkungsweisen und Bedeutung für die Stadtentwicklung", in *Jahrbuch des Departments für Raumplanung der TU Wien 2018*, ed. by J. Suitner, S. J. Dangschat and R. Giffinger, vol. 6, 129–144.

Glaeser, E. L. 2011. *Triumph of the City: How Our Greatest Invention Makes Us Richer, Smarter, Greener, Healthier, and Happier.* New York, NY: Penguin Press.

Gray, J. N., D. Pessel and P. P. Varaiya. 1972. "A Critique of Forrester's Model of an Urban Area", in *IEEE – Transactions on Systems, Man, and Cybernetics (SMC-2)* 2, 139–144. https://doi.org/10.1109/TSMC.1972.4309083.

Haasnoot, M., A. Warren and J. H. Kwakkel 2019. "Dynamic Adaptive Policy Pathways (DAPP)", in *Decision Making under Deep Uncertainty*, ed. by V. Marchau, W. Walker, P. Bloemen and S. Popper Cham: Springer, 71–92. https://doi.org/10.1007/978-3-030 05252-2_4.

Hamarat, C., J. H. Kwakkel and E. Pruyt 2013. "Adaptive Robust Design under deep uncertainty", in *Technological Forecasting and Social Change* 80 (3), 408–418. https://doi.org/10.1016/j.techfore.2012.10.004.

Hartikainen, A., J.-P. Pitkänen, A. Riihelä, J. Räsänen, I. Sacs, A. Sirkiä and A. Uteng 2019. *Whimpact: Insights from the world's first Mobility-as-a-Service (MaaS) system.* N.p.: Ramboll. https://ramboll.com/-/media/files/rfi/publications/Ramboll_whimpact-2019 (6/7/2020).

Haselmayer, M. and M. Jenny 2017. "Sentiment analysis of political communication: combining a dictionary approach with crowdcoding", in *Quality and Quantity* 51, 2623–2646. https://doi.org/10.1007/s11135-016-0412-4.

HSL 2016. "HSL Government Approves Model Contract for Travel Chain Cooperation HSL". https://www.hsl.fi/uutiset/2016/hsln-hallitus-hyvaksyi-sopimusmallin-matkaketjuyhteistyosta-9317 (23/1/2020).

Helmond, A. 2015. "The Platformization of the Web: Making Web Data Platform Ready", in *Social Media + Society* (1) 2. https://doi.org/10.1177/2056305115603080.

House of Lords Select Committee on Communications 2019. *Regulating in a Digital World. 2nd Report of Session 2017–19.* https://www.regulation.org.uk/library/2019-HoL-Regulating_in_a_Digital_World.pdf (6/7/2020).

ITU 2020. "International Telecommunication Union, United Nations Specialized Agency for information and communication technologies", homepage. https://www.itu.int/en/Pages/default.aspx (6/7/2020).

Jessop B. 2004. "Hollowing out the 'nation state' and multi-level governance", in *A Handbook of Comparative Social Policy*, ed. by P. Kennett. Cheltenham/Northampton, MA: Edward Elgar, 11–27.

Jittrapirom, P., V. Caiati, A.-M. Feneri, S. Ebrahimigharehbaghi, M. J. Alonso González and J. Narayan 2017a. "Mobility as a Service: A Critical Review of Definitions, Assessments of Schemes, and Key Challenges", in *Urban Planning 2* (2), 13–25. https://doi.org/10.17645/up.v2i2.931.

Jittrapirom, P., H. Knoflacher and M. Mailer 2017b. "The conundrum of the motorcycle in the mix of sustainable urban transport", in *Transportation Research Procedia* (25), 4869–4890. https://doi.org/10.1016/j.trpro.2017.05.365.

Jittrapirom, P., V. Marchau, R. van der Heijden and H. Meurs 2018a. "Dynamic adaptive policymaking for implementing Mobility-as-a Service (MaaS)", in *Research in Transportation Business & Management* (27), 46–55. https://doi.org/10.1016/j.rtbm.2018.07.001.

Jittrapirom, P., V. Marchau, R. van der Heijden and H. Meurs 2018b. "Future implementation of mobility as a service (MaaS): Results of an international Delphi study", in *Travel Behaviour and Society*, 1–59. https://doi.org/10.1016/j.tbs.2018.12.004.

Jittrapirom, P., W. van Neerven, K. Martens, D. Trampe and H. Meurs 2019. "The Dutch elderly's preferences toward a smart demand-responsive transport service", in *Research in Transportation Business & Management* (30). https://doi.org/10.1016/j.rtbm.2019.100383.

Kamargianni, M., W. Li, M. Matyas and A. Schäfer 2016. "A Critical Review of New Mobility Services for Urban Transport", in *Transportation Research Procedia* (14), 3294–3303. https://doi.org/10.1016/j.trpro.2016.05.277.

Kamargianni, M., M. Matyas, W. Li and A. Schäfer 2015. *Feasibility Study for "Mobility as a Service" concept in London.* London: UCL Energy Institute. https://doi.org/10.13140/RG.2.1.3808.1124.

Kitchin, R. 2014. "The real-time city? Big data and smart urbanism", in *GeoJournal* (79), 1–14. https://doi.org/10.1007/s10708-013-9516-8.

Kitchin, R. and G. McArdle 2016. "What makes Big Data, Big Data? Exploring ontological characteristics of 26 datasets", in *Big Data and Society* (3) 1. https://doi.org/10.1177/2053951716631130.

Kitchin, R., T. P. Lauriault and G. McArdle (eds.) 2017. *Data and the City.* London: Routledge.

Knoflacher, H. 2007. "Success and failures in urban transport planning in Europe – understanding the transport system", in *Sadhana* 32, 293–307. https://doi.org/10.1007/s12046-007-0026-6.

Kölbl, R., M. Niegl and H. Knoflacher 2008. "A strategic planning methodology", in *Transport Policy* 15 (5), 273–282. https://doi.org/10.1016/j.tranpol.2008.07.001.

Langley, P. and A. Leyshon 2017. "Platform capitalism: The intermediation and capitalisation of digital economic circulation", in *Finance and Society* 3 (1), 11–31.

Lempert, R. J. 2019. "Robust Decision Making (RDM)", in *Decision Making under Deep Uncertainty*, ed. by V. Marchau, W. Walker, P. Bloemen and S. Popper. Cham: Springer, 23–51. https://doi.org/10.1007/978-3-030-05252-2_2.

Leszczynski, A. 2016. "Speculative futures: Cities, data, and governance beyond smart urbanism", in *Environment and Planning A: Economy and Space* 48 (9), 1691–1708. https://doi.org/10.1177/0308518X16651445.

Leszczynski, A. 2020. "Glitchy vignettes of platform urbanism", in *Environment and Planning D: Society and Space* (38) 2, 189–208. https://doi.org/10.1177/0263775819878721.

Li, X. and P. Gong 2016. "Urban growth models: progress and perspective", in *Science Bulletin* (61) 21, 1637–1650. https://doi.org/10.1007/s11434-016-1111-1.

Mazzucatto, M. 2013. *The Entrepreneurial State: Debunking Public vs. Private Sector Myths.* London: Anthem Press.

MaaSifest (n.d.). "Mobility as a Service is the future for Mobility". http://www.connekt.nl/initiatief/mobility-as-a-service/ (6/7/2020).

May, A. D. 2003. *A Decision Makers' Guidebook: Developing Sustainable Urban Land Use and Transport Strategies*. Brussels: European Commission. https://www.researchgate.net/publication/241745361_A_Decision_Makers%27_Guidebook (6/7/2020).

Moore, M. H. 1995. *Creating Public Value: Strategic Management in Government*. Cambridge, MA/London: Harvard University Press.

OECD 2019. "Regulatory effectiveness in the era of digitalisation", June 2019. https://www.oecd.org/gov/regulatory-policy/Regulatory-effectiveness-in-the-era-of-digitalisation.pdf (6/7/2020).

Offenhuber, D. and C. Ratti (eds.) 2014. *Decoding the City: Urbanism in the Age of Big Data*. Basel: Birkhäuser.

Pangbourne, K., M. N. Mladenović, D. Stead and D. Milakis 2019. "Questioning mobility as a service: Unanticipated implications for society and governance", in *Transportation Research Part A: Policy and Practice* (131), 35–49. https://doi.org/10.1016/j.tra.2019.09.033.

Pangbourne, K., D. Stead, M. Mladenović and D. Milakis 2018. "The Case of Mobility as a Service: A Critical Reflection on Challenges for Urban Transport and Mobility Governance", in *Governance of the Smart Mobility Transition*, 33–48. https://doi.org/10.1108/978-1-78754-317-120181003.

Pieterse, E. 2008. *City Futures: Confronting the Crisis of Urban Development*. Cape Town: UCT Press.

Pinter, R. 2008. *Information Society: From Theory to Political Practice*. Budapest: Gondolat uj Madátum.

Pojani, D. and D. Stead 2015. "Sustainable Urban Transport in the Developing World: Beyond Megacities", in *Sustainability* 7 (6), 7784–7805. https://doi.org/10.3390/su7067784.

Polis 2017. "Mobility as a Service: Implications for urban and regional transport". https://www.polisnetwork.eu/wp-content/uploads/2017/12/polis-maas-discussion-paper-2017-final_.pdf (6/7/2020).

Priisalu, J. and R. Ottis 2017. "Personal control of privacy and data: Estonian experience", in *Health and Technology* (7), 441–451. https://doi.org/10.1007/s12553-017-0195-1.

Rudolph, S. 2019. *Digitale Medien, Partizipation und Ungleichheit. Eine Studie zum sozialen Gebrauch des Internets*. Wiesbaden: Springer VS.

Scott, J. C. 1998. *Seeing like a State: How Certain Schemes to Improve the Human Condition Have Failed*. Berkeley, CA: Yale University Press.

Shelton, T., M. Zook and A. Wiig 2015. "The 'actually existing smart city'", in *Cambridge Journal of Regions, Economy and Society* (8) 1, 13–25. https://doi.org/10.1093/cjres/rsu026.

SMILE 2014. "Pilot operation". http://smile-einfachmobil.at/pilotbetrieb_en.html (14/2/2020).

Söderström, O. 2014. "Smart city as corporate storytelling", in *City* (18) 3, 307–320. https://doi.org/10.1080/13604813.2014.906716.

Söderström, O. and A.-C. Mermet 2020. "When Airbnb Sits in the Control Room: Platform Urbanism as Actually Existing Smart Urbanism in Reykjavík", in *Frontiers in Sustainable Cities* (2) 15. https://doi.org/10.3389/frsc.2020.00015.

Strowel, A. and W. Vergote 2016. "Digital Platforms: To Regulate or Not To Regulate?", in *European Commission Newsroom*. https://ec.europa.eu/information_society/newsroom/image/document/2016-7/uclouvain_et_universit_saint_louis_14044.pdf (6/7/2020).

Srnicek, N. 2017. *Platform Capitalism*. Cambridge/Malden: Polity Press.

Tiwana, A. 2014. *Platform Ecosystems: Aligning Architecture, Governance, and Strategy*. Amsterdam/Boston, MA/Heidelberg/London/New York, NY/Oxford/Paris/San Diego, CA/San Francisco, CA/Singapore/Sydney/Tokyo: Morgan Kaufmann.

Townsend, A. M. 2013. *Smart Cities: Big Data, Civic Hackers, and the Quest for a New Utopia*. New York, NY/London: W. W. Norton & Company.

Pas, J. W. G. M. van der, J. H. Kwakkel and B. van Wee 2012. "Evaluating Adaptive Policymaking using expert opinions", in *Technological Forecasting and Social Change* 79 (2), 311–325. https://doi.org/10.1016/j.techfore.2011.07.009.

Vayla 2018. "The role of the Finnish Transport Agency in Finnish public transport". https://vayla.fi/web/en/transport-system/public-transport#.Xik1VGgzZaQ (23/1/2020).

Walker, W. E., V. A. W. J. Marchau and J. H. Kwakkel 2019. "Dynamic Adaptive Planning (DAP)", in *Decision Making under Deep Uncertainty*, ed. by V. Marchau, W. Walker, P. Bloemen and S. Popper. Cham: Springer, 53–69. https://doi.org/10.1007/978-3-030-05252-2_3.

Wong, Y. Z., D. A. Hensher and C. Mulley 2020. "Mobility as a service (MaaS): Charting a future context", in *Transportation Research Part A: Policy and Practice* (131), 5–19. https://doi.org/10.1016/j.tra.2019.09.030.

Zipper, D. 2019. "There's No App for Getting People Out of Their Cars", *Bloomberg CityLab*, 13/11/2019, https://www.citylab.com/perspective/2019/11/mobility-app-transit-options-ridesharing-bike-car-ownership/601858/ (14/2/2020).

Printed in the United States
by Baker & Taylor Publisher Services